Description of Input and Examples for PHREEQC Version 3—A Computer Program for Speciation, Batch-Reaction, One-Dimensional Transport, and Inverse Geochemical Calculations

By David L. Parkhurst and C.A.J. Appelo[1]

Chapter 43 of
Section A, Groundwater
Book 6, Modeling Techniques

[1]Hydrochemical Consultant
Valeriusstraat 11
1071 MB Amsterdam, NL
appt@hydrochemistry.eu
http://www.hydrochemistry.eu

Techniques and Methods 6–A43

U.S. Department of the Interior
U.S. Geological Survey

U.S. Department of the Interior
KEN SALAZAR, Secretary

U.S. Geological Survey
Marcia K. McNutt, Director

U.S. Geological Survey, Denver, Colorado: 2013

This and other USGS information products are available at http://store.usgs.gov/
U.S. Geological Survey
Box 25286, Denver Federal Center
Denver, CO 80225

To learn about the USGS and its information products visit http://www.usgs.gov/
1-888-ASK-USGS

Suggested citation:
Parkhurst, D.L., and Appelo, C.A.J., 2013, Description of input and examples for PHREEQC version 3—A computer program for speciation, batch-reaction, one-dimensional transport, and inverse geochemical calculations: U.S. Geological Survey Techniques and Methods, book 6, chap. A43, 497 p., available only at http://pubs.usgs.gov/tm/06/a43/.

Contents

Figures

Figures showing:

Tables

Conversion Factors and Abbreviations

SI to Inch/Pound

Multiply	By	To obtain
Length		
centimeter (cm)	0.3937	inch (in.)
millimeter (mm)	0.03937	inch (in.)
meter (m)	3.281	foot (ft)
kilometer (km)	0.6214	mile (mi)
meter (m)	1.094	yard (yd)
Area		
square meter (m^2)	0.0002471	acre
square centimeter (cm^2)	0.001076	square foot (ft^2)
square meter (m^2)	10.76	square foot (ft^2)
Volume		
liter (L)	1.057	quart (qt)
liter (L)	0.2642	gallon (gal)
cubic meter (m^3)	264.2	gallon (gal)
cubic meter (m^3)	0.0002642	million gallons (Mgal)
cubic meter (m^3)	35.31	cubic foot (ft^3)
cubic meter (m^3)	1.308	cubic yard (yd^3)
Flow rate		
meter per second (m/s)	3.281	foot per second (ft/s)
square meter per second (m^2/s)	10.76	square foot per second (ft^2/s)
cubic meter per second (m^3/s)	35.31	cubic foot per second (ft^3/s)

Mass		
gram (g)	0.03527	ounce, avoirdupois (oz)
kilogram (kg)	2.205	pound avoirdupois (lb)
Pressure		
kilopascal (kPa)	0.009869	atmosphere, standard (atm)
kilopascal (kPa)	0.01	bar
Density		
kilogram per cubic meter (kg/m^3)	0.06242	pound per cubic foot (lb/ft^3)
Energy		
joule (J)	0.0000002	kilowatthour (kWh)

Temperature in degrees Celsius (°C) may be converted to degrees Fahrenheit (°F) as follows:
$$°F=(1.8×°C)+32$$

Abbreviations

[For abbreviations of elements and element valence states, see Table 3]

atmosphere	atm
atmosphere cubic centimeter per mole per kelvin	atm cm^3 mol^{-1} /K^{-1}
atmosphere cubic centimeter per mole squared	atm cm^3 mol^{-2}
bar cubic centimeter per calorie	bar cm^3/cal
calorie kelvin per mol	cal K mol^{-1}
calorie kelvin per mole per bar	cal K mol^{-1} bar^{-1}
calorie per mole	cal/mol
calorie per mole per bar	cal mol^{-1} bar^{-1}
coulomb per mole	C/mol
cubic centimeter per calorie	cm^3/cal
cubic centimeter per mole	cm^3/mol
cubic decimeter per mole	dm^3/mol
cubic meter	m^3
cubic meter per mole	m^3/mol
day	d
degree Celsius	°C
electrostatic unit of charge	esu
equation of state	EOS
equivalent	eq

equivalent per cubic meter	eq/m^3
equivalent per kilogram	eq/kg
equivalent per kilogram water	eq/kgw
equivalent per liter	eq/L
erg per kelvin	erg/K
farad per square meter	F/m^2
gram	g
gram per cubic centimeter	g/cm^3
gram per equivalent	g/eq
gram per liter	g/L
gram formula weight	gfw
gram per liter per hour	$g\ L^{-1}h^{-1}$
gram per mole	g/mol
hour	h
joule	J
joule per mole	J/mol
joule per volt per equivalent	$J\ V^{-1}eq^{-1}$
joule per kelvin per mole	$JK^{-1}mol^{-1}$
joule per mole per meter	$J\ mol^{-1}m^{-1}$
kelvin	K
kilocalorie	kcal
kilocalorie per mole	kcal/mol
kilogram	kg
kilogram per liter	kg/L
kilogram solution	kgs
kilogram water	kgw
kilogram per cubic meter	kg/m^3
kilojoule per mole	kJ/mol
kilojoule per degree Celsius per kilogram	$kJ°C^{-1}kg^{-1}$
kilojoule per degree Celsius per meter per second	$kJ°C^{-1}m^{-1}s^{-1}$
kilopascals	kPa
liter	L
liter per gram	L/g
liter per mole	L/mol
liter per degree kelvin per mole	$L\ K^{-1}mol^{-1}$
microgram	μg
microgram per gram	μg/g
microgram per liter	μg/L
microgram per mole	μg/mol
micromole	μmol
micromole per kilogram water	μmol/kgw
microsiemens per centimeter	μS/cm
megapascals	MPa

meter	m
meter per second	m/s
milliequivalent	meq
milliequivalent per gram	meq/g
milliequivalent per kilogram	meq/kg
milliequivalent per kilogram water	meq/kgw
milliequivalent per 100 gram	meq/100 g
milligram per kilogram water	mg/kgw
milligram per liter	mg/L
millimole	mmol
millimole per square centimeter per second	$mmol\ cm^{-2}s^{-1}$
millimolar (millimole per liter)	mM
millimole per kilogram water	mmol/kgw
mole	mol
mole per kilogram	mol/kg
mole per kilogram water	mol/kgw
mole per liter	mol/L
mole per liter per hour	$mol\ L^{-1}h^{-1}$
mole per liter per second	$mol\ L^{-1}s^{-1}$
mole per cubic meter	mol/m^3
mole per meter to the fourth power	mol/m^4
mole per mole	mol/mol
mole per second	mol/s
mole per square centimeter per second	$mol\ cm^{-2}s^{-1}$
mole per square meter per second	$mol\ m^{-2}s^{-1}$
parts per billion	ppb
parts per million	ppm
parts per thousand	ppt
second	s
siemens per meter	S/m
siemens square meter per mole	$S\ m^2\ mol^{-1}$
square meter	m^2
square meter per gram	m^2/g
square meter per mole	m^2/mol
square meter per second	m^2/s
square meter per second per volt	$m^2\ s^{-1}V^{-1}$
square nanometer	nm^2
tracer diffusion coefficient in water	Dw
volt	V
year	yr

Description of Input and Examples for PHREEQC Version 3—A Computer Program for Speciation, Batch-Reaction, One-Dimensional Transport, and Inverse Geochemical Calculations

By David L. Parkhurst and C.A.J. Appelo

Abstract

PHREEQC version 3 is a computer program written in the C and C++ programming languages that is designed to perform a wide variety of aqueous geochemical calculations. PHREEQC implements several types of aqueous models: two ion-association aqueous models (the Lawrence Livermore National Laboratory model and WATEQ4F), a Pitzer specific-ion-interaction aqueous model, and the SIT (Specific ion Interaction Theory) aqueous model. Using any of these aqueous models, PHREEQC has capabilities for (1) speciation and saturation-index calculations; (2) batch-reaction and one-dimensional (1D) transport calculations with reversible and irreversible reactions, which include aqueous, mineral, gas, solid-solution, surface-complexation, and ion-exchange equilibria, and specified mole transfers of reactants, kinetically controlled reactions, mixing of solutions, and pressure and temperature changes; and (3) inverse modeling, which finds sets of mineral and gas mole transfers that account for differences in composition between waters within specified compositional uncertainty limits.

Many new modeling features were added to PHREEQC version 3 relative to version 2. The Pitzer aqueous model (*pitzer.dat* database, with keyword **PITZER**) can be used for high-salinity waters that are beyond the range of application for the Debye-Hückel theory. The Peng-Robinson equation of state has been implemented for calculating the solubility of gases at high pressure. Specific volumes of aqueous species are calculated as a function of the dielectric properties of water and the ionic strength of the solution, which allows calculation of pressure effects on chemical reactions and the density of a solution. The specific conductance and the density of a solution are calculated and printed in the output file. In addition to Runge-Kutta integration, a stiff ordinary differential equation solver (CVODE) has been included for kinetic calculations with multiple rates that occur at widely different time scales.

Surface complexation can be calculated with the CD-MUSIC (Charge Distribution MUltiSIte Complexation) triple-layer model in addition to the diffuse-layer model. The composition of the electrical double layer of a surface can be estimated by using the Donnan approach, which is more robust and faster than the alternative Borkovec-Westall integration. Multicomponent diffusion, diffusion in the electrostatic double layer on a surface, and transport of colloids with simultaneous surface complexation have been added to the transport module.

A series of keyword data blocks has been added for isotope calculations—ISOTOPES, CALCULATE_VALUES, ISOTOPE_ALPHAS, ISOTOPE_RATIOS, and NAMED_EXPRESSIONS. Solution isotopic data can be input in conventional units (for example, permil, percent modern carbon, or tritium units) and the numbers are converted to moles of isotope by PHREEQC. The isotopes are treated as individual components (they must be defined as individual master species) so that each isotope has its own set of aqueous species, gases, and solids. The isotope-related keywords allow calculating equilibrium fractionation of isotopes among the species and phases of a system. The calculated isotopic compositions are printed in easily readable conventional units.

New keywords and options facilitate the setup of input files and the interpretation of the results. Keyword data blocks can be copied (keyword COPY) and deleted (keyword DELETE). Keyword data items can be altered by using the keyword data blocks with the _MODIFY extension and a simulation can be run with all reactants of a given index number (keyword RUN_CELLS). The definition of the complete chemical state of all reactants of PHREEQC can be saved in a file in a raw data format (DUMP and _RAW keywords). The file can be read as part of another input file with the INCLUDE$ keyword. These keywords facilitate the use of IPhreeqc, which is a module implementing all PHREEQC version 3 capabilities; the module is designed to be used in other programs that need to implement geochemical calculations; for example, transport codes.

Charting capabilities have been added to some versions of PHREEQC. Charting capabilities have been added to Windows distributions of PHREEQC version 3. (Charting on Linux requires installation of Wine.) The keyword data block USER_GRAPH allows selection of data for plotting and manipulation of chart appearance. Almost any results from geochemical simulations (for example, concentrations, activities, or saturation indices) can be retrieved by using Basic language functions and specified as data for plotting in USER_GRAPH. Results of transport simulations can be plotted against distance or time. Data can be added to a chart from tab-separated-values files.

All input for PHREEQC version 3 is defined in keyword data blocks, each of which may have a series of identifiers for specific types of data. This report provides a complete description of each keyword data block and its associated identifiers. Input files for 22 examples that demonstrate most of the capabilities of PHREEQC version 3 are described and the results of the example simulations are presented and discussed.

Introduction

PHREEQC version 3 is a computer program for simulating chemical reactions and transport processes in natural or polluted water, in laboratory experiments, or in industrial processes. The program is based on equilibrium chemistry of aqueous solutions interacting with minerals, gases, solid solutions, exchangers, and sorption surfaces, which accounts for the original acronym—pH-REdox-EQuilibrium, but the program has evolved to include the capability to model kinetic reactions and 1D (one-dimensional) transport. Rate equations are completely user-specifiable in the form of Basic statements. Kinetic and equilibrium reactants can be interconnected, for example, by linking the number of surface sites to the amount of a kinetic reactant that is consumed (or produced) in a model period. A 1D transport algorithm simulates dispersion and diffusion; solute movement in dual porosity media; and multicomponent diffusion, where species have individual, temperature-dependent diffusion coefficients, but ion fluxes are modified to maintain charge balance during transport. A powerful inverse modeling capability allows identification of reactions that account for observed water compositions along a flowline or in the time course of an experiment. Extensible chemical databases allow application of the reaction, transport, and inverse-modeling capabilities to almost any chemical reaction that is recognized to influence rainwater, soil-water, groundwater, and surface-water quality.

PHREEQC evolved from the Fortran program PHREEQE (Parkhurst and others, 1980). PHREEQE was capable of simulating a variety of geochemical reactions for a system, including:

- Mixing of waters,
- Addition of net irreversible reactions to solution,
- Dissolving and precipitating phases to achieve equilibrium with the aqueous phase, and
- Effects of changing temperature.

PHREEQE calculated concentrations of elements, molalities and activities of aqueous species, pH, pe (negative log of the conventional activity of the electron), saturation indices, and mole transfers of phases to achieve equilibrium as a function of specified reversible and irreversible geochemical reactions.

PHREEQC version 1 (Parkhurst, 1995) was a completely new program written in the C programming language that implemented all of the capabilities of PHREEQE and added many capabilities that were not available in PHREEQE, including:

- Ion-exchange equilibria,

- Surface-complexation equilibria,

- Fixed-pressure gas-phase equilibria

- Advective transport, and

- Geochemical inverse modeling.

Other improvements relative to PHREEQE included complete accounting for elements in solids and the aqueous and gas phase, mole balance on hydrogen and oxygen to account for the mass of water in the aqueous phase, identification of the stable phase assemblage from a list of candidate phases, use of redox couples for definition of redox state in speciation calculations, and a more robust non-linear equation solver.

PHREEQC version 2 was a modification of PHREEQC version 1. All of the capabilities and most of the code for version 1 were retained in version 2 and several new capabilities were added, including:

- Kinetically controlled reactions,

- Solid-solution equilibria,

- Fixed-volume gas-phase equilibria,

- Variation of the number of exchange or surface sites in proportion to a mineral or kinetic reactant,

- Diffusion or dispersion in 1D transport,

- 1D transport coupled with diffusion into stagnant zones, and

- Isotope mole balance in inverse modeling.

The numerical method was modified to use several sets of convergence parameters in an attempt to avoid convergence problems. User-defined quantities could be written to the primary output file and (or) to a file suitable for importation into a spreadsheet, and solution compositions could be defined in a format compatible with spreadsheet programs.

PHREEQC version 3 extends PHREEQC version 2 with new features based on experience gained while simulating the results from laboratory experiments and field investigations. Furthermore, the code has been generalized into a computer object (IPhreeqc) to facilitate its use by other software programs that need to calculate chemical reactions or the distribution of chemicals in various phases.

PHREEQC can be used as a speciation program to calculate saturation indices, the distribution of aqueous species, and the density and specific conductance of a specified solution composition. For calculating solute activities, PHREEQC uses ion-association, Pitzer, or SIT (Specific ion Interaction Theory) equations to account for the nonideality of aqueous solutions. Analytical data for mole balances can be defined for any valence state or combination of valence states for an element. Distribution of redox elements among their valence states can be based on a specified pe or any redox couple for which data are available. PHREEQC allows the concentration of an element to be adjusted to obtain equilibrium (or a specified saturation index or gas partial pressure) with a specified phase, or to obtain charge balance. Solution compositions can be specified with a variety of concentration units.

In batch-reaction calculations, PHREEQC is oriented toward system equilibrium rather than just aqueous equilibrium. For an equilibrium calculation, all of the moles of each element in the system are distributed among the aqueous phase, pure phases, solid solutions, gas phase, exchange sites, and surface sites to attain system equilibrium. Non-equilibrium reactions can also be modeled, including aqueous-phase mixing, user-specified changes in the elemental totals of the system, and any kind of kinetically controlled reaction. Mole balances on hydrogen and oxygen allow the calculation of pe and the mass of water in the aqueous phase, which allows water-producing or -consuming reactions to be modeled correctly. Temperature effects can be modeled with the reaction enthalpy (Van't Hoff equation) or with a polynomial for the equilibrium constant. Pressure effects can be simulated by entering molar volumes of solids and parameters for defining the specific volume of aqueous species as a function of temperature, pressure, and ionic strength with a Redlich-type equation (for example, Redlich and Meyer, 1964). The solubility of gases in gas mixtures at (very) high pressures can be calculated with the Peng-Robinson equation of state (Peng and Robinson, 1976). The parameters for calculating the specific volume of aqueous species, the Peng-Robinson parameters for gases, and molar volumes of minerals have been added to the databases *phreeqc.dat, Amm.dat,* and *pitzer.dat*.

Sorption and desorption can be modeled as surface complexation reactions or as (charge neutral) ion exchange reactions. PHREEQC has two models for surface complexation. One surface complexation model is based on the Dzombak and Morel (1990) database for complexation of heavy metal ions on hydrous ferric oxide (Hfo, or commonly referred to as ferrihydrite). Ferrihydrite, like many other oxy-hydroxides, binds metals and protons on strong and weak sites and develops a charge depending on the ions sorbed. The model uses the Gouy-Chapman equation to relate surface charge and potential. The other surface complexation model is CD-MUSIC (Charge Distribution MUltiSIte Complexation), which also allows multiple binding sites for each surface. In addition, the charge, the potential, and even the sorbed species can be distributed over the Stern layer and the Helmholtz layer in this model (Hiemstra and Van Riemsdijk, 1996). The CD-MUSIC model has more options to fit experimental data and was developed for sorption on goethite. In both models, the surface charge can be neutralized by an electrical double layer (EDL) on the surface. The composition of the EDL can be calculated by explicit integration of the Poisson-Boltzmann equation (Borkovec and Westall, 1983), or by averaging for a Donnan volume (Appelo and Wersin, 2007). Surface complexation constants for two of the databases distributed with the program (*phreeqc.dat* and *wateq4f.dat*) are taken from Dzombak and Morel (1990); surface complexation constants for the other databases distributed with the program (*minteq.dat* and *minteq.v4.dat*) are taken from MINTEQA2 (Allison and others, 1990; U.S. Environmental Protection Agency, 1998).

Ion exchange can be modeled with the Gaines-Thomas convention (the equivalent fraction of the exchangeable cation is used for activity of the exchange species), the Gapon convention (equivalent fraction of exchange sites occupied by a cation is used for activity of the exchange species), or the Vanselow convention (mole-fraction of the exchangeable cations is used for activity of the exchange species). The equilibrium constants for the Gaines-Thomas model as listed in Appelo and Postma (2005) are included in several of the databases distributed with the program (*Amm.dat, iso.dat, llnl.dat, phreeqc.dat, pitzer.dat,* and *wateq4f.dat*).

Kinetically controlled reactions can be defined in a general way by using an embedded Basic interpreter. Rate expressions written in the Basic language can be included in the input file, and the program uses the Basic interpreter to calculate rates, which can depend on any parameter of the chemical model. Multiple rates can be integrated simultaneously by using Runge-Kutta explicit or the CVODE implicit (stiff) equation solver (Cohen and Hindmarsh, 1996). Formulations for ideal, multicomponent and nonideal, binary solid and liquid solutions are available. The equilibrium compositions of nonideal, binary solid solutions can be calculated even if miscibility gaps exist, and the equilibrium composition of ideal solid and liquid solutions that have two or more components also can be calculated. It is possible to precipitate solid solutions from supersaturated conditions with no preexisting solid, and to dissolve solid solutions completely. Both fixed-pressure gas-phase (fixed-pressure gas bubbles) and fixed-volume gas-phase compositions can be included in the calculations.

It is possible to define independently any number of solution compositions, gas phases, or pure-phase, solid-solution, exchange, or surface-complexation assemblages. Batch reactions allow any combination of solution (or mixture of solutions), gas phase, and assemblages to be brought together, any irreversible reactions can be added, and equilibrium is calculated for the resulting system. (Equilibrium is identical to the minimum Gibbs energy for the system.) If kinetic reactions are defined, then the kinetic reactions are integrated with an automatic time-stepping algorithm while system equilibrium is maintained for the equilibrium reactions that are defined.

PHREEQC provides a numerically efficient method for simulating the movement of solutions through a column or 1D flow path with or without the effects of dispersion. The initial composition of the aqueous, gas, and solid phases within the column are specified and the changes in composition due to advection and dispersion and (or) diffusion (Appelo and Postma, 2005) coupled with reversible and irreversible chemical reactions within the column can be modeled. For simulating colloidal transport, surfaces can be given a diffusion coefficient and transported as solutes through the column. For modeling a dual porosity medium, stagnant zones can be incorporated in the column. Multicomponent diffusion, a process where each solute diffuses according to its own diffusion coefficient, can be included in advective transport simulations or as a stand-alone diffusion process. In the multicomponent diffusion process, diffusion in the EDL and in the interlayers of clay minerals can be included, and the diffusion coefficients can be coupled to porosity changes that may result from mineral dissolution and precipitation, thus providing a framework for simulating experiments with clays and clay rocks. A stagnant-zone option can be used for modeling (multicomponent) diffusion in three dimensions by using explicit finite-difference equations to define mixing among the stagnant cells. A simple advective-reactive transport simulation option with reversible and irreversible chemical reactions is retained from version 1.

Inverse modeling attempts to account for the chemical changes that occur as water evolves along a flow path. Assuming two water analyses represent starting and ending water compositions along a flow path, inverse modeling is used to calculate the moles of minerals and gases that must enter or leave solution to account for the differences in composition. Inverse models that mix two or more waters to form a final water also can be calculated. PHREEQC allows uncertainty limits to be defined for all analytical data, such that inverse models are constrained to satisfy mole balance for each element and valence state as well as charge balance for each solution, while adjustments to the analytical data are constrained to be within the specified uncertainty limits. Isotope mole-balance equations with associated uncertainty limits can be specified, but inverse modeling does not include Rayleigh fractionation processes.

The input to PHREEQC is completely free format and is based on chemical symbolism. Balanced equations, written in chemical symbols, are used to define aqueous species, exchange species, surface-complexation species, solid solutions, and pure phases, which eliminates all use of index numbers to identify elements or species. The C programing language allows dynamic allocation of computer memory, so there are no limitations on array sizes, string lengths, or numbers of entities, such as solutions, phases, sets of phases, exchangers, solid solutions, or surfaces that can be defined to the program. The graphical user interface PhreeqcI (Charlton and Parkhurst, 2002) provides input screens for all of the features of version 2 and most of the features of version 3, including charting. Another graphical user interface with charting options, PHREEQC for Windows, has been written by Vincent Post (Post, 2012). The free-format structure of the data, the use of order-independent keyword data blocks, and the relatively simple syntax facilitate the generation of input files with a standard editor.

A new capability in PHREEQC version 3—the INCLUDE$ keyword—allows files to be inserted into input and database files. The point of insertion can, but does not have to correspond to the end of keyword data blocks. Inserted files may in turn insert other files, so that a collection of files may be merged into one stream for PHREEQC database and (or) input files. The merging is done "on-the-fly", so that it is possible to write a file with a SELECTED_OUTPUT data block that is subsequently included in the same run.

Charting capabilities similar to those in PHREEQC for Windows have been added to the Windows distributions of PHREEQC version 3. Charting is possible for Linux, but requires installation of Wine. The keyword data block USER_GRAPH allows selection of data for plotting and manipulation of chart appearance. Almost any results from geochemical simulations (for example, concentrations, activities, or saturation indices) can be retrieved by using Basic language functions and specified as data for plotting in USER_GRAPH. Results of transport simulations can be plotted against distance or time.

Program Limitations

PHREEQC is a general geochemical program and is applicable to many hydrogeochemical environments. However, several limitations need to be considered.

Aqueous Model

One limitation of the aqueous model is lack of internal consistency in the data in the databases. The database *pitzer.dat* defines the most consistent aqueous model; however, it includes only a limited number of elements. All of the other databases are compendia of logarithms of equilibrium constants (log Ks) and enthalpies of reaction that have been taken from various literature sources. No systematic attempt has been made to determine the aqueous model that was used to develop the individual log Ks or whether the aqueous models defined by the current database files are consistent with the original experimental data. The database files provided with the program should be considered to be preliminary. Careful selection of aqueous species and thermodynamic data is left to the users of the program.

Ion Exchange

The default ion-exchange formulation assumes that the thermodynamic activity of an exchange species is equal to its equivalent fraction. Optionally, the equivalent fraction can be multiplied by a Debye-Hückel activity coefficient and (or) an "active fraction" coefficient to define the activity of an exchange species (Appelo, 1994a). Other formulations use other definitions of activity (mole fraction instead of equivalent fraction, for example) and may be included in the database with appropriate rewriting of species or solid solutions. No attempt has been made to include other or more complicated exchange models. In many field studies, ion-exchange modeling requires experimental data on material from the study site for appropriate model application.

Surface Complexation

Davis and Kent (1990) reviewed surface-complexation modeling and note theoretical problems with the use of molarity as the standard state for sorbed species. PHREEQC uses mole fraction for the activity of surface species instead of molarity. This change in standard state has no effect on monodentate surface species but does affect multidentate species significantly. Other uncertainties occur in determining the number of sites, the surface area, the composition of sorbed species, and the appropriate log Ks. In many field studies, surface-complexation modeling requires experimental data on material from the study site for appropriate model application.

Solid Solutions

PHREEQC uses a Guggenheim approach for determining activities of components in nonideal, binary solid solutions (Glynn and Reardon, 1990). Ternary nonideal solid solutions are not implemented. It is possible to model two or more component solid solutions by assuming ideality. However, the assumption of ideality is usually an oversimplification, except possibly for isotopes of the same element.

Transport Modeling

An explicit finite difference algorithm is included for calculations of 1D advective-dispersive transport, and optionally, diffusion in stagnant zones. The algorithm may show numerical dispersion when the grid is coarse. The magnitude of numerical dispersion also depends on the nature of the modeled reactions; numerical dispersion may be large in many cases—linear exchange, surface complexation, diffusion into stagnant zones, among others—but may be small when chemical reactions counteract the effects of dispersion. It is recommended that modeling be performed stepwise, starting with a coarse grid to obtain results rapidly and to investigate the hydrochemical reactions, and finishing with a finer grid to assess the effects of numerical dispersion on both reactive and conservative species.

Inverse Modeling

Inclusion of uncertainties in the process of identifying inverse models is a major advance over previous inverse modeling programs. However, the numerical method has shown some inconsistencies in results due to the way the solver handles small numbers. The option to change the tolerance used by the solver (**-tol** in **INVERSE_MODELING** data block) is an attempt to remedy this problem. Some versions of PHREEQC have an option to use an extended precision solver in inverse calculations, but this option has not proved to be effective. The inability to make Rayleigh fractionation calculations for isotopes in precipitating minerals is a major limitation.

Purpose and Scope

The purpose of this report is to describe the input and provide example calculations for the program PHREEQC version 3. The report includes a discussion of the versions of PHREEQC that are available, a list of the types of calculations that can performed, a complete description of the keyword data blocks that comprise the input for the program, and presentation of a series of examples of input files and model results that demonstrate many of the capabilities of the program.

Versions of PHREEQC

PHREEQC is available for a number of different software environments. Batch and library (or DLL) versions are available for Windows and Linux. Graphical user interfaces and a COM (Component Object Model) version are available only for Windows operating systems. All versions are available at the Web site *http://wwwbrr.cr.usgs.gov/projects/GWC_coupled/phreeqc*.

Batch Versions

Batch (or stand-alone) versions of PHREEQC have been compiled for Windows and Linux operating systems. These versions are executed from a command line with one to four arguments: (1) input file, (2) output file, (3) database file, and (4) screen-output file. The input file is required and the rest are optional, but all preceding arguments are required to enter the third or fourth argument.

A batch version with charting capabilities is distributed for Windows operating systems. (This version can be run with Wine on Linux and OS2 operating systems.) It also is possible to install Wine and related software to compile PHREEQC with charting capabilities on Linux operating systems, but that version is not distributed presently (2012).

Graphical User Interfaces

The program PhreeqcI is a graphical user interface to PHREEQC that runs on Windows operating systems. It provides data entry screens for most keyword data blocks with a description of each input data item. The interface allows input to be defined, simulations run, output viewed, and charts generated. An input tree allows all the keyword input for multiple files to be viewed and selected for running or editing. Data entry for keyword data blocks related to isotopes (CALCULATE_VALUES, ISOTOPE_ALPHAS, ISOTOPE_RATIOS) has not been implemented.

The program PHREEQC for Windows (PfW) written by Vincent Post (Post, 2012) is a graphical user interface that allows building and running PHREEQC input files. It provides templates for each keyword data block and identifier of PHREEQC version 2 and USER_GRAPH, which can be edited to define the desired input. PfW has charting capabilities equivalent to PHREEQC version 3 (excluding the PLOT_XY function).

The general-purpose editor Notepad++ has been adapted for writing, editing, and running PHREEQC input files, including charting. Notepad++ provides the following capabilities:

- Syntax highlighting,

- Autocompletion of keywords, identifiers, and any other already present word in the text,

- Calltips for Basic functions and keywords,

- Colored numbers (| and l are differently colored!),

- Parenthesis matching: print $(1 + 9.0) / (9 + 1.0)$,

- Commenting or uncommenting multiple lines at once,

- Column editor for easy Basic line renumbering,

- Shortcuts: for example, Ctrl+F6 runs PHREEQC on the current file,

- File recognition of extensions *.ppi*, *.pqi*, *.phrq*, *.phr*, *.dat*, and *.out*.

The Notepad++ version adapted for PHREEQC version 3 can be downloaded from *http://www.hydrochemistry.eu/downl.html* (accesed December 19, 2012).

PHREEQC Modules for Use with Scripting and Programming Languages

PHREEQC modules have been developed that allow PHREEQC to be included in many software environments (Charlton and Parkhurst, 2011). A limited set of methods implement the full reaction capabilities of PHREEQC for each module. These input methods use strings or files to define geochemical calculations in exactly the same formats used by PHREEQC. Output methods provide a table of user-selected model results, such as concentrations, activities, saturation indices, or densities. The PHREEQC modules may be used in scripting languages to fit parameters; to plot PHREEQC results for field, laboratory, or theoretical investigations; or to develop new models that include simple or complex geochemical calculations.

PHREEQC version 3 has been implemented as a C++ class, and a derived class, called IPhreeqc, includes the methods that make IPhreeqc easy to use in other software. IPhreeqc has been compiled in libraries for Linux and Windows that allow PHREEQC to be called from C++, C, and Fortran. A Microsoft COM (Component Object Model) version of PHREEQC has been implemented, which allows IPhreeqc to be used by any software that can interface with a COM server—for example, Excel, Visual Basic, Python, or MATLAB.

Types and Sequence of Calculations

After reading a database file of thermodynamic data, PHREEQC reads an input file until it reaches an END keyword or the end of the file. At that point, PHREEQC begins a simulation by processing any new thermodynamic data that have been read, creating updated lists of elements, phases, and aqueous, exchange, and surface species. After processing these new data, up to 12 types of calculations or data management operations may be performed in the following order: (1) initial solution or speciation calculations, (2) initial exchange calculations, (3) initial surface calculations, (4) initial gas-phase calculations, (5) batch-reaction calculations, (6) inverse-modeling calculations, (7) advective-transport calculations, (8) advective-dispersive transport calculations, (9) cell batch-reaction calculations, (10) copy operations, (11) dump operations, and (12) delete operations. After all of these calculations and operations are completed, PHREEQC reads the input file to the next END keyword and starts another simulation. All the simulations together are termed a "run".

Initial Solution or Speciation Calculations

Generally, a chemical analysis provides total concentrations of elements in solution. A speciation calculation distributes these totals among aqueous species by using an aqueous model; the results of the speciation calculation are the activities of all of the aqueous species. The activities can be used for calculating saturation indices for minerals, relative to the water. Speciation modeling requires only a SOLUTION data block for each water analysis for which saturation indices are to be calculated. Alternatively, the SOLUTION_SPREAD data block can be used to define multiple solutions in the same data block—one row for each solution—and speciation calculations are performed for each analysis. Example 1 demonstrates speciation calculations.

Initial Exchange Calculations

One type of reactant used by PHREEQC is an assemblage of one or more ion exchangers. The elemental composition of the exchange assemblage can be defined explicitly or implicitly with the EXCHANGE data block. If the composition is defined explicitly, that is by specifying the number of moles of each element on exchange sites, then there is no need for an initial exchange calculation. If the composition of the exchange assemblage is defined implicitly, a calculation is performed to determine the composition of the exchanger that is in equilibrium with a specified solution composition. In this calculation, the composition of the solution does not change, but the composition of the exchanger is adjusted until it is in equilibrium with the solution. Examples 13, 14, 19, and 21 demonstrate this calculation in Examples.

Initial Surface Calculations

One type of reactant used by PHREEQC is an assemblage of one or more surfaces, each with one or more types of complexation sites. Similar to ion exchange, the elemental composition of the surface assemblage can be defined explicitly or implicitly with the **SURFACE** data block. If the composition is defined explicitly, that is by specifying the number of moles of each element on surface sites, then there is no need for an initial surface calculation. If the composition of the surface assemblage is defined implicitly, a calculation is performed to determine the composition of the surface that is in equilibrium with a specified solution composition. In this calculation, the composition of the solution does not change, but the compositions of the surfaces are adjusted until they are in equilibrium with the solution. Examples 14, 19, and 21 demonstrate this calculation in Examples.

Initial Gas-Phase Calculations

A gas phase is defined by the number of moles of gas components that are present. The gas phase may have either constant pressure or constant volume. The number of moles of each component in a constant-pressure gas phase can be defined by specifying the partial pressures of all the gas components, in which case no initial gas-phase calculation is done. An initial gas-phase calculation is done when the identifier **-equilibrate** is included in a fixed-volume gas-phase definition. In this case, the partial pressure of each gas component in the solution is taken as the partial pressure in the gas phase. From the partial pressures and the volume of the gas phase, the moles of each component will be calculated by the ideal gas law or (approximately) with the Peng-Robinson equation of state. The composition of the solution is not changed during an initial gas-phase calculation. Example 7 demonstrates this calculation in Examples.

Batch-Reaction Calculations

Batch-reaction calculations simulate reactions occurring in a beaker and can involve equilibrium and irreversible reactions. Equilibrium reactions are defined by specifying a solution or mixture of solutions to be put in the beaker along with a pure-phase assemblage, an exchange assemblage, a multicomponent gas phase, a solid-solution assemblage, and (or) a surface assemblage. The solution or mixture is brought to equilibrium with the reactants. Furthermore, irreversible reactions can be specified, including addition or removal of specified reactants, pressure changes, temperature changes, and (or) kinetic reactions, where the reaction rate can depend on solution composition or other, user-definable parameters. Conceptually, the irreversible reactions are added and equilibrium is calculated for the system (see examples 2, 3, 4, 5, 6, 7, 8, 10, 19, 20, 21, and 22 in the Examples). Kinetic reactions are integrated for a specified time step by calculating equilibrium following each of a series of irreversible reactions that depend on the evolving composition of the solution (see examples 6, 9, and 15 in Examples).

Initial conditions for batch reactions are defined with **SOLUTION, SOLUTION_SPREAD, EQUILIBRIUM_PHASES, EXCHANGE, GAS_PHASE, SOLID_SOLUTIONS**, and **SURFACE** data blocks. Irreversible reactions are defined with the following data blocks: **MIX**, for mixing of solutions; **REACTION**, for adding or removing fixed amounts of specified reactants; **KINETICS** and **RATES**, for defining kinetic reactions; **REACTION_PRESSURE**, for changing the pressure at which the batch reaction occurs; and **REACTION_TEMPERATURE**, for changing the temperature at which the batch reaction occurs.

Different sets of keyword data blocks can be defined within one simulation, each set being identified by the number or range of numbers which follow the keyword. In the subsequent batch reaction, a set may be included either implicitly or explicitly. For an implicit calculation, a solution or mixture (**SOLUTION** or **MIX** keywords) must be defined within the simulation, and the first of each keyword set (defined before the **END**) will be included in the calculation. That is, the first solution (or mixture) will be used along with the first of each of the data blocks **EQUILIBRIUM_PHASES, EXCHANGE, GAS_PHASE, SOLID_SOLUTIONS, SURFACE, KINETICS, REACTION, REACTION_PRESSURE**, and **REACTION_TEMPERATURE**. For an explicit calculation, "**USE** *keyword number*" defines a set that is to be used regardless of position within the input lines (see examples 3, 6, 7, 8, and 9 in Examples). "**USE** *keyword* **none**" eliminates a reactant that was implicitly defined (see example 8 in Examples). If the composition of the solution, pure-phase assemblage, exchange assemblage, gas phase, solid-solution assemblage, or surface assemblage has changed after the batch-reaction calculation, it can be saved with the **SAVE** keyword. **RUN_CELLS** invokes all the defined keyword data blocks automatically for specified cell numbers, as explained in Cell Batch-Reaction Calculations.

Inverse-Modeling Calculations

Inverse modeling is used to deduce the geochemical reactions that account for the change in chemical composition of water along a flow path. At least two chemical analyses of water at different points along the flow path are needed, as well as a set of phases that are potentially reactive along the flow path. From the analyses and phases, mole-balance models are calculated. A mole-balance model is a set of mole transfers of phases and reactants that accounts for the change in composition along the flow path. Normally, only **SOLUTION** or **SOLUTION_SPREAD** data blocks and an **INVERSE_MODELING** keyword data block are needed for inverse modeling calculations. However, additional reactant phases may need to be defined with **PHASES** or **EXCHANGE_SPECIES** data blocks (see examples 16, 17, and 18 in Examples).

Advective-Transport Calculations

Advective-transport calculations are used to simulate advection and chemical reactions as water moves through a 1D column (**ADVECTION** data block). The column is divided into a number of cells, n, which is defined by the user. The cells are numbered 1 through n, and these cells initially contain solutions with identifying numbers 1 through n. A solution composition for each of these integers must have been defined by **SOLUTION** (or **SOLUTION_SPREAD**) data blocks or the **SAVE** keyword. The cells also may contain other reversible or irreversible reactants. For a given cell number, i, if a phase assemblage, exchange assemblage, solid-solution assemblage, surface assemblage, gas phase, mixture, reaction, reaction-pressure, or reaction-temperature data block with identifying number i has been defined, then it is present in cell i during the advective-transport calculation. Thus, the initial conditions and the set of reactants can be defined individually for each cell, which provides flexibility to simulate a variety of chemical conditions throughout the column (see examples 11 and 14 in Examples).

The infilling solution for the column is always solution number 0. Advection is modeled by "shifting" solution 0 to cell 1, the solution in cell 1 to cell 2, and so on. At each shift, kinetic reactions are integrated in each cell, while maintaining equilibrium with any gas phase or solid-phase assemblages that are present in the cell. To facilitate definition of the initial conditions, the keywords **EQUILIBRIUM_PHASES**, **EXCHANGE, GAS_PHASE, KINETICS, MIX, REACTION, REACTION_PRESSURE, REACTION_TEMPERATURE, SOLID_SOLUTIONS, SOLUTION**, and **SURFACE** allow simultaneous definition for a range of cell numbers. The **SAVE** keyword also permits a solution, gas-phase, or assemblage numbers to be saved to a range of numbers simultaneously.

Advective-Dispersive Transport Calculations

Analogous to purely advective transport, advective-dispersive 1D and diffusive 2- or 3D transport can be modeled with the TRANSPORT data block. Dispersivities can be specified for cells, and either a single diffusion coefficient is used for all chemical species or each species can be given its own temperature-dependent diffusion coefficient. Like purely advective transport calculations, a column of *n* cells is defined, but also dispersion/diffusion parameters, boundary conditions, direction of flow, cell lengths, and advective time step can be provided. It also is possible to model double porosity by including the relevant information (example 13). The infilling solution depends on the direction of flow and may be solution number 0 or *n+1*. For each shift (advection or diffusion time step) a number of dispersion/diffusion mixing steps are performed, the number depending on numerical stability criteria. For each shift and dispersion step, kinetic reactions are integrated for each cell while maintaining equilibrium with any gas phase or assemblages that are present in the cell (see examples 11, 12, 13, 15, 19, and 21, in Examples).

Cell Batch-Reaction Calculations

Cell batch-reaction calculations are a special case of batch-reaction calculations. Whereas any reactant with any identification number can be reacted in a batch-reaction calculation, only reactants with the same identification number are used in a cell batch-reaction calculation. For a cell batch-reaction calculation, the identification number (cell number) is specified, and any reactants with that identification number are brought together and reacted. The reactants may include any defined by the EQUILIBRIUM_PHASES, EXCHANGE, GAS_PHASE, KINETICS, MIX, REACTION, REACTION_PRESSURE, REACTION_TEMPERATURE, SOLID_SOLUTIONS, SOLUTION, and SURFACE data blocks. If both a MIX and SOLUTION are defined with the specified cell number, then the MIX reactant is used. The USE data block has no effect on cell batch-reaction calculations. The compositions of the EQUILIBRIUM_PHASES, EXCHANGE, GAS_PHASE, KINETICS, SOLUTION, and SURFACE reactants are automatically saved after a cell batch-reaction calculation and the SAVE data block has no effect.

COPY Operations

The **COPY** operation allows any numbered reactant to be replicated, including reactants defined with **EQUILIBRIUM_PHASES, EXCHANGE, GAS_PHASE, KINETICS, MIX, REACTION, REACTION_PRESSURE, REACTION_TEMPERATURE, SOLID_SOLUTIONS, SOLUTION,** and **SURFACE** data blocks. A reactant can be copied to a single new instance with a specified identification number or to multiple new instances with a contiguous range of identification numbers. If a reactant is copied to an identification number, and a reactant of that type already exists with that number, the existing reactant is overwritten.

DUMP Operations

The **DUMP** operation allows any numbered reactant to be written to a file in a form that is readable by PHREEQC (**_RAW** data blocks), including reactants defined with **EQUILIBRIUM_PHASES, EXCHANGE, GAS_PHASE, KINETICS, MIX, REACTION, REACTION_PRESSURE, REACTION_TEMPERATURE, SOLID_SOLUTIONS, SOLUTION,** and **SURFACE** data blocks. The file can be used to reestablish the state of a PHREEQC calculation in a subsequent PHREEQC run. IPhreeqc modules can retrieve **DUMP** information as a string, which can be used to copy the reactant definitions to other IPhreeqc modules.

DELETE Operations

The **DELETE** operation allows any numbered reactant to be deleted from the current PHREEQC run, including reactants defined with **EQUILIBRIUM_PHASES, EXCHANGE, GAS_PHASE, KINETICS, MIX, REACTION, REACTION_PRESSURE, REACTION_TEMPERATURE, SOLID_SOLUTIONS, SOLUTION,** and **SURFACE** data blocks. The reactants to be deleted can be identified by individual or ranges of identification numbers. In some circumstances, such as cell batch-reaction calculations, it may be useful to delete a reactant that is no longer wanted. For most batch-reaction calculations, memory is sufficiently available in modern computers so that it is not necessary to delete reactants that are no longer used. However, if IPhreeqc modules are used for simulations with hundreds of thousands of cells, then **DELETE** operations may be needed for efficient management of computer memory.

Description of Data Input

The input for PHREEQC is arranged by keyword data blocks. Each data block begins with a line that contains the keyword (and possibly additional data) followed by additional lines containing data related to the keyword. The keywords that define the input data for running the program are listed in table 1. Keywords and their associated data are read from a database file at the beginning of a run to define the elements, exchange reactions, surface complexation reactions, mineral phases, gas components, and rate expressions. Any data items read from the database file can be redefined by keyword data blocks in the input file. After the database file is read, data are read from the input file until the first END keyword is encountered, after which the specified calculations are performed. The process of reading data from the input file until an END, followed by doing the calculations, is repeated until the end of the input file is encountered. The set of calculations, defined by keyword data blocks terminated by an END, is termed a "simulation". A "run" is a series of one or more simulations that are contained in the same input data file and calculated during the same invocation of the program PHREEQC.

Each simulation may contain one or more of seven types of speciation, batch-reaction, and transport calculations: (1) initial solution speciation, (2) determination of the composition of an exchange assemblage in equilibrium with a fixed solution composition, (3) determination of the composition of a surface assemblage in equilibrium with a fixed solution composition, (4) determination of the composition of a fixed-volume gas phase in equilibrium with a fixed solution composition, (5) calculation of chemical composition as a result of batch reactions, which include mixing; kinetically controlled reactions; net addition or removal of elements from solution, termed "net stoichiometric reaction"; variation in temperature and pressure; equilibration with assemblages of pure phases, exchangers, surfaces, and (or) solid solutions; and equilibration with a gas phase at a fixed total pressure or fixed volume, (6) advective-reactive transport, or (7) advective-dispersive-reactive transport. The combination of capabilities allows the modeling of complex geochemical reactions and transport processes during one or more simulations.

In addition to speciation, batch-reaction, and transport calculations, the code may be used for inverse modeling, by which net chemical reactions are deduced that account for composition differences between an initial water or a mixture of initial waters and a final water.

Table 1. List of keyword data blocks.

Keyword data block	Function
ADVECTION	Specify parameters for advective-reactive transport, no dispersion
CALCULATE_VALUES	Define Basic functions
COPY	Make copies of keyword data blocks with new identifying numbers
DATABASE	Specify the database for the simulations
DELETE	Delete specified reactants
DUMP	Write complete descriptions of specified reactants to file (or to a string for IPhreeqc modules)
END	Demarcate end of a simulation
EQUILIBRIUM_PHASES	Define assemblage of minerals and gases to react with an aqueous solution
EXCHANGE	Define exchange assemblage composition
EXCHANGE_MASTER_SPECIES	Identify exchange sites and corresponding exchange master species
EXCHANGE_SPECIES	Define association half-reaction and thermodynamic data for exchange species
GAS_PHASE	Define a gas-phase composition
INCLUDES	Insert a file into the input or database file
INCREMENTAL_REACTIONS	Define whether reaction increments are incremental or cumulative
INVERSE_MODELING	Specify solutions, reactants, and parameters for mole-balance modeling
ISOTOPES	Identify isotopes of elements and define the absolute isotopic ratio of standards
ISOTOPE_ALPHAS	Specify fractionation factors to appear in the output file
ISOTOPE_RATIOS	Specify isotope ratios to appear in the output file
KINETICS	Specify kinetic reactions and define parameters
KNOBS	Define parameters for numerical method and printing debugging information
LLNL_AQUEOUS_MODEL_PARAMETERS	Specify activity coefficient parameters for the Lawrence Livermore National Laboratory aqueous model
MIX	Define mixing fractions of aqueous solutions
NAMED_EXPRESSIONS	Assigns a name to an analytical expression for an equilibrium constant or isotope fractionation factor
PHASES	Define dissociation reactions and thermodynamic data for minerals and gases
PITZER	Specify the parameters of a Pitzer specific-ion-interaction aqueous model
PRINT	Select data blocks to be printed to the output file
RATES	Define rate equations with Basic language statements
REACTION	Specify irreversible reactions
REACTION_PRESSURE	Specify pressure(s) for batch reactions
REACTION_TEMPERATURE	Specify temperature(s) for batch reactions
RUN_CELLS	Specify a reaction simulation that includes all reactants of a given identification number
SAVE	Save results of batch reactions for use in subsequent simulations
SELECTED_OUTPUT	Print specified quantities to a user-defined file

Table 1. List of keyword data blocks.—Continued

Keyword data block	Function
SIT	Specify the parameters of a SIT (Specific ion Interaction Theory) aqueous model
SOLID_SOLUTIONS	Define the composition of a solid-solution assemblage
SOLUTION	Define the composition of an aqueous solution
SOLUTION_MASTER_SPECIES	Identify elements and corresponding aqueous master species
SOLUTION_SPECIES	Define association reaction and thermodynamic data for aqueous species
SOLUTION_SPREAD	Define one or more aqueous solution compositions using a tab-delimited format (Alternative input format for **SOLUTION**)
SURFACE	Define the composition of an assemblage of surfaces
SURFACE_MASTER_SPECIES	Identify surface sites and corresponding surface master species
SURFACE_SPECIES	Define association reaction and thermodynamic data for surface species
TITLE	Specify a text string to be printed in the output file
TRANSPORT	Specify parameters for advective-dispersive-reactive transport, optionally with dual porosity
USE	Select aqueous solution or other reactants that define batch reactions
USER_GRAPH	Specify data and parameters for a user-defined X–Y plot
USER_PRINT	Print user-defined quantities to the output file
USER_PUNCH	Print user-defined quantities to the selected-output file

Conventions for Data Input

PHREEQC was designed to eliminate some of the input errors due to complicated data formatting in Fortran-type input files. Data for the program are free format; spaces or tabs may be used to delimit input fields (except SOLUTION_SPREAD, which is delimited only with tabs); blank lines are ignored. Keyword data blocks within a simulation may be entered in any order. However, data elements entered on a single line are order specific. As much as possible, the program is case insensitive. However, chemical formulas are case sensitive.

The following conventions are used for data input to PHREEQC:

Keywords—Input data blocks are identified with an initial keyword. This word must be spelled exactly, although case is not important. Several of the keywords have synonyms. For example, **PURE_PHASES** is a synonym for EQUILIBRIUM_PHASES.

Identifiers—Identifiers are options that may be used within a keyword data block. Identifiers may have two forms: (1) they may be spelled completely and exactly (case insensitive) or (2) they may be preceded by a hyphen and then only enough characters to uniquely define the identifier are needed. The form with the hyphen is always acceptable and is recommended. Usually, the form without the hyphen is acceptable, but in some cases the hyphen is needed to indicate the word is an identifier rather than an identically spelled keyword; these cases are noted in the definitions of the identifiers in the following sections. In this report, the form with the hyphen is used except for identifiers of the SOLUTION keyword and the identifiers **log_k** and **delta_h**. The hyphen in the identifier <u>never</u> implies that the negative of a quantity is entered.

Chemical equations—For aqueous, exchange, and surface species, chemical reactions must be *association* reactions, with the defined species occurring in the first position after the equal sign. For phases, chemical reactions must be *dissolution* reactions with the formula for the defined phase occurring in the first position on the left-hand side of the equation. Additional terms on the left-hand side are allowed. All chemical equations must contain an equal sign, "=". In addition, left- and right-hand sides of all chemical equations must balance in numbers of atoms of each element and total charge. All equations are checked for these criteria at runtime, unless they are specifically excepted. Nested parentheses in chemical formulas are acceptable. Spaces and tabs within chemical equations are ignored. Waters of hydration and other chemical formulas (that are normally represented by a " · ", as in the formula for gypsum, $CaSO_4 \cdot 2H_2O$) are designated with a colon (":") in PHREEQC (thus, $CaSO_4:2H_2O$), but only one colon per formula is permitted.

Element names—Two forms of element names are available (1) those beginning with an alphabetic character and (2) those beginning with a square bracket. For form 1, an element formula, wherever it is used, must begin with a capital letter and may be followed by one or more lowercase letters or underscores, "_". Numbers are not permitted, except in parentheses for defining the redox state. In general, element names are simply the chemical symbols for elements, which have a capital letter and zero or one lower case letter. It is sometimes useful to define other entities as elements, which allows mole balance and mass-action equations to be applied. Thus, "Fulvate" is an acceptable element name, and it would be possible to define metal binding constants in terms of metal-Fulvate complexes.

Form 2 of element names is less restrictive than form 1. Within the square brackets, any combination of alphanumeric characters and the characters plus, minus, equal, colon, decimal point, and underscore can be used. The form-2 element name is case dependent, but upper and lower case characters can be used in any position. The *iso.dat* database makes extensive use of the square-bracket form for element names by using the mass number and chemical symbol for minor-isotope definitions, such as [13C], [15N], and [34S].

Charge on a chemical species—The charge on a species may be defined by the proper number of pluses or minuses following the chemical formula or by a single plus or minus followed by an integer designating the charge. Either of the following is acceptable, Al+3 or Al+++. However, Al3+ would be interpreted as a molecule with three aluminum atoms and a charge of plus one.

Valence states—Redox elements that exist in more than one valence state in solution are identified for definition of solution composition by the element name followed by a valence in parentheses. Thus, sulfur that exists as sulfate is defined as S(6) and total sulfide (H_2S, HS^-, and others) is identified by S(-2). The valence may include a decimal point. The valence number is for identification purposes only and does not otherwise affect the calculations.

log *K* and temperature dependence—The identifier **log_k** is used to define the log *K* at 25 °C for a reaction. The temperature dependence for log *K* may be defined by the Van't Hoff expression or by an analytical expression. The identifier **delta_h** is used to give the standard enthalpy of reaction at 25 °C for a chemical reaction, which is used in the Van't Hoff equation. By default the units of the standard enthalpy are kilojoule per mole (kJ/mol). Optionally, for each reaction the units may be defined to be kilocalorie per mole (kcal/mol). An analytical expression for the temperature dependence of log *K* for a reaction may be defined with the **-analytical_expression** identifier. Up to six numbers may be given, which are the coefficients for

the equation: $\log_{10} K = A_1 + A_2 T + \dfrac{A_3}{T} + A_4 \log_{10} T + \dfrac{A_5}{T^2} + A_6 T^2$, where T is in kelvin. A log *K* is defined either

with **log_k** or **-analytical_expression** (default **log_k** is zero); the enthalpy is optional (default is zero). If present, an analytical expression is used in preference to the **log_k** and enthalpy values for calculation of the log *K* at the specified temperature.

Pressure dependence of log *K*—Pressure dependency of reaction constants for species, and the pressure-dependent solubilities of minerals and gases, are calculated from the volume change of the reaction. The molar volume of solids and parameters for calculating the molal volume of aqueous species are defined in *Amm.dat*, *phreeqc.dat,* and *pitzer.dat*.

Comments—The "#" character delimits the beginning of a comment in the input file. All characters in the line that follow this character are ignored. If the entire line is a comment, the line is not echoed to the output file. If the comment follows input data on a line, the entire line, including the comment, is echoed to the output file. The "#" is useful for adding comments explaining the source of various data or describing the problem setup. In addition, it is useful for temporarily removing lines from an input file.

Logical line separator—A semicolon (";") is interpreted as a logical end-of-line character. This allows multiple logical lines to be entered on the same physical line. For example, solution data could be entered as:

```
pH 7.0; pe 4.0; temp 25.0
```

on one line. The semicolon should not be used in character fields, such as the title or other comment or description fields.

Logical line continuation—A backslash ("\") at the end of a line may be used to merge two physical lines into one logical line. For example, a long chemical equation could be entered as:

```
Ca0.165Al2.33Si3.67O10(OH)2 + 12 H2O = \
0.165Ca+2 + 2.33 Al(OH)4- + 3.67 H4SiO4 + 2 H+
```

on two lines. The program would interpret this sequence as a balanced equation entered on a single logical line. For a line to be logically continued, the backslash must be the last character in the line except for white space.

Repeat count—An asterisk ("*") can be used to indicate a repeat count for the data item that follows the asterisk. The format is an integer followed directly by the asterisk, which is followed directly by a numeric value. For example "4*1.0" is the same as entering four values of 1.0 ("1.0 1.0 1.0 1.0"). Repeat counts can be used for specifying data for the identifiers **-length** and **-dispersivity** in the **TRANSPORT** data block and for specifying reaction steps in the **REACTION** and **KINETICS** data blocks.

Range of integers—A hyphen ("-") can be used to indicate a range of integers for the keywords with an identification number (for example, **SOLUTION** 2-5). It is also possible to define a range of cell numbers for the identifiers **-print_cells** and **-punch_cells** in the **ADVECTION** and **TRANSPORT** data blocks and in the options for the **COPY**, **DELETE**, **DUMP**, and **RUN_CELLS** data blocks. A range of integers is given in the form *m-n*, where *m* and *n* are positive integers, *m* is less than *n*, and the two numbers are separated by a hyphen without intervening spaces.

Special characters—A summary of all of the special characters used in PHREEQC formatting is given in table 2.

Reducing Chemical Equations to a Standard Form

The numerical algorithm of PHREEQC requires that chemical equations be written in a particular form. Internally, every equation must be written in terms of a minimum set of chemical species; essentially, one species for each element or valence state of an element. For the program PHREEQE, these species were called "master species" and the reactions for all aqueous complexes had to be written using only these species. PHREEQC also needs reactions in terms of master species; however, the program contains the logic to rewrite the input equations into this form. Thus, it is possible to enter an association reaction and log K for an aqueous species in terms of any aqueous species in the database (not just master species), and PHREEQC will rewrite the equation to the proper internal form.

Table 2. Summary of special characters for input data files.

Special character	Use
-	When preceding a character string, a hyphen indicates an identifier (option) for a keyword.
-	Indicates a range of cell numbers for keyword data blocks (for example, **SOLUTION** 2-5), for identifiers **-print_cells** and **-punch_cells** in the ADVECTION and TRANSPORT data blocks, and for identifiers in the COPY, DELETE, DUMP, and RUN_CELLS data blocks.
:	In a chemical equation, ":" replaces "·" in a formula like $CaSO_4 \cdot 2H_2O$.
[ ]	Used to define element names including numeric and a limited set of special characters (+-._:).
( )	The redox state of an element is defined by a valence enclosed by parentheses following an element name.
#	Comment character, all characters following # are ignored.
;	Logical line separator.
\	Line continuation if "\" is the last non-white-space character of a line.
*	Can be used to indicate a repeat count for **-length** and **-dispersivity** values in the TRANSPORT data block and steps in the REACTION and KINETICS blocks.

PHREEQC also will rewrite reactions for phases, exchange complexes, and surface complexes. Reactions are required to be dissolution reactions for phases and association reactions for aqueous, exchange, or surface complexes. Dissolution reactions for phases allow inclusion of names of solids and gases in the equations, provided they are appended with the strings "(s)" and "(g)"; for example,

```
CaCO2[18O](s) + H2O(l) = H2[18O](aq) + Calcite(s).
```

The string "(l)" can be appended to the water formula and "(aq)" to aqueous species for clarity, but they are not required. The "(s)" and "(g)" suffixes cause the program to look in the list of phases to find equations that can be used to reduce the original equation to an equation that contains exclusively aqueous species. This capability to use solids and gases in chemical reactions for phases was implemented primarily to simplify the definition of equations for isotopic solid and gas components. The log Ks for these isotopic species often depend on the log K for the predominant isotopic species (solid or gas) offset by a fractionation factor and (or) a symmetry-derived log K. The inclusion of gases and solids in the equations for isotopic solids and gases is a straightforward method to define these dependencies of the isotopic species equilibrium constant on the equilibrium constant for the predominant isotopic species. In the example given here, the equilibrium constant for the single oxygen-18 form of calcium carbonate solid depends on the equilibrium constant of the pure carbon-12, oxygen-16 form of calcite, which is specified by "Calcite(s)" in the example equation and refers to the equation and log K defined for the calcite phase.

There is one major restriction on the rewriting capabilities for aqueous species. PHREEQC calculates mole balances on individual valence states or combinations of valence states of an element for initial solution calculations. It is necessary for PHREEQC to be able to determine the valence state of an element in a species from the chemical equation that defines the species. To do this, the program requires that only one aqueous species of an element valence state is defined by the electron half-reaction that relates it to another valence state. The aqueous species defined by this half-reaction is termed a "secondary master species"; there must be a one-to-one correspondence between valence states and secondary master species and the coefficient of the newly defined species must be one. In addition, there must be one "primary master species" for each element, such that reactions for all aqueous species for an element can be rewritten in terms of the primary master species. The equation for the primary master species is simply an identity reaction. If the element is a redox element, the primary master species must also be a secondary master species. For example, to be able to calculate mole balances on total iron, total ferric iron, or total ferrous iron, a primary master species must be defined for Fe (iron) and secondary master species must be defined for Fe(+3) (ferric iron) and Fe(+2) (ferrous iron). In the default databases, the primary master species for Fe is Fe^{+2}, the secondary master species for Fe(+2) is Fe^{+2}, and the secondary master species for Fe(+3) is Fe^{+3}. The correspondence between master species and elements and element valence states is defined by the SOLUTION_MASTER_SPECIES data block, which for iron in *phreeqc.dat* is as follows:

```
SOLUTION_MASTER_SPECIES
Fe                    Fe+2    0.0    Fe              55.847
Fe(+2)                Fe+2    0.0    Fe
Fe(+3)                Fe+3   -2.0    Fe
```

The line with "Fe" (without parentheses) defines the primary master species, and the last two lines, which have parentheses following "Fe", define the secondary master species. The chemical equations for the master species and all other aqueous species are defined by the SOLUTION_SPECIES data block.

Conventions for Documentation

The descriptions of keywords and their associated input are now described in alphabetical order as listed in table 1. Several formatting conventions are used to help the user interpret the input requirements. In this report, keywords are always capitalized and bold. Words in bold must be included literally when creating input files (although upper and lower case are interchangeable and optional spellings may be permitted). "Identifiers" are additional keywords that apply only within a given keyword data block; they can be considered to be sub-keywords or options. Although identifiers are case independent, lowercase bold is used in this report for all identifiers except **pH**, **-Donnan**, **-multi_D**, and **-interlayer_D**, for which mixed case is used. "**temperature**" is an identifier for SOLUTION input. Each identifier may have two forms: (1) the identifier word spelled exactly ("**temperature**", in this case), or (2) a hyphen followed by a sufficient number of characters to define the identifier uniquely (for example, **-t** for temperature in SOLUTION the data block.). The form with the hyphen is recommended. Words in *italics* are input values that are variable and depend on user selection of appropriate values. Items in brackets ([ ]) are optional input fields. Mutually exclusive input fields are enclosed in parentheses and separated by the word "or". In general, the optional fields in a line must be entered in the specified order, but it is sometimes possible to omit intervening fields. For clarity, commas sometimes are used to delimit input fields in the explanations of data input; however, commas are not allowed in the input data file except in Basic programs; in all other cases, only white space (spaces and tabs) may be used to delimit fields in input files. Where applicable, default values for input fields are stated.

Getting Started

When the program PHREEQC is invoked, two files are used to define the thermodynamic model and the types of calculations that will be done, the database file and the input file. The database file is read once (to the end of the file or until an END keyword is encountered) at the beginning of the program. The input file is then read and processed simulation by simulation (as defined by END keywords) until the end of the file. The formats for the keyword data blocks are the same for either the input file or the database file.

The database file is used to define static data for the thermodynamic model. Although any keyword data block can occur in the database file, normally, the file contains the keyword data blocks: EXCHANGE_MASTER_SPECIES, EXCHANGE_SPECIES, PHASES, RATES, SOLUTION_MASTER_SPECIES, SOLUTION_SPECIES, SURFACE_MASTER_SPECIES, and SURFACE_SPECIES. These keyword data blocks define rate expressions, master species, and the stoichiometric and thermodynamic properties of all of the aqueous phase species, exchange species, surface species, and pure phases.

Nine database files are provided with the program: (1) *phreeqc.dat*, a database file derived from PHREEQE (Parkhurst and others, 1980), which is consistent with *wateq4f.dat*, but has a smaller set of elements and aqueous species (table 3); (2) *Amm.dat* is the same as *phreeqc.dat*, except that ammonia redox state has been decoupled from the rest of the nitrogen system; that is, ammonia has been defined as a separate component; (3) *wateq4f.dat*, a database file derived from WATEQ4F (Ball and Nordstrom, 1991); (4) *llnl.dat*, a database file derived from databases for EQ3/6 and Geochemist's Workbench that uses thermodynamic data compiled by the Lawrence Livermore National Laboratory; (5) *minteq.dat*, a database derived from the databases for the program MINTEQA2 (Allison and others, 1990); (6) *minteq.v4.dat*, a database derived from MINTEQA2 version 4 (U.S. Environmental Protection Agency, 1998); (7) *pitzer.dat*, a database for the specific-ion-interaction model of Pitzer (Pitzer, 1973) as implemented in PHRQPITZ (Plummer and others, 1988); (8) *sit.dat*, a database implementing the Specific ion Interaction Theory (SIT) as described by Grenthe and others (1997); and (9) *iso.dat*, a partial implementation of the individual component approach to isotope calculations as described by Thorstenson and Parkhurst (2002, 2004). The elements and element valence states, corresponding notation, and default formula used to convert mass concentration to mole concentration units in the database *phreeqc.dat* are listed in table 3. Other databases may use different sets of elements, different notation for the element names, or different default conversion formulas.

The input data file is used (1) to define the types of calculations that are to be done, and (2) if necessary, to modify the data read from the database file. If new elements and aqueous species, exchange species, surface species, or phases need to be included in addition to those defined in the database file, or if the stoichiometry, log K, or activity coefficient information from the database file needs to be modified for a given run, then the keywords mentioned in the previous paragraph can be included in the input file. The data read for these keyword data blocks in the input file will augment or supersede the data read from the database file. In many cases, the thermodynamic model defined in the database will not be modified, and the above keywords will not be used in the input data file.

The place to start is with the simplest input file, which contains only a **SOLUTION** data block containing the dissolved concentrations of elements. With this input file, PHREEQC will perform a speciation calculation and calculate saturation indices for the solution. More complex calculations will calculate new solution compositions as a function of reactions. Reactions can be understood as occurring in a beaker, where a solution (as defined by a **SOLUTION** data block) is placed in the beaker, and then additional reactants are added. The reactants are defined with the keywords **EQUILIBRIUM_PHASES**, **EXCHANGE, GAS_PHASE, KINETICS, REACTION, SOLID_SOLUTIONS**, and **SURFACE.** One or more of these reactants may be added to the beaker, and then system equilibrium is calculated, which results in mole transfers into and out of solution, and new pH and element concentrations. The pressure and temperature of the reaction may be defined with **REACTION_PRESSURE** and

Table 3. Elements and element valence states included in default database *phreeqc.dat*, including PHREEQC notation and default formula for gram formula weight.
[For alkalinity, formula for gram equivalent weight is given]

Element or element valence state	PHREEQC notation	Formula used for default gram formula weight
Alkalinity	Alkalinity	$Ca_{0.5}(CO_3)_{0.5}$
Aluminum	Al	Al
Barium	Ba	Ba
Boron	B	B
Bromide	Br	Br
Cadmium	Cd	Cd
Calcium	Ca	Ca
Carbon	C	HCO_3
Carbon(IV)	C(4)	HCO_3
Carbon(-IV), methane	C(-4)	CH_4
Chloride	Cl	Cl
Copper	Cu	Cu
Copper(II)	Cu(2)	Cu
Copper(I)	Cu(1)	Cu
Fluoride	F	F
Hydrogen(0), dissolved hydrogen	H(0)	H
Iron	Fe	Fe
Iron(II)	Fe(2)	Fe
Iron(III)	Fe(3)	Fe
Lead	Pb	Pb
Lithium	Li	Li
Magnesium	Mg	Mg
Manganese	Mn	Mn
Manganese(II)	Mn(2)	Mn
Manganese(III)	Mn(3)	Mn
Nitrogen	N	N
Nitrogen(V), nitrate	N(5)	N
Nitrogen(III), nitrite	N(3)	N
Nitrogen(0), dissolved nitrogen	N(0)	N
Nitrogen(-III), ammonia	N(-3)	N
Oxygen(0), dissolved oxygen	O(0)	O
Phosphorus	P	P
Potassium	K	K
Silica	Si	SiO_2
Sodium	Na	Na
Strontium	Sr	Sr
Sulfur	S	SO_4
Sulfur(VI), sulfate	S(6)	SO_4
Sulfur(-II), sulfide	S(-2)	S
Zinc	Zn	Zn

REACTION_TEMPERATURE. So, the design of PHREEQC is fairly intuitive. You must choose the composition of a starting solution and then decide which types of reactants you need to add to the beaker to model your system. Transport reactions are simply defined by a series of beakers, each containing a set of reactants, and water flows and mixes from one beaker to the next and equilibrates with the reactants in each beaker in sequence.

Units

The concentrations of elements in solution and the mass of water in the solution are defined through the SOLUTION or SOLUTION_SPREAD data block. Internally, all concentrations are converted to molality and the number of moles of each element in solution (including hydrogen and oxygen) is calculated from the molalities and the mass of water. Thus, internally, a solution is simply a list of elements and the number of moles of each element.

PHREEQC allows each reactant to be defined independently. In particular, reactants (EQUILIBRIUM_PHASES, EXCHANGE, GAS_PHASE, KINETICS, REACTION, SOLID_SOLUTIONS, and SURFACE) are defined in terms of moles, without reference to a volume or mass of water. Systems are defined by combining a solution with a set of reactants that react either reversibly (EQUILIBRIUM_PHASES, EXCHANGE, GAS_PHASE, SOLID_SOLUTIONS, and SURFACE) or irreversibly (KINETICS or REACTION). Essentially, all of the moles of elements in the solution and the reversible reactants are combined, the moles of irreversible reactants are added (or removed), and a new system equilibrium is calculated. Only after system equilibrium is calculated is the mass of water in the system known, and only then the molalities of all entities can be calculated.

For transport calculations, each cell is a system that is defined by the solution and all the reactants contained in keywords that bear the same number as the cell number. The system for the cell initially is defined by the moles of elements that are present in the solution and the moles of each reactant. The compositions of all these entities evolve as the transport calculations proceed.

Keywords

The following sections describe the data input requirements for the program. Each type of data is input through a specific keyword data block. Most keywords are listed in alphabetical order within this section of the report; however, a set of keywords most pertinent to model developers is described in Appendix A. Each keyword data block may have a number of identifiers, many of which are optional. Identifiers may be entered in any order; the line numbers given in examples for the keyword data blocks are for identification purposes only. Default values for identifiers are used if the identifier is omitted.

ADVECTION

This keyword data block is used to specify the number of cells and the number of "shifts" for an advection simulation. Advection simulations are used to model one-dimensional advective or "plug" flow with reactions. No dispersion or diffusion is simulated and no cells with immobile water are allowed. However, all chemical processes modeled by PHREEQC may be included in an advection simulation. The **TRANSPORT** data block may be used to model additional physical processes, such as dispersion, diffusion, and exchange with cells containing immobile water.

Example data block

```
Line  0: ADVECTION
Line  1:      -cells 5
Line  2:      -shifts 25
Line  3:      -time_step 1 yr
Line  4:      -initial_time 1000
Line  5:      -print_cells 1-3 5
Line  6:      -print_frequency 5
Line  7:      -punch_cells 2-5
Line  8:      -punch_frequency 5
Line  9:      -warnings false
```

Explanation

Line 0: **ADVECTION**

ADVECTION is the keyword for the data block. No other data are input on the keyword line.

Line 1: **-cells** *cells*

-cells—Identifier for number of cells in the advection simulation. Optionally, **cells** or **-c[ells]**.

cells—Number of cells in the one-dimensional column to be used in the advection simulation. Default is 1.

Line 2: **-shifts** *shifts*

-shifts—Identifier for the number of shifts or time steps in the advection simulation. Optionally, **shifts** or **-sh[ifts]**.

shifts—Number of times the solution in each cell will be shifted to the next higher numbered cell. Default is 1.

Line 3: **-time_step** *time_step* [*unit*]

> **-time_step**—Identifier for time step associated with each advective shift. The identifier is required if kinetic reactions (**KINETICS** data blocks) are part of the advection simulation and optional for other advection simulations. If **-time_step** is defined, then the values for time printed to the selected-output file will be *initial_time* + *shift_number* × *time_step*, where *shift_number* represents the sequence number of a shift; if **-time_step** is not defined, the value of time printed to the selected-output file will be the advection shift number. Once **-time_step** is defined, the time step will be used for all subsequent advection simulations until it is redefined. Optionally, **timest**, **-t**[**imest**], **time_step**, or **-t**[**ime_step**].

> *time_step*—The time associated with each advective shift. Kinetic reactions will be integrated for this period of time for each advective shift. Default is 0 s (second).

> *unit*—Optional time unit may be **second**, **minute**, **hour**, **day**, **year**, or an abbreviation of one of these units. The *time_step* is converted to seconds after reading the data block; all internal calculations, Basic functions, and output times are in seconds. Default is second.

Line 4: **-initial_time** *initial_time*

> **-initial_time**—Identifier to set the time at the beginning of an advection simulation. The identifier **-initial_time** has effect only if **-time_step** has been set in this or a previous **ADVECTION** data block. The identifier sets the initial value of the variable controlled by **-time** in **SELECTED_OUTPUT** data block. Optionally, **initial_time** or **-i**[**nitial_time**].

> *initial_time*—Time (seconds) at the beginning of the advection simulation. Default is the cumulative time including all preceding **ADVECTION** simulations for which **-time_step** has been defined and all preceding **TRANSPORT** simulations.

Line 5: **-print_cells** *list of cell numbers*

> **-print_cells**—Identifier to select cells for which results will be written to the output file. If **-print_cells** is not included, results for all cells will be written to the output file. Once **-print_cells** is defined, the list of cells will be used for all subsequent advection simulations until the list is redefined. Optionally, **print_cells** or **-pr**[**int_cells**]. Note the hyphen is required in **-print** to avoid a conflict with the keyword **PRINT**.

> *list of cell numbers*—Printing to the output file will occur only for these cell numbers. The list of cell numbers must be delimited by spaces or tabs and may be continued on the succeeding line(s). A range of cell numbers may be included in the list in the form *m-n*, where *m* and *n*

are positive integers, m is less than n, and the two numbers are separated by a hyphen without intervening spaces. Default 1-*cells*.

Line 6: **-print_frequency** *print_modulus*

-print_frequency—Identifier to select shifts for which results will be written to the output file. Once defined, the print frequency will be used for all subsequent advection simulations until it is redefined. Optionally, **print_frequency, -print_f[requency], output_frequency, -o[utput_frequency]**.

print_modulus—Printing to the output file will occur for advection shifts that are evenly divisible by *print_modulus*. Default is 1.

Line 7: **-punch_cells** *list of cell numbers*

-punch_cells—Identifier to select cells for which results will be written to the selected-output file. If **-punch_cells** is not included, results for all cells will be written to the selected-output file. Once defined, the list of cells will be used for all subsequent advection simulations until the list is redefined. Optionally, **punch, punch_cells, -pu[nch_cells], selected_cells**, or **-selected_c[ells]**.

list of cell numbers—Printing to the selected-output file will occur only for these cell numbers. The list of cell numbers must be delimited by spaces or tabs and may be continued on the succeeding line(s). A range of cell numbers may be included in the list in the form *m-n*, where *m* and *n* are positive integers, *m* is less than *n*, and the two numbers are separated by a hyphen without intervening spaces. Default 1-*cells*.

Line 8: **-punch_frequency** *punch_modulus*

-punch_frequency—Identifier to select shifts for which results will be written to the selected-output file. Once defined, the punch frequency will be used for all subsequent advection simulations until it is redefined. Optionally, **punch_frequency, -punch_f[requency], selected_output_frequency, -selected_o[utput_frequency]**.

punch_modulus—Printing to the selected-output file will occur for advection shifts that are evenly divisible by *punch_modulus*. Default is 1.

Line 9: **-warnings** [(*True* or *False*)]

-warnings—Identifier enables or disables printing of warning messages for advection calculations. In some cases, advection calculations could produce many warnings, which are not errors. Once it is determined that the warnings are not due to erroneous input, disabling the warning messages can avoid generating large output files. Default is **true** at startup. Optionally, **warnings**, **warning**, or **-w[arnings]**.

(*True* or *False*)—If true, warning messages are printed to the screen and the output file; if false, warning messages are not printed to the screen or the output file. The value set with **-warnings** is retained in subsequent advection simulations until changed. If neither **true** nor **false** is entered on the line, **true** is assumed. Optionally, **t[rue]** or **f[alse]**.

Notes

The capabilities available through the **ADVECTION** data block are a simplified version of a more complete formulation of 1D advective-dispersive-reactive transport that is presented by Appelo and Postma (2005) and implemented in the TRANSPORT data block. **ADVECTION** allows flow only in the forward direction (from lower to higher numbered cells) and does not simulate dispersion, stagnant-zone transport, multicomponent diffusion, or surface transport. Calculations using the **ADVECTION** keyword are sufficient for initial investigations, and in comparison to other problems that include dispersion, the calculations are fast. The TRANSPORT data block allows modeling of the additional processes of diffusion, dispersion, and diffusion into stagnant zones. The transport capabilities of the **ADVECTION** keyword in PHREEQC versions 2 and 3 are equivalent to the capabilities of the TRANSPORT keyword in PHREEQC version 1.

In the Example data block given in this section, a column of five cells (*cells*) is modeled and 5 pore volumes of filling solution are moved through the column (*shifts/cells* is 5). Unless kinetic reactions are modeled, no explicit definition of time is required, only the number of shifts. Also, no distance is explicitly specified for advection calculations, only the number of cells.

The **-time_step** identifier is required if kinetic reactions (KINETICS data block) are defined for at least one cell in the column. If kinetic reactions are defined, then an integration is performed for each cell that has kinetic reactions for each advective shift. Kinetic reactions significantly increase the run time of a simulation because the integration of the rates of reaction imposes 1 to 6 (or possibly more) additional batch-reaction calculations for each cell that has kinetic reactions for each advective shift. The total time modeled in the Example data block simulation is 25 yr (*time_step* × *shifts*).

By default, the composition of the solution, pure-phase assemblage, exchange assemblage, gas phase, solid-solution assemblage, surface assemblage, and kinetic reactants are printed for each cell for each shift. Use of **-print_cells** and **-print_frequency** will limit the data written to the output file. The **-print_cells** identifier restricts printing in the output file to the specified cells; in the Example data block, results for cells 1, 2, 3, and 5 are printed to the output file. The identifier **-print_frequency** restricts printing in the output file to those advection shifts that are evenly divisible by *print_modulus*. In the Example data block, results are printed to the output file after each integer pore volume (5 shifts). Data written to the output file can be further limited with the keyword PRINT (see **-reset false**). The USER_PRINT data block can be used to calculate quantities to be printed to the output file.

If the SELECTED_OUTPUT data block has been defined, then data specified in the SELECTED_OUTPUT and USER_PUNCH data blocks are written to the selected-output file. Use of **-punch_cells** and **-punch_frequency** in the **ADVECTION** data block will limit what is written to the selected-output file. The **-punch_cells** identifier restricts printing to the selected-output file to the specified cells; in the Example data block, results for cells 2, 3, 4, and 5 are printed to the selected-output file. The identifier **-punch_frequency** restricts printing to the selected-output file to those advection shifts that are evenly divisible by *punch_modulus*. In the Example data block, results are printed to the selected-output file after each integer pore volume (5 shifts). All printing to the selected-output file can be switched on or off through the **-selected_output** identifier of the keyword PRINT.

Most of the information for advection calculations must be entered with other keywords. This advection calculation assumes that solutions with numbers 0 through 5 have been defined by using the SOLUTION, COPY, or SAVE data blocks. Solution 0 is the infilling solution and solutions 1 through 5 are the initial solutions in the cells of the column. Other reactants may be defined for each of the cells. Pure-phase assemblages may be defined with EQUILIBRIUM_PHASES, COPY, or SAVE, with the number of the assemblage corresponding to the cell number. Likewise, an exchange assemblage, gas phase, solid-solution assemblage, or surface assemblage can be defined for each cell through EXCHANGE, GAS_PHASE, SOLID_SOLUTIONS, SURFACE, COPY, or SAVE data blocks, with the identifying number corresponding to the cell number. Note that ranges of numbers can be used (for example SOLUTION 1-5) to define multiple solutions, pure-phase assemblages, exchange assemblages, gas phases, solid-solution assemblages, or surface assemblages and that COPY and SAVE allow a range of numbers to be used.

The REACTION data block can be used to define a stoichiometric reaction that applies to a cell at each shift, with the reaction number corresponding to the cell number. This capability is not very useful because it represents only zero-order kinetics, and the reaction rate is constant throughout the advection simulation. The KINETICS data block provides a better definition of time-varying reactions for individual cells.

The **MIX** keyword can be used with **ADVECTION** modeling to define simplistic dispersion or lateral inflow to the column. At each shift, solution 0 is moved to cell 1, any stoichiometric reaction or mixing for cell 1 is added, kinetic reactions are integrated while maintaining equilibrium with the contents of cell 1; solution 1 (before mixing and reaction) is moved to cell 2, reaction or mixing for cell 2 is added, kinetic reactions are integrated while maintaining equilibrium with the contents of cell 2; and so on until solution *cells-1* is moved to cell *cells*. The moles of pure phases and kinetic reactants, and the compositions of the exchange assemblage, surface assemblage, and gas phase in each cell are updated with each shift, but only after mixing for the next cell has been accomplished.

Example problems

The keyword **ADVECTION** is used in example problems 11 and 14.

Related keywords

COPY, EQUILIBRIUM_PHASES, EXCHANGE, GAS_PHASE, KINETICS, MIX, PRINT, REACTION, REACTION_PRESSURE, REACTION_TEMPERATURE, SAVE, SELECTED_OUTPUT, SOLID_SOLUTIONS, SOLUTION, SURFACE, TRANSPORT, USER_PRINT, and USER_PUNCH.

CALCULATE_VALUES

This keyword data block allows the definition of Basic functions that can be used in other PHREEQC Basic programs or to display isotopic compositions as defined in the **ISOTOPE_ALPHAS** and **ISOTOPE_RATIOS** data blocks. Isotope ratios are the ratio of moles of a minor isotope to moles of a major isotope in an aqueous, gas, or mineral species. Isotope alphas are the ratio of two isotope ratios, which is the fractionation factor between two species. The data block is used primarily to display results from the isotopic simulations, but also can be used to write Basic functions to simplify Basic programming for **RATES**, **USER_GRAPH**, **USER_PRINT**, and **USER_PUNCH**.

Example data block

```
Line 0: CALCULATE_VALUES
Line 1: R(D)
Line 2: -start
Line 3:    10  ratio = -9999.999
Line 3a:   20  if (TOT("D") <= 0) THEN GOTO 100
Line 3b:   30  total_D = TOT("D")
Line 3c:   40  total_H = TOT("H")
Line 3d:   50  ratio = total_D/total_H
Line 3e:   100 SAVE ratio
Line 4: -end
```

Explanation

Line 0: **CALCULATE_VALUES**

CALCULATE_VALUES is the keyword for the data block. No other data are input on the keyword line.

Line 1: *name of Basic function*

name of Basic function—Alphanumeric character string that defines the name of the Basic function; no spaces are allowed. The function defined in this example would be evaluated in other PHREEQC Basic programs by using CALC_VALUE("R(D)").

Line 2: **-start**

-start—Identifier marks the beginning of a Basic program. Optional.

Line 3: *numbered Basic statement*

numbered Basic statement—A valid Basic language statement that must be numbered. The statements are evaluated in numerical order. The statement "SAVE *expression*" must be included in the list of statements, where the value of *expression* is the result that is returned from the Basic function. Statements and functions that are available through the Basic interpreter are listed in The Basic Interpreter (tables 7 and 8).

Line 4: **-end**

> **-end**—Identifier marks the end of a Basic function. Note the hyphen is required to avoid a conflict with the keyword **END**.

Notes

The **CALCULATE_VALUES** data block is used to write Basic functions. These functions have no arguments and return a single numeric result defined in the line with the SAVE command. The functions may be called from other PHREEQC Basic programs by using the function CALC_VALUE(*"function_name"*), where *function_name* is the name of a function defined in a **CALCULATE_VALUES** data block.

The **CALCULATE_VALUES** data block is used primarily to implement isotopic calculations as described in Thorstenson and Parkhurst (2002, 2004), but could be used for other purposes as well. The database *iso.dat* contains many instances of Basic functions in **CALCULATE_VALUES** data blocks. Basic functions have been written to calculate isotope ratios (moles of minor isotope divided by moles of major isotope) for a large number of isotopes and aqueous, gas, and mineral species. The example given at the beginning of this section calculates the ratio of deuterium to hydrogen in an aqueous solution. The functions also are used to calculate ratios of isotope ratios (alphas), which are equal to the fractionation factors among species. These functions are then specified in the ISOTOPE_ALPHAS and ISOTOPE_RATIOS data blocks to enable listing isotopic ratios and alphas in the output file. The functions also can be used in Basic programs in RATES, USER_GRAPH, USER_PRINT, and USER_PUNCH data blocks.

Example problems

The keyword **CALCULATE_VALUES** is used in the database *iso.dat*.

Related keywords

ISOTOPE_ALPHAS, ISOTOPE_RATIOS, RATES, USER_GRAPH, USER_PRINT, and USER_PUNCH.

COPY

This keyword data block is used to make copies of any of the numbered reactants, which include equilibrium-phase assemblages, exchange assemblages, gas phases, kinetic reactions, mix definitions, reactions, reaction-pressure definitions, reaction-temperature definitions, solid-solution assemblages, solutions, or surface assemblages. These reactants are usually defined by the EQUILIBRIUM_PHASES, EXCHANGE, GAS_PHASE, KINETICS, MIX, REACTION, REACTION_PRESSURE, REACTION_TEMPERATURE, SOLID_SOLUTIONS, SOLUTION, and SURFACE data blocks, but may be defined or modified with the SAVE, _RAW, or _MODIFY (see Appendix A) data blocks.

Example data block

```
Line 0:   COPY equilibrium_phases 2 3-5
Line 0a:  COPY exchange 1 11
Line 0b:  COPY gas_phase 1 11
Line 0c:  COPY kinetics 2 3-5
Line 0d:  COPY mix 1 11
Line 0e:  COPY reaction 1 11
Line 0f:  COPY reaction_pressure 25 15
Line 0g:  COPY reaction_temperature 2 3-5
Line 0h:  COPY solid_solution 1 11
Line 0i:  COPY solution 1 11
Line 0j:  COPY surface 1 11
Line 0k:  COPY cell 1 21
```

Explanation

Line 0: **COPY** *reactant source_number destination_number_range*

COPY is the keyword for the data block.

reactant—The word "**cell**", or one of the 10 reactants that can be identified by an integer—**equilibrium_phases, exchange, gas_phase, kinetics, mix, reaction, reaction_pressure, reaction_temperature, solid_solution, solution,** or **surface.**

source_number—An integer designating the reactant to be copied. If *reactant* is **cell**, all reactants identified by *source_number* will be copied.

destination_number_range—A single number or a range of numbers designated by an integer followed by a hyphen, followed by an integer, with no intervening spaces. A copy of the source reactant will be made for each of the numbers in the range. If *reactant* is **cell**, all reactants identified by *source_number* will be copied for each of the numbers in the range.

Notes

The **COPY** operations are done after all reaction, advection, and transport calculations for a simulation, but before the DUMP and DELETE operations. If the reactant numbered *source_number* does not exist, the copy request is ignored. The *source_number* reactant will be copied so that after the copy operation, reactants will exist with each of the numbers designated by *destination_number_range*. If **cell** is designated for *reactant*, then for each reactant numbered *source_number*, a new copy will be generated for each number in the range given by *destination_number_range*. Unlike DELETE and DUMP, only a single number range is allowed for **COPY**. If a reactant with a specified number exists before the copy operation, that reactant will be overwritten.

Example problems

The keyword **COPY** is used in example problems 11, 12, and 15.

Related keywords

DUMP, DELETE, EQUILIBRIUM_PHASES, EXCHANGE, GAS_PHASE, KINETICS, MIX, REACTION, REACTION_PRESSURE, REACTION_TEMPERATURE, SOLID_SOLUTIONS, SOLUTION, and SURFACE.

DATABASE

This keyword data block is used to specify a database for the simulations.

Example data block

```
Line 0: DATABASE ../../database/pitzer.dat
```

Explanation

Line 0: **DATABASE** *database_file_name*

DATABASE is the keyword for the data block.

database_file_name—File name for the database. If the database is not in the working directory, then a path name relative to the working directory or an absolute path name must be given.

Notes

DATABASE must be the first keyword data block in an input file. It may be preceded by comment lines, but not by other keyword data blocks. The file specified in the **DATABASE** data block is used for the simulations regardless of other default or command-line-argument definition of the database file.

Example problems

The keyword **DATABASE** is used in example problems 14, 15, 17, and 20.

DELETE

This keyword data block is used to delete reactants. Any reactant that is identified by an integer can be deleted, which includes reactants defined by EQUILIBRIUM_PHASES, EXCHANGE, GAS_PHASE, KINETICS, MIX, REACTION, REACTION_PRESSURE, REACTION_TEMPERATURE, SOLID_SOLUTIONS, SOLUTION, and SURFACE data blocks.

Example data block

```
Line  0: DELETE
Line  1:        -equilibrium_phases
Line  2:        -exchange          2 4
Line  3:                           3 5
Line  4:        -gas_phase         2-4 5
Line  5:        -kinetics          3-5 2
Line  6:        -mix               2 3
Line  6a:       -mix               4 5
Line  7:        -reaction          2-3 4 5
Line  8:        -reaction_pressure 2 3 4-5
Line  9:        -reaction_temperature
Line  3a:                          2 3 4 5
Line 10:        -solid_solution
Line  3b:                          5 4 3 2
Line 11:        -solution          2 3 4 5
Line 12:        -surface
Line  3c:                          2-3 4-5
Line  3d:                          5
Line 13:        -cells             2-5
Line 14:        -all
```

Explanation

Line 0: **DELETE**

DELETE is the keyword for the data block. No other data are input on the keyword line.

Line 1: **-equilibrium_phases** *list_of_ranges*

-equilibrium_phases—Identifier indicates that equilibrium-phase assemblages will be deleted. Optionally, **-e[quilibrium_phases]**; note that the hyphen is necessary to distinguish the identifier from the keyword EQUILIBRIUM_PHASES.

list_of_ranges—List of number ranges. The number ranges may be a single integer or a range defined by an integer, a hyphen, and an integer, without intervening spaces.

Equilibrium-phase assemblages identified by any of the numbers in the list will be deleted. If *list_of_ranges* is empty, all equilibrium-phase-assemblage definitions will be deleted.

Line 2: **-exchange** *list_of_ranges*

-exchange—Identifier indicates that exchange assemblages will be deleted. Optionally, **-ex**[**change**]; note that the hyphen is necessary to distinguish the identifier from the keyword **EXCHANGE**.

list_of_ranges—List of number ranges. The number ranges may be a single integer or a range defined by an integer, a hyphen, and an integer, without intervening spaces. Exchange assemblages identified by any of the numbers in the list will be deleted. If *list_of_ranges* is empty, all exchange-assemblage definitions will be deleted.

Line 3: *list_of_ranges*

list_of_ranges—List of number ranges. The number ranges may be a single integer or a range defined by an integer, a hyphen, and an integer, without intervening spaces. Line 3 can be used to begin a list of ranges or to continue a list of ranges.

Line 4: **-gas_phase** *list_of_ranges*

-gas_phase—Identifier indicates that gas phases will be deleted. Optionally, **-g**[**as_phase**]; note that the hyphen is necessary to distinguish the identifier from the keyword **GAS_PHASE**.

list_of_ranges—List of number ranges. The number ranges may be a single integer or a range defined by an integer, a hyphen, and an integer, without intervening spaces. Gas phases identified by any of the numbers in the list will be deleted. If *list_of_ranges* is empty, all gas-phase definitions will be deleted.

Line 5: **-kinetics** *list_of_ranges*

-kinetics—Identifier indicates that kinetic reactants will be deleted. Optionally, **-k**[**inetics**]; note that the hyphen is necessary to distinguish the identifier from the keyword **KINETICS**.

list_of_ranges—List of number ranges. The number ranges may be a single integer or a range defined by an integer, a hyphen, and an integer, without intervening spaces. Kinetics reactants identified by any of the numbers in the list will be deleted. If *list_of_ranges* is empty, all kinetic-reaction definitions will be deleted.

Line 6: **-mix** *list_of_ranges*

-mix—Identifier indicates that solution mix definitions will be deleted. Optionally, **-m**[**ix**]; note that the hyphen is necessary to distinguish the identifier from the keyword **MIX**.

list_of_ranges—List of number ranges. The number ranges may be a single integer or a range defined by an integer, a hyphen, and an integer, without intervening spaces. Mix definitions identified by any of the numbers in the list will be deleted. If *list_of_ranges* is empty, all mix definitions will be deleted.

Line 7: **-reaction** *list_of_ranges*

-reaction—Identifier indicates that reactions will be deleted. Optionally, **-r[eaction]**; note that the hyphen is necessary to distinguish the identifier from the keyword REACTION.

list_of_ranges—List of number ranges. The number ranges may be a single integer or a range defined by an integer, a hyphen, and an integer, without intervening spaces. Reactions identified by any of the numbers in the list will be deleted. If *list_of_ranges* is empty, all reaction definitions will be deleted.

Line 8: **-reaction_pressure** *list_of_ranges*

-reaction_pressure—Identifier indicates that reaction-pressure definitions will be deleted. Optionally, **-reaction_p[ressure]**, **pressure**, or **-pr[essure]**; note that the hyphen is necessary to distinguish the identifier from the keyword REACTION_PRESSURE.

list_of_ranges—List of number ranges. The number ranges may be a single integer or a range defined by an integer, a hyphen, and an integer, without intervening spaces. Reaction-pressure definitions identified by any of the numbers in the list will be deleted. If *list_of_ranges* is empty, all reaction-pressure definitions will be deleted.

Line 9: **-reaction_temperature** *list_of_ranges*

-reaction_temperature—Identifier indicates that reaction-temperature definitions will be deleted. Optionally, **-reaction_[temperature]**, **temperature**, or **-t[emperature]**; note that the hyphen is necessary to distinguish the identifier from the keyword REACTION_TEMPERATURE.

list_of_ranges—List of number ranges. The number ranges may be a single integer or a range defined by an integer, a hyphen, and an integer, without intervening spaces. Reaction-temperature definitions identified by any of the numbers in the list will be deleted. If *list_of_ranges* is empty, all reaction-temperature definitions will be deleted.

Line 10: **-solid_solution** *list_of_ranges*

> **-solid_solution**—Identifier indicates that solid-solution assemblages will be deleted. Optionally, **-soli[d_solutions]**; note that the hyphen is necessary to distinguish the identifier from the keyword **SOLID_SOLUTIONS**.
>
> *list_of_ranges*—List of number ranges. The number ranges may be a single integer or a range defined by an integer, a hyphen, and an integer, without intervening spaces. Solid-solution assemblages identified by any of the numbers in the list will be deleted. If *list_of_ranges* is empty, all solid-solution assemblage definitions will be deleted.

Line 11: **-solution** *list_of_ranges*

> **-solution**—Identifier indicates that solutions will be deleted. Optionally, **-s[olution]**; note that the hyphen is necessary to distinguish the identifier from the keyword **SOLUTION**.
>
> *list_of_ranges*—List of number ranges. The number ranges may be a single integer or a range defined by an integer, a hyphen, and an integer, without intervening spaces. Solutions identified by any of the numbers in the list will be deleted. If *list_of_ranges* is empty, all solution definitions will be deleted.

Line 12: **-surface** *list_of_ranges*

> **-surface**—Identifier indicates that surface assemblages will be deleted. Optionally, **-su[rfaces]**; note that the hyphen is necessary to distinguish the identifier from the keyword **SURFACE**.
>
> *list_of_ranges*—List of number ranges. The number ranges may be a single integer or a range defined by an integer, a hyphen, and an integer, without intervening spaces. Surface assemblages identified by any of the numbers in the list will be deleted. If *list_of_ranges* is empty, all surface-assemblage definitions will be deleted.

Line 13: **-cells** *list_of_ranges*

> **-cells**—Identifier indicates that all reactants identified by a specified number will be deleted, including equilibrium-phase assemblages, exchanger assemblages, gas phases, kinetic reactants, mix definitions, reaction definitions, reaction-temperature definitions, solid-solution assemblages, solutions, and surface assemblages. Optionally, **-c[ells]**.
>
> *list_of_ranges*—List of number ranges. The number ranges may be a single integer or a range defined by an integer, a hyphen, and an integer, without intervening spaces. Reactants of any type that are identified by any of the numbers in the list will be deleted. If *list_of_ranges* is empty, all numbered reactants of all types will be deleted (same as **-all**).

Line 14: **-all**

-all—All numbered reactants of all types will be deleted.

Notes

The **DELETE** data block allows reactants to be deleted. All definitions in the example at the beginning of this section, except the first, delete reactants numbered 2 through 5. The first definition, **-equilibrium_phases** with no *list_of_ranges*, causes all equilibrium-phase-assemblage definitions to be deleted. The **DELETE** operations follow the COPY and DUMP operations and are the last operations of a simulation.

Usually, it is not necessary to use the **DELETE** data block in PHREEQC simulations because new definitions of reactants automatically overwrite old definitions and all reactants are deleted at the end of a run. The **DELETE** data block may be useful to remove reactants from cells between advection or transport calculations, to conserve memory during a run, or for reinitializing IPhreeqc modules (Charlton and Parkhurst, 2011).

Related keywords

COPY, DUMP, EQUILIBRIUM_PHASES, EXCHANGE, GAS_PHASE, KINETICS, MIX, REACTION, REACTION_PRESSURE, REACTION_TEMPERATURE, SOLID_SOLUTIONS, SOLUTION, and SURFACE.

DUMP

This keyword data block is used to write complete definitions of reactants to a specified file. The reactants are written in a "raw" format that saves the exact chemical state of each specified reactant. The raw data blocks are intended to be used intact to reinitialize simulations at the current state of the calculations or to transfer reactants from one IPhreeqc module (Charlton and Parkhurst, 2011) to another.

Example data block

```
Line  0:  DUMP
Line  1:      -file                myfile.dmp
Line  2:      -append              true
Line  3:      -equilibrium_phases
Line  4:      -exchange            2 4
Line  5:                           3 5
Line  6:      -gas_phase           2-4 5
Line  7:      -kinetics            3-5 2
Line  8:      -mix                 2 3
Line 8a:      -mix                 4 5
Line  9:      -reaction            2-3 4 5
Line 10:      -reaction_pressure   2 3 4 5
Line 11:      -reaction_temperature 5
Line 5a:                           2 3 4
Line 12:      -solid_solution
Line 5b:                           5 4 3 2
Line 13:      -solution            2 3 4 5
Line 14:      -surface
Line 5c:                           2-3 4-5
Line 5d:                           5
Line 15:      -cells               2-5
Line 16:      -all
```

Explanation

Line 0: **DUMP**

DUMP is the keyword for the data block. No other data are input on the keyword line.

Line 1: **-file** *file_name*

-file—Identifier is used to specify the name of the file to which the dump data are written. Optionally, **file** or **-f**[**ile**].

file_name— Name of file where dump data are written. File names must conform to operating system conventions. Default is **dump.out**.

Line 2: **-append** [(*True* or *False*)]

> **-append**—Identifier is used to specify whether the dump data will overwrite existing data or will be appended to the end of the file, if the file exists. Default is **false** at startup; any data in the dump file are overwritten. Optionally, **append** or -a[**ppend**].

> (*True* or *False*)—A value of **true** indicates that the dump data will be appended to the end of the dump file. A value of **false** indicates that any data in the dump file will be overwritten. If neither **true** nor **false** is entered on the line, **true** is assumed. Optionally, **t**[**rue**] or **f**[**alse**].

Line 3: **-equilibrium_phases** *list_of_ranges*

> **-equilibrium_phases**—Identifier indicates that equilibrium-phase assemblages will be dumped. Optionally, **-e**[**quilibrium_phases**]; note that the hyphen is necessary to distinguish the identifier from the keyword EQUILIBRIUM_PHASES.

> *list_of_ranges*—List of number ranges. The number ranges may be a single integer or a range defined by an integer, a hyphen, and an integer, without intervening spaces. Equilibrium-phase assemblages identified by any of the numbers in the list will be dumped. If *list_of_ranges* is empty, all equilibrium-phase-assemblage definitions will be dumped.

Line 4: **-exchange** *list_of_ranges*

> **-exchange**—Identifier indicates that exchange assemblages will be dumped. Optionally, **-ex**[**change**]; note that the hyphen is necessary to distinguish the identifier from the keyword EXCHANGE.

> *list_of_ranges*—List of number ranges. The number ranges may be a single integer or a range defined by an integer, a hyphen, and an integer, without intervening spaces. Exchange assemblages identified by any of the numbers in the list will be dumped. If *list_of_ranges* is empty, all exchange-assemblage definitions will be dumped.

Line 5: *list_of_ranges*

> *list_of_ranges*—List of number ranges. The number ranges may be a single integer or a range defined by an integer, a hyphen, and an integer, without intervening spaces. Line 5 can be used to begin a list of ranges or to continue a list of ranges.

Line 6: **-gas_phase** *list_of_ranges*

> **-gas_phase**—Identifier indicates that gas phases will be dumped. Optionally, **-g**[**as_phase**]; note that the hyphen is necessary to distinguish the identifier from the keyword GAS_PHASE.

list_of_ranges—List of number ranges. The number ranges may be a single integer or a range defined by an integer, a hyphen, and an integer, without intervening spaces. Gas phases identified by any of the numbers in the list will be dumped. If *list_of_ranges* is empty, all gas-phase definitions will be dumped.

Line 7: **-kinetics** *list_of_ranges*

-kinetics—Identifier indicates that kinetic reactants will be dumped. Optionally, **-k[inetics]**; note that the hyphen is necessary to distinguish the identifier from the keyword KINETICS.

list_of_ranges—List of number ranges. The number ranges may be a single integer or a range defined by an integer, a hyphen, and an integer, without intervening spaces. Kinetic reactants identified by any of the numbers in the list will be dumped. If *list_of_ranges* is empty, all kinetic-reaction definitions will be dumped.

Line 8: **-mix** *list_of_ranges*

-mix—Identifier indicates that mix definitions will be dumped. Optionally, **-m[ix]**; note that the hyphen is necessary to distinguish the identifier from the keyword MIX.

list_of_ranges—List of number ranges. The number ranges may be a single integer or a range defined by an integer, a hyphen, and an integer, without intervening spaces. Mix definitions identified by any of the numbers in the list will be dumped. If *list_of_ranges* is empty, all mix definitions will be dumped.

Line 9: **-reaction** *list_of_ranges*

-reaction—Identifier indicates that reactions will be dumped. Optionally, **-r[eaction]**; note that the hyphen is necessary to distinguish the identifier from the keyword REACTION.

list_of_ranges—List of number ranges. The number ranges may be a single integer or a range defined by an integer, a hyphen, and an integer, without intervening spaces. Reactions identified by any of the numbers in the list will be dumped. If *list_of_ranges* is empty, all reaction definitions will be dumped.

Line 10: **-reaction_pressure** *list_of_ranges*

-reaction_pressure—Identifier indicates that reaction-pressure definitions will be dumped. Optionally, **-reaction_p[ressure]**, **pressure**, or **-pr[essure]**; note that the hyphen is necessary to distinguish the identifier from the keyword REACTION_PRESSURE.

list_of_ranges—List of number ranges. The number ranges may be a single integer or a range defined by an integer, a hyphen, and an integer, without intervening spaces.

Reaction-pressure definitions identified by any of the numbers in the list will be dumped. If *list_of_ranges* is empty, all reaction-pressure definitions will be dumped.

Line 11: **-reaction_temperature** *list_of_ranges*

-reaction_temperature—Identifier indicates that reaction-temperature definitions will be dumped. Optionally, **-reaction_[temperatures]**, **temperature**, or **-t[emperature]**; note that the hyphen is necessary to distinguish the identifier from the keyword REACTION_TEMPERATURE.

list_of_ranges—List of number ranges. The number ranges may be a single integer or a range defined by an integer, a hyphen, and an integer, without intervening spaces. Reaction-temperature definitions identified by any of the numbers in the list will be dumped. If *list_of_ranges* is empty, all reaction-temperature definitions will be dumped.

Line 12: **-solid_solution** *list_of_ranges*

-solid_solution—Identifier indicates that solid-solution assemblages will be dumped. Optionally, **-soli[d_solutions]**; note that the hyphen is necessary to distinguish the identifier from the keyword SOLID_SOLUTIONS.

list_of_ranges—List of number ranges. The number ranges may be a single integer or a range defined by an integer, a hyphen, and an integer, without intervening spaces. Solid-solution assemblages identified by any of the numbers in the list will be dumped. If *list_of_ranges* is empty, all solid-solution-assemblage definitions will be dumped.

Line 13: **-solution** *list_of_ranges*

-solution—Identifier indicates that solutions will be dumped. Optionally, **-s[olution]**; note that the hyphen is necessary to distinguish the identifier from the keyword SOLUTION.

list_of_ranges—List of number ranges. The number ranges may be a single integer or a range defined by an integer, a hyphen, and an integer, without intervening spaces. Solutions identified by any of the numbers in the list will be dumped. If *list_of_ranges* is empty, all solution definitions will be dumped.

Line 14: **-surface** *list_of_ranges*

-surface—Identifier indicates that surface assemblages will be dumped. Optionally, **-su[rfaces]**; note that the hyphen is necessary to distinguish the identifier from the keyword SURFACE.

list_of_ranges—List of number ranges. The number ranges may be a single integer or a range defined by an integer, a hyphen, and an integer, without intervening spaces. Surface assemblages identified by any of the numbers in the list will be dumped. If *list_of_ranges* is empty, all surface-assemblage definitions will be dumped.

Line 15: **-cells** *list_of_ranges*

-cells—Identifier indicates that all reactants identified by a specified number will be dumped, including equilibrium-phase assemblages, exchanger assemblages, gas phases, kinetic reactants, mix definitions, reaction definitions, reaction-temperature definitions, solid-solution assemblages, solutions, and surface assemblages. Optionally, **cell**, **cells**, or **-c[ells]**.

list_of_ranges—List of number ranges. The number ranges may be a single integer or a range defined by an integer, a hyphen, and an integer, without intervening spaces. Reactants of any type that are identified by any of the numbers in the list will be dumped.

Line 16: **-all**

-all—Identifier indicates that all reactants will be dumped, including equilibrium-phase assemblages, exchanger assemblages, gas phases, kinetic reactants, mix definitions, reaction definitions, reaction-temperature definitions, solid-solution assemblages, solutions, and surface assemblages. Optionally, **all** or **-al[l]**.

Notes

The **DUMP** data block allows a complete description of reactants to be written to a file in the raw formats. The unedited raw formats can be read directly by PHREEQC. The dump file can be included in an input file with keyword INCLUDES or can be read by an IPhreeqc module. All definitions of number ranges in the example at the beginning of this section, except for the first, dump reactants numbered 2 through 5. The first definition, **-equilibrium_phases** with no *list_of_ranges*, causes all equilibrium-phase assemblages to be dumped. The **-cells** identifier dumps reactants of all types that are identified by numbers in the *list_of_ranges.* The **-all** identifier dumps all reactants that are defined prior to or within the current simulation. The **DUMP** operations follow the COPY operations and immediately precede the DELETE operations, which are the last operations of a simulation.

The raw format in which reactant data are written is not intended to be edited. Many of the fields are mandatory and deletion of them will cause an error when a raw data block is processed by PHREEQC. Furthermore, the raw formats may change. If additional capabilities are added to PHREEQC, the raw data formats will likely contain additional information to implement the new capabilities. Thus, data formatted in the raw formats may not be compatible with future versions of PHREEQC.

The **DUMP** data block may be useful to save the state of a transport or any other simulation to be able to restart calculations from that point; it is also possible to use the saved composition of a cell as the starting point for additional calculations. (Keyword TRANSPORT also has a dump option for writing files at transport steps that are definable with identifier **-dump_frequency**.) When using IPhreeqc modules, **DUMP** can be used to write reactant compositions so that they can be transferred to other IPhreeqc modules; the **DUMP** data (string or file) can be read by another IPhreeqc module.

Related keywords

COPY, DELETE, EQUILIBRIUM_PHASES, EQUILIBRIUM_PHASES_RAW, EXCHANGE, EXCHANGE_RAW, GAS_PHASE, GAS_PHASE_RAW, KINETICS, KINETICS_RAW, MIX, REACTION, REACTION_RAW, REACTION_PRESSURE, REACTION_PRESSURE_RAW, REACTION_TEMPERATURE, REACTION_TEMPERATURE_RAW, RUN_CELLS, SOLID_SOLUTIONS, SOLID_SOLUTIONS_RAW, SOLUTION, SOLUTION_RAW, SURFACE, and SURFACE_RAW.

END

This keyword has no associated data. It ends the data input for a simulation. After this keyword is read by the program, the calculations described by the input for the simulation are performed and the results printed. Additional simulations may follow in the input file, each in turn will be terminated with an **END** keyword or the end of the file.

Example problems

The keyword **END** is used in all example problems, 1 through 22.

EQUILIBRIUM_PHASES

This keyword data block is used to define the amounts of an assemblage of pure phases that can react reversibly with the aqueous phase. When the phases included in this keyword data block are brought into contact with an aqueous solution, each phase will dissolve or precipitate to achieve equilibrium or will dissolve completely. Pure phases include minerals with fixed composition and gases with fixed partial pressures. Two types of input are available: in one type, the phase itself reacts to equilibrium (or a specified saturation index or log gas partial pressure); in the other type, an alternative reaction occurs to the extent necessary to reach equilibrium (or a specified saturation index or log gas partial pressure) with the specified pure phase.

Example data block

```
Line 0:    EQUILIBRIUM_PHASES 1 Define amounts of phases in assemblage.
Line 1a:      Chalcedony  0.0      0.0
Line 1b:      CO2(g)      -3.5     1.0
Line 1c:      Gibbsite(c) 0.0      KAlSi3O8  1.0  dissolve_only
Line 1d:      Calcite     0.0      Gypsum    1.0  precipitate_only
Line 1e:      pH_Fix      -5.0     HCl       10.0
Line 2:                 -force_equality
```

Explanation

Line 0: **EQUILIBRIUM_PHASES** [*number*] [*description*]

EQUILIBRIUM_PHASES is the keyword for the data block. Optionally, **EQUILIBRIUM**, **EQUILIBRIA**, **PURE_PHASES**, **PURE**.

number—A positive number designates the phase assemblage and its composition. A range of numbers may also be given in the form *m-n*, where *m* and *n* are positive integers, *m* is less than *n*, and the two numbers are separated by a hyphen without intervening spaces. Default is 1.

description—Optional comment that describes the phase assemblage.

Line 1: *phase name* [*saturation index* [(*alternative formula* or *phase*)] [*amount* [(**dis** or **pre**)]]]

phase name—Name of a phase. The phase must be defined with PHASES input, either in the database file or in the current or previous simulations of the run. The name must be spelled identically to the name used in PHASES input (except for case).

saturation index—Target saturation index for the pure phase in the aqueous phase (Line 1a); for gases, this number is the log of the partial pressure (Line 1b). The target saturation index (or log partial pressure) may not be attained if the amount of the phase in the assemblage is insufficient. Default is 0.0.

alternative formula—Chemical formula that is added (or removed) to attain the target saturation index (or log partial pressure). By default, the mineral defined by *phase name* dissolves or precipitates to attain the target saturation index. If *alternative formula* is entered, *phase name* does not react; the stoichiometry of *alternative formula* is added or removed from the aqueous phase to attain the target saturation index for *phase name*. *Alternative formula* must be a legitimate chemical formula composed of elements defined to the program. Line 1c indicates that the stoichiometry given by *alternative formula*, $KAlSi_3O_8$ (potassium feldspar), will be added or removed from the aqueous phase until gibbsite equilibrium is attained. *Alternative formula* and *alternative phase* are mutually exclusive fields.

alternative phase—The chemical formula defined for *alternative phase* is added (or removed) to attain the target saturation index (or log partial pressure). By default, the mineral defined by *phase name* dissolves or precipitates to attain the target saturation index. If *alternative phase* is entered, *phase name* does not react; the stoichiometry of the *alternative phase* is added or removed from the aqueous phase to attain the target saturation index for *phase name*. *Alternative phase* must be defined through **PHASES** input (either in the database file or in the present or previous simulations). Line 1d indicates that the phase gypsum will be added to or removed from the aqueous phase until calcite equilibrium is attained. *Alternative formula* and *alternative phase* are mutually exclusive fields.

amount—Moles of the phase in the phase assemblage or moles of the alternative reaction. This number of moles defines the maximum amount of the mineral or gas that can dissolve. It may be possible to dissolve the entire amount without reaching the target saturation index, in which case the solution will have a smaller saturation index for this phase than the target saturation index. If *amount* is equal to zero (as in Line 1a), then the phase cannot dissolve, but will precipitate if the solution becomes supersaturated with the phase. Default is 10.0 mol (moles).

dissolve_only or **precipitate_only**—Optionally, the phase, *alternative formula*, or *alternative phase* can be required only to dissolve or only to precipitate. **Dissolve_only** indicates that the phase, *alternative formula*, or *alternative phase* will dissolve if moles are present and if the saturation index of *phase name* is less than the target *saturation_index* (Line 1c); the phase, *alternative formula*, or *alternative phase* cannot precipitate. **Precipitate_only** indicates that the phase, *alternative formula*, or *alternative phase* will precipitate if the saturation index of *phase name* is greater than the target *saturation_index* (Line 1d); the phase, *alternative formula*, or *alternative phase* cannot dissolve. Optionally, **d[issolve_only]** or **p[recipitate_only]**.

Line 2: **-force_equality** [(*True* or *False*)]

-force_equality—Identifier is used to include the immediately preceding phase in the list of equality constraints. Normally, to be able to find the stable phase assembly, equilibrium phases are included as inequality constraints; each phase is forced to have a saturation index of less than or equal to its target saturation index. However, if a fixed pH or other specific phase boundary is required, using the **force_equality** identifier will force that phase to attain its target saturation index. Default is **false** if **-force_equality** is not defined. Optionally, **force_equality** or **-f[orce_equality]**.

(*True* or *False*)—A value of **true** indicates that the immediately preceding phase will be forced to attain its target saturation index. A value of **false** indicates that the preceding phase will attain its target saturation index if it is part of the stable phase assemblage. If neither **true** nor **false** is entered on the line, **true** is assumed. Optionally, **t[rue]** or **f[alse]**.

Notes

If just one number is included on Line 1, it is assumed to be the target saturation index (or log partial pressure of a gas), and the amount of the phase defaults to 10.0 mol. If two numbers are included on the line, the first is the target saturation index and the second is the amount of the phase present. Line 1 may be repeated to define all pure phases that are assumed to react reversibly.

It is possible to include a pure phase that has an amount of zero (Line 1a). In the example, chalcedony cannot dissolve, but it can precipitate if the solution is supersaturated with chalcedony, either by initial conditions, through dissolution of pure phases, or through other specified reactions (mixing, stoichiometric or kinetic reactions). However, if chalcedony does precipitate and the equilibrium phase assemblage is saved, then it is possible in a subsequent simulation to dissolve the chalcedony that forms. If **dissolve_only** (only the initial "d" is required) or **precipitate_only** (only the initial "p" is required) are specified in the final field on the line defining the phase, the phase can only dissolve or precipitate, regardless of whether the equilibrium phases are saved and reacted again.

It is possible to maintain constant pH conditions by specification of an *alternative formula* and a special phase (PHASES input). Line 1e would maintain a pH of 5.0 (log activity of H+ of -5.0) by adding HCl, provided a phase named "pH_Fix" were defined with reaction $H^+ = H^+$ and log $K = 0.0$. (Note: If the acid, HCl, is specified and, in fact, a base is needed to attain pH 5.0, it is possible that the program will fail to find a solution to the algebraic equations.) In some cases, where a pH_Fix phase is defined along with other phases in the **EQUILIBRIUM_PHASES** definition, it is possible that the attempt to find a stable phase assemblage will result in not attaining the specified pH. In this case, it is useful to include the **-force_equality** identifier for the pH_Fix phase to force the desired pH to be attained (Line 2).

The number of exchange sites can be related to the moles of a phase that are present in an **EQUILIBRIUM_PHASES** phase assemblage (see EXCHANGE). As the moles of the phase increase or decrease, the number of exchange sites will increase or decrease. Likewise, the number of surface sites can be related to the moles of a phase that are present in an **EQUILIBRIUM_PHASES** phase assemblage (see SURFACE).

For batch reactions, after a pure-phase assemblage has reacted with the solution, it is possible to save the resulting assemblage composition (that is, the identity, target saturation index, and moles of each phase) with the SAVE keyword. If the new composition is not saved, the assemblage composition will remain the same as it was before the batch reaction. After it has been defined or saved, the assemblage may be used in subsequent simulations by the USE keyword. TRANSPORT and ADVECTION calculations automatically update the pure-phase assemblage and SAVE has no effect during these calculations.

Example problems

The keyword **EQUILIBRIUM_PHASES** is used in example problems 2, 3, 5, 6, 7, 8, 9, 10, 14, 17, and 21.

Related keywords

ADVECTION, COPY, DELETE, DUMP, EXCHANGE, PHASES, SAVE **equilibrium_phases**, SURFACE, TRANSPORT, and USE **equilibrium_phases**.

EXCHANGE

This keyword data block is used to define the amount and composition of an assemblage of exchangers. The initial composition of the exchange assemblage can be defined in two ways: (1) explicitly, by listing the composition of each exchange component; or (2) implicitly, by specifying that each exchanger is in equilibrium with a solution of fixed composition. The exchange master species, stoichiometries, and log Ks for the exchange reactions are defined with the keywords EXCHANGE_MASTER_SPECIES and EXCHANGE_SPECIES. The number of exchange sites can be fixed, can be related to the amount of a phase in an equilibrium-phase assemblage, or can be related to the amount of a kinetic reactant.

Example data block 1

```
Line 0:   EXCHANGE 10 Measured exchange composition
Line 1a:       CaX2      0.3
Line 1b:       MgX2      0.2
Line 1c:       NaX       0.5
Line 2a:       CaY2      Ca-montmorillonite  equilibrium_phase 0.165
Line 2b:       NaZ       Kinetic_clay        kinetic_reactant  0.1
Line 3:       -exchange_gammas                true
```

Explanation 1

Line 0: **EXCHANGE** [*number*] [*description*]

EXCHANGE is the keyword for the data block.

number—A positive number designates the exchange assemblage and its composition. A range of numbers may also be given in the form *m-n*, where *m* and *n* are positive integers, *m* is less than *n*, and the two numbers are separated by a hyphen without intervening spaces. Default is 1.

description—Optional comment that describes the exchanger.

Line 1: *exchange formula, amount*

exchange formula—Exchange species including stoichiometry of exchange ion and exchanger.

amount—Quantity of exchange species (mol).

Line 2: *exchange formula, name,* [(**equilibrium_phase** or **kinetic_reactant**)], *exchange_per_mole*

exchange formula—Exchange species including stoichiometry of exchange ion and exchange site(s). The *exchange formula* must be charge balanced; if no exchange ions are included in the formula, then the exchange site must be uncharged.

name—Name of the pure phase or kinetic reactant that has this kind of exchange site. If *name* is a phase, the amount of the phase in an EQUILIBRIUM_PHASES data block with the same number as this exchange number (10, in the Example data block) will be used to determine the number of exchange sites. If *name* is a kinetic reactant, the amount of the reactant in a KINETICS data block with the same number as this exchange number (10, in the Example data block) will be used to determine the number of exchange sites. Some care is needed in defining the stoichiometry of the exchange species if the exchangeable ions are related to a phase or kinetic reactant. The assumption is that some of the ions in the pure phase or kinetic reactant are available for exchange and these ions are defined through one or more entries of Line 2. The stoichiometry of the phase (defined in a PHASES data block) or kinetic reactant (defined in a KINETICS data block) must contain sufficient amounts of the exchangeable ions. From the Example data block (Line 2a) there must be at least 0.165 mol of calcium per mole of Ca-montmorillonite. From the Example data block (Line 2b) there must be at least 0.1 mol of sodium per mole of the reactant "kinetic_clay".

equilibrium_phase or **kinetic_reactant**—If **equilibrium_phase** is used, the *name* on the line is a phase defined in an EQUILIBRIUM_PHASES data block. If **kinetic_reactant** is used, the name on the line is the rate name for a kinetic reactant defined in a KINETICS data block. Optionally, **e[quilibrium_phase]** or **k[inetic_reactant]**. Default is **equilibrium_phase**.

exchange_per_mole—Number of moles of the exchange species per mole of phase or kinetic reactant, unitless (mol/mol).

Line 3: **-exchange_gammas** [(*True* or *False*)]

-exchange_gammas—This identifier selects whether exchange activity coefficients are assumed to be equal to aqueous activity coefficients when using the Pitzer or SIT aqueous model. This option has no effect when using ion-association aqueous models. Default is **true** if **-exchange_gammas** is not included. Optionally, **exchange_gammas** or **-ex[change_gammas]**.

(*True* or *False*)—When using the Pitzer or SIT aqueous model, a value of **true** indicates that the aqueous activity coefficient for an ion will be used as the activity coefficient for the corresponding exchange species. A value of **false** indicates that activity of an exchange species will be equal to the equivalent fraction. If neither **true** nor **false** is entered on the line, **true** is assumed. Optionally, t[**rue**] or f[**alse**].

Notes 1

Line 1 may be repeated to define the entire composition of each exchanger. This Example data block defines the amount and composition of three exchangers, X, Y, and Z. Line 2 should be entered only once for each type of exchange site. The total number of exchange sites of X is 1.5 mol and the total concentrations of calcium, magnesium, and sodium on exchanger X are 0.3, 0.2, and 0.5 mol, respectively. When the composition of the exchanger is defined explicitly, such as in this Example data block, the exchanger will almost certainly not be in equilibrium with any of the solutions that have been defined. Any batch reaction that includes an explicitly defined exchanger will produce a reaction that causes change in solution and exchange composition.

Exchanger Y is related to the amount of Ca-montmorillonite in EQUILIBRIUM_PHASES 10, where 10 is the same number as the exchange-assemblage number. If m represents the moles of Ca-montmorillonite in EQUILIBRIUM_PHASES 10, then the number of moles of exchangeable component CaY_2 is $0.165m$, and the total number of exchange sites (Y) is $0.33m$ (0.165×2). The stoichiometry of Ca must be at least 0.165 in the formula for Ca-montmorillonite. During batch-reaction simulations the exchange composition, including the moles of Ca exchanged, will change depending on competing species defined in EXCHANGE_SPECIES. In addition, the moles of Ca-montmorillonite in EQUILIBRIUM_PHASES 10 may change, in which case the total moles of the exchange sites (Y) will change.

Exchanger Z is related to the amount of a kinetic reactant that dissolves and precipitates according to a rate expression named "kinetic_clay". The formula for the kinetic reactant is defined in KINETICS 10, where 10 is the same number as the exchange-assemblage number. If m represents the moles of kinetic_clay in KINETICS 10, then the number of moles of exchangeable sodium (NaZ) is $0.1m$, which is equal to the total number of exchange sites. The stoichiometry of Na must be at least 0.1 in the formula for the kinetic reactant. The exchange composition will change during reaction calculations, depending on competing species defined in EXCHANGE_SPECIES. In addition, the moles of kinetic_clay in KINETICS 10 may change, in which case the total moles of the exchange sites (Z) will change.

The **-exchange_gammas** identifier selects whether exchange-species activity coefficients are set equal to aqueous activity coefficients for the Pitzer (*pitzer.dat* database) or SIT (*sit.dat* database) aqueous models. If **-exchange_gammas** is set to true, the activity coefficient for an exchange species is set equal to the activity coefficient of the corresponding aqueous species and is multiplied times the equivalent fraction of the exchange species to obtain the activity. If **-exchange_gammas** is set to false, the activity of an exchange species is equal to its equivalent fraction. For ion-association aqueous models (databases *phreeqc.dat*, *wateq4f.dat*, *llnl.dat*, *minteq.dat*, among others), exchange-species activity coefficient parameters (which are the same parameters as aqueous species) are defined in the EXCHANGE_SPECIES data block.

Example data block 2

```
Line  0: EXCHANGE 1 Exchanger in equilibrium with solution 1
Line 1a:      X        1.0
Line 1b:      Xa       0.5
Line  2:      CaY2     Ca-montmorillonite   equilibrium_phase  0.165
Line  3:      -equilibrate with solution 1
Line  4:      -exchange_gammas             true
```

Explanation 2

Line 0: **EXCHANGE** [*number*] [*description*]

Same as Example data block 1.

Line 1: *exchange_site moles*

exchange_site—Only the name of the exchange site needs to be entered.

moles—Quantity of exchange site (mol).

Line 2: *exchange formula, name,* [(**equilibrium_phase** or **kinetic_reactant**)], *exchange_per_mole* (same as Example data block 1).

Line 3: **-equilibrate** *number*

-equilibrate—This string at the beginning of the line indicates that the exchange assemblage is defined to be in equilibrium with a given solution composition. Optionally, **equil**, **equilibrate**, **equilibrium**, **-e[quilibrate]**, or **-e[quilibrium]**.

number—Solution number with which the exchange assemblage is to be in equilibrium. Any alphabetic characters following the identifier and preceding an integer ("with solution" in Line 1) are ignored.

Line 4: **-exchange_gammas** [(*True* or *False*)]

-exchange_gammas—Same as Example data block 1.

Notes 2

The order of Lines 1, 2, 3, and 4 is not important. Line 3 should occur only once within the data block. Lines 1 and 2 may be repeated to define the amounts of other exchangers, if more than one exchanger is present in the assemblage. Example data block 2 requires PHREEQC to make a calculation to determine the composition of the exchange assemblage. The calculation will be performed before any batch-reaction calculations to determine the concentrations of each exchange component [such as CaX_2, MgX_2, or NaX (from the default database) provided calcium, magnesium, and sodium are present in solution] that would exist in equilibrium with the specified solution (solution 1 in this Example data block). The composition of the solution will not change during this calculation. When an exchange assemblage (as defined in Example data block 1 or Example data block 2) is placed in contact with a solution during a batch reaction, both the exchange composition and the solution composition will adjust to reach a new equilibrium.

The exchange ions given by the formulas in Lines 2 are not used in the initial exchange-composition calculation. However, the definition of the exchange ions is important for batch-reaction and transport calculations if the number of exchange sites is related to a pure phase or kinetic reactant. As the reactant, either a pure phase or a kinetic reactant, dissolves or precipitates, the number of exchange sites varies. Any new sites are initially filled with the exchangeable ions given in Line 2. When exchange sites are removed (for example, when a pure phase dissolves) then the net effect is to subtract from the pure phase formula the amount of the exchange ions defined in Line 2 and add an equivalent amount of ions to the solution. As an example, suppose some Ca-montmorillonite precipitates. Initially, calcium is in the exchange positions, but sodium replaces part of the calcium on the exchanger. When the montmorillonite dissolves again, the calcium in the formula for the phase is added to solution, the exchange ion (calcium from Line 2) is removed from solution, and the sodium and calcium from the exchanger is added to solution; the net effect is dissolution of (Na, Ca)-montmorillonite. Note that equilibrium for Ca-montmorillonite always uses the same mass-action equation, which includes only calcium, even though the composition of the phase is changing. Note also that this formulation implies that a pure Na-montmorillonite can never be attained because calcium must always be present to attain equilibrium with Ca-montmorillonite.

It is possible to realize a complete exchange of sodium and calcium by defining Y without cations under **EXCHANGE**, and a new equilibrium with only the structural ions of montmorillonite under PHASES. The combined reaction of exchanger and equilibrium phase must be electrically neutral. In the Example data block, the montmorillonite would be defined with a positive charge deficit of 0.165. When montmorillonite forms, the exchange sites Y increase in proportion and take cations from solution to exactly balance the charge deficit. Note that log_k for montmorillonite is adjusted by $\log_{10}(0.001^{0.165})$ to account for an estimated contribution of 1 mmol/kgw (millimole per kilogram water) Ca in solution. Yet another possibility is to use the capabilities of the SOLID_SOLUTIONS data block to define a variable composition solid solution between calcium and sodium montmorillonite end members.

```
EXCHANGE 1 Exchanger in equilibrium with solution 1
    Y Montmorillonite equilibrium_phase  0.165
    -equilibrate with solution 1
PHASES
    -no_check          # must use no_check because of unbalanced equation
Montmorillonite        # Montmorillonite has 0.165 mol Y-/mol
Al2.33Si3.67O10(OH)2 + 12 H2O = 2.33 Al(OH)4- + 3.67 H4SiO4 + 2 H+
    log_k    -44.532 #Assume a_ca = 0.001 at equilibrium
    delta_h  58.373  kcal
```

An exchanger can be defined with a fixed number of sites initially, but through special definition of a kinetic reactant, the number of sites can vary with reaction progress. Changes in the number of exchange sites can be included in the KINETICS keyword, under -formula. The combination of exchanger and kinetic reaction must be neutral.

```
EXCHANGE 1
    # Z+ is related to Goethite, initial amount is 0.2 * m_go = 0.02
    Z   0.02
    -equil 1
KINETICS 1
    # Z has a charge of +1.0, Fe(OH)2+ sorbs anions.
    -formula  FeOOH 0.8  Fe(OH)2 0.2  Z  -0.2
    m    0.1
```

After a batch reaction has been simulated, it is possible to save the resulting exchange assemblage composition with the SAVE keyword. If the new composition is not saved, the exchange assemblage composition will remain the same as it was before the batch reaction. After it has been defined or saved, the exchange assemblage can be used in subsequent simulations through the USE keyword. TRANSPORT and ADVECTION calculations automatically update the pure-phase assemblage and SAVE has no effect during these calculations.

Example problems

The keyword EXCHANGE is used in example problems 11, 12, 13 14, 19, and 21.

Related keywords

ADVECTION, COPY, DELETE, DUMP, EQUILIBRIUM_PHASES,
EXCHANGE_MASTER_SPECIES, EXCHANGE_SPECIES, KINETICS, SAVE **exchange**,
TRANSPORT, and USE **exchange**.

EXCHANGE_MASTER_SPECIES

This keyword data block is used to define the correspondence between the name of an exchange site and an exchange species that is used as the master species in calculations. Normally, this data block is included in the database file and only additions and modifications are included in the input file.

Example data block

```
Line  0:  EXCHANGE_MASTER_SPECIES
Line 1a:      X                 X-
Line 1b:      Xa                Xa-
Line 1c:      [exSite]          [exSite]-
```

Explanation

Line 0: EXCHANGE_MASTER_SPECIES

Keyword for the data block. No other data are input on the keyword line.

Line 1: *exchange name, exchange master species*

exchange name—Name of an exchange site, X, Xa, and [exSite] in this Example data block. Two forms for exchange names are available: (1) exchange names that begin with a capital letter followed by zero or more lower case letters and underscores ("_") and no numbers; and (2) exchange names that are enclosed in square brackets (see Line 1c) and use any combination of alphanumeric characters and the characters plus (+), minus (-), equal (=), colon (:), decimal point (.), and underscore (_). In general, the exchange names using form 1 have a capital letter and zero or more lower case letters. Exchange names using form 2 also are case dependent, but upper and lower case characters can be used in any position.

exchange master species—Formula for the master exchange species, X^-, Xa^-, and $[exSite]^-$ in this Example data block.

Notes

All half-reactions for the exchanger (X, Xa, and [exSite] in this Example data block) must be written in terms of the master exchange species (X^-, Xa^-, and $[exSite]^-$ in this Example data block). Each exchange master species must be defined by an identity reaction with log K of 0.0 in EXCHANGE_SPECIES input. Any additional exchange species are defined with association reactions in EXCHANGE_SPECIES input.

Example problems

The keyword **EXCHANGE_MASTER_SPECIES** is used in the *Amm.dat*, *iso.dat*, *llnl.dat*, *phreeqc.dat*, *pitzer.dat*, and *wateq4f.dat* databases.

Related keywords

EXCHANGE, EXCHANGE_SPECIES, SAVE **exchange**, and USE **exchange**.

EXCHANGE_SPECIES

This keyword data block is used to define a half-reaction and relative log K for each exchange species. Normally, this data block is included in the database file and only additions and modifications are included in the input file.

Example data block

```
Line  0:   EXCHANGE SPECIES
Line 1a:      X- = X-
Line 2a:          log_k          0.0
Line 1b:      X- + Na+ = NaX
Line 2b:          log_k          0.0
Line  3:          -gamma         4.      0.075   0.1
Line 1c:      2X- + Ca+2 = CaX2
Line 2c:          log_k          0.8
Line  4:          -davies
Line 1d:      Xa- = Xa-
Line 2d:          log_k          0.0
Line 1e:      Xa- + Na+ = NaXa
Line 2e:          log_k          0.0
Line 1f:      2Xa- + Ca+2 = CaXa2
Line 2f:          log_k          2.0
```

Explanation

Line 0: **EXCHANGE_SPECIES**

Keyword for the data block. No other data are input on the keyword line.

Line 1: *Association reaction*

Association reaction for exchange species. The defined species must be the first species to the right of the equal sign. The association reaction must precede any identifiers related to the exchange species. Master species have an identity reaction (Lines 1a and 1d).

Line 2: **log_k** log K

log_k—Identifier for log K at 25 °C. Optionally, **-log_k**, **logk**, **-l[og_k]**, or **-l[ogk]**.

log K—Log K at 25 °C for the reaction. Unlike log K for aqueous species, the log K for exchange species is implicitly relative to a reference exchange species. In the default database file, sodium (NaX) is used as the reference and the reaction $X^- + Na^+ = NaX$ is given a log K of 0.0 (Line 2b). By subtracting the reaction for NaX in Line 1b twice from the reaction for

CaX_2 in Line 1c, it follows that log K for the reaction in Line 2c is numerically equal to log K for the reaction $2NaX + Ca^{+2} = CaX_2 + 2Na^+$. The identity reaction for a master species has log K of 0.0 (Lines 2a and 2d); reactions for reference species also have log K of 0.0 (Lines 2b and 2e). Default is 0.0.

Line 3: **-gamma** *Debye-Hückel a, Debye-Hückel b, active_fraction_coefficient*

-gamma—Indicates WATEQ Debye-Hückel equation will be used to calculate an activity coefficient for the exchange species if the aqueous model is an ion-association model (see **-exchange_gammas** in the EXCHANGE data block for information about activity coefficients when using the Pitzer or SIT aqueous models). If **-gamma** or **-davies** is not input for an exchange species, the activity of the species is equal to its equivalent fraction. If **-gamma** is entered, then an activity coefficient of

the form of WATEQ (Truesdell and Jones, 1974), $\log\gamma = \dfrac{-Az_e^2\sqrt{\mu}}{1 + Ba^o\sqrt{\mu}} + b\mu$, is multiplied

times the equivalent fraction to obtain activity for the exchange species. In this equation, γ is the activity coefficient, μ is ionic strength (mol/L [mole per liter], assumed to be equal to mol/kgw [mole per kilogram water]), A and B are constants at a given temperature and pressure, z_e is the number of equivalents of exchanger in the exchange species, and a^o and b are ion-specific parameters. Optionally, **gamma** or **-g[amma]**.

Debye-Hückel a—Parameter a^o in the WATEQ activity-coefficient equation.

Debye-Hückel b—Parameter b in the WATEQ activity-coefficient equation.

active_fraction_coefficient—Parameter for changing log_k as a function of the exchange sites occupied (Appelo, 1994a). The active-fraction model is useful for modeling sigmoidal exchange isotherms and proton exchange on organic matter (see *http://www.hydrochemistry.eu/exmpls/a_f.html*, accessed June 25, 2012).

Line 4: **-davies**

-davies—Indicates the Davies equation will be used to calculate an activity coefficient. If **-gamma** or **-davies** is not input for an exchange species, the activity of the species is equal to its equivalent fraction. If **-davies** is entered, then an activity coefficient of the form of the

Davies equation, $\log\gamma = -Az_e^2\left(\dfrac{\sqrt{\mu}}{1+\sqrt{\mu}} - 0.3\mu\right)$, is multiplied times the equivalent fraction to obtain activity for the exchange species. In this equation, γ is the activity coefficient, μ is ionic strength, A is a constant at a given temperature, and z_e is the number of equivalents of exchanger in the exchange species. Optionally, **davies** or **-d[avies]**.

Notes

Lines 1 and 2 may be repeated as necessary to define all of the exchange reactions, with Line 1 preceding Line 2 for each exchange species. One identity reaction that defines the exchange master species (in the Example data block, Lines 1a and 2a, 1d and 2d) and one reference half-reaction are needed for each exchanger. The identity reaction has a log K of 0.0. The reference half-reaction for each exchanger also will have a log K of 0.0 (in the Example data block, Lines 1b and 2b, 1e and 2e); in the default database file the reference half-reaction is $Na^+ + X^- = NaX$. Multiple exchangers may be defined simply by defining multiple exchange master species and additional half-reactions involving these master species, as in this Example data block.

Activities of exchange species may be expressed as equivalent or mole fractions of the species (Gaines-Thomas or Vanselow convention, respectively), or as fractions of the exchange sites occupied (Gapon convention). All three conventions can be used in PHREEQC (see *http://www.hydrochemistry.eu/pub/ap_pa02.pdf*, accessed June 25, 2012). In the databases, the Gaines-Thomas convention is used.

Cation exchange experiments with heterovalent exchange in which the salinity of the solutions is varied (for example, exchange of $2Na^+$ for Ca^{2+} at varying Cl^- concentrations) can be modeled better when exchange is calculated with molal concentrations for solute species instead of activities. This implies that the activity coefficients of solute cations and exchangeable species are the same, perhaps because a large part of cation exchange in soils and sediments takes part in the electrostatic double layer. Accordingly, PHREEQC permits the activity coefficient for exchangeable species to be defined in the same way as the solute species. The **-gamma** identifier allows the equivalent fraction to be multiplied by an activity coefficient by using the WATEQ Debye-Hückel equation. Similarly, when using the *llnl.dat* database, **-llnl_gamma** can be used to multiply the equivalent fraction by the activity coefficient that is defined according to the conventions of the *llnl.dat* database. The Davies equation can be used to calculate the activity coefficient of the exchange species by specifying the **-davies** identifier. The use of these equations is strictly empirical and is motivated by the observation that these activity corrections provide a better fit to some experimental data.

Temperature dependence of log K can be defined with the standard enthalpy of reaction (**-delta_h**) using the Van't Hoff equation or with an analytical expression (**-analytical_expression**). Sometimes it is useful to offset a log K from zero for parameter fitting, or to account for dependencies among log K values, in which case the **-add_log_k** identifier can be used to add the value defined by a named analytical expression (NAMED_EXPRESSIONS) to the log K of the exchange species. See SOLUTION_SPECIES for examples.

The identifier **-no_check** can be used to disable checking charge and elemental balances (see SOLUTION_SPECIES) and allows the Gapon exchange convention to be used (See *http://www.hydrochemistry.eu/a&p/6/exch_phr.pdf*, accessed June 25, 2012).

Example problems

The keyword **EXCHANGE_SPECIES** is used in example problems 12, 13, 18, and 21. See also the databases *Amm.dat*, *iso.dat*, *llnl.dat*, *phreeqc.dat*, *pitzer.dat*, and *wateq4f.dat*.

Related keywords

EXCHANGE, EXCHANGE_MASTER_SPECIES, SAVE **exchange**, SOLUTION_SPECIES, and USE **exchange**.

GAS_PHASE

This keyword data block is used to define the composition of a fixed-total-pressure or a fixed-volume multicomponent gas phase. The thermodynamic properties of the gas components are defined with **PHASES** input. If the critical pressure and temperature are defined for a gas component with **PHASES**, the Peng-Robinson equation of state (EOS) will be used for calculating the relation between pressure and molar volume, and fugacity coefficients will be calculated for the gases. If the critical temperature and pressure are not defined, the ideal gas law will be used. Ideal gases and Peng-Robinson gases cannot be mixed in a GAS_PHASE. A GAS_PHASE data block is not needed if fixed partial pressures of gas components are desired; use **EQUILIBRIUM_PHASES** instead. The gas phase defined with this keyword data block subsequently may be equilibrated with an aqueous phase in combination with pure-phase, surface, exchange, and solid-solution assemblages in batch-reaction calculations. Either Henry's law (ideal gases) or the Peng-Robinson EOS (nonideal gases) is used for calculating the solubility of the gases. As a consequence of batch reactions, a fixed-pressure gas phase may exist or not, depending on the sum of the partial pressures of the dissolved gases in solution. A fixed-volume gas phase always contains some amount of each gas component that is present in solution. The initial composition of a fixed-pressure gas phase is defined by the partial pressures of each gas component. The initial composition of a fixed-volume gas may be defined by the partial pressures of each gas component or may be defined to be that which is in equilibrium with a fixed-composition aqueous phase. When the Peng-Robinson EOS is used and the **GAS_PHASE** has a pressure higher than about 10 atmospheres, the initial gas-phase composition calculated for a fixed-composition aqueous phase is only an approximation of the true gas composition.

Example data block 1

```
Line 0:    GAS_PHASE 1-5  Air
Line 1:         -fixed_pressure
Line 2:         -pressure      1.001
Line 3:         -volume        1.0
Line 4:         -temperature   25.0
Line 5a:        CH4(g)         0.0
Line 5b:        CO2(g)         0.000316
Line 5c:        O2(g)          0.2
Line 5d:        N2(g)          0.78
```

Explanation 1

Line 0: **GAS_PHASE** [*number*] [*description*]

> **GAS_PHASE** is the keyword for the data block.

> *number*—A positive number designates the gas phase and its composition. A range of numbers may also be given in the form *m-n*, where *m* and *n* are positive integers, *m* is less than *n*, and the two numbers are separated by a hyphen without intervening spaces. Default is 1.

> *description*—Optional comment that describes the gas phase.

Line 1: **-fixed_pressure**

> **-fixed_pressure**—Identifier defining the gas phase to have a fixed total pressure; that is, a gas bubble. A fixed-pressure gas phase is the default if neither the **-fixed_pressure** nor the **-fixed_volume** identifier is used. Optionally **fixed_pressure** or **-fixed_p[ressure]**.

Line 2: **-pressure** *pressure*

> **-pressure**—Identifier defining the fixed pressure of the gas phase that applies during all batch-reaction and transport calculations. Optionally **pressure** or **-p[ressure]**.

> *pressure*—The pressure of the gas phase, in atm (atmosphere). Default is 1.0 atm.

Line 3: **-volume** *volume*

> **-volume**—Identifier defining the <u>initial</u> volume of the fixed-pressure gas phase. Optionally, **volume** or **-v[olume]**.

> *volume*—The <u>initial</u> volume of the fixed-pressure gas phase, in liters. The ideal gas law or the Peng-Robinson EOS is used to calculate the initial moles, *n*, of each gas component in the fixed-pressure gas phase. Default is 1.0 L (liter).

Line 4: **-temperature** *temp*

> **-temperature**—Identifier defining the <u>initial</u> temperature of the gas phase. Optionally, **temperature** or **-t[emperature]**.

> *temp*—The <u>initial</u> temperature of the gas phase, in °C (degree Celsius). The *temp* along with *volume* and *partial pressure* are used to calculate the initial moles of each gas component in the fixed-pressure gas phase. Default is 25.0 °C.

Line 5: *phase name, partial pressure*

> *phase name*—Name of a gas component. A phase with this name must be defined by PHASES input in the database or input file.

partial pressure—<u>Initial</u> partial pressure of this component in the gas phase (atm). The *partial pressure* along with *volume* and *temp* are used to calculate the initial moles of this gas component in the fixed-pressure gas phase.

Notes 1

Line 5 must be repeated as necessary to define all of the components initially present in the fixed-pressure gas phase as well as any components which may subsequently enter the gas phase. The initial moles of a gas component that is defined to have a positive partial pressure in **GAS_PHASE** input will be computed using either the ideal gas law, $n = PV/RT$, where n is the moles of the gas, P is the defined partial pressure (Line 5), V is the initial volume, given by **-volume**, R is the gas constant (0.08207 L K^{-1}mol^{-1}, liter per degree kelvin per mole), and T is given by **-temperature** (converted to kelvin), or the Peng-Robinson EOS (see keyword **PHASES** for the equations). Thus, in Example data block 1 and with the *wateq4f.dat* database, which does not define critical temperatures and pressures, the moles of all gases are calculated by $n = (0.000316 + 0.2 + 0.78) \times 1.0 / (298 \times 0.02807) = 0.04$ mol. If this gas phase reacts with a solution with a very small amount of water so that n does not change (that is, the dissolution of gas is negligible), the volume becomes $V = 0.04 \times (298 \times 0.02807) / 1.001 = 0.979$ L. It is likely that the sum of the partial pressures of the defined gases will not be equal to the pressure given by **-pressure**. However, when the **GAS_PHASE** reacts with a solution during a batch-reaction simulation, the moles of gases and volume of the gas phase will be adjusted so that each component is in equilibrium with the solution while the total pressure (sum of the partial pressures) is that specified by **-pressure**. It is possible that the gas phase disappears if the sum of the partial pressures of dissolved gases is less than the pressure given by **-pressure**.

A gas component may be defined to have initial partial pressure of zero. In this case, no moles of that component will be present initially, but the component may enter the gas phase when in contact with a solution that contains that component. If no gas phase exists initially, the initial partial pressures of all components should be set to 0.0; a gas phase may subsequently form if batch reactions cause the sum of the partial pressures of the gas components to exceed *pressure*.

Example data block 2

```
Line 0:   GAS_PHASE 1-5  Find composition from solution 1
Line 1:        -fixed_volume
Line 2:        -volume         1.0
Line 3:        -temperature    25.0
```

```
Line 4a:      CH4(g)        0.0
Line 4b:      CO2(g)        0.000316
Line 4c:      O2(g)         0.2
Line 4d:      N2(g)         0.78
```

Explanation 2

Line 0: **GAS_PHASE** [*number*] [*description*]

GAS_PHASE is the keyword for the data block.

number—a positive number designates the gas phase and its composition. A range of numbers may also be given in the form *m-n*, where *m* and *n* are positive integers, *m* is less than *n*, and the two numbers are separated by a hyphen without intervening spaces. Default is 1.

description—Optional comment that describes the gas phase.

Line 1: **-fixed_volume**

-fixed_volume—Identifier defining the gas phase to be one that has a fixed volume (not a gas bubble). A fixed-pressure gas phase is the default if neither the **-fixed_pressure** nor the **-fixed_volume** identifier is used. Optionally **fixed_volume** or **-fixed_v**[**olume**].

Line 2: **-volume** *volume*

-volume—Identifier defining the volume of the fixed-volume gas phase, which applies for all batch-reaction or transport calculations. Optionally, **volume** or **-v**[**olume**].

volume—The volume of the fixed-volume gas phase, in liters. Default is 1.0 L.

Line 3: **-temperature** *temp*

-temperature—Identifier defining the <u>initial</u> temperature of the gas phase. Optionally, **temperature** or **-t**[**emperature**].

temp—The <u>initial</u> temperature of the gas phase, in °C. Default is 25.0 °C.

Line 4: *phase name, partial pressure*

phase name—Name of a gas component. A phase with this name must be defined by PHASES input in the database or input file.

partial pressure—<u>Initial</u> partial pressure of this component in the gas phase, in atm. The *partial pressure* along with *volume* and *temp* are used to calculate the initial moles of this gas component in the fixed-volume gas phase.

Notes 2

Line 4 may be repeated as necessary to define all the components initially present in the fixed-volume gas phase, as well as any components which may subsequently enter the gas phase. The initial moles of a gas component with a positive partial pressure will be computed using either the ideal gas law, $n = PV/(RT)$, where n is the moles of the gas, P is the defined partial pressure (Line 4), V is given by **-volume**, R is the gas constant, and T is given by **-temperature** (converted to kelvin), or the Peng-Robinson EOS. When the gas phase reacts with a solution during a batch-reaction simulation, the total pressure, the partial pressures of the gas components in the gas phase, and the partial pressures of the gas components in the aqueous phase will be adjusted so that equilibrium is established for each component. A constant-volume gas phase always exists unless all of the gas components are absent from the system. The identifier **-pressure** is not used for a fixed-volume gas phase.

A gas component may be defined to have an initial partial pressure of zero. In this case, no moles of that component will be present initially, but the component will enter the gas phase when in contact with a solution containing the component.

Example data block 3

```
Line 0:    GAS_PHASE 1-5  Air
Line 1:         -fixed_volume
Line 2:         -equilibrate with solution 10
Line 3:         -volume        1.0
Line 4a:      CH4(g)
Line 4b:      CO2(g)
Line 4c:      O2(g)
Line 4d:      N2(g)
```

Explanation 3

Line 0: **GAS_PHASE** [*number*] [*description*]

GAS_PHASE is the keyword for the data block.

number—A positive number designates the gas phase and its composition. A range of numbers may also be given in the form *m-n*, where *m* and *n* are positive integers, *m* is less than *n*, and the two numbers are separated by a hyphen without intervening spaces. Default is 1.

description—Optional comment that describes the gas phase.

Line 1: **-fixed_volume**

> **-fixed_volume**—Identifier defining the gas phase to be one that has a fixed volume (not a gas bubble). A fixed-pressure gas phase is the default if neither the **-fixed_pressure** nor the **-fixed_volume** identifier is used. Optionally **fixed_volume** or **-fixed_v[olume]**.

Line 2: **-equilibrate** *number*

> **-equilibrate**—Identifier indicates that the fixed-volume gas phase is defined to be in equilibrium with a solution of a fixed composition. This identifier may only be used with the **-fixed_volume** identifier. Optionally, **equil**, **equilibrium**, **-e[quilibrium]**, **equilibrate**, **-e[quilibrate]**.

> *number*—Solution number with which the fixed-volume gas phase is to be in equilibrium. Any alphabetic characters following the identifier and preceding an integer ("with solution" in Line 2) are ignored.

Line 3: **-volume** *volume*

> **-volume**—Identifier defining the volume of the fixed-volume gas phase, which applies for all batch-reaction or transport calculations. Optionally, **volume** or **-v[olume]**.

> *volume*—The volume of the fixed-volume gas phase, L. Default is 1.0 L.

Line 4: *phase name*

> *phase name*—Name of a gas component. A phase with this name must be defined by PHASES input in the database or input file.

Notes 3

Line 4 may be repeated as necessary to define all of the components that may be present in the fixed-volume gas phase. The **-equilibrate** identifier specifies that the initial moles of the gas components are to be calculated by equilibrium with solution 10. This calculation is termed an "initial gas-phase-composition calculation". During this calculation, the composition of solution 10 does not change, only the moles of each component in the gas phase are calculated. This calculation is approximate for a Peng-Robinson GAS_PHASE due to the fugacity coefficient, which is used for calculating the activity of the gas in the solubility equation. Alternatively, for Peng-Robinson gases, keyword GAS_PHASE_MODIFY may be used, but this is still approximate for a gas-mixture at high pressure. A constant-volume gas phase always exists unless all of the gas components are absent from the system. When the **-equilibrate** identifier is used, the identifiers **-pressure** and **-temperature** are not needed and initial partial pressures for each gas component need not be specified; the partial pressures for the gas components are calculated from the partial pressures in solution and the temperature is equal to the solution temperature. The **-equilibrate** identifier cannot be used with a fixed-pressure gas phase.

A gas component may have an initial partial pressure of zero because the solution with which the gas phase is in equilibrium does not contain that gas component. In this case, no moles of that component will be present initially, but the component may enter the gas phase when the gas is in contact with another solution that does contain that component.

After a batch reaction has been simulated, it is possible to save the resulting gas-phase composition with the SAVE keyword. If the new composition is not saved, the gas-phase composition will remain the same as it was before the batch reaction. After it has been defined or saved, the gas phase can be used in subsequent simulations through the USE keyword. TRANSPORT and ADVECTION calculations automatically update the gas-phase composition and SAVE has no effect during these calculations.

Example problems

The keyword **GAS_PHASE** is used in example problems 7 and 22.

Related keywords

ADVECTION, COPY, DELETE, DUMP, EQUILIBRIUM_PHASES, GAS_PHASE_MODIFY, PHASES, SAVE **gas_phase**, TRANSPORT, and USE **gas_phase**.

INCLUDE$

This keyword is used to insert the contents of another file into the input or database file. The inserted file may extend the data block of the preceding keyword and (or) add additional keyword data blocks. Files that are inserted may contain further **INCLUDE$** statements. The files are included dynamically, which means that an input file can write a file with DUMP or USER_PUNCH and subsequently include that file into the input stream.

Example

Input file:

```
SOLUTION
        pH      6
INCLUDE$ A
END
```

File A:

```
        Na      2
        S(6)    1
INCLUDE$ B
```

File B:

```
EQUILIBRIUM_PHASES
        Calcite
```

Is equivalent to the following input:

```
SOLUTION
        pH      6
        Na      2
        S(6)    1
EQUILIBRIUM_PHASES
        Calcite
END
```

Notes

The **INCLUDE$** keyword is used to include a file into the input file. The inclusion is done as PHREEQC is processing the input file and running simulations. Thus, it is possible to use a DUMP or SELECTED_OUTPUT data block to write a file that is included at a later point in the run. The keyword may be used in database files or input files.

INCLUDE$

Example problems

The keyword **INCLUDE$** is used in example problems 8, 20, and 21.

Related keywords

DUMP and SELECTED_OUTPUT.

INCREMENTAL_REACTIONS

This keyword data block is included mainly to speed up batch-reaction calculations that include kinetic reactions (KINETICS keyword). The keyword has no effect on transport calculations. By default (**INCREMENTAL_REACTIONS false**), for each time t_i is given by **-steps** in the KINETICS keyword data block, rates of kinetic reactions are integrated from time 0 to t_i. This default repeats the integration over early times for each reaction step even though the early times may be the most central processing unit (CPU) intensive part of the integration. If **INCREMENTAL_REACTIONS** is set to true, the values of t_i are the incremental times for which to integrate the rates; each kinetic calculation

(denoted by i) integrates over the time interval from $\displaystyle\sum_{n=0}^{i-1} t_n$ to $\displaystyle\sum_{n=0}^{i} t_n$. **INCREMENTAL_REACTIONS** has a similar effect for **-steps** in the REACTION data block.

Example data block

```
Line 0:   INCREMENTAL_REACTIONS true
```

Explanation

Line 0: **INCREMENTAL_REACTIONS** [(*True* or *False*)]

> **INCREMENTAL_REACTIONS** is the keyword for the data block. If value is **true**, reaction steps for REACTION and time steps for KINETICS data blocks are incremental amounts of reaction and time that add to previous reaction steps. If the value is **false**, reaction steps and time steps are total amounts of reaction and time, independent of previous reaction steps. Initial setting at the beginning of the run is **false**. If neither **true** nor **false** is entered on the line, **true** is assumed. Optionally, **t**[**rue**] or **f**[**alse**].

Notes

Frequently, kinetic reactions are faster at early times and slower at later times. The integration of kinetic reactions for the early times is CPU intensive because the rates must be evaluated at many time subintervals to achieve an accurate integration of the rate equations when reactions are fast. If the time steps in the KINETICS data block are 0.1, 1, 10, and 100 s (seconds) and the time steps are not incremental (default at initialization of a run), then the kinetic reactions will be integrated from 0 to 0.1, 0 to 1, 0 to 10, and 0 to

100 s; the early part of the reactions (0 to 0.1 s) must be integrated for each specified time. By using incremental time steps, the kinetic reactions will be integrated from 0 to 0.1, 0.1 to 1.1, 1.1 to 11.1, and 11.1 to 111.1 s; the results from the previous time step are used as the starting point for the next time step, and integrating over the same early time interval is avoided.

If the time steps in the **KINETICS** data block are defined as "**-steps** 100 **in** 2 **steps**" and **INCREMENTAL_REACTIONS false**, then the kinetic reactions will be integrated from 0 to 50 and from 0 to 100 s. By using **INCREMENTAL_REACTIONS true**, the kinetic reactions will be integrated from 0 to 50 and from 50 to 100 s. Although the calculation procedure differs, results of calculations using the "**in**" form of data input should be the same for **INCREMENTAL_REACTIONS true** or **false**.

For consistency, the **INCREMENTAL_REACTIONS** keyword also has an effect on the interpretation of steps defined in the **REACTION** data block. If the steps in the **REACTION** data block were 0.1, 1, 10, and 100 mmol (millimole), then by default, solution compositions would be calculated after a total of 0.1, 1, 10, and 100 mmol of reaction had been added to the initial solution. By using incremental reaction steps, solution compositions would be calculated after a total of 0.1, 1.1, 11.1, and 111.1 mmol of reaction had been added.

If the reaction steps in the **REACTION** data block are defined as "**-steps** 1 **in** 2 **steps**" and **INCREMENTAL_REACTIONS false** (default), then the solution composition will be calculated after 0.5 mol of reaction are added to the initial solution and after 1 mol of reaction has been added to the initial solution. By using **INCREMENTAL_REACTIONS true**, the solution composition will be calculated after 0.5 mol of reaction are added to the initial solution and again after an additional 0.5 mol of reaction are added to the reacted solution. Although the calculation procedure differs, results of calculations using the "**in**" form of data input should be the same for **INCREMENTAL_REACTIONS true** or **false**.

If **INCREMENTAL_REACTIONS true**, REACTION is defined with a list of steps, and more batch-reaction steps (maximum number of steps defined in KINETICS, REACTION, REACTION_PRESSURE, and REACTION_TEMPERATURE) than REACTION steps are defined; then, the last reaction step is repeated for the additional batch-reaction steps. Thus the reaction continues to be added to solution during the final batch-reaction steps. If no additional reaction is desired in these final batch-reaction steps, then additional reaction amounts equal to zero should be entered in the REACTION data block. Similarly, if more batch-reaction steps are defined than kinetic steps, the final time step from the KINETICS data block will be used for the final batch-reaction steps.

If "**in**" is used in **-steps** in the REACTION data block and the number of batch-reaction steps is greater than the number of steps defined in the REACTION data block, then the reaction step is zero for REACTION in the remaining batch-reaction steps. Likewise, if "**in**" is used in **-steps** in the KINETICS data block, and the number of batch-reaction steps is greater than the number steps defined in the KINETICS data block, then the time step for kinetic reactions in the remaining batch-reaction steps will be zero.

The incremental approach is not implemented for the MIX keyword. If a MIX data block is used, then solutions are mixed only once before any reaction or kinetic steps. REACTION_PRESSURE and REACTION_TEMPERATURE steps are always nonincremental.

Example problems

The keyword **INCREMENTAL_REACTIONS** is used in example problems 6, 9, 17, 20, and 22.

Related keywords

KINETICS, MIX, REACTION, REACTION_PRESSURE, and REACTION_TEMPERATURE.

INVERSE_MODELING

This keyword data block is used to specify the information needed for an inverse modeling calculation. Inverse modeling attempts to determine sets of mole transfers of phases that account for changes in water chemistry between one or a mixture of initial waters and a final water. Isotope mole balance, but not isotope fractionation, can be included in the calculations. The data block includes definition of the solutions, phases, and uncertainty limits used in the calculations.

Example data block

```
Line  0:   INVERSE_MODELING 1
Line  1:       -solutions              10 3 5
Line  2:       -uncertainty            0.02    0.04
Line  3:       -phases
Line  4a:          Calcite             force   pre    13C    -1.0    1
Line  4b:          Anhydrite           force   dis    34S    13.5    2
Line  4c:          CaX2
Line  4d:          NaX
Line  5:       -balances
Line  6a:          pH                  0.1
Line  6b:          Ca                  0.01        -0.005
Line  6c:          Alkalinity          -1.0e-6
Line  6d:          Fe                  0.05        0.1        0.2
Line  7:       -isotopes
Line  8a:          13C                 0.05        0.1        0.05
Line  8b:          34S                 1.0
Line  9:       -range                  10000
Line 10:       -minimal
Line 11:       -tolerance              1e-10
Line 12:       -force_solutions        true        false
Line 13:       -uncertainty_water      0.55  # moles (~1%)
Line 14:       -mineral_water          false
Line 15:       -lon_netpath            prefix
Line 16:       -pat_netpath            prefix
Line 17:       -multiple_precision     false
Line 18:       -mp_tolerance           1e-25
Line 19:       -censor_mp              1e-12
```

Explanation

Line 0: **INVERSE_MODELING** [*number*] [*description*]

INVERSE_MODELING is the keyword for the data block.

number—A positive number designates the following inverse-modeling definition. Default is 1.

description—Optional comment that describes the inverse-modeling calculation.

Line 1: **-solutions**, *list of solution numbers*

> **-solutions**—Identifier that indicates a list of solution numbers follows on the same line. Optionally, **sol** or **-s[olutions]**. Note the hyphen is required to avoid conflict with the keyword SOLUTION.

> *list of solution numbers*—List of solution numbers to use in mole-balance calculations. At least two solution numbers are required and these solutions must be defined by SOLUTION (or SOLUTION_SPREAD) input or by SAVE after a batch-reaction calculation in the current or previous simulations. The final solution number is listed last; all but the final solution are termed "initial solutions". If more than one initial solution is listed, the initial solutions are assumed to mix to form the final solution. The mixing proportions of the initial solutions are calculated in the modeling process. In the Example data block (Line 1), solution 5 is to be made by mixing solutions 10 and 3 in combination with phase mole transfers.

Line 2: **-uncertainty**, *list of uncertainty limits*

> **-uncertainty**—Identifier that indicates a list of default uncertainty limits for each solution follows on the same line. The uncertainty limits defined with **-uncertainty** do not apply to pH; default for pH is 0.05 pH units and may be changed with the **-balances** identifier. If **-uncertainty** is not entered, the program uses 0.05. The default uncertainty limits can be overridden for individual elements or element valence states using the **-balances** identifier. Optionally, **uncertainty, uncertainties, -u[ncertainty]**, or **-u[ncertainties]**.

> *list of uncertainty limits*—List of default uncertainty limits that are applied to each solution in the order given by **-solutions**. The first uncertainty limit in the list is applied to all the element and element valence states in the first solution listed in **-solutions**. The second uncertainty limit in the list is applied to all the element and element valence states in the second solution listed in **-solutions**, and so on. If fewer uncertainty limits are entered than the number of solutions, the final uncertainty limit in the list is used for the remaining solutions. Thus, if only one uncertainty limit is entered, it is applied to all solutions. The uncertainty limit may have two forms: (1) if the uncertainty limit is positive, it is interpreted as a fraction to be used to calculate the uncertainty limit for each element or element valence state; a value of 0.02 indicates that an uncertainty limit of 2 percent of the moles of each element in solution

will be used, and (2) if the uncertainty limit is negative, it is interpreted as an absolute value in moles to use for each mole-balance constraint. The second form is rarely used in **-uncertainty** input. In this Example data block, the default uncertainty limit for the first solution is set to 0.02, which indicates that the concentration of each element in the first solution (solution 10) is allowed to vary up to plus or minus 2 percent, and a default uncertainty limit of 4 percent will be applied to each element and valence state in the second solution (solution 3) and in all remaining solutions (solution 5 in this case).

Line 3: **-phases**

-phases—Identifier that indicates a list of phases to be used in inverse modeling follows on succeeding lines. Optionally, **phase, phase_data, -p[hases]**, or **-p[hase_data]**. Note the hyphen is required in **-phases** to avoid conflict with the keyword PHASES.

Line 4: *phase name* [**force**] [(**dissolve** or **precipitate**)] [*list of isotope name, isotope ratio, isotope uncertainty limit*]

phase name—Name of a phase to be used in inverse modeling. The phase must be defined in PHASES input or it must be a charge-balanced exchange species defined in EXCHANGE_SPECIES input. Any phases and exchange species defined in the database file or in the current or previous simulations are available for inverse modeling. Only the chemical reaction in PHASES or EXCHANGE_SPECIES input is important; the log K is not used in inverse-modeling calculations.

force—The phase is included ("forced") to be in the range calculation (see Line 9) whether or not the phase mole transfer is nonzero. This will give another degree of freedom to the range calculation for models that do not include the phase and the resulting range of mole transfers may be larger. The order of this option following the phase name is not important. Optionally, **f[orce]**.

dissolve or **precipitate**—The phase may be constrained only to enter the aqueous phase, "**dissolve**", or leave the aqueous phase, "**precipitate**". Any set of initial letters from these two words are sufficient to define a constraint.

list of isotope name, isotope ratio, isotope uncertainty limit—Isotopic information for the phase may be defined for one or more isotopes by appending (to Line 4) triplets of *isotope name, isotope ratio, isotope uncertainty limit.*

isotope name—Isotope name written with mass number first followed by element name with no intervening spaces.

isotope ratio—Isotope ratio for this isotope of this element (*isotope name*) in the phase, frequently permil, but percent or other units can be used. Units must be consistent with the units in which this isotope of the element is defined in SOLUTION input.

isotope uncertainty limit—Uncertainty limit for isotope ratio in the phase. Units must be consistent with the units for *isotope ratio* and units in which this isotope of this element is defined in SOLUTION input.

Line 5: **-balances**

-balances—Identifier that indicates a list of names of elements or element-valence-states follow on succeeding lines. Optionally, **bal**, **balance**, **balances**, or **-b[alances]**.

Line 6: *element or valence state name* [*list of uncertainty limits*]

element or valence state name—Name of an element or element valence state to be included as a mole-balance constraint in inverse modeling. The identifier **-balances** is used for two purposes: (1) to include mole-balance equations for elements not contained in any of the phases (**-phases**); and (2) to override the uncertainty limits defined with **-uncertainty** (or the default uncertainty limits) for elements, element valences states, or pH. Mole-balance equations for all elements that are found in the phases of **-phases** input are automatically included in inverse modeling with the default uncertainty limits defined by the **-uncertainties** identifier; mole-balance equations for all valence states of redox elements are included if the element is in any of the phases of **-phases**.

list of uncertainty limits—List of uncertainty limits for the specified element, element valence-state constraint, or pH. It is possible to input an uncertainty limit for *element or valence state name* for each solution used in inverse modeling (as defined by **-solutions**). If fewer uncertainty limits are entered than the number of solutions, the final uncertainty limit in the list is used for the remaining solutions. Thus, if only one uncertainty limit is entered, it is used for the given element or element valence state for all solutions. The uncertainty limit for pH must be given in standard units. Thus, the uncertainty limit in pH given on Line 6a is 0.1 pH units for all solutions. The uncertainty limits for elements and element valence states (but not for pH) may have two forms: (1) if the uncertainty limit is positive, it is interpreted as a fraction that when multiplied times the moles in solution gives the uncertainty limit in moles—a value of 0.02 would indicate an uncertainty limit of 2 percent of the moles in solution; and (2) if the uncertainty limit is negative, it is interpreted as an absolute value in moles to use for the solution in the mole-balance equation for *element* or *valence state name*. In the Example data block, Line 6b, the uncertainty limit for calcium in solution 10 is 1 percent of the moles of calcium in solution 10. The uncertainty limit for calcium in solution 3 and solution 5 is 0.005 mol. The uncertainty limit for iron (Line 6d) is 5 percent in solution 10, 10 percent in solution 3, and 20 percent in solution 5.

Line 7: **-isotopes**

-isotopes—Identifier that specifies mole balances be included in the calculations for the isotopes listed on succeeding lines. Optionally, **isotopes** or **-i**[**sotopes**].

Line 8: *isotope_name, list of uncertainty limits*

isotope_name—Name of an isotope for which mole balance is desired. The name must be written with mass number first followed by element name or redox state with no intervening spaces.

list of uncertainty limits—List of uncertainty limits for the specified isotope for the solutions used in inverse modeling (as defined by **-solutions**). If fewer uncertainty limits are entered than the number of solutions, the final uncertainty limit in the list is used for the remaining solutions. Thus, if only one uncertainty limit is entered, it is used for the given isotope for all solutions. In the Example data block (Line 8), the uncertainty limit for carbon-13 (Line 8a) is 0.05 permil in solution 10, 0.1 permil in solution 3, and 0.05 permil in solution 5. The uncertainty limit for sulfur-34 (Line 8b) is 1 permil in all solutions. Units of the uncertainty limits for an isotope must be consistent with units used to define the isotope in SOLUTION input and with the units used to define isotope values under the **-phases** identifier (Line 4).

Line 9: **-range** [*maximum*]

-range—Identifier that specifies that ranges in mole transfer for each phase in each model should be calculated. The range in mole transfer for a phase is the minimum and maximum mole transfers that can be attained for a given inverse model by varying element concentrations within their uncertainty limits. Any phase with the **force** option will be included for each range calculation even if the inverse model does not contain this phase. Optionally, **range**, **ranges**, or **-r**[**anges**].

maximum—The maximum value for the range is calculated by minimizing the difference between the value of *maximum* and the calculated mole transfer of the phase or the solution fraction. The minimum value of the range is calculated by minimizing the difference between the negative of the value of *maximum* and the calculated mole transfer of the phase or the solution fraction. In some evaporation problems, the solution fraction could be greater than 1000 (over 1,000-fold evaporative concentration). In these problems, the default value is not large enough and a larger value of maximum should be entered. Default is 1000.

Line 10: **-minimal**

-minimal—Identifier that specifies that models be reduced to the minimum number of phases that can satisfy all of the constraints within the specified uncertainty limits. Note that two minimal models may have different numbers of phases; minimal models imply that no model with any proper subset of phases and solutions could be found. The **-minimal** identifier minimizes the number of calculations that will be performed and produces the models that contain the most essential geochemical reactions. However, models that are not minimal may also be of interest, so the use of this option is left to the discretion of the user. In the interest of expediency, it is suggested that models are first identified using the **-minimal** identifier, checked for plausibility and geochemical consistency, and then rerun without the **-minimal** identifier. Optionally, **minimal**, **minimum**, **-m[inimal]**, or **-m[inimum]**.

Line 11: **-tolerance** *tol*

-tolerance—Identifier that indicates a tolerance for the optimizing solver is to be given. Optionally, **tolerance** or **-t[olerance]**.

tol—Tolerance used by the optimizing solver. The value of *tol* should be greater than the greatest calculated mole transfer or solution fraction multiplied by 1×10^{-15}. The default value is adequate unless very large mole transfers (greater than 1,000 mol) or solution fractions (greater than 1,000-fold evaporative concentration) occur. In these cases, a larger value of *tol* may be needed. Essentially, a value less than *tol* is treated as zero. Thus, the value of *tol* should not be too large, or significantly different concentrations will be treated as equal. Uncertainty limits less than *tol* are assumed to be zero. Default is approximately 1×10^{-10} for default compilation, but may be smaller if the program is compiled by using long double precision.

Line 12: **-force_solutions** *list of* (*True* or *False*)

-force_solutions—Identifier that indicates one or more solutions will be forced to be included in all range calculations. If **-force_solutions** is not included, the default is false for all solutions; no solutions are forced to be included in the range calculations. Optionally, **force_solution**, **force_solutions**, or **-force_[solutions]**.

list of (*True* or *False*)—**True** values include initial solutions in all range calculations. It is possible to input a **true** or **false** value for each initial solution used in inverse modeling. If fewer values are entered than the number of initial solutions (**-solutions** identifier), then the final value in the list is used for the remaining initial solutions. Thus, if only one **true** or **false** value is entered, it is used for all initial solutions. In the Example data block (Line 12), solution 10 will be included in all range calculations for all models; even if a model does not include solution 10 (mixing fraction of zero), the range calculation will allow for nonzero mixing fractions of solution 10 in calculating the minimum and maximum mole transfers of phases. Solutions 3 and 5 will be included in range calculations only for models that have a nonzero mixing fraction for these solutions.

Line 13: **-uncertainty_water** *moles*

-uncertainty_water—Identifier for uncertainty term in the water-balance equation. For completeness in the formulation of inverse modeling, an uncertainty term can be added to the water balance equation. The sum of the moles of water derived from each initial solution must balance the moles of water in the final solution plus or minus *moles* of water. Optionally, **uncertainty_water**, **u_water**, **-uncertainty_[water]**, or **-u_[water]**.

moles—Uncertainty term for the water-balance equation. Default is 0.0 mol.

Line 14: **-mineral_water** [(*True* or *False*)]

-mineral_water—Identifier to include or exclude water derived from minerals in the water-balance equation. Normally, water from minerals should be included in the water-balance equation. Sometimes unreasonable models are generated that create all the water in solution by dissolution and precipitation of minerals. Setting **-mineral_water** to **false** removes the terms for water derived from minerals from the water-balance equation, which eliminates these unreasonable models. However, removing these terms may introduce errors in some models by ignoring water derived from minerals (for example, water from dissolution of gypsum) that should be considered in the water-balance equation. Default is **true** if **-mineral_water** is not included. Optionally, **mineral_water** or **-mine[ral_water]**.

(*True* or *False*)—**True** includes terms for water derived from minerals in the water-balance equation, **false** excludes these terms from the equation. If neither **true** nor **false** is entered on the line, **true** is assumed. Optionally, **t[rue]** or **f[alse]**.

Line 15: **-lon_netpath** *prefix*

-lon_netpath—At the beginning of an inverse-modeling calculation, all solutions that have been defined to PHREEQC are written to a file named *prefix*.**lon** (indicating a Netpath "lon" file format). The file contains the solution compositions (with concentrations converted to moles per kilogram water) in a format that is readable by DBXL. DBXL is distributed with NetpathXL (Parkhurst and Charlton, 2008). Optionally, **lon_netpath** or **-l[on_netpath]**.

prefix—The alphanumeric string is used to generate a file name.

Line 16: **-pat_netpath** *prefix*

-pat_netpath—A Netpath model file is written for each inverse model that is found by PHREEQC. The model files are named *prefix-n*.**mod**, where *n* is the sequence number of the model. In addition, a file is written with the name *prefix*.**pat** (indicating a Netpath "pat" file format); it contains the compositions of the solutions associated with each model. The solution compositions for each model include the concentration adjustments calculated by the PHREEQC inverse model. The model and **.pat** files are readable with NetpathXL (Parkhurst and Charlton, 2008). Optionally, **pat_netpath** or **-pa[t_netpath]**.

prefix—The alphanumeric string used to generate file names for model files and the corresponding **.pat** file.

Line 17: **-multiple_precision** [(*True* or *False*)]

-multiple_precision—Invokes multiple-precision version of Cl1, the simplex optimization routine (provided PHREEQC has been compiled with the INVERSE_CL1MP preprocessor directive). Use of the multiple-precision version of Cl1 has not proven to be significantly better than the default version. Default is **false** if Line 17 is not included. Optionally, **multiple_precision** or **-mu[ltiple_precision]**.

(*True* or *False*)—**True** uses the multiple-precision version of Cl1; **false** uses the default precision version of Cl1. If neither **true** nor **false** is entered on the line, **true** is assumed. Optionally, **t[rue]** or **f[alse]**.

Line 18: **-mp_tolerance**

   -mp_tolerance—Identifier that indicates a tolerance for the multiple-precision version of the optimizing solver is to be given. Optionally, **mp_tolerance** or **-mp[_tolerance]**.

   mp_tol—Tolerance used by the multiple-precision version of the optimizing solver. Uncertainty limits less than *mp_tol* are assumed to be zero. Default is 1×10^{-12}.

Line 19: **-censor_mp** *value*

   -censor_mp—Identifier that indicates coefficients in the inverse-modeling matrix will be censored (set to zero). Optionally, **censor_mp** or **-c[ensor_mp]**.

   value—As calculations occur in the linear-equation array, elements less than *value* are set to zero. If *value* is zero, no censoring occurs. Default is 1×10^{-20}.

Notes

Writing of inverse models to the output file can be enabled or disabled with the **-inverse** identifier in the PRINT data block. Inverse models can be written to the selected-output file by including the **-inverse** identifier in the SELECTED_OUTPUT data block. For each model that is found the following values are written to the selected-output file: (1) the sum of residuals, sum of each residual divided by its uncertainty limit, and the maximum fractional error; (2) for each solution—the mixing fraction, minimum mixing fraction, and maximum mixing fraction; and (3) for each phase in the list of phases (**-phase** identifier)—the mole transfer, minimum mole transfer, and maximum mole transfer. Mixing fractions and mole transfers are zero for solutions and phases not included in the model. Minimum and maximum values are 0.0 unless the **-range** calculation is performed. The result of printing to the selected-output file is columns of numbers, where each row represents a mole-balance model.

The numerical method for inverse modeling *requires* consideration of the uncertainties related to aqueous concentrations. Uncertainties related to mineral compositions may be equally important, but they are not automatically considered. To consider uncertainties in mineral compositions, it is possible to include two (or more) phases (under **-phases** identifier and definitions in PHASES data block) that represent end member compositions for minerals. The inverse modeling calculation will attempt to find models considering the entire range of mineral composition. Usually, each model that is found will include only one or the other of the end members, but any mixture of inverse models, which in this case would represent mixtures of the end members, is also a valid inverse model.

The possibility of evaporation or dilution can be included in inverse modeling by including water as one of the phases under the **-phases** identifier [H2O(g) for databases distributed with program]. The mole transfer of this phase will affect only the water-balance equation. If the mole transfer is positive, dilution is simulated; if negative, evaporation is simulated (see example 17 in Examples).

If **-uncertainty** is not included, a default uncertainty limit of 0.05 (5 percent) is used for elements and 0.05 for pH. Default uncertainty limits, specified by **-uncertainty**, will almost always be specified as positive numbers, indicating fractional uncertainty limits. A default uncertainty limit specified by a negative number, indicating a fixed molal uncertainty limit for all elements in solution, is usually not reasonable because of wide ranges in concentrations among elements present in solution.

No mole-balance equation is used for pH and the uncertainty limit in pH only affects the mole balance on alkalinity. Alkalinity is assumed to co-vary with pH and carbon, and an equation relating the uncertainty term for alkalinity and the uncertainty terms for pH and carbon is included in the inverse model (see "Equations and Numerical Methods for Inverse Modeling" in Parkhurst and Appelo, 1999).

All phase names and phase stoichiometries must be defined through PHASES or EXCHANGE_SPECIES input. Lines 4c and 4d are included to allow ion-exchange reactions in the inverse model; exchange species with the names CaX_2 and NaX are among the exchange species defined in the default database and are thus available for use in inverse modeling when this database is used. In the Example data block and in the example problems (16, 17, and 18), the composition of the phases is assumed to be relatively simple. In real systems, the composition of reactive phases—for example pyroxenes, amphiboles, or alumino-silicate glasses—may be complex. Application of inverse modeling in these systems will require knowledge of specific mineral compositions or appropriate simplification of the mineral stoichiometries.

By default, mole-balance equations for every element that occurs in the phases listed in **-phases** input are included in the inverse-modeling formulation. If an element is redox active, then mole-balance equations for all valence states of that element are included. The **-balances** identifier is necessary to define (1) uncertainty limits for pH, elements, or element valence states that are different from the default uncertainty limits or (2) mole-balance equations for elements not included in the phases. Mole-balance equations for alkalinity and electrons are always included in the inverse model. In some solutions, such as pure water or pure sodium chloride solutions, the alkalinity may be small (less than 1×10^{-7} eq [equivalent]) in both initial and final solutions. In this case, it may be necessary to use large (relative to 1×10^{-7} eq) uncertainty limits

($+1.0$ or -1×10^{-6}) to obtain a mole balance on alkalinity. For most natural waters, alkalinity will not be small in both solutions and special handling of the alkalinity uncertainty will not be necessary (note alkalinity is a negative number in acid solutions). Uncertainty limits for electrons are never used because it is always assumed that no free electrons exist in an aqueous solution.

If isotope mole balances are used, then (1) isotopic values for the aqueous phases must be defined through the SOLUTION data block, (2) the **-isotopes** identifier must be used in the INVERSE_MODELING data block to specify the isotopes for which mole balances are desired (and, optionally, the uncertainty limits in isotopic values associated with each solution), and (3) for each phase listed below the **-phases** identifier of the INVERSE_MODELING data block, isotopic values and uncertainty limits must be defined for each isotope that is contained in the phase. In addition, each phase that contains isotopes must be constrained either to dissolve or to precipitate. Default uncertainty limits for isotopes are given in table 4.

Table 4. Default uncertainty limits for isotopes.
[PDB, Pee Dee Belemnite; CDT, Cañon Diablo Troilite, VSMOW, Vienna Standard Mean Ocean Water]

Isotope	Default uncertainty limit
^{13}C	1 permil PDB
$^{13}C(4)$	1 permil PDB
$^{13}C(-4)$	5 permil PDB
^{34}S	1 permil CDT
$^{34}S(5)$	1 permil CDT
$^{34}S(-2)$	5 permil CDT
^{2}H	1 permil VSMOW
^{18}O	0.1 permil VSMOW
^{87}Sr	0.01 ratio

The options **-minimal** and **-range** affect the speed of the calculations. The fastest calculation is one that includes the **-minimal** identifier and does not include **-range**. The slowest calculation is one that does not include **-minimal** and does include **-range**.

The **force** option for a phase in **-phases** and the **-force_solutions** identifier affects only the range calculation; it does not affect the number of models that are found. When the **-range** identifier is specified and a model is found by the numerical method, then the model is augmented by any phase for which **force** is specified and by any solution for which **-force_solutions** is **true**; the range calculation is performed with the augmented model. The effect of these options is to calculate wider ranges for mole transfers for some models. If every phase and every solution were forced to be in the range calculation, then the results of the range calculation would be the same for every model and the results would be the maximum possible ranges of mole transfer for any models that could be derived from the given set of solutions, phases, and uncertainty limits.

Data interchange from PHREEQC to NetpathXL (Plummer and others, 1991, 1994; Parkhurst and Charlton, 2008) is available through the **-lon_netpath** and **-pat_netpath** identifiers. By using **-lon_netpath**, solutions defined to PHREEQC (through the SOLUTION, SOLUTION_SPECIES, SOLUTION_SPREAD, or SAVE data blocks) can be written in a format readable by DBXL, which is distributed with NetpathXL. DBXL in turn can write data as an Excel file that can be used by NetpathXL for inverse modeling.

The **-pat_netpath** identifier allows PHREEQC inverse models to be recreated in NetpathXL. This feature is useful for inverse modeling of isotopes. The inverse model of PHREEQC has capabilities to account for uncertainties in element concentrations, but has a limited capability for modeling isotope evolution (in forward models, isotopes can be fractionated with kinetics or the capabilities included in *iso.dat*). NETPATH (Plummer and others, 1991, 1994) as implemented in NetpathXL has a complete formulation for inverse modeling with isotopes that includes fractionation processes. The **-pat_netpath** identifier allows inverse models that include adjustments for uncertainties to be imported into NetpathXL. Model files, as defined in NETPATH, are exported from PHREEQC along with a **.pat** file that includes solution compositions as adjusted by the PHREEQC inverse modeling calculation. These model files and **.pat** file will recreate the PHREEQC inverse model in NetpathXL. In addition, data can be translated from NetpathXL Excel files to PHREEQC input files by using the program DBXL.

The numerical method for inverse modeling with PHREEQC occasionally fails, presumably because of ill-conditioned matrices for the linear equations. A higher precision version of the optimization solver Cl1 was implemented to try to improve the numerical stability of the solver. Unfortunately, results with the higher precision solver have not been significantly better than the default precision solver. Two parameters are available to adjust the numerical method with the high precision solver, **-mp_tolerance** and **-censor_mp**. It is possible that using the higher precision solver with these parameters will result in a solution to an inverse modeling problem that is not possible with the default precision solver.

Example problems

The keyword **INVERSE_MODELING** is used in example problems 16, 17, and 18.

Related keywords

EXCHANGE_SPECIES, PHASES, PRINT, SELECTED_OUTPUT, SOLUTION, and SAVE.

ISOTOPES

This keyword data block is used to identify isotopes of elements and to define the absolute ratio of the minor isotope to the major isotope in the isotope standard. This keyword data block is used to implement the treatment of isotopes as individual thermodynamic components (Thorstenson and Parkhurst, 2002, 2004). The **ISOTOPES** data block is used in the database file *iso.dat* and is unlikely to be used in any other context.

Example data block

```
Line 0: ISOTOPES
Line 1: H
Line 2:        -isotope      D       permil 155.76e-6      # VSMOW
Line 2a:       -isotope      T       TU     1e-18
Line 1a: H(0)
Line 2b:       -isotope      D(0)    permil 155.76e-6      # VSMOW
Line 2c:       -isotope      T(0)    TU     1e-18
```

Explanation

Line 0: **ISOTOPES**

ISOTOPES is the keyword for the data block. No other data are input on the keyword line.

Line 1: (*element* or *element redox state*)

element or *element redox state*—Name of an element or element redox state that has two or more isotopes of environmental interest. The element or redox state must be defined in SOLUTION_MASTER_SPECIES.

Line 2: **-isotope**, (*isotope name* or *isotope redox state*), *units*, *ratio*

-isotope—Identifier used to define an isotope of an element or element redox state. Optionally, **isotope** or **-i**[**sotope**].

isotope name or *isotope redox state*—An isotope that has been defined as an element or element redox state in SOLUTION_MASTER_SPECIES. The isotope is an isotope of the element or element redox state defined in the preceding Line 1.

units—Units of measurement for the isotope. Legal units are permil, pct (percent), pmc (percent modern carbon), tu (tritium units), and pci/L (picocurie per liter).

ratio—Absolute mole ratio in the standard of the (minor) isotope to the predominant isotope.

Notes

Reaction calculations with isotopes are performed by assuming each isotope is a separate thermodynamic component. Thus, in addition to the principle isotope of an element, which typically is named by the standard element nomenclature (for example, C for carbon), each isotope also is defined as an element in a **SOLUTION_MASTER_SPECIES** data block. The isotope name is usually formed by placing the element name prefixed by the isotopic number in brackets (for example, [13C] for carbon-13), or by special names like D for deuterium and T for tritium.

The individual component approach for isotopes posits that each aqueous species containing a minor isotope can have a slightly different equilibrium constant than the major isotope species and that the difference can be related to symmetry numbers and fractionation factors. Likewise for heterogeneous reactions between the solution and a gas phase or solid phases, minor-isotope gas or solid components have slightly different equilibrium constants than the major isotope versions. Equilibrium constants must be defined for each isotopic gas and solid component. Heterogeneous fractionation is calculated as an equilibrium process between solution and a gas phase (**GAS_PHASE**) and (or) between solution and solid solutions (**SOLID_SOLUTIONS**). Kinetic fractionation can be calculated by using slightly different rates of reaction for minor isotopic components than for major isotope components.

The **ISOTOPES** data block describes which isotopes are related to which elements. In the Example data block given in this section, the elements and redox states of D and T are related to the element H and the redox state H(0). The **ISOTOPES** data block also defines the units of measurement for each isotope and the absolute ratio in the standard of the isotope to the predominant isotope. This ratio is used to convert the isotopic measurement from the units of the standard into moles of isotope in solution. Once the number of moles of an isotope in solution is known, an isotope is treated exactly the same as any other element. For example, the aqueous model for deuterium is defined with **SOLUTION_SPECIES** data block and is nearly the same as the aqueous model for H, with the exception that the equilibrium constants are slightly different. The differences in equilibrium constants can be related to fractionation factors. The **NAMED_EXPRESSIONS** data block is used to simplify the definition of the relationship between fractionation factors and equilibrium constants. Additional keyword data blocks (**CALCULATE_VALUES**, **ISOTOPE_ALPHAS**, **ISOTOPE_RATIOS**) are available by which molar concentrations can be converted back to standard isotopic units for output.

Example problems

The keyword **ISOTOPES** is used in the *iso.dat* database.

Related keywords

CALCULATE_VALUES, ISOTOPE_ALPHAS, ISOTOPE_RATIOS, NAMED_EXPRESSIONS, SOLUTION_MASTER_SPECIES, and SOLUTION_SPECIES.

ISOTOPE_ALPHAS

This keyword data block is used to enable printing of isotopic fractionation factors, referred to as alphas, to the output file. A Basic function defined in CALCULATE_VALUES is used to calculate the fractionation factor from the current isotopic composition of species or phases and an analytical expression for a fractionation factor is evaluated by a definition in NAMED_EXPRESSIONS. These two values and related data are printed in the output file under the heading "Isotope Alphas". The **ISOTOPE_ALPHAS** data block is used in the database file *iso.dat* and is unlikely to be used in any other context.

Example data block

```
Line 0:  ISOTOPE ALPHAS
Line 1:      Alpha_D_OH-/H2O(l)      Log_alpha_D_OH-/H2O(l)
Line 2:      Alpha_T_OH-/H2O(l)      Log_alpha_T_OH-/H2O(l)
```

Explanation

Line 0: **ISOTOPE_ALPHAS**

ISOTOPE_ALPHAS is the keyword for the data block. No other data are input on the keyword line.

Line 1: *calculate_values_function named_expression*

calculate_values_function—The name of a calculate values function (CALCULATE_VALUES data block) that evaluates a fractionation factor based on the isotopic compositions of species or phases.

named_expression—The name of a named expression (NAMED_EXPRESSIONS data block) that evaluates an analytical expression for a fractionation factor between species or phases.

Notes

This keyword data block is used to implement the treatment of isotopes as individual thermodynamic components (Thorstenson and Parkhurst, 2000, 2004). If R is defined to be the ratio of the number of moles of the minor isotope to the number of moles of the predominant isotope in a species or phase, then the fractionation factor, or alpha, is the ratio of R in one species or phase to R in another species or phase. In the Example data block given in this section, the fractionation factors are calculated for deuterium (D) and tritium (T) between hydroxide ion and liquid water. Analytical expressions for fractionation factors are defined in the database through the use of the NAMED_EXPRESSIONS data block and are incorporated into equilibrium constants for species and phases in SOLUTION_SPECIES and PHASES data blocks. The fractionation factor based on solution and phase composition can be calculated by Basic functions that are defined in the CALCULATE_VALUES data block. At equilibrium, fractionation factors derived from the composition of the solution and other phases should equal the fractionation factor derived from the named expression, just as the ion-activity product of a phase should equal the equilibrium constant at equilibrium. This correspondence between composition-derived and analytical fractionation factors is printed in the output file under the heading "Isotope Alphas". The **ISOTOPE_ALPHAS** data block only defines quantities to print and by itself does not affect the equilibrium distribution of species in a simulation.

The use of CALCULATE_VALUES functions to evaluate isotope alphas may be expensive in terms of computer time. If **-isotope_alphas** is true (PRINT data block), all isotope alphas defined in the database or the input file are evaluated for each reaction calculation, even if the relevant isotopes are not in the reaction system. The Basic function SUM_SPECIES, which is used in many of the isotope alpha calculations, is especially time consuming. Minimizing the number of isotope alphas that are defined, minimizing the use of the SUM_SPECIES function in the CALCULATE_VALUES programs, and setting **-isotope_alphas false** in a PRINT data block will decrease execution times for isotopic calculations.

Example problems

The keyword **ISOTOPE_ALPHAS** is used in the *iso.dat* database.

Related keywords

CALCULATE_VALUES, ISOTOPE_RATIOS, and NAMED_EXPRESSIONS.

ISOTOPE_RATIOS

This keyword data block is used to enable printing of isotopic ratios in species or phases to the output file. A Basic function defined in CALCULATE_VALUES is used to calculate an isotope ratio, which is then printed in the output file under the heading "Isotope Ratios". The **ISOTOPE_RATIOS** data block is used in the database file *iso.dat* and is unlikely to be used in any other context.

Example data block

```
Line  0:  ISOTOPE_RATIOS
Line  1:       R(D)_H2O(l)            D
Line  1a:      R(T)_H2O(l)            T
Line  1b:      R(D)_OH-               D
Line  1c:      R(T)_OH-               T
```

Explanation

Line 0: **ISOTOPE_RATIOS**

> **ISOTOPE_RATIOS** is the keyword for the data block. No other data are input on the keyword line.

Line 1: *calculate_values_function isotope*

> *calculate_values_function*—The name of a calculate values function (CALCULATE_VALUES data block) that evaluates an isotopic ratio based on the isotopic compositions of species or phases.
>
> *isotope*—The name of the isotope used in calculating the isotope ratio.

Notes

This keyword data block is used to implement the treatment of isotopes as individual thermodynamic components (Thorstenson and Parkhurst, 2000, 2004). An isotopic ratio, R, is defined to be the ratio of the number of moles of the minor isotope to the number of moles of the predominant isotope in a species or phase. A fractionation factor is defined as the ratio of two Rs. In the Example data block given in this section, isotopic ratios are calculated for deuterium (D) and tritium (T) in liquid water and in the hydroxide ion. The isotopic ratios based on solution and phase compositions are calculated by Basic functions defined in the CALCULATE_VALUES data block. For example, the CALCULATE_VALUES function that defines the deuterium to ^{1}H ratio in hydroxide is as follows:

```
R(D)_OH-
    -start
10 ratio = -9999.999
20 if (TOT("D") <= 0) THEN GOTO 100
30 total_D = sum_species("*{O,[18O]}D*","D")
40 total_H = sum_species("*{O,[18O]}H*","H")
50 if (total_H <= 0) THEN GOTO 100
60 ratio = total_D/total_H
100 save ratio
    -end
```

Results of evaluating the Basic functions specified in the **ISOTOPE_RATIOS** data block are printed in the output file under the heading "Isotope Ratios". The **ISOTOPE_RATIOS** data block only defines quantities to print and by itself does not affect the equilibrium distribution of species in a simulation.

The use of CALCULATE_VALUES functions to evaluate isotope ratios may be expensive in terms of computer time. If **-isotope_ratios** is true (PRINT data block), isotope ratios are evaluated for each isotope in the reaction system. The Basic function SUM_SPECIES, which is used in many of the isotope ratio calculations, is especially time consuming. Minimizing the number of isotope ratios that are defined in the database and input file, minimizing the use of the SUM_SPECIES function in the CALCULATE_VALUES programs, and setting **-isotope_ratios false** in a PRINT data block will decrease execution times for isotopic calculations.

Example problems

The keyword **ISOTOPE_RATIOS** is used in the *iso.dat* database.

Related keywords

CALCULATE_VALUES, ISOTOPE_ALPHAS, and NAMED_EXPRESSIONS.

KINETICS

This keyword data block is used to specify kinetic reactions and parameters for batch-reaction and reactive-transport calculations. Mathematical expressions for the rates of the kinetic reactions are defined with the **RATES** data block. The rate equations are integrated over a time step by either a Runge-Kutta method or by an implicit stiff-equation solver, which is more robust and faster when kinetic reactions have widely varying rates. Both methods estimate the error of the integration and use appropriate time subintervals to maintain the errors within specified tolerances for each time interval.

Example data block 1

```
Line 0:   KINETICS 1 Define 3 explicit time steps
Line 1a: Pyrite
Line 2a:      -formula      FeS2 1.0 FeAs2 0.001
Line 3a:      -m            1e-3
Line 4a:      -m0           1e-3
Line 5a:      -parms        3.0   0.67   .5   -0.11
Line 6a:      -tol          1e-9
Line 1b: Calcite
Line 3b:      -m            7.e-4
Line 4b:      -m0           7.e-4
Line 5b:      -parms        5.0      0.3
Line 6b:      -tol          1.e-8
Line 1c: Organic_C
Line 2c:      -formula      CH2O(NH3)0.1 0.5
Line 3c:      -m            5.e-3
Line 4c:      -m0           5.e-3
Line 6c:      -tol          1.e-8
Line 7:  -steps            100 200 300 day
Line 8:  -step_divide      100
Line 9:  -runge_kutta      6
Line 10: -cvode            false
Line 11: -bad_step_max     500
Line 12: -cvode_order      5
Line 13: -cvode_steps      100
```

Explanation 1

> Line 0: **KINETICS** [*number*] [*description*]
>
> **KINETICS** is the keyword for the data block.

number—A positive number designates the set of kinetic reactions. A range of numbers may also be given in the form *m-n*, where *m* and *n* are positive integers, *m* is less than *n*, and the two numbers are separated by a hyphen without intervening spaces. Default is 1.

description—Optional comment that describes the kinetic reactions.

Line 1: *rate name*

rate name—Name of a rate expression. The *rate name* and its associated rate expression must be defined within a RATES data block, either in the default database file or in the current or previous simulations of the run. The name must be spelled identically to the name used in RATES input (except for case).

Line 2: **-formula** *list of formula [stoichiometric coefficient]*

By default, the *rate name* is assumed to be the name of a phase that has been defined in a PHASES data block, and the formula for that phase is then used for the stoichiometry of the reaction (for example, the definitions for calcite in the Example data block above). However, kinetic reactions are not restricted to mineral phases, any set of elements produced or consumed by the kinetic reaction (relative to the aqueous phase) can be specified through a list of doublets *formula* and *stoichiometric coefficient* (Lines 2a and 2c). Optionally, **formula** or **-f**[ormula].

formula—Chemical formula or the name of a phase to be added by the kinetic reaction. If a chemical formula is used, it must begin with a capital letter and contain element symbols and stoichiometric coefficients (Line 2a). A phase name may be entered independent of case. Each *formula* must be a charge-balanced combination of elements. (An exception may be for defining exchangers or surfaces related to kinetic reactants). Any charge (+/-) in the formula is ignored. The formula may be considered as adding or removing native elements from the system in the given stoichiometry.

stoichiometric coefficient—Defines the mole transfer coefficient for *formula* per mole of reaction progress (the value for SAVE that is calculated in the RATES Basic program). The product of the coefficient times the moles of reaction progress gives the mole transfer for *formula* relative to the aqueous solution; a negative stoichiometric coefficient and a positive value for reaction progress gives a negative mole transfer, which removes reactants from the aqueous solution. In Line 2a, each mole of reaction dissolves 1.0 mol of FeS_2 and 0.001 mol of $FeAs_2$ into the aqueous solution; Line 2c demonstrates the use of parentheses in a

Description of Data Input 105

formula; each mole of reaction (as defined for SAVE in **RATES**) adds 0.5 mol of the specified formula (*stoichiometric coefficient* is 0.5). Thus, 1 mole of reaction will add 0.5 mol of CH_2O and 0.05 mol of NH_3 to the aqueous solution to simulate the degradation of nitrogen-containing organic matter. Default is 1.0, unitless (mol/mol).

Line 3: **-m** *moles*

> *moles*—Current moles of reactant. As reactions occur, the *moles* will increase or decrease. Default is equal to *initial moles* if *initial moles* is defined, or 1.0 mol if *initial moles* is not defined. Optionally, **m** or **-m**.

Line 4: **-m0** *initial moles*

> *initial moles*—Initial moles of reactant. This identifier is useful if the rate of reaction is dependent on grain size. Formulations for this dependency often include the ratio of the amount of reactant remaining to the amount of reactant initially present. The quantity *initial moles* does not change as the kinetic reactions proceed. Frequently, the quantity *initial moles* is equal to *moles* at the beginning of a kinetic reaction. Default is equal to *moles* if *moles* is defined, or 1.0 mol if *moles* is not defined. Optionally, **m0** or **-m0**

Line 5: **-parms** *list of parameters*

> *list of parameters*—A list of numbers may be entered that can be used in the rate expressions; for example, constants, exponents, or half saturation constants. In the rate expression defined with the **RATES** keyword, these numbers are available to the Basic interpreter in the array *PARM*; *PARM(1)* is the first number entered, *PARM(2)* the second, and so on. Optionally, **parms**, **-p[arms]**, **parameters**, or **-p[arameters]**.

Line 6: **-tol** *tolerance*

> *tolerance*—Tolerance for integration procedure (mol). For each integration time interval, the absolute difference between two estimates of the integral of the rate expression must be less than this tolerance or the time interval is automatically reduced. The value of *tolerance* is related to the concentration differences that are considered significant for the elements in the reaction. Smaller concentration differences that are considered significant require smaller tolerances. Numerical accuracy of the kinetic integration can be tested by decreasing the tolerance to determine if results change significantly. Default is 1×10^{-8} mol. Optionally, **tol** or **-t[ol]**.

Line 7: **-steps** *list of time steps* [*unit*]

> *list of time steps*—Time steps over which to integrate the rate expressions (s). The **-steps** identifier is used only during batch-reaction calculations; it is not needed for transport calculations. By default, the list of time steps are considered to be independent times, all starting from zero. The Example data block would produce results after 100, 200, and 300 d of reaction. However, the INCREMENTAL_REACTIONS keyword can be used to make the time steps incremental so that the results of the previous time step are the starting point of the new time step. For incremental time steps, the Example data block would produce results after 100, 300, and 600 d. Default is 1.0 s. Optionally, **steps,-s[teps]**, **time_steps**, or **-ti[me_steps]**.

> *unit*—Optional time unit may be **second**, **minute**, **hour**, **day**, **year**, or an abbreviation of one of these units. The *time steps* are converted to seconds after reading the data block; all internal calculations, Basic functions, and output times are in seconds. Default is second.

Line 8: **-step_divide** *step_divide*

> *step_divide*—This parameter affects integration by the Runge-Kutta solver. If *step_divide* is greater than 1.0, the first time interval of each integration is set to *time step* / *step_divide*; at least two time intervals must be integrated to reach the total time of *time step*—0 to *time step* / *step_divide* and *time step* / *step_divide* to *time step*. If *step_divide* is less than 1.0, then *step_divide* is the maximum moles of reaction that can be added during a kinetic integration subinterval. Frequently reaction rates are fast initially, thus requiring small time intervals to produce an accurate integration of the rate expressions. The Runge-Kutta method will adapt to these fast rates when the integration fails the **-tolerance** criterion, but it may require several reductions in the length of the initial time interval for the integration to meet the criterion; *step_divide* > 1 can be used to make the initial time interval of each integration sufficiently small to satisfy the criterion, which may speed the overall calculation time. However, the smaller time interval will apply to all integrations throughout the simulation, even if reaction rates are slow later in the simulation. Using an appropriate *step_divide* < 1 can also cause sufficiently small initial time intervals when rates are fast, but will not require small time intervals later in the simulation if rates are slow; however, the appropriate value for *step_divide* < 1 is not easily known and usually must be found by trial and error. The default maximum moles of reaction is 0.1 mol during a time subinterval. Normally,

-step_divide is not used unless run times are long and it is apparent that each integration requires several time intervals. The status line, which is printed to the screen, notes the number of integration intervals that fail the **-tolerance** criterion as "bad" and the number of integration intervals that pass the criterion as "OK". Optionally, **step_divide** or **-step_[divide]**.

Line 9: **-runge_kutta (1, 2, 3, or 6)**

(**1, 2, 3,** or **6**)—This parameter affects integration by the Runge-Kutta solver. It designates the preferred number of time subintervals to use when integrating rates and is related to the order of the integration method. A value of **6** specifies that a 5th order embedded Runge-Kutta method, which requires six intermediate rate evaluations, will be used for all integrations. For values of **1, 2,** or **3**, the program will try to limit the rate evaluations to this number. If the **-tolerance** criterion is not satisfied among the evaluations or over the full integration interval, the method will automatically revert to the Runge-Kutta method of order 5. A value of **6** means that the 5th order method will be used exclusively. Values of **1** or **2** are mainly expedient when it is known that the rate is nearly constant in time. Default is **3**. Optionally, **rk, -r[k], runge_kutta** or **-r[unge_kutta]**.

Line 10: **-cvode** [(*True* or *False*)]

-cvode—Specifies whether to use the explicit Runge-Kutta method or the implicit CVODE method (Cohen and Hindmarsh, 1996) to integrate the kinetic rate equations. Default is **false** if **-cvode** is not included; Runge-Kutta method is used. Optionally, **cvode** or **-c[vode]**.

(*True* or *False*)—A value of **true** indicates the CVODE implicit integration method is used; **false** indicates the explicit Runge-Kutta integration method is used. If neither **true** nor **false** is entered on the line, **true** is assumed. Optionally, **t[rue]** or **f[alse]**.

Line 11: **-bad_step_max** *tries*

-bad_step_max—Defines the maximum number of attempts at integrating a set of kinetic reactions over a time step. For the Runge-Kutta method, it is the maximum number of times the integration fails in integrating over a time interval. For the CVODE method, it is the number of times that CVODE is invoked in integrating over a time interval. Default is 500 if **-bad_step_max** is not included. Optionally, **bad_step_max** or **-b[ad_step_max]**.

tries—The maximum number of integration attempts.

Line 12: **-cvode_order** (*1, 2, 3, 4,* or *5*)

> **-cvode_order**—Specifies the number of terms to use in the extrapolation of rates when using the CVODE method. Default is 5. Optionally, **cvode_order** or **-cvode_o[rder]**.
>
> (*1, 2, 3, 4,* or *5*) –Number of terms used in the extrapolation of rates.

Line 13: **-cvode_steps** *steps*

> **-cvode_steps**—Specifies the maximum number of steps that will be taken during one invocation of CVODE. Default is 100 if **-cvode_steps** is not included. Optionally, **cvode_steps** or **-cvode_[steps]**.
>
> *steps*—Maximum number of steps.

Example data block 2

```
Line 0:   KINETICS 1 Define 3 equal time steps
Line 1a: Calcite
Line 3a:     -m       7.e-4
Line 5a:     -parms   5      0.3
Line 7:  -steps       300 day in 3 steps
```

Explanation 2

> Line 0: **KINETICS** [*number*] [*description*]
>
> Same as Example data block 1.
>
> Line 1: *rate name*
>
> Same as Example data block 1.
>
> Line 3: **-m** *moles*
>
> Same as Example data block 1.
>
> Line 5: **-parms** *list of parameters*
>
> Same as Example data block 1.
>
> Line 7: **-steps** *total time* [*unit*] [**in** *steps*]
>
> > *total time*—Total time over which to integrate kinetic reactions, in seconds. The total time may be divided into a number of calculations given by *steps*. The **-steps** identifier is used only in batch-reaction calculations; it is not needed for transport calculations. Default is 1.0 s. Optionally, **steps,-s[teps]**, **time_steps**, or **-ti[me_steps]**.

unit—Optional time unit may be **second, minute, hour, day, year**, or an abbreviation of one of these units. The *total time* is converted to seconds after reading the data block; all internal calculations, Basic functions, and output times are in seconds. Default is second.

in *steps*—"**in**" indicates that the *total time* will be divided into *steps* number of steps. INCREMENTAL_REACTIONS has no effect on the output for Example data block 2; results will be printed after 100, 200, and 300 d of reaction. However, INCREMENTAL_REACTIONS does affect the computational method. If INCREMENTAL_REACTIONS is **false** the reactions will be integrated over the time intervals from 0 to 100, 0 to 200, and 0 to 300 d. If INCREMENTAL_REACTIONS is **true** the reactions will be integrated over the time intervals from 0 to 100, 100 to 200, and 200 to 300 d.

Notes

Both **KINETICS** and REACTION data blocks are used to model irreversible reactions. REACTION can only be used to define specified amounts of stoichiometric reactions. For kinetic batch reactions or advective or advective-dispersive transport calculations, the fixed stoichiometric reaction of REACTION is added at each kinetic, advective, or transport time step, regardless of the length of the time step. **KINETICS** is used to define time-dependent kinetic reactions. To use **KINETICS**, a mathematical rate expression based on the solution composition must be defined and this expression is used to calculate the rate of reaction at any point in time. The RATES data block is used to define a set of general rate expressions that may apply over the entire modeling domain. The **KINETICS** data block is used to identify the subset of general rate expressions that apply to a given batch reaction or to specified cells of transport calculations. The **KINETICS** data block also is used to define specific parameters for the rate expression, such as the moles of reactant initially present in a cell, spatially varying coefficients, or cell-specific exponents for the rate equation. In advective (ADVECTION data block) and advective-dispersive transport (TRANSPORT data block) calculations, the number(s) assigned with the **KINETICS** keyword defines the cell(s) to which the kinetic reactions apply.

Two integration methods are available in PHREEQC, an explicit fifth-order Runge-Kutta method and an implicit CVODE method (Cohen and Hindmarsh, 1996). The Runge-Kutta method is likely to be faster for non-stiff ordinary differential equations; typically, these are differential equations that have similar time scales. CVODE is more robust for stiff sets of equations in which kinetic reactants have rates that vary widely for the same element. The **-bad_steps_max** parameter applies to both integration methods, but has slightly different meanings depending on the method. With the Runge-Kutta method, an attempt is made to integrate over a time substep. If that integration fails to keep errors less than specified tolerances, the integration is repeated with a smaller time step. For the Runge-Kutta method, the **-bad_steps_max** parameter is the maximum number of times that a new attempt is made after the integration fails to meet error tolerances. For the CVODE method, the parameter **-cvode_steps** determines the number of steps that can be taken with one invocation of CVODE. If the integration has not completed the entire time step, CVODE is reinvoked. For the CVODE method, the **-bad_steps_max** parameter is the number of times that CVODE can be invoked to complete the integration over a time step. The **-cvode_order** parameter specifies the number of terms to use in the extrapolation of rates within the CVODE method. Although less than five terms may be specified, experience has not shown that limiting the number of terms is helpful either to achieve or to accelerate a successful integration.

For a batch-reaction calculation, the number of reaction steps is the maximum number of steps defined in any of the following keyword data blocks: **KINETICS**, REACTION, REACTION_PRESSURE, and REACTION_TEMPERATURE. When the maximum number of steps is greater than the number of steps defined in **KINETICS**, then if INCREMENTAL_REACTIONS is false (cumulative reaction steps), the reactions are integrated for the time specified by the final time step for each of the additional steps; if INCREMENTAL_REACTIONS is true (incremental reaction steps), kinetic reactions are not included in the additional steps.

The SAVE data block does not apply to kinetic reactions, and a "SAVE **kinetics** *n*" statement results in an error message. The number of moles of a kinetic reaction is updated continuously during a simulation that includes kinetic reactions. Thus, unlike other reactants, a **KINETICS** data block definition will be changed automatically by a batch reaction; other reactant compositions that vary in a batch reaction must be saved with the SAVE data block to be updated to the new compositions.

Example problems

The keyword **KINETICS** is used in example problems 6, 9, and 15.

Related keywords

ADVECTION, COPY, DELETE, DUMP, INCREMENTAL_REACTIONS, PHASES, RATES, REACTION, and TRANSPORT.

KNOBS

This keyword data block is used to redefine parameters that affect convergence of the numerical method during speciation, batch-reaction, and transport calculations. The first six identifiers listed can be used to modify the numerical method to try to obtain a numerical solution to the nonlinear equations. The remaining identifiers produce long, uninterpretable output files, which are of little use to the user.

Example data block

```
Line  0:    KNOBS
Line  1:        -iterations            150
Line  2:        -convergence_tolerance 1e-8
Line  3:        -tolerance             1e-14
Line  4:        -step_size             10.
Line  5:        -pe_step_size          5.
Line  6:        -diagonal_scale        true
Line  7:        -debug_diffuse_layer   true
Line  8:        -debug_inverse         true
Line  9:        -debug_model           true
Line 10:        -debug_prep            true
Line 11:        -debug_set             true
Line 12:        -logfile               true
```

Explanation

Line 0: **KNOBS**

KNOBS is the keyword for the data block. Optionally, **DEBUG**.

Line 1: **-iterations** *iterations*

-iterations—Allows changing the maximum number of iterations. Default is 100 at startup. Optionally, **iterations** or **-i[terations]**.

iterations—Positive integer limiting the maximum number of iterations used to solve the set of algebraic equations for a single calculation. Values greater than 200 are not usually effective.

Line 2: **-convergence_tolerance** *convergence_tolerance*

-convergence_tolerance—Changes the convergence criterion used to determine when the algebraic equations have been solved. For an element mole-balance equation, convergence is satisfied when mole balance is within *convergence_tolerance* times the total moles of the element (*convergence_tolerance* · T_m). When the **-high_precision** identifier of

SELECTED_OUTPUT is true, the convergence criterion is set to the smaller of *convergence_tolerance* and 1×10^{-12}. Default is 1×10^{-8} at startup. Optionally, **convergence_tolerance** or **-c[onvergence_tolerance]**.

convergence_tolerance—Tolerance for determining convergence in the nonlinear equation solver.

Line 3: **-tolerance** *tolerance*

-tolerance—Allows changing the tolerance used by the optimization solver (subroutine cl1) to determine numbers equal to zero. This is <u>not</u> the convergence criterion used to determine when the algebraic equations have been solved. At starutp, default is 1×10^{-15} (or possibly smaller if the program is compiled with long double precision). Optionally, **tolerance** or **-t[olerance]**.

tolerance—Positive, decimal number used by the optimization solver (subroutine cl1). All numbers smaller than this number are treated as zero. This number should approach the value of the least significant decimal digit that can be interpreted by the computer. The value of tolerance should be on the order of 1×10^{-12} to 1×10^{-15} for most computers and most simulations.

Line 4: **-step_size** *step_size*

-step_size—Allows changing the maximum step size. At startup, default is 100; activities of master species may change by up to 2 orders of magnitude in a single iteration. Optionally, **step_size** or **-s[tep_size]**.

step_size—Positive, decimal number limiting the maximum, multiplicative change in the activity of an aqueous master species on each iteration.

Line 5: **-pe_step_size** *pe_step_size*

-pe_step_size—Allows changing the maximum step size for the activity of the electron. Optionally, **pe_step_size** or **-p[e_step_size]**.

pe_step_size—Positive, decimal number limiting the maximum, multiplicative change in the conventional activity of electrons on each iteration. Normally, *pe_step_size* should be smaller than the *step_size* because redox species are particularly sensitive to changes in pe. Default is 10; that is, a_{e^-} may change by up to 1 order of magnitude in a single iteration or pe may change by up to 1 unit.

Line 6: **-diagonal_scale** [(*True* or *False*)]

> **-diagonal_scale**—Allows changing the default method for scaling equations. Invoking this alternative method of scaling causes any mole-balance equations with the diagonal element (approximately the total concentration of the element or element valence state in solution) less than 1×10^{-10} to be scaled by the reciprocal of the diagonal element. Default is **false** at startup. Optionally, **diagonal_scale** or **-d[iagonal_scale]**.

> (*True* or *False*)—A value of **true** indicates the alternative scaling method is to be used; **false** indicates the alternative scaling method will not be used. If neither **true** nor **false** is entered on the line, **true** is assumed. Optionally, **t[rue]** or **f[alse]**.

Line 7: **-debug_diffuse_layer** [(*True* or *False*)]

> **-debug_diffuse_layer**—Includes debugging prints for diffuse layer calculations. This identifier applies only when **-diffuse_layer** is used in the SURFACE data block. If this option is set to **true**, values of the *g* function—the surface excess—are printed for each value of charge for aqueous species; the charge(s) for which the value of *g* has not converged are printed; and the number of iterations needed for the integration, by which *g* values are calculated, are printed. Default is **false** at startup. Optionally, **debug_diffuse_layer** or **-debug_d[iffuse_layer]**.

> (*True* or *False*)—A value of **true** indicates the debugging information will be included in the output file; **false** indicates debugging information will not be printed. If neither **true** nor **false** is entered on the line, **true** is assumed. Optionally, **t[rue]** or **f[alse]**.

Line 8: **-debug_inverse** [(*True* or *False*)]

> **-debug_inverse**—Includes debugging prints for subroutines called by subroutine *inverse_models*. If this option is set to **true**, a large amount of information about the process of finding inverse models is printed. The program will print the following for each set of equations and inequalities that are attempted to be solved by the optimizing solver: a list of the unknowns, a list of the equations, the array that is to be solved, any nonnegativity or nonpositivity constraints on the unknowns, the solution vector, and the residual vector for the linear equations and inequality constraints. The printout is long and not very useful. Default is **false** at startup. Optionally, **debug_inverse** or **-debug_i[nverse]**.

(*True* or *False*)—A value of **true** indicates the debugging information will be included in the output file; **false** indicates debugging information will not be printed. If neither **true** nor **false** is entered on the line, **true** is assumed. Optionally, **t[rue]** or **f[alse]**.

Line 9: **-debug_model** [(*True* or *False*)]

-debug_model—Includes debugging prints for subroutines called by subroutine *model*. If this option is set to **true**, a large amount of information about the Newton-Raphson iterations is printed. The program will print some or all of the following at each iteration: the array that is solved, the solution vector calculated by the solver, the residuals of the linear equations and inequality constraints, the values of all of the master unknowns and their change, the moles of each pure phase and phase mole transfers, the moles of each element in the system minus the amount in pure phases and the change in this quantity. The printout is long and not very useful. If the numerical method does not converge in *iterations*-1 iterations (default is after 99 iterations), this printout is automatically begun and sent to the log file *phreeqc.log*. Default is **false** at startup. Optionally, **debug_model** or **-debug_m[odel]**.

(*True* or *False*)—A value of **true** indicates the debugging information will be included in the output file; **false** indicates debugging information will not be printed. If neither **true** nor **false** is entered on the line, **true** is assumed. Optionally, **t[rue]** or **f[alse]**.

Line 10: **-debug_prep** [(*True* or *False*)]

-debug_prep—Includes debugging prints for subroutine *prep*. If this option is set to **true**, the chemical equation and log K for each species and phase, as rewritten for the current calculation, are written to the output file. The printout is long and not very useful. Default is **false** at startup. Optionally, **debug_prep** or **-debug_p[rep]**.

(*True* or *False*)—A value of **true** indicates the debugging information will be included in the output file; **false** indicates debugging information will not be printed. If neither **true** nor **false** is entered on the line, **true** is assumed. Optionally, **t[rue]** or **f[alse]**.

Line 11: **-debug_set** [(*True* or *False*)]

-debug_set—Includes debugging prints for subroutines called by subroutine *set*. If this option is set to **true**, the initial revisions of the master unknowns (see equation 84, Parkhurst and Appelo, 1999), which occur in subroutine *set*, are printed for each element or element valence state that fails the initial convergence criteria. The initial revisions occur before the Newton-Raphson method is invoked and attempt to provide good estimates of the master unknowns to the Newton-Raphson method. Default is **false** at startup. Optionally, **debug_set** or **-debug_s[et]**.

(*True* or *False*)—A value of **true** indicates the debugging information will be included in the output file; **false** indicates debugging information will not be printed. If neither **true** nor **false** is entered on the line, **true** is assumed. Optionally, **t[rue]** or **f[alse]**.

Line 12: **-logfile** [(*True* or *False*)]

-logfile—Prints information to a file named *phreeqc.log*. If this option is set to **true**, information about each calculation will be written to the log file. The information includes the number of iterations in revising the initial estimates of the master unknowns, the number of Newton-Raphson iterations, and the iteration at which any infeasible solution was encountered while solving the system of nonlinear equations. (An infeasible solution occurs if no solution to the equality and inequality constraints can be found.) At each iteration, the identity of any species that exceeds 30 mol (an unreasonably large number) is written to the log file and noted as an "overflow". Any basis switches are noted in the log file. The information about infeasible solutions and overflows can be useful for altering other parameters defined through the **KNOBS** data block, as described below. Default is **false** at startup. Optionally, **logfile** or **-l[ogfile]**.

(*True* or *False*)—A value of **true** indicates log information will be written to the file, *phreeqc.log*; **false** indicates log information will not be written. If neither **true** nor **false** is entered on the line, **true** is assumed. Optionally, **t[rue]** or **f[alse]**.

Notes

Convergence problems are less frequent with PHREEQC version 3 than with previous versions of PHREEQE and PHREEQC; however, they may still occur. The main causes of nonconvergence appear to be (1) calculation of very large molalities in intermediate iterations, (2) accumulation of roundoff errors in simulations involving very small concentrations of elements in solution, and (3) loss of precision in problems with no redox buffering. The first cause can be identified by "overflow" messages at iteration 1 or greater that appear in the file *phreeqc.log* (see **-logfile** above). This problem can usually be eliminated by decreasing the maximum allowable step sizes from the default values. The second and third causes of nonconvergence can be identified by messages in *phreeqc.log* that indicate "infeasible solutions". The remedy to these problems is an ongoing investigation, but altering **-tolerance** or **-diagonal_scaling** sometimes fixes the problem, and it should be noted that the program attempts several combinations of these parameters automatically before terminating the calculations. Additional iterations (**-iterations**) beyond 200 usually do not solve nonconvergence problems. A trick that is sometimes helpful with nonconvergence is to include the following fictitious aqueous species that has a concentration of about 1×10^{-9} and produces terms in the charge-, hydrogen-, and oxygen-balance equations of a magnitude great enough for the solver to solve the equations:

```
SOLUTION_SPECIES
H2O + 0.01e- = H2O-0.01
     log_k   -9.0
```

If the numerical method does not converge with the original set of convergence parameters (either default or user specified), 12 additional sets of parameters are tried automatically to obtain convergence: (1) *iterations* is doubled and smaller values for *step_size* and *pe_step_size* are used; (2) *iterations* is doubled and *tol* is decreased by a factor of 10.0; (3) *iterations* is doubled and *tol* is increased by a factor of 10.0; (4) *iterations* is doubled and the value of *diagonal_scale* is switched from false to true or from true to false; (5) *iterations* is doubled, *diagonal_scale* is switched, and *tol* is decreased by a factor of 10.0; (6) *iterations* is doubled and pure phase columns are scaled; (7) *iterations* is doubled, pure phase columns are scaled, and *diagonal_scale* is switched from false to true or from true to false; (8) *iterations* is doubled and the scaling is increased by a factor of 10.0; (9) only equality constraints are solved for the first five iterations; (10) additional inequality constraints are added to ensure new concentrations are positive; (11) *iterations* is doubled and *tol* is decreased by a factor of 100.0; (12) *iterations* is doubled and *tol* is decreased by a factor of 1,000.0.

Example problems

The keyword **KNOBS** is used in example problems 20 and 21.

LLNL_AQUEOUS_MODEL_PARAMETERS

This keyword data block is used to define aqueous model parameters for the Lawrence Livermore National Laboratory aqueous model. The parameters are the temperature-dependent values of the Debye-Hückel A, B, and B-dot parameters and the expression for the activity coefficient of aqueous carbon dioxide as a function of temperature and ionic strength. The **LLNL_AQUEOUS_MODEL_PARAMETERS** data block is used in the database file *llnl.dat*, which is the only context in which it can be used.

Example data block

```
Line 0: LLNL_AQUEOUS_MODEL_PARAMETERS
Line 1: -temperature
Line 2:        0.0100        25.0000        60.0000       100.0000
Line 2a      150.0000       200.0000       250.0000       300.0000
Line 3: -dh_a
Line 4:        0.4939         0.5114         0.5465         0.5995
Line 4a:       0.6855         0.7994         0.9593         1.2180
Line 5: -dh_b
Line 6:        0.3253         0.3288         0.3346         0.3421
Line 6a:       0.3525         0.3639         0.3766         0.3925
Line 7: -bdot
Line 8:        0.0374         0.0410         0.0438         0.0460
Line 8a:       0.0470         0.0470         0.0340         0.0000
Line 9: -co2_coefs
Line 10:      -1.0312       0.0012806
Line 10a:    255.9           0.4445
Line 10b:     -0.001606
```

Explanation

Line 0: **LLNL_AQUEOUS_MODEL_PARAMETERS**

LLNL_AQUEOUS_MODEL_PARAMETERS is the keyword for the data block. No other data are input on the keyword line.

Line 1: **-temperature**

-temperature—Begins a block of data that defines a temperature grid. Values of the other parameters are specified at each of the temperatures in the grid. Optionally, **temperature**, **temperatures**, or **-t[emperatures]**.

Line 2: *list_of_temperatures*

 list_of_temperatures—Temperatures, °C, for the temperature grid. Any number of temperatures can be given, but values of the parameters defined by **-dh_a**, **-dh_b**, and **-bdot** must be defined for each temperature.

Line 3: **-dh_a**

 -dh_a—Begins a block of data that defines the Debye-Hückel *A* parameter at each temperature in the temperature grid. Optionally, **adh**, **dh_a**, **-a[dh]**, or **-d[h_a]**.

Line 4: *dh_a_values*

 dh_a_values—Values of Debye-Hückel *A* for each temperature in the temperature grid.

Line 5: **-dh_b**

 -dh_b—Begins a block of data that defines the Debye-Hückel *B* parameter at each temperature in the temperature grid. Optionally, **bdh**, **dh_b**, **-b[dh]**, or **-dh_b**.

Line 6: *dh_b_values*

 dh_b_values—Values of Debye-Hückel *B* for each temperature in the temperature grid.

Line 7: **-bdot**

 -bdot—Begins a block of data that defines the Debye-Hückel *B*-dot parameter at each temperature in the temperature grid. Optionally, **bdot**, **b_dot**, **-bdo[t]**, or **-b_[dot]**.

Line 8: *dh_bdot_values*

 dh_bdot_values—Values of Debye-Hückel *B*-dot for each temperature in the temperature grid.

Line 9: **-co2_coefs**

 -co2_coefs—Begins a block of data that defines the parameters in the expression for the activity coefficient of CO_2(aq) as a function of temperature and ionic strength. Optionally, **c_co2**, **co2_coefs**, **-c[_co2]**, or **-c[o2_coefs]**.

Line 10: *C, F, G, E, H*

C, F, G, E, H—Parameters in the expression $\ln\gamma_{CO_2} = \left(C + FT + \frac{G}{T}\right)I - (E + HT)\left(\frac{I}{I+1}\right)$.

Notes

The Lawrence Livermore National Laboratory aqueous model (Daveler and Wolery, 1992) uses this expression for the log (base 10) of an activity coefficient: $\log\gamma_i = \dfrac{A_\gamma z_i^2 \sqrt{I}}{1 + \mathring{a}_i B_\gamma \sqrt{I}} + \dot{B}I$, where A_γ is the Debye-Hückel A parameter corresponding to **-dh_a**, B_γ is the Debye-Hückel B parameter corresponding to **-dh_b**, $\dot{B}$ is the Debye-Hückel B-dot parameter corresponding to **-bdot**, $\mathring{a}_i$ is the hard core diameter, which is specific to each aqueous species, and I is the ionic strength. The **LLNL_AQUEOUS_MODEL_PARAMETERS** data block defines A_γ, B_γ, and $\dot{B}$ as functions of temperature. A temperature grid is defined with **-temperatures** and values of each of these parameters are specified at each point in the temperature grid. For a given simulation temperature, values of the parameters are interpolated linearly between points on the temperature grid. The ion-specific parameter $\mathring{a}_i$ is defined in the **SOLUTION_SPECIES** data block with the identifier **-llnl_gamma**.

The activity coefficient of aqueous carbon dioxide is defined as a function of ionic strength based on the parameterization of Drummond (1981). The formula for the natural log of the activity coefficient is $\ln\gamma_{CO_2} = \left(C + FT + \dfrac{G}{T}\right)I - (E + HT)\left(\dfrac{I}{I+1}\right)$, where T is temperature in kelvin; I is ionic strength; and $C, F, G, E,$ and H are defined in order with the **-co2_coefs** identifier. The activity coefficient calculated from this formula can be applied to specific neutral species by using the identifier **-CO2_llnl_gamma** in the **SOLUTION_SPECIES** data block.

Example problems

The **LLNL_AQUEOUS_MODEL_PARAMETERS** data block is used in the *llnl.dat* database.

Related keywords

SOLUTION_SPECIES.

MIX

This keyword data block is used to mix together two or more aqueous solutions. Mixing may be used alone, in combination with additional reactions, or during advection or transport calculations. All applications of MIX result in a batch-reaction calculation that produces aqueous equilibrium, including redox equilibrium.

Example data block

```
Line  0:  MIX 2 Mixing solutions 5, 6, and 7.
Line 1a:      5     1.1
Line 1b:      6     0.5
Line 1c:      7     0.3
```

Explanation

Line 0: **MIX** [*number*] [*description*]

MIX is the keyword for the data block.

number—A positive number designates the following mixing parameters. Default is 1.

description—Optional comment that describes the mixture.

Line 1: *solution number, mixing fraction*

solution number—Defines a solution to be part of the mixture.

mixing fraction—Decimal number that is multiplied times the moles of each element in the specified solution; the mixture is the sum of each solution times its mixing fraction. Mixing fractions may be greater than 1.0.

Notes

In mixing, each solution is multiplied by its mixing fraction and a new solution is calculated by summing over all of the fractional solutions. In the Example data block, if the moles of sodium in solutions 5, 6, and 7 were 0.1, 0.2, and 0.3, the moles of sodium in the mixture would be $0.1 \times 1.1 + 0.2 \times 0.5 + 0.3 \times 0.3 = 0.3$. The moles of all elements are multiplied by the mixing fraction of the solution, including elements hydrogen and oxygen. Thus, the mass of water is effectively multiplied by the same fractions. In the Example data block, if all solutions have 1 kg of water, the total mass of water in the mixture is approximately $1.1 + 0.5 + 0.3 = 1.9$ kg, and the concentration of sodium would be approximately 0.16 mol/kgw (0.3/1.9). The charge imbalance of each solution is multiplied by the mixing fraction, and all the imbalances are then summed to calculate the charge imbalance of the mixture. The temperature of the mixture is approximated by multiplying each solution temperature by its mixing fraction, summing these numbers, and dividing by the sum of the mixing fractions. Other intensive properties of the mixture are calculated in the same way as temperature. This approach for calculating the temperature of mixtures is an approximation because enthalpies of reaction are ignored. For example, heat generated by mixing a strong acid with a strong base is not considered.

This formulation of mixing can be used to approximate constant volume processes if the sum of the mixing fractions is 1.0 and all of the solutions have the same mass of water. The calculations are only approximate in terms of mixing volumes because the summation is made in terms of moles (or mass) and no consideration is given to the partial molar volumes of solutes. Similarly, the formulation for mixing can approximate processes with varying volume; for example, a titration.

Mixing results in a batch-reaction calculation, which produces aqueous equilibrium, including redox equilibrium. **SOLUTION**s may be defined with redox disequilibrium by defining concentrations of individual valence states of elements. When **SOLUTION**s are mixed, all valence states of elements react to redox equilibrium. Thus, even if a single solution in redox disequilibrium is mixed with a mixing fraction of 1.0 (which will not change the total concentrations of elements), redox reactions will occur among the valence states of elements, which in turn will change the pH and pe of the solution.

When multiple batch-reaction steps are defined in **KINETICS**, **REACTION**, **REACTION_PRESSURE**, or **REACTION_TEMPERATURE**, and if **INCREMENTAL_REACTIONS** is false (cumulative reaction steps), then each batch-reaction step uses the same mixing factors; if **INCREMENTAL_REACTIONS** is true (incremental reaction steps), then the mixing fractions are applied during the first batch-reaction step only.

Example problems

The keyword **MIX** is used in example problems 3, 4, 13, and 21.

Related keywords

INCREMENTAL_REACTIONS, **SOLUTION**, **SAVE solution**, **USE solution**, and **USE mix**.

NAMED_EXPRESSIONS

This keyword data block is used to assign names to expressions for equilibrium constants and fractionation factors. The named expressions are useful to simplify definition of thermodynamic data for isotopic species in **SOLUTION_SPECIES** and **PHASES** data blocks, and are used in the **ISOTOPE_ALPHAS** data block to provide printing of fractionation factors to the output file. The **NAMED_EXPRESSIONS** data block is used in the database file *iso.dat* and *llnl.dat* databases.

Example data block

```
Line  0:   NAMED_EXPRESSIONS
Line  1:   Log_alpha_13C_CO2(aq)/CO2(g)
Line  2:    -ln_alpha1000    -0.9    0.0    0.0    0.0    .0063e6
Line 1a:   Log_alpha_13C_HCO3-/CO2(aq)
Line  3:    -add_logk          Log_alpha_13C_HCO3-/CO2(g)    1
Line 3a:    -add_logk          Log_alpha_13C_CO2(aq)/CO2(g)   -1
```

Explanation

Line 0: **NAMED_EXPRESSIONS**

NAMED_EXPRESSIONS is the keyword for the data block. No other data are input on the keyword line. Optionally, **NAMED_LOG_K**, **NAMED_EXPRESSIONS**, or **NAMED_ANALYTICAL_EXPRESSIONS**.

Line 1: *Name*

Name—Name for the expression.

Line 2: **-ln_alpha1000** $A_1, A_2, A_3, A_4, A_5, A_6$

-ln_alpha1000—An analytical expression for a fractionation factor is defined. The six coefficients are applied to the expression

$$1000\ln\alpha = A_1 + A_2 T + \frac{A_3}{T} + A_4 \log_{10} T + \frac{A_5}{T^2} + A_6 T^2,$$ where T is temperature in kelvin. If less than six parameters are defined, the undefined parameters are assumed to be zero. This identifier may be used only once in the definition of a named expression. Optionally, **ln_alpha1000** or **-ln[_alpha1000]**.

$A_1, A_2, A_3, A_4, A_5, A_6$—Coefficients for the analytical expression for a fractionation factor.

Line 3: **-add_logk** *named_expression, coefficient*

> **-add_logk**—The *named_expression* is used in calculating the value for *Name* from the preceding Line 1. This identifier may be used multiple times in the definition of a named expression. Optionally, **add_logk**, **add_log_k**, **-ad[d_logk]** or **-ad[d_log_k]**.
>
> *named_expression*—Name of an expression defined in a **NAMED_EXPRESSIONS** data block.
>
> *coefficient*—The *coefficient* is multiplied by the value of the *named_expression* when calculating the value of *Name*.

Notes

The **NAMED_EXPRESSIONS** data block is implemented to avoid multiple definitions of analytical expressions, to simplify combining expressions for fractionation factors, and to preserve relationships among isotopic fractionation factors and equilibrium constants. It is used to implement the individual isotope equilibrium constant approach of Thorstenson and Parkhurst (2000, 2004).

Fractionation factors are commonly reported with analytical expressions for $1000\ln(\alpha)$. The **-ln_alpha1000** identifier allows definition of a fractionation factor in this form. However, the analytical expression is immediately converted to an expression for the log base 10 fractionation factor. All combinations of named expressions are performed with log base 10 operations.

The **-add_logk** identifier is used to sum up multiple expressions to produce a new expression. In the Example data block, the fractionation factor between HCO_3^- and $CO_{2(aq)}$ is defined as a combination of the fractionation factor between HCO_3^- and $CO_{2(g)}$ and the fractionation factor between $CO_{2(g)}$ and $CO_{2(aq)}$. It is good practice to define the analytical expressions for fractionation factors only once, and then use the **-add_logk** identifier to add it to form other expressions. In this way, the relationships among fractionation factors are preserved, even if the analytical expression is changed.

The database *iso.dat* uses the **NAMED_EXPRESSIONS** data block to implement isotope fractionation factors. Within that data block, the only identifiers used are **-ln_alpha1000** and **-add_logk**. However, for completeness, it is also possible to use the identifiers **-analytical_expression**, **-log_k**, and **-delta_h** to define named expressions. Descriptions of these identifiers can be found in the PHASES and SOLUTION_SPECIES data blocks.

Named expressions are used to define the equilibrium constants for isotopic species in the SOLUTION_SPECIES and PHASES data blocks of the *iso.dat* database. Named expressions are used in the ISOTOPE_ALPHAS data block to print values of analytical expressions to the output file. The values of the named expressions for the current temperature can be obtained with the Basic function LK_NAMED("*name*").

Example problems

The **NAMED_EXPRESSIONS** data block is used in the *iso.dat* and *llnl.dat* databases.

Related keywords

ISOTOPE_ALPHAS, PHASES, and SOLUTION_SPECIES.

PHASES

This keyword data block is used to define a name, chemical reaction, log K, and temperature dependence of log K for each gas component and mineral that can be used for speciation, batch-reaction, transport, or inverse-modeling calculations. In addition, molar volumes can be defined for solids, and the critical temperature and pressure and the acentric factor can be defined for gases. Normally, this data block is included in the database file and only additions and modifications are included in the input file.

Example data block 1

```
Line  0:  PHASES
Line 1a: Gypsum
Line 2a:      CaSO4:2H2O = Ca+2 + SO4-2 + 2H2O
Line 3a:      log_k      -4.58
Line 4a:      delta_h    -0.109
Line  5:      -analytical_expression 68.2401 0.0 -3221.51  -25.0627
Line  6:      -Vm        73.9 cm3/mol
Line 1b: O2(g)
Line 2b:      O2 = O2
Line 3b:      log_k      -2.96
Line 4b:      delta_h     1.844
Line  7:      -T_c       154.6
Line  8:      -P_c       49.80
Line  9:      -Omega      0.021
```

Explanation 1

Line 0: **PHASES**

Keyword for the data block. No other data are input on the keyword line.

Line 1: *Phase name*

phase name—Alphanumeric name of phase; no spaces are allowed.

Line 2: *Dissolution reaction*

Dissolution reaction for phase to aqueous species. Any aqueous species, including e⁻, may be used in the dissolution reaction. The chemical formula for the defined phase must be the first chemical formula on the left-hand side of the equation. The dissolution reaction must precede any identifiers related to the phase. The stoichiometric coefficient for the phase in the chemical reaction must be 1.0.

Line 3: **log_k** *log K*

> **log_k**—Identifier for log *K* at 25 °C. Optionally, **-log_k**, **logk**, **-l[og_k]**, or **-l[ogk]**.
>
> *log K*—Log *K* at 25 °C for the reaction. Default is 0.0.

Line 4: **delta_h** *enthalpy, [units]*

> **delta_h**—Identifier for enthalpy of reaction at 25 °C. Optionally, **-delta_h**, **deltah**, **-d[elta_h]**, or **-d[eltah]**.
>
> *enthalpy*—Enthalpy of reaction at 25 °C for the reaction. Default is 0.0.
>
> *units*—Units may be calories, kilocalories, joules, or kilojoules per mole. Only the energy unit is needed (per mole is implied) and abbreviations of these units are acceptable. Explicit definition of units for all enthalpy values is recommended. The enthalpy of reaction is used in the Van't Hoff equation to determine the temperature dependence of the equilibrium constant. Internally, all enthalpy calculations are performed in the units of kJ/mol. Default units are kJ/mol.

Line 5: **-analytical_expression** $A_1, A_2, A_3, A_4, A_5, A_6$

> **-analytical_expression**—Identifier for coefficients for an analytical expression for the temperature dependence of log *K*. If defined, the analytical expression takes precedence over **log_k** and the Van't Hoff equation to determine the temperature dependence of the equilibrium constant. Optionally, **analytical_expression, a_e, ae, -a[nalytical_expression], -a[_e], -a[e]**.
>
> $A_1, A_2, A_3, A_4, A_5, A_6$—Six values defining log *K* as a function of temperature in the expression $\log_{10}K = A_1 + A_2 T + \dfrac{A_3}{T} + A_4 \log_{10} T + \dfrac{A_5}{T^2} + A_6 T^2$, where *T* is kelvin. Coefficients are defined in order from A_1 to A_6; if less than six parameters are defined, the undefined parameters are set to zero.

Line 6: **-Vm** *molar_volume [units]*

> **-Vm**—Identifier for the molar volume of the solid phase.
>
> *molar_volume*, the molecular weight divided by the density of the solid at 25 °C. In the example for gypsum 172.18 (g/mol, gram per mole) / 2.33 (g/cm^3, gram per cubic centimeter) = 73.9 cm^3/mol (cubic centimeter per gram). Default is 0 cm^3/mol.

units—Units may be cm^3/mol, dm^3/mol (cubic decimeter per mole), or m^3/mol (cubic meter per mole). Default is cm^3/mol.

Line 7: **-T_c** *critical temperature*

-T_c—Identifier for the critical temperature of the gas.

critical temperature—Temperature, K (kelvin). Default is 0 K.

Line 8: **-P_c** *critical pressure*

-P_c—Identifier for the critical pressure of the gas.

critical pressure—Pressure, atm. Default is 0 atm.

Line 9: **-Omega** *acentric factor*

-Omega—Identifier for the acentric factor of the gas.

acentric factor—Acentric factor dimensionless. Default is 0.

Notes 1

The set of Lines 1 and 2 must be entered in order, and either Line 3 (**log_k**) or Line 5 (**-analytical_expression**) should be entered for each phase (default log K is 0.0). The analytical expression (**-analytical_expression**) takes precedence over **log_k** and the Van't Hoff equation (**delta_H**) to determine the temperature dependence of the equilibrium constant. Lines 3 to 5, Line 6 for a solid, and Lines 7 to 9 for a gas may be entered as needed in any order. The equations for the phases may be written in terms of any aqueous chemical species, including e$^-$.

The molar volume of a solid is, together with the volumes of the solute species, used to calculate the pressure dependence of **log_k**:

$$\log K_P = \log K_{P=1} - \frac{\Delta V_r}{2.303 \times RT}(P-1)$$,

where P is pressure (atm), ΔV_r is the volume change of the reaction (cm^3/mol), R is the gas constant (82.06 atm cm^3 mol^{-1} K^{-1}, atmosphere cubic centimeter per mole per kelvin), and T is the temperature (K).

The critical temperature and pressure and the acentric factor of a gas are used to calculate the equation of state according to Peng and Robinson (1976):

$$P = \frac{RT}{V_m - b} - \frac{a\alpha}{V_m^2 + 2bV_m - b^2}$$,

where V_m is the molar volume of the gas (cm³/mol), b is the minimal volume of the gas (cm³/mol), a is the Van der Waals attraction factor (atm cm³ mol⁻², atmosphere cubic centimeter per mole squared) and α is a dimensionless function of reduced temperature and acentric factor. With P and Vm known, the fugacity coefficient of a gas can be calculated as:

$$\ln(\varphi) = \left(\frac{PV_m}{RT} - 1\right) - \ln\left(\frac{P(V_m - b)}{RT}\right) + \frac{a\,\alpha}{2.828\,b\,RT}\,\ln\left(\frac{V_m + 2.414b}{V_m - 0.414b}\right),$$

where φ is the fugacity coefficient. The product of the fugacity coefficient and the gas pressure yields the activity of the gas which can be entered in the law of mass action.

For a multicomponent gas phase, the weighted sums are taken of a, b, and α:

$b_{sum} = \Sigma(x_i\,b)$, where x_i is mole-fraction of gas i,

$a\alpha_{sum} = \Sigma_i(\ \Sigma_j(x_i\,x_j\,(a_i\alpha_i\,a_j\alpha_j)^{0.5})\ )\ (1 - k_{ij})$, and

k_{ij} = binary interaction coefficient.

As a result, the fugacity coefficient of a gas in a mixture will be different from the fugacity coefficient of the pure gas at the same pressure.

The identifier **-no_check** can be used to disable checking charge and elemental balances (see **SOLUTION_SPECIES**). The use of **-no_check** is not recommended, except in cases where the phase is only to be used for inverse modeling. Even in this case, equations defining phases should be charge balanced. The identifier also can be used to define the mineral formula for an exchanger with an explicit charge imbalance (see explanation under **EXCHANGE**).

Example data block 2

```
Line 0:   PHASES
Line 1:   CH3D(g)
Line 2:   CH3D(g) + H2O(l) = CH4(g) + HDO(aq)
Line 3:       log_k          -0.301029995663
Line 4:       -add_logk      Log_alpha_D_CH4(g)/H2O(l)      -1.0
Line 1a: Ca[34S]O4:2H2O
Line 2a:    Ca[34S]O4:2H2O + SO4-2 = [34S]O4-2 + Gypsum(s)
Line 4a:       -add_logk      Log_alpha_34S_Gypsum/SO4-2      -1.0
Line 1b: Gypsum
Line 2b:    CaSO4:2H2O = Ca+2 + SO4-2 + 2 H2O
Line 3a:       log_k          -4.580
```

Explanation 2

Line 0: **PHASES**

Keyword for the data block. No other data are input on the keyword line.

Line 1: *Phase name*

phase name—Alphanumeric name of phase; no spaces are allowed.

Line 2: *Dissolution reaction*

Dissolution reaction for phase. In implementing isotope calculations, the dissolution reaction was generalized to allow, in addition to any aqueous species, solid and gas species. To distinguish solids and gases (defined in the **PHASES** data block) from aqueous species, "(s)" and "(g)" are appended to the species in the equation. For clarity, "H2O(l)" can be used in an equation to designate liquid water, but it is equivalent to using "H2O". The chemical formula for the defined phase must be the first chemical formula on the left-hand side of the equation. The stoichiometric coefficient for the defined phase in the chemical reaction must be 1.0. The dissolution reaction must precede any identifiers related to the phase.

Line 3: **log_k** *log K*

log_k—Identifier for log K at 25 °C. Optionally, **-log_k**, **logk**, **-l[og_k]**, or **-l[ogk]**.

log K—Log K at 25 °C for the reaction. Default is 0.0.

Line 4: **-add_logk** *named_expression, coefficient*

-add_logk—The value of the *named_expression* is multiplied by the *coefficient* and added to the log K for the phase. This identifier may be used multiple times in the definition of the equilibrium constant for the phase. Optionally, **add_logk**, **add_log_k**, **-ad[d_logk]** or **-ad[d_log_k]**.

named_expression—Name of an expression defined in a **NAMED_EXPRESSIONS** data block.

coefficient—The *coefficient* is multiplied by the value of the *named_expression* and added to the equilibrium constant for the phase.

Notes 2

Example data block 2 demonstrates capabilities that were added during the implementation of isotopic calculations as described by Thorstenson and Parkhurst (2000, 2004). First, to simplify the definition of equilibrium constants for the isotopic variants of a mineral or gas, the dissociation reaction was generalized to allow solids and gases in the equation. Solids are identified by an appended "(s)" and gases are identified by an appended "(g)". The original definition of the solid or gas may or may not have the appended string in its name; PHREEQC attempts to find the name in the list of phases with and without the appended string. Note in this example that "Gypsum(s)" is used in Line 2a, but "Gypsum" is defined as the name of the phase in Line 1b.

When specifying individual equilibrium constants for isotopic phases, a single fractionation factor may appear in the expressions of the equilibrium constants for multiple isotopic forms of a phase. To avoid many manipulations with analytical expressions in the definition of equilibrium constants, the NAMED_EXPRESSIONS data block allows assigning a name to the expression for a fractionation factor. This named expression can then be used in calculating the equilibrium constants for phases defined in the **PHASES** data block by use of the **-add_logk** identifier. In the definition of CH3D(g) in the Example data block 2, the fractionation factor for deuterium between methane and liquid water is added to an equilibrium constant derived from the symmetry of the molecule to define the equilibrium constant for the dissociation reaction.

Example problems

The keyword **PHASES** is used in example problems 1, 6, 7, 8, 9, 10, 16, 18, and 21. It is also found in all of the database files.

Related keywords

EQUILIBRIUM_PHASES, EXCHANGE, INVERSE_MODELING, KINETICS, NAMED_EXPRESSIONS, REACTION, SAVE **equilibrium_phases**, and USE **equilibrium_phases**.

PITZER

This keyword data block is used to define specific-ion-interaction parameters for the Pitzer aqueous model. The **PITZER** data block is used in the database file *pitzer.dat*, which is derived from the database for the program PHRQPITZ. Details of the implementation of the aqueous model and the sources for data in the *pitzer.dat* database can be found in Plummer and others (1988).

Example data block 1

```
Line  0: PITZER
Line  1:     -macinnes      true
Line  2:     -use_etheta    true
Line  3:     -redox         true
```

Explanation 1

Line 0: **PITZER**

PITZER is the keyword for the data block. No other data are input on the keyword line.

Line 1: **-macinnes** [(*True* or *False*)]

-macinnes—Identifier determines whether activity coefficients printed in the output are scaled by the MacInnes assumption that the activity coefficients of Cl^- and K^+ are equal (see Plummer and others, 1988). Default is **true** at startup. Optionally, **macinnes** or **-m[acinnes]**.

(*True* or *False*)—A value of **true** indicates that activity coefficients printed in the output are scaled by the MacInnes assumption; **false** indicates that the activity coefficients printed in the output are unscaled. If neither **true** nor **false** is entered on the line, **true** is assumed. Optionally, **t[rue]** or **f[alse]**.

Line 2: **-use_etheta** [(*True* or *False*)]

-use_etheta—Identifier determines whether nonsymmetric mixing coefficients are used in the calculations (see Plummer and others, 1988, for a description of the coefficients $^E\theta_{ij}(I)$ and $^E\theta'_{ij}(I)$). Default is **true** at startup. Optionally, **use_etheta** or **-u[se_etheta]**.

(*True* or *False*)—A value of **true** indicates that the nonsymmetric mixing coefficients will be calculated; **false** indicates that the nonsymmetric mixing coefficients will be set to 1.0. If neither **true** nor **false** is entered on the line, **true** is assumed. Optionally, **t[rue]** or **f[alse]**.

Line 3: **-redox** [(*True* or *False*)]

-**redox**—Identifier determines whether a redox-related equation is included in the calculations. This option is useful only if using a database that contains at least one redox couple. Note that the *pitzer.dat* database does not include any redox couples and thus cannot simulate any redox reactions. Default is **false** at startup. Optionally, **redox** or **-r[edox]**.

(*True* or *False*)—A value of **true** indicates that a redox-related equation will be included; **false** indicates that a redox-related equation is not included. If neither **true** nor **false** is entered on the line, **true** is assumed. Optionally, **t[rue]** or **f[alse]**.

Notes 1

The identifiers of Example data block 1 are the only options likely to be included in a PHREEQC input file. Of these three identifiers, only **-macinnes** is likely to be used. Scaling activity coefficients by the MacInnes assumption may help in comparing activity coefficients to activity coefficients from other sources. However, the value of the identifier makes no difference in the calculations; it only affects the values of the activity coefficients that are printed to the output file. The **-use_etheta** identifier was added to simplify the results so that a user could compare activity coefficients with simple calculations without having to calculate the nonsymmetric mixing coefficients (**-use_etheta** false). Note that calculations without the nonsymmetric mixing coefficients are not meaningful in the Pitzer model. By default, the **-redox** identifier is false because the *pitzer.dat* database contains no elements defined with more than one redox state. A database with multiple redox states for an element is needed to make use of the **-redox** identifier.

Example data block 2

```
Line 0:  PITZER
Line 1:  -B0
Line 2:      Na+     Cl-     0.0765  -777.03  -4.4706  0.008946 -3.3158e-6  0
Line 3:  -B1
Line 2:      Na+     Cl-     0.2664  0        0        6.1608e-5 1.0715e-6
Line 4:  -B2
Line 2:      Mg+2    SO4-2   -37.23  0        0        -0.253
Line 5:  -C0
Line 2:      Na+     Cl-     0.00127 33.317   0.09421  -4.655e-5
Line 6:  -THETA
Line 7:      K+      Na+     -0.012
Line 8:  -LAMBDA
Line 9:      CO2     Na+     0.1
Line 10: -PSI
```

```
Line 11:       Na+      K+       Cl-       -0.0018
Line 12: -ZETA
Line 13:       B(OH)3   H+       Cl-       -0.0102
Line 14: -MU
Line 15:       NH3      NH3      CO3-2     0.000625
Line 16: -ETA
Line 17:       CO2      Na+      K+        0.0
Line 18: -alphas
Line 19:       Fe+2     Cl-      2         1
Line 19a:      Fe+2     SO4-2    1.559     5.268
```

Explanation 2

Line 0: **PITZER**

PITZER is the keyword for the data block. No other data are input on the keyword line.

Line 1: **-B0**

-B0—Identifier begins a block of data that defines $\beta_{MX}^{(0)}$ cation-anion interaction parameters for the Pitzer aqueous model (see Plummer and others, 1988).

Line 2: *cation anion A_0, A_1, A_2, A_3, A_4, A_5*

cation anion—A cation-anion pair.

A_0, A_1, A_2, A_3, A_4, A_5—Coefficients for the temperature dependence of the Pitzer parameter. The expression for a Pitzer parameter is as follows:

$$P = A_0 + A_1\left(\frac{1}{T} - \frac{1}{T_r}\right) + A_2 \ln\left(\frac{T}{T_r}\right) + A_3(T - T_r) + A_4(T^2 - T_r^2) + A_5\left(\frac{1}{T^2} - \frac{1}{T_r^2}\right), \text{ where } P \text{ is the}$$

parameter, T is the temperature in kelvin, T_r is the reference temperature (298.15 K), and ln is the natural log. If less than six parameters are defined, the undefined parameters are assumed to be zero.

Line 3: **-B1**

-B1—Identifier begins a block of data that defines $\beta_{MX}^{(1)}$ cation-anion interaction parameters for the Pitzer aqueous model (see Plummer and others, 1988).

Line 4: **-B2**

-B2—Identifier begins a block of data that defines $\beta_{MX}^{(2)}$ cation-anion interaction parameters for the Pitzer aqueous model (see Plummer and others, 1988).

Line 5: **-C0**

> **-C0**—Identifier begins a block of data that defines C_{MX}^{ϕ} cation-anion interaction parameters for the Pitzer aqueous model (see Plummer and others, 1988).

Line 6: **-THETA**

> **-THETA**—Identifier begins a block of data that defines θ_{ij} cation-cation and anion-anion interaction parameters for the Pitzer aqueous model (see Plummer and others, 1988).

Line 7: (*cation cation* or *anion anion*) A_0, A_1, A_2, A_3, A_4, A_5

> (*cation cation* or *anion anion*)—A cation-cation pair of ions or an anion-anion pair of ions.
>
> A_0, A_1, A_2, A_3, A_4, A_5—Coefficients for the temperature dependence of the Pitzer parameter (see Line 2).

Line 8: **-LAMBDA**

> **-LAMBDA**—Identifier begins a block of data that defines λ_{nc} neutral-cation or λ_{na} neutral-anion interaction parameters for the Pitzer aqueous model (see Plummer and others, 1988).

Line 9: (*neutral cation* or *neutral anion*) A_0, A_1, A_2, A_3, A_4, A_5

> (*neutral cation* or *neutral anion*)—A neutral-cation pair of species or a neutral-anion pair of species.
>
> A_0, A_1, A_2, A_3, A_4, A_5—Coefficients for the temperature dependence of the Pitzer parameter (see Line 2).

Line 10: **-PSI**

> **-PSI**—Identifier begins a block of data that defines $\psi_{aa'c}$ anion-anion-cation (where a and a' are dissimilar) or $\psi_{cc'a}$ cation-cation-anion (where c and c' are dissimilar) interaction parameters for the Pitzer aqueous model (see Plummer and others, 1988).

Line 11: (*cation cation anion* or *anion anion cation*) A_0, A_1, A_2, A_3, A_4, A_5

> (*cation cation anion* or *anion anion cation*)—A cation-cation-anion triple of ions or an anion-anion-cation triple of ions.
>
> A_0, A_1, A_2, A_3, A_4, A_5—Coefficients for the temperature dependence of the Pitzer parameter (see Line 2).

Line 12: **-ZETA**

> **-ZETA**—Identifier begins a block of data that defines neutral-cation-anion (ζ_{Mna}, ζ_{cnX}, or ζ_{Nca} where M and c represent cations, and a and X represent anions, and N and n represent neutral species) interaction parameters for the Pitzer aqueous model (see Clegg and Whitfield, 1991; Clegg and Whitfield, 1995, p. 2404 corrects the coefficient of the zeta term in the 1991 paper).

Line 13: *neutral cation anion* $A_0, A_1, A_2, A_3, A_4, A_5$

> *neutral cation anion*—A neutral-cation-anion triple of species.

> $A_0, A_1, A_2, A_3, A_4, A_5$—Coefficients for the temperature dependence of the Pitzer parameter (see Line 2).

Line 14: **-MU**

> **-MU**—Identifier begins a block of data that defines $\mu_{Nnn'}$ neutral-neutral-neutral (where *n and n'* represent dissimilar neutral species) interaction parameters for the Pitzer aqueous model (see Clegg and Whitfield, 1991).

Line 15: *neutral neutral neutral* $A_0, A_1, A_2, A_3, A_4, A_5$

> *neutral neutral neutral*—A neutral-neutral-neutral triple of species.

> $A_0, A_1, A_2, A_3, A_4, A_5$—Coefficients for the temperature dependence of the Pitzer parameter (see Line 2).

Line 16: **-ETA**

> **-ETA**—Identifier begins a block of data that defines neutral-anion-anion and neutral-cation-cation (η_{Mnc}, η_{Xna}, $\eta_{Ncc'}$, or $\eta_{Naa'}$, where M and c represent cations, and a and X represent anions, and N and n represent neutral species, and prime indicated dissimilar species) interaction parameters for the Pitzer aqueous model (see Clegg and Whitfield, 1991; Clegg and Whitfield, 1995, p. 2404 corrects the coefficient of the eta term in the 1991 paper).

Line 17: (*neutral cation cation* or *neutral anion anion*) $A_0, A_1, A_2, A_3, A_4, A_5$

> (*neutral cation cation* or *neutral anion anion*)—A neutral-cation-cation or neutral-anion-anion triple of species.

$A_0, A_1, A_2, A_3, A_4, A_5$—Coefficients for the temperature dependence of the Pitzer parameter (see Line 2).

Line 18: **-alphas**

-alphas—Identifier begins a block of data that defines alpha parameters for the Pitzer aqueous model that override default values for specific cation-anion pairs. For any electrolyte containing a monovalent ion, a single α parameter with the default value of 2.0 is used in the calculation of B_{MX}^{ϕ}, B_{MX}, and B'_{MX}. For electrolytes containing two polyvalent ions, two parameters, α_1 and α_2, are used in the calculation of B_{MX}^{ϕ}, B_{MX}, and B'_{MX}. For 2-2 electrolytes, the defaults are $\alpha_1 = 1.4$ and $\alpha_2 = 12.0$. For 3-2 and 4-2 electrolytes, the defaults are $\alpha_1 = 2.0$ and $\alpha_2 = 50.0$ (see Plummer and others, 1988).

Line 19: *cation anion* α_1 α_2

cation anion—A cation-anion pair.

α_1—Value of the α_1 parameter. For electrolytes with at least one monovalent ion, α_1 is interpreted as α.

α_2—Value of the α_2 parameter. For electrolytes with at least one monovalent ion, α_2 is not used.

Notes 2

The identifiers of Example data block 2 are used to define a Pitzer aqueous model. Examples of most of these identifiers are found in the Pitzer database, *pitzer.dat*. If definition or modification of a Pitzer aqueous model is undertaken, then a complete description of the aqueous model can be found in Plummer and others (1988) and Clegg and Whitfield (1991), as amended by Clegg and Whitfield (1995, p. 2404), among other sources. Symbols used in this report for the Pitzer parameters are consistent with most descriptions of the Pitzer approach. When modifying a Pitzer aqueous interaction parameter, care is needed to ensure thermodynamic consistency among all of the parameters.

Most Pitzer parameters are defined for a pair or triple of species; the order in which these species are defined is not important. If the same type of parameter with the same set of species is redefined, even if the order of the species is different, then the previous definition is removed and replaced with the new definition.

If a **PITZER** data block is read in the database file or the input file, then the Pitzer aqueous model is used for the simulations. Only one aqueous model can be used in a PHREEQC run; it is an error to read both a **PITZER** data block and a SIT data block.

Example problems

The **PITZER** data block is used in the *pitzer.dat* database.

Related keywords

SIT.

This keyword data block is used to select which results are written to the output file. In addition, this data block enables or disables writing results to the selected-output file and writing a status line to the screen, which monitors the type of calculation being performed.

Example data block

```
Line  0:   PRINT
Line  1:        -reset               false
Line  2:        -eh                  true
Line  3:        -echo_input          true
Line  4:        -equilibrium_phases  true
Line  5:        -exchange            true
Line  6:        -gas_phase           true
Line  7:        -headings            true
Line  8:        -initial_isotopes    true
Line  9:        -inverse_modeling    true
Line 10:        -isotope_alphas      true
Line 11:        -isotope_ratios      true
Line 12:        -kinetics            true
Line 13:        -other               true
Line 14:        -saturation_indices  true
Line 15:        -solid_solutions     true
Line 16:        -species             true
Line 17:        -surface             true
Line 18:        -totals              true
Line 19:        -user_print          true
Line 20:        -alkalinity          false
Line 21:        -dump                true
Line 22:        -censor_species      1e-8
Line 23:        -selected_output     false
Line 24:        -status              false
Line 25:        -user_graph          true
Line 26:        -warnings            200
```

Explanation

Line 0: **PRINT**

Keyword for the data block. No other data are input on the keyword line.

Line 1: **-reset** [(*True* or *False*)]

  -reset—Changes all print options listed on Lines 2 through 19 to **true** or **false**. If used, this identifier should be the first identifier of the data block. Individual print options may follow. Optionally, **reset** or **-r[eset]**.

  (*True* or *False*)—If true, all data blocks described on Lines 2 through 19 are printed to the output file; if false, these data blocks are excluded from the output file. If neither **true** nor **false** is entered on the line, **true** is assumed. Optionally, **t[rue]** or **f[alse]**.

Line 2: **-eh** [(*True* or *False*)]

  -eh—Prints eh values calculated from redox couples to the output file for initial solution calculations. Default is **true** at startup. Optionally, **eh**.

  (*True* or *False*)—If true, eh values calculated from redox couples are printed to the output file; if false, eh values are not printed. If neither **true** nor **false** is entered on the line, **true** is assumed. Optionally, **t[rue]** or **f[alse]**.

Line 3: **-echo_input** [(*True* or *False*)]

  -echo_input—Prints non-comment lines from the input file to the output file. Default is **true** at startup. Optionally, **echo_input** or **-ec[ho_input]**.

  (*True* or *False*)—If true, input lines are echoed to the output file; if false, input lines are not echoed to the output file. If neither **true** nor **false** is entered on the line, **true** is assumed. Optionally, **t[rue]** or **f[alse]**.

Line 4: **-equilibrium_phases** [(*True* or *False*)]

  -equilibrium_phases—Prints the compositions of equilibrium-phase assemblages to the output file. Default is **true** at startup. Optionally, **equilibria**, **equilibrium**, **pure**, **-eq[uilibrium_phases]**, **-eq[uilibria]**, **-p[ure_phases]**, or **-p[ure]**. Note the hyphen is required to avoid a conflict with the keyword EQUILIBRIUM_PHASES; the same is true for the synonym **PURE_PHASES**.

  (*True* or *False*)—If true, compositions of equilibrium-phase assemblages are printed to the output file; if false, compositions of equilibrium-phase assemblages are not printed to the output file. If neither **true** nor **false** is entered on the line, **true** is assumed. Optionally, **t[rue]** or **f[alse]**.

Line 5: **-exchange** [(*True* or *False*)]

> **-exchange**—Prints the compositions of exchange assemblages to the output file. Default is **true** at startup. Optionally, **-ex**[**change**]. Note the hyphen is required to avoid a conflict with the keyword EXCHANGE.

> (*True* or *False*)—If true, compositions of exchange assemblages are printed to the output file; if false, compositions of exchange assemblages are not printed to the output file. If neither **true** nor **false** is entered on the line, **true** is assumed. Optionally, **t**[**rue**] or **f**[**alse**].

Line 6: **-gas_phase** [(*True* or *False*)]

> **-gas_phase**—Prints the compositions of gas phases to the output file. Default is **true** at startup. Optionally, **-g**[**as_phase**]. Note the hyphen is required to avoid a conflict with the keyword GAS_PHASE.

> (*True* or *False*)—If true, compositions of gas phases are printed to the output file; if false, compositions of gas phases are not printed to the output file. If neither **true** nor **false** is entered on the line, **true** is assumed. Optionally, **t**[**rue**] or **f**[**alse**].

Line 7: **-headings** [(*True* or *False*)]

> **-headings**—Prints the titles and headings that identify the beginning of each type of calculation to the output file. Default is **true** at startup. Optionally, **heading**, **headings**, or **-h**[**eadings**].

> (*True* or *False*)—If true, headings are printed to the output file; if false, headings are not printed to the output file. If neither **true** nor **false** is entered on the line, **true** is assumed. Optionally, **t**[**rue**] or **f**[**alse**].

Line 8: **-initial_isotopes** [(*True* or *False*)]

> **-initial_isotopes**—Prints the molalities of isotopic elements to the output file for initial solution calculations. Default is **true** at startup. Optionally, **initial_isotopes** or **-ini**[**tial_isotopes**].

> (*True* or *False*)—If true, molalities of isotopic elements are printed to the output file for initial solution calculations; if false, molalities of isotopic elements are not printed to the output file for initial solution calculations. If neither **true** nor **false** is entered on the line, **true** is assumed. Optionally, **t**[**rue**] or **f**[**alse**].

Line 9: **-inverse_modeling** [(*True* or *False*)]

> **-inverse_modeling**—Prints the results of inverse modeling to the output file. Default is **true** at startup. Optionally, **inverse** or **-i**[**nverse_modeling**]. Note the hyphen is required to avoid a conflict with the keyword INVERSE_MODELING.

(*True* or *False*)—If true, results of inverse modeling are printed to the output file; if false, results of inverse modeling are not printed to the output file. If neither **true** nor **false** is entered on the line, **true** is assumed. Optionally, **t**[**rue**] or **f**[**alse**].

Line 10: **-isotope_alphas** [(*True* or *False*)]

-isotope_alphas—Prints isotope fractionation factors (as defined by the ISOTOPE_ALPHAS data block) to the output file. Default is **true** at startup. Optionally, **-is**[**otope_alphas**]. Note the hyphen is required to avoid a conflict with the keyword ISOTOPE_ALPHAS.

(*True* or *False*)—If true, isotope fractionation factors are printed to the output file; if false, isotope fractionation factors are not printed to the output file. If neither **true** nor **false** is entered on the line, **true** is assumed. Optionally, **t**[**rue**] or **f**[**alse**].

Line 11: **-isotope_ratios** [(*True* or *False*)]

-isotope_ratios—Prints isotope ratios (as defined by the ISOTOPE_RATIOS data block) to the output file. Default is **true** at startup. Optionally, **-isotope_r**[**atios**]. Note the hyphen is required to avoid a conflict with the keyword ISOTOPE_RATIOS.

(*True* or *False*)—If true, isotope ratios are printed to the output file; if false, isotope ratios are not printed to the output file. If neither **true** nor **false** is entered on the line, **true** is assumed. Optionally, **t**[**rue**] or **f**[**alse**].

Line 12: **-kinetics** [(*True* or *False*)]

-kinetics—Prints the compositions of kinetic-reaction assemblages to the output file. Default is **true** at startup. Optionally, **-k**[**inetics**]. Note the hyphen is required to avoid a conflict with the keyword KINETICS.

(*True* or *False*)—If true, the compositions of kinetic-reaction assemblages are printed to the output file; if false, the compositions of kinetic-reaction assemblages are not printed to the output file. If neither **true** nor **false** is entered on the line, **true** is assumed. Optionally, **t**[**rue**] or **f**[**alse**].

Line 13: **-other** [(*True* or *False*)]

-other—Controls all printing to the output file not controlled by any of the other identifiers, including lines that identify the solution or mixture, exchange assemblage, solid-solution assemblage, surface assemblage, pure-phase assemblage, kinetic reaction, and gas phase to be used in each calculation; and description of the stoichiometric reaction. Default is **true** at startup. Optionally, **other**, **-o**[**ther**], **use**, or **-u**[**se**].

(*True* or *False*)—If true, output items controlled by the **-other** identifier are printed to the output file; if false, output items controlled by the **-other** identifier are not printed to the output file. If neither **true** nor **false** is entered on the line, **true** is assumed. Optionally, **t[rue]** or **f[alse]**.

Line 14: **-saturation_indices** [(*True* or *False*)]

-saturation_indices—Prints saturation indices to the output file. Default is **true** at startup. Optionally, **-si**, **si**, **saturation_indices**, or **-sa[turation_indices]**.

(*True* or *False*)—If true, saturation indices are printed to the output file; if false, saturation indices are not printed to the output file. If neither **true** nor **false** is entered on the line, **true** is assumed. Optionally, **t[rue]** or **f[alse]**.

Line 15: **-solid_solutions** [(*True* or *False*)]

-solid_solutions—Prints the compositions of solid-solution assemblages to the output file. Default is **true** at startup. Optionally, **-so[lid_solutions]**. Note the hyphen is required to avoid a conflict with the keyword SOLID_SOLUTIONS.

(*True* or *False*)—If true, the compositions of solid-solution assemblages are printed to the output file; if false, the compositions of solid-solution assemblages are not printed to the output file. If neither **true** nor **false** is entered on the line, **true** is assumed. Optionally, **t[rue]** or **f[alse]**.

Line 16: **-species** [(*True* or *False*)]

-species—Prints the distributions of aqueous species (including molality, activity, and activity coefficient) to the output file. Default is **true** at startup. Optionally, **species** or **-sp[ecies]**.

(*True* or *False*)—If true, the distributions of aqueous species are printed to the output file; if false, the distributions of aqueous species are not printed to the output file. If neither **true** nor **false** is entered on the line, **true** is assumed. Optionally, **t[rue]** or **f[alse]**.

Line 17: **-surface** [(*True* or *False*)]

-surface—Prints the compositions of surface assemblages to the output file. Default is **true** at startup. Optionally, **-su[rface]**. Note the hyphen is required to avoid a conflict with the keyword SURFACE.

(*True* or *False*)—If true, the compositions of surface assemblages are printed to the output file; if false, the compositions of surface assemblages are not printed to the output file. If neither **true** nor **false** is entered on the line, **true** is assumed. Optionally, **t[rue]** or **f[alse]**.

Line 18: **-totals** [(*True* or *False*)]

>   **-totals**—Prints the total molalities of elements (or element valence states in initial solutions), pH, pe, temperature, and other solution characteristics to the output file. Default is **true** at startup. Optionally, **totals** or **-t**[**otals**].

>   (*True* or *False*)—If true, the total molalities of elements and other solution characteristics are printed to the output file; if false, the total molalities of elements and other solution characteristics are not printed to the output file. If neither **true** nor **false** is entered on the line, **true** is assumed. Optionally, **t**[**rue**] or **f**[**alse**].

Line 19: **-user_print** [(*True* or *False*)]

>   **-user_print**—Prints the information defined in a USER_PRINT data block to the output file. Default is **true** at startup. Optionally, **-u**[**ser_print**]. Note the hyphen is required to avoid a conflict with the keyword USER_PRINT.

>   (*True* or *False*)—If true, the information defined in a USER_PRINT data block is printed to the output file; if false, the information defined in a USER_PRINT data block is not printed to the output file. If neither **true** nor **false** is entered on the line, **true** is assumed. Optionally, **t**[**rue**] or **f**[**alse**].

Line 20: **-alkalinity** [(*True* or *False*)]

>   **-alkalinity**—Prints listings of the species that contribute to alkalinity to the output file. Default is **false** at startup. Optionally, **alkalinity** or **-a**[**lkalinity**].

>   (*True* or *False*)—If true, listings of the species that contribute to alkalinity are printed to the output file; if false, listings of the species that contribute to alkalinity are not printed to the output file. If neither **true** nor **false** is entered on the line, **true** is assumed. Optionally, **t**[**rue**] or **f**[**alse**].

Line 21: **-dump** [(*True* or *False*)]

>   **-dump**—Controls writing dump files. Default is **true** at startup. Optionally, **dump** or **-d**[**ump**].

>   (*True* or *False*)—If true, dump files are written as specified in DUMP and TRANSPORT data blocks. If false, dump files are not written. If neither **true** nor **false** is entered on the line, **true** is assumed. Optionally, **t**[**rue**] or **f**[**alse**].

Line 22: **-censor_species** *fraction*

>   **-censor_species**—Sets a criterion for exclusion of species with small molalities from the distribution of species blocks of the output file (see **-species**). When *fraction* is 0, all species

of each element or element redox state are printed. When fraction is a small number greater than zero, if the molality of an element or element redox state in a species is less than *fraction* times the total molality of the element or element redox state, then the species is excluded from the distribution of species for that element or element redox state. Default is 0.0 at startup. Optionally, **censor_species** or **-c**[**ensor_species**].

fraction—Small number greater than zero (1×10^{-8}, for example).

Line 23: **-selected_output** [(*True* or *False*)]

-selected_output—Controls printing of information defined in SELECTED_OUTPUT and USER_PUNCH data blocks to the selected-output file. This identifier has no effect unless the SELECTED_OUTPUT data block is included in the input file. If a SELECTED_OUTPUT data block is included, **-selected_output** enables or disables printing to the selected-output file. This print-control option is not affected by **-reset**. Default is **true** at startup. Optionally, **-se**[**lected_output**]. Note the hyphen is required to avoid a conflict with the keyword SELECTED_OUTPUT.

(*True* or *False*)—If true, printing to the selected-output file is enabled; if false, printing to the selected-output file is disabled. If neither **true** nor **false** is entered on the line, **true** is assumed. Optionally, **t**[**rue**] or **f**[**alse**].

Line 24: **-status** [(*True* or *False* or *time_interval*)]

-status—Controls printing of information that monitors calculations to the screen. When set to **true**, a status line is printed to the screen identifying the simulation number and the type of calculation that is being processed by the program. When set to **false**, no status line is printed to the screen. When set to an integer number, the printout will be suspended for that number of milliseconds. This print-control option is not affected by **-reset**. Default is **true** at startup. Optionally, **status** or **-st**[**atus**].

(*True* or *False* or *time_interval*)—**True** enables printing the status line to the screen; **false** disables printing the status line; and *time_interval* sets the frequency for refreshing the status line (milliseconds).

Line 25: **-user_graph** [(*True* or *False*)]

-user_graph—Enables plotting graphs defined by the USER_GRAPH data blocks. Default is **true** at startup. Optionally, **-user_g**[**raph**]. Note the hyphen is required to avoid a conflict with the keyword USER_GRAPH.

(*True* or *False*)—If true, plotting of graphs defined by the USER_GRAPH data blocks is enabled; if false, plotting of graphs defined by the USER_GRAPH data blocks is disabled. If neither **true** nor **false** is entered on the line, **true** is assumed. Optionally, **t[rue]** or **f[alse]**.

Line 26: **-warnings** *count*

-warnings—Sets a limit to the number of warnings that are printed to the screen and the output file. Default is 100 at startup; up to 100 warnings are printed. Optionally, **warning**, **warnings**, or **-w[arnings]**.

count—Maximum number of warnings written to the screen and the output file.

Notes

By default, all print options are set to **true** at the beginning of a run, with the exception of **-alkalinity**. Once set by the keyword data block **PRINT**, options remain in effect until the end of the run or until changed in another **PRINT** data block.

Unlike most PHREEQC input, the order in which the identifiers are entered is important when using the **-reset** identifier. For the identifiers controlled by **-reset**, any identifier set before **-reset** in the data block will be reset when **-reset** is encountered. Thus, **-reset** should be the first identifier in the data block. Using **-reset false** will eliminate all printing to the output file except the echoing of the input file and the printing of warning and error messages.

For long TRANSPORT and ADVECTION calculations with KINETICS, printing the status line [**-status true** (default)] may cause a significant increase in run time. This has been the case on some Macintosh systems. If printing to the screen is unbuffered, the program must wait for the status line to be written before continuing calculations, which slows overall execution time. In this case, setting **-status false** may speed up run times. Alternatively, the time interval for updating the status line may be set to be an integer number of milliseconds. For example, **-status** 500 will suspend any printout to the status line for 500 milliseconds while computations continue unhindered. With a set time interval, the on-screen status line may not show the actual final status of the program when it reports "Done".

The identifiers **-species** and **-saturation_indices** control the longest output data blocks in the output file and are the most likely to be selectively excluded from long computer runs. Use of the **-censor_species** identifier also will decrease the size of the output file and simplify the results. If transport calculations are made, the output file could become very large unless some or all of the output is excluded though the **PRINT** data block (**-reset false**). Alternatively, the output in transport calculations may be limited by using the

-**print_cells** and -**print_frequency** identifiers in the ADVECTION and TRANSPORT data block. For transport calculations, the SELECTED_OUTPUT data block usually is used to produce a compact file of selected results.

Example problems

The keyword **PRINT** is used in example problems 6, 10, 12, 13, 14, 15, 19, 20, and 21.

Related keywords

ADVECTION: -**print_cells** and -**print_frequency**, SELECTED_OUTPUT, TRANSPORT: -**print_cells** and -**print_frequency**, USER_GRAPH, USER_PRINT, and USER_PUNCH.

RATES

This keyword data block is used to define mathematical rate expressions for kinetic reactions. General rate formulas are defined in the **RATES** data block and specific kinetic parameters for batch reaction or transport are defined in the **KINETICS** data block.

Example data block

```
Line 0:   RATES
Line 1:       Calcite
Line 2:       -start
Basic: 1    rem M = current number of moles of calcite
Basic: 2    rem M0 = number of moles of calcite initially present
Basic: 3    rem PARM(1) = A/V, cm^2/L
Basic: 4    rem PARM(2) = exponent for M/M0
Basic: 10   si_cc = SI("Calcite")
Basic: 20   if (M <= 0 and si_cc < 0) then goto 200
Basic: 30     k1 = 10^(0.198 - 444.0 / TK )
Basic: 40     k2 = 10^(2.84 - 2177.0 / TK)
Basic: 50     if TC <= 25 then k3 = 10^(-5.86 - 317.0 / TK )
Basic: 60     if TC > 25 then k3  = 10^(-1.1 - 1737.0 / TK )
Basic: 70     t = 1
Basic: 80     if M0 > 0 then t = M/M0
Basic: 90     if t = 0 then t = 1
Basic: 100    area = PARM(1) * (t)^PARM(2)
Basic: 110    rf = k1*ACT("H+")+k2*ACT("CO2")+k3*ACT("H2O")
Basic: 120    rem 1e-3 converts mmol to mol
Basic: 130    rate = area * 1e-3 * rf * (1 - 10^(2/3*si_cc))
Basic: 140    moles = rate * TIME
Basic: 200 SAVE moles
Line 3:       -end
Line 1a:      Pyrite
Line 2a:      -start
Basic: 1    rem  PARM(1) = log10(A/V, 1/dm)
Basic: 2    rem  PARM(2) = exp for (M/M0)
Basic: 3    rem  PARM(3) = exp for O2
Basic: 4    rem  PARM(4) = exp for H+
Basic: 10   if (M <= 0) then goto 200
Basic: 20   if (SI("Pyrite") >= 0) then goto 200
Basic: 30     lograte = -10.19 + PARM(1) + PARM(2)*LOG10(M/M0)
Basic: 40     lograte = lograte + PARM(3)*LM("O2") + PARM(4)*LM("H+")
Basic: 50     moles = (10^lograte) * TIME
Basic: 60     if (moles > M) then moles = M
Basic: 200 SAVE moles
Line 3a:      -end
```

Explanation

Line 0: **RATES**

RATES is the keyword for the data block. No other data are input on the keyword line.

Line 1: *name of rate expression*

name of rate expression—Alphanumeric character string that identifies the rate expression; no spaces are allowed.

Line 2: **-start**

-start—Identifier marks the beginning of a Basic program by which the moles of reaction for a time subinterval are calculated.

Basic: *numbered Basic statement*

numbered Basic statement—A valid Basic language statement that must be numbered. The statements are evaluated in numerical order. The sequence of statements must extrapolate the rate of reaction over the time subinterval given by the internally defined variable TIME. There must be a statement "**SAVE** *expression*", where the value of *expression* is the moles of reaction that are transferred during time subinterval TIME. Statements and functions that are available through the Basic interpreter are listed in the section on the Basic interpreter. Parameters defined in the KINETICS data block also are available through the Basic array PARM.

Line 3: **-end**

-end—Identifier marks the end of a Basic program by which the number of moles of a reaction for a time subinterval is calculated. Note the hyphen is required to avoid a conflict with the keyword **END**.

Notes

A Basic interpreter (David Gillespie, Synaptics, Inc., San Jose, Calif., written commun., 1997) distributed with the Linux operating system (Free Software Foundation, Inc.) is embedded in PHREEQC. The Basic interpreter is used during the integration of the kinetic reactions to evaluate the moles of reaction progress for a time subinterval. A Basic program for each kinetic reaction must be included in the input or database file. Each program must stand alone with its own set of variables and numbered statement lines. There is no conflict in using the same variable names or line numbers in separate rate programs.

It is possible to transfer data among rates with the special Basic statements PUT and GET (see The Basic Interpreter). The programs are used to calculate the instantaneous rate of reaction and extrapolate that rate for a time subinterval given by the variable "TIME" (calcite, line 140; pyrite line 50). TIME is an internally generated and variable time substep, and its value cannot be changed. The total moles of reaction must be returned to the main program with a SAVE command (line 200 in each example). Note that moles of reaction are returned, not the rate of the reaction. Moles are counted positive when the solution concentration of the reactant increases.

Table 5. Description of Basic program for calcite kinetics given in example for **RATES** data block.

Line number	Function
1–4	Comments.
10	Calculate calcite saturation index.
20	If undersaturated and no moles of calcite, exit; moles=0 by default.
30–60	Calculate temperature dependence of constants k1, k2, and k3.
70–90	Calculate ratio of current moles of calcite to initial moles of calcite; set ratio to 1 if no moles of calcite are present.
100	Calculate surface area.
110	Calculate forward rate.
130	Calculate overall rate, factor of 1e–3 converts rate to moles from millimoles.
140	Calculate moles of reaction over time interval given by TIME. Note that the multiplication of the rate by TIME must be present in one of the Basic lines.
200	Return moles of reaction for time subinterval with "SAVE". A SAVE statement must always be present in a rate program.

The first example estimates the rate of calcite dissolution or precipitation on the basis of a rate expression from Plummer and others (1978) (see also equations 101 and 106, Parkhurst and Appelo, 1999). The forward rate is given by

$$R_f = k_1[H^+] + k_2[CO_{2(aq)}] + k_3[H_2O] \,, \tag{1}$$

where square brackets indicate activity and k_1, k_2, and k_3 are functions of temperature (Plummer and others, 1978). In a pure calcite-water system with fixed P_{CO_2}, the overall rate for calcite (forward rate minus backward rate) is approximated by

$$R_{Calcite} = R_f \left[1 - \left(\frac{IAP}{K_{Calcite}} \right)^{\frac{2}{3}} \right], \tag{2}$$

where $R_{Calcite}$ is mmol cm^{-2}s^{-1}(millimole per square centimeter per second). Equation 2 is implemented in Basic for the first example above. Explanations of the Basic lines for this rate expression are given in table 5.

The second example is for the dissolution of pyrite in the presence of dissolved oxygen from Williamson and Rimstidt (1994):

$$R_{Pyrite} = 10^{-10.19}(O_{2(aq)})^{0.5}(H^+)^{-0.11},$$ (3)

where parentheses indicate molality. This rate is based on detailed measurements in solutions of varying compositions and shows a square root dependence on the molality of oxygen and a small dependence on pH. This rate is applicable only for dissolution in the presence of oxygen and will be incorrect near equilibrium when oxygen is depleted. Explanations of the Basic lines for this rate expression are given in table 6.

Table 6. Description of Basic program for pyrite dissolution kinetics given in example for **RATES** data block.

Line number	Function
1–4	Comments.
10	Checks that pyrite is still available, otherwise exits with value of moles=0 by default.
20	Checks that the solution is undersaturated (the rate is for dissolution only), otherwise exits with value of moles=0.
30, 40	Calculate log of the rate of pyrite dissolution.
50	Calculate the moles of pyrite dissolution over time interval given by TIME.
60	Limits pyrite dissolution to remaining moles of pyrite.
200	Return moles of reaction for time subinterval with SAVE. A SAVE statement must always be present in a rate program.

Some special statements and functions have been added to the Basic interpreter to allow access to quantities that may be needed in rate expressions. These functions are listed in The Basic Interpreter, table 8. Standard Basic statements that are implemented in the interpreter are listed in The Basic Interpreter, table 7. Upper or lower case may be used for statement, function, and variable names. String variable names must end with the character "$".

The PRINT command in Basic programs is useful for debugging rate expressions. It can be used to write quantities to the output file to check that rates are calculated correctly. However, the PRINT command will write to the output file every time a rate is evaluated, which may be many times per time step. The sequence of information from PRINT statements in **RATES** definitions may be difficult to interpret because of the automatic time-step adjustment of the integration method.

Example problems

The keyword **RATES** is used in example problems 6, 9, and 15. It is also found in the *Amm.dat*, *llnl.dat*, *phreeqc.dat*, and *wateq4f.dat* databases.

Related keywords

ADVECTION, KINETICS, and TRANSPORT.

REACTION

This keyword data block is used to define irreversible reactions that transfer specified amounts of elements to or from the aqueous solution during batch-reaction calculations. **REACTION** steps are specified explicitly and do not depend on solution composition or time. The KINETICS and RATES data blocks should be used to model the rates of irreversible reactions that evolve with time and vary with solution composition.

Example data block 1

```
Line 0:   REACTION 5 Add sodium chloride and calcite to solution.
Line 1a:       NaCl    2.0
Line 1b:       Calcite  0.001
Line 2:        0.25    0.5     0.75    1.0  moles
```

Explanation 1

Line 0: **REACTION** [*number*] [*description*]

REACTION is the keyword for the data block.

number—A positive number designates this stoichiometric reaction definition. A range of numbers also may be given in the form *m-n*, where *m* and *n* are positive integers, *m* is less than *n*, and the two numbers are separated by a hyphen without intervening spaces. Default is 1.

description—Optional comment that describes the stoichiometric reaction.

Line 1: (*phase name* or *formula*), [*relative stoichiometry*]

phase name or *formula*—If a *phase name* is given, the program uses the stoichiometry of that phase as defined by PHASES input; otherwise, *formula* is a chemical formula to be used in the stoichiometric reaction. Additional lines can be used to define additional reactants.

relative stoichiometry—Amount of this reactant relative to other reactants; it is a molar ratio between reactants. In the Example data block, the reaction contains 2,000 times more NaCl (Line 1a) than calcite (line 1b). Default is 1.0 unitless (mol/mol).

Line 2: *list of reaction amounts*, [*units*]

list of reaction amounts—A separate calculation will be made for each listed amount. If INCREMENTAL_REACTIONS is **false** (default), Example data block 1 performs the calculation as follows: the first step adds 0.25 mol of reaction (assuming *units* are "moles") to the initial solution; the second step adds 0.5 mol of reaction to the initial solution; the third 0.75 mol; and the fourth 1.0 mol; each reaction step begins with the same initial solution and adds only the amount of reaction specified. If INCREMENTAL_REACTIONS keyword is **true**, the calculations are performed as follows: the first step adds 0.25 mol of reaction and the intermediate results are saved as the starting point for the next step; then 0.5 mol of reaction are added and the intermediate results saved; then 0.75 mol; then 1.0 mol; the total amount of reaction added to the initial solution is 2.5 mol. The total amount of each reactant added at any step in the reaction is the reaction amount times the relative stoichiometric coefficient of the reactant. Additional lines may be used to define all reactant amounts.

units—Units may be moles, millimoles, or micromoles. Units must follow all reaction amounts. Default is moles.

If Line 2 is not entered, the default is one step of 1.0 mol.

Example data block 2

```
Line 0: REACTION 5 Add sodium chloride and calcite to reaction solution.
Line 1a:      NaCl       2.0
Line 1b:      Calcite    0.001
Line 2:       1.0 moles in 4 steps
```

Explanation 2

Line 0: **REACTION** [*number*] [*description*]

Same as Example data block 1.

Line 1: (*phase name* or *formula*), [*relative stoichiometry*]

Same as Example data block 1.

Line 2: *reaction amount* [*units*] [**in** *steps*]

reaction amount—A single reaction amount is entered. This amount of reaction will be added in *steps* steps.

units—Same as Example data block 1.

in *steps*—"**in**" indicates that the stoichiometric reaction will be divided into *steps* number of steps. If INCREMENTAL_REACTIONS is **false** (default), Example data block 2 performs the calculations as follows: the first step adds 0.25 mol of reaction to the initial solution; the second step adds 0.5 mol of reaction to the initial solution; the third 0.75 mol; and the fourth 1.0 mol. If INCREMENTAL_REACTIONS keyword is **true**, the calculations are performed as follows: each of the four steps adds 0.25 mol of reaction and the intermediate results are saved as the starting point for the next step.

If Line 2 is not entered, the default is one step of 1.0 mol.

Notes

The **REACTION** data block is used to increase or decrease solution concentrations by specified amounts of reaction. If the product of *reaction amount* and *relative stoichiometry* is positive, then the *phase name* or *formula* will be added to the solution; if the product is negative, the *phase name* or *formula* will be removed from the solution. The specified reactions are added to or removed from solution without regard to equilibrium, time, or reaction kinetics. Irreversible reactions that evolve in time or depend on concentration must be modeled with the KINETICS and RATES keywords.

Example data block 1 with INCREMENTAL_REACTIONS **false** and Example data block 2 with INCREMENTAL_REACTIONS **true** or **false** will generate the same solution compositions after 0.25, 0.5, 0.75, and 1.0 mol of reaction have been added. Example data block 1 with INCREMENTAL_REACTIONS **true** generates results after 0.25, 0.75, 1.5, and 2.5 mol of reaction have been added.

If a phase name is used to define the stoichiometry of a reactant, that phase must have been defined by PHASES input in the database or in the input data file. If negative relative stoichiometries or negative reaction amounts are used, it is possible to remove more of an element than is present in the system, which results in negative concentrations. Negative concentrations will cause the calculations to fail. It is possible to "evaporate" a solution by removing H_2O or dilute a solution by adding H_2O. If more reaction steps are defined in the KINETICS, REACTION_PRESSURE, or REACTION_TEMPERATURE data blocks than in **REACTION**, then the final reaction amount defined by **REACTION** will be repeated for the additional steps. Suppose only one reaction step of 1.0 mol is specified in a **REACTION** data block and two temperature steps are specified in a REACTION_TEMPERATURE data block. If INCREMENTAL_REACTIONS is **false**, then the total amount of reaction added by the end of step 1 and step 2 is the same, 1.0 mol. However, if INCREMENTAL_REACTIONS is **true**, the total amount of reaction added by the end of step 1 will be 1.0 mol and by the end of step 2 will be 2.0 mol.

Example problems

The keyword **REACTION** is used in example problems 4, 5, 6, 7, 10, 17, 19, 20, and 22.

Related keywords

INCREMENTAL_REACTIONS, KINETICS, PHASES, RATES, REACTION_PRESSURE, and REACTION_TEMPERATURE.

REACTION_PRESSURE

This keyword data block is used to define pressure during batch-reaction steps. This data block can also be used to specify the pressure in a cell or range of cells during advective-transport calculations (ADVECTION) and advective-dispersive transport calculations (TRANSPORT).

Example data block 1

```
Line 0: REACTION_PRESSURE 1 Three explicit reaction pressures.
Line 1:    1.0    250.5    500.0
```

Explanation 1

Line 0: **REACTION_PRESSURE** [*number*] [*description*]

REACTION_PRESSURE is the keyword for the data block.

number—Positive number or a range of numbers to designate this pressure definition. A range of numbers may be given in the form *m-n*, where *m* and *n* are positive integers, *m* is less than *n*, and the two numbers are separated by a hyphen without intervening spaces. Default is 1.

description—Optional comment that describes the pressure data.

Line 1: *list of pressures*

list of pressures—A list of pressures (atm) that will be applied to batch-reaction calculations. More lines may be used to supply additional pressures. One batch-reaction calculation will be performed for each listed pressure.

Example data block 2

```
Line 0: REACTION_PRESSURE 1 Three implicit reaction pressures.
Line 1:    1.0    500.0 in 3 steps
```

Explanation 2

Line 0: **REACTION_PRESSURE** [*number*] [*description*]

Same as Example data block 1.

Line 1: *pres$_1$, pres$_2$,* **in** *steps*

pres$_1$—Pressure of first reaction step, atm.

pres$_2$—Pressure of final reaction step, atm.

in *steps*—"**in**" indicates that the pressure will be calculated for each of *steps* number of steps. The pressure at each step, *i*, will be calculated by the formula

$$pres_i = pres_1 + \frac{(i-1)}{(steps-1)}(pres_2 - pres_1)$$; if *steps* = 1, then the pressure of the batch reaction will be $pres_1$. Example data block 2 performs exactly the same calculations as Example data block 1. If more batch-reaction steps are defined by KINETICS, REACTION, or REACTION_TEMPERATURE input, the pressure of the additional steps will be $pres_2$.

Notes

If more batch-reaction steps are defined in KINETICS, REACTION, or REACTION_TEMPERATURE than pressure steps in **REACTION_PRESSURE**, then the final pressure will be used for all of the additional batch-reaction steps. INCREMENTAL_REACTIONS keyword has no effect on the **REACTION_PRESSURE** data block. The default pressure of a reaction step is equal to the pressure of the initial solution or the mixing-fraction-averaged pressure of a mixture. **REACTION_PRESSURE** input can be used even if there is no REACTION input. The method of calculation of pressure steps using "**in**" is slightly different than that for reaction steps. If *n* pressure steps are defined with "**in** *n*" in a **REACTION_PRESSURE** data block, then the pressure of the first reaction step is equal to *pres₁*; pressures in the remaining steps change in *n*-1 equal increments. In contrast, if *n* reaction steps are defined with "**in** *n*" in a REACTION data block, then the reaction is added in *n* equal increments.

In an advective-transport calculation (ADVECTION), if **REACTION_PRESSURE** *n* is defined (or a range is defined *n-m*), and *n* is less than or equal to the number of cells in the simulation, then the first pressure in the data block of **REACTION_PRESSURE** *n* is used as the pressure in cell *n* (or cells *n-m*) for all shifts in the advective-transport calculation. In advective-dispersive transport simulations (TRANSPORT), the initial equilibration also occurs at the first pressure of **REACTION_PRESSURE** *n* in cell *n*.

Example problems

The keyword **REACTION_PRESSURE** is used in example problem 2.

Related keywords

ADVECTION, KINETICS, REACTION, REACTION_TEMPERATURE, and TRANSPORT.

REACTION_TEMPERATURE

This keyword data block is used to define temperature during batch-reaction steps. This data block can also be used to specify the temperature in a cell or range of cells during advective-transport calculations (ADVECTION) and the initial temperature for a cell or range of cells in advective-dispersive transport calculations (TRANSPORT).

Example data block 1

```
Line 0: REACTION_TEMPERATURE 1 Three explicit reaction temperatures.
Line 1:    15.0    25.0    35.0
```

Explanation 1

Line 0: **REACTION_TEMPERATURE** [*number*] [*description*]

REACTION_TEMPERATURE is the keyword for the data block.

number—Positive number or a range of numbers to designate this temperature definition. A range of numbers may be given in the form *m-n*, where *m* and *n* are positive integers, *m* is less than *n*, and the two numbers are separated by a hyphen without intervening spaces. Default is 1.

description—Optional comment that describes the temperature data.

Line 1: *list of temperatures*

list of temperatures—A list of temperatures (°C) that will be applied to batch-reaction calculations. More lines may be used to supply additional temperatures. One batch-reaction calculation will be performed for each listed temperature.

Example data block 2

```
Line 0: REACTION_TEMPERATURE 1 Three implicit reaction temperatures.
Line 1:    15.0    35.0 in 3 steps
```

Explanation 2

Line 0: **REACTION_TEMPERATURE** [*number*] [*description*]

Same as Example data block 1.

Line 1: *temp₁, temp₂,* **in** *steps*

$temp_1$—Temperature of first reaction step, °C.

$temp_2$—Temperature of final reaction step, °C.

in *steps*—"**in**" indicates that the temperature will be calculated for each of *steps* number of steps. The temperature at each step, *i*, will be calculated by the formula

$$temp_i = temp_1 + \frac{(i-1)}{(steps-1)}(temp_2 - temp_1)$$; if *steps* = 1, then the temperature of the batch reaction will be $temp_1$. Example data block 2 performs exactly the same calculations as Example data block 1. If more batch-reaction steps are defined by KINETICS, REACTION, or REACTION_PRESSURE input, the temperature of the additional steps will be $temp_2$.

Notes

If more batch-reaction steps are defined in KINETICS, REACTION, or REACTION_PRESSURE than temperature steps in **REACTION_TEMPERATURE**, then the final temperature will be used for all of the additional batch-reaction steps. INCREMENTAL_REACTIONS keyword has no effect on the **REACTION_TEMPERATURE** data block. The default temperature of a reaction step is equal to the temperature of the initial solution or the mixing-fraction-averaged temperature of a mixture. **REACTION_TEMPERATURE** input can be used even if there is no REACTION input. The method of calculation of temperature steps using "**in**" is slightly different than that for reaction steps. If *n* temperature steps are defined with "**in** *n*" in a **REACTION_TEMPERATURE** data block, then the temperature of the first reaction step is equal to $temp_1$; temperatures in the remaining steps change in *n*-1 equal increments. In contrast, if *n* reaction steps are defined with "**in** *n*" in a REACTION data block, then the reaction is added in *n* equal increments.

In an advective-transport calculation (ADVECTION), if **REACTION_TEMPERATURE** *n* is defined (or a range is defined *n-m*), and *n* is less than or equal to the number of cells in the simulation, then the first temperature in the data block of **REACTION_TEMPERATURE** *n* is used as the temperature in cell *n* (or cells *n-m*) for all shifts in the advective-transport calculation. In advective-dispersive transport simulations (TRANSPORT), the initial equilibration also occurs at the first temperature of **REACTION_TEMPERATURE** *n* in cell *n*. However, depending on the setting of *temperature retardation factor* (**-thermal_diffusion** in the TRANSPORT data block), an exchange of heat may take place that will cause the temperature of the cell to change as transport progresses.

Example problems

The keyword **REACTION_TEMPERATURE** is used in example problems 2 and 22.

Related keywords

ADVECTION, KINETICS, REACTION, REACTION_PRESSURE, and TRANSPORT.

RUN_CELLS

This keyword data block is used to run reaction simulations for a specified set of cells. For a specified cell number, *n*, all reactants numbered *n* are reacted together and the resulting reactant compositions are saved to the same cell number. If multiple steps have been defined in **KINETICS** *n*, **REACTION** *n*, **REACTION_PRESSURE** *n,* or **REACTION_TEMPERATURE** *n* data blocks, multiple calculations will be done for each cell. It is possible to specify the starting time and the time step for cells that have kinetic reactants; these time values defined in **RUN_CELLS** supersede the definitions in the **KINETICS** data block.

Example data block

```
Line 0: RUN_CELLS
Line 1:     -cells  1 2
Line 2:             5-6
Line 2a:            7
Line 3:     -start_time    100 day
Line 4:     -time_step     10  day
```

Explanation

Line 0: **RUN_CELLS**

RUN_CELLS is the keyword for the data block. No other data are input on the keyword line.

Line 1: **-cells** *list of cell numbers*

-cells—Identifier for a list of cells to be run. Optionally, **cell**, **cells**, or **-c[ells]**.

list of cell numbers—A list of cell numbers. Each item of the list may be a single cell number or a range of cell numbers defined by two integers separated by a hyphen, with no intervening spaces.

Line 2: *list of cell numbers*

list of cell numbers—The list of cell numbers for the **-cells** identifier may be continued on multiple lines.

Line 3: **-start_time** *time* [*unit*]

-start_time—Identifier defining a start time for cells that have kinetic reactants. Optionally, **start_time** or **-s[tart_time]**.

time—Time at the beginning of the simulation for a cell that has kinetic reactants, s.

unit—Optional time unit may be **second**, **minute**, **hour**, **day**, **year**, or an abbreviation of one of these units. The *time* is converted to seconds after reading the data block; all internal calculations, Basic functions, and output times are in seconds. Default is second.

Line 4: **-time_step** *time_step* [*unit*]

-time_step—Identifier defining a time step for cells that have kinetic reactants. Optionally, **time_step** or **-t**[**ime_step**].

time_step—Time step for the simulation for a cell that has kinetic reactants, s.

unit—Optional time unit may be **second**, **minute**, **hour**, **day**, **year**, or an abbreviation of one of these units. The *time_step* is converted to seconds after reading the data block; all internal calculations, Basic functions, and output times are in seconds. Default is second.

Notes

The **RUN_CELLS** data block is a streamlined method for running a simulation that uses all reactants that have been defined with the same identification number (cell number). The calculation for a cell defined in the **-cells** data block is equivalent to a series of USE and SAVE data blocks for all of the reactants with a specified cell number. If both a solution and a mix definition exist for a cell number, the mix definition is used in preference to the solution. If multiple steps have been defined in KINETICS, REACTION, REACTION_PRESSURE, or REACTION_TEMPERATURE data blocks, then multiple calculations will be done for that cell.

It is possible to specify an initial time for a kinetic integration by using the **-start_time** identifier and to override the time step from the KINETICS data block by using the **-time_step** identifier. If **-time_step** is not defined or a KINETICS data block is not defined for the cell, then the calculations occur exactly as they would by a series of USE and SAVE data blocks. If **-time_step** is defined, then the kinetic reaction will be integrated over an interval of *time_step*. If n_{max} is the maximum number of steps defined for the cell with KINETICS, REACTION, REACTION_PRESSURE, or REACTION_TEMPERATURE data blocks, then the kinetic reaction will be divided into n_{max} equal increments; the results are equivalent to defining "**-step** *time_step* **in** n_{max} steps" in the KINETICS data block. The **time_step** will be used for all subsequent **RUN_CELLS** calculations or until it is changed with another **-time_step** definition in a **RUN_CELLS** data block.

The **RUN_CELLS** data block simplifies the definition of repetitive reactions in batch calculations. It also is intended to be used when an IPhreeqc module implements geochemical reactions in a reactive-transport model. For example, if a transport model has a set of cells numbered 1 through n, and the chemical reactants for those cells are saved as identification numbers 1 through n in an IPhreeqc module, then a simple sequential calculation can be implemented. The transport code is used to transport elemental concentrations conservatively, and the concentrations for solutions in the IPhreeqc module are updated with a series of SOLUTION_MODIFY data blocks. Geochemical reactions are calculated by the data block: **RUN_CELLS**; **-cells** 1-n. The new compositions of solutions and reactants are automatically stored in the IPhreeqc module and new solution concentrations are retrieved by extracting data defined by SELECTED_OUTPUT or by the output from a DUMP data block. The new solution concentrations then are used to begin a new conservative transport step.

Example problems

The keyword **RUN_CELLS** is used in example problem 20.

Related keywords

DUMP, SAVE, SELECTED_OUTPUT, SOLUTION_SPECIES, and USE.

SAVE

This keyword data block is used to save the composition of a solution, exchange assemblage, gas phase, equilibrium-phase assemblage, solid-solution assemblage, or surface assemblage following a batch-reaction calculation. The composition is stored internally in computer memory and can be retrieved subsequently with the USE keyword during the remainder of the computer run.

Example data block

```
Line 0a:  SAVE equilibrium_phases 2
Line 0b:  SAVE exchange 2
Line 0c:  SAVE gas_phase 2
Line 0d:  SAVE solid_solution 1
Line 0e:  SAVE solution 2
Line 0f:  SAVE surface 1
```

Explanation

Line 0: **SAVE** *keyword*, *number*

SAVE is the keyword for the data block.

keyword—One of six keywords with an index number, **equilibrium_phases**, **exchange**, **gas_phase**, **solid_solution**, **solution**, or **surface**. Options for **equilibrium_phases**: **equilibrium**, **equilibria**, **pure_phases**, or **pure**.

number—User defined positive integer to be associated with the respective composition. A range of numbers may also be given in the form *m-n*, where *m* and *n* are positive integers, *m* is less than *n*, and the two numbers are separated by a hyphen without intervening spaces.

Notes

SAVE affects only the internal storage of chemical-composition information during the current run; it does not save information between PHREEQC runs. To save results to a permanent file, see SELECTED_OUTPUT or DUMP. The **SAVE** data block applies only at the end of batch-reaction calculations and has no effect following initial solution, initial exchange-composition, initial surface-composition, initial gas-phase-composition, transport, run cells, or inverse calculations. During batch-reaction calculations, the compositions of the solution, exchange assemblage, gas phase, pure-phase assemblage, solid-solution assemblage, and surface assemblage vary to attain equilibrium. The compositions that exist at the end of a batch reaction are <u>not</u> automatically saved (unless RUN_CELLS is used); however, the compositions may be saved explicitly for use in subsequent simulations within the run by using the **SAVE** keyword. The **SAVE** keyword must be used for each type of composition that is to be saved (solution, exchange assemblage, gas phase, pure-phase assemblage, solid-solution assemblage, or surface assemblage). **SAVE** assigns *number* to the corresponding composition. If one of the compositions is saved in a *number* that already exists, the old composition is deleted. There is no need to save the compositions unless they are to be used in subsequent simulations within the run. ADVECTION, TRANSPORT, and RUN_CELLS calculations automatically save results after each calculation and the **SAVE** keyword has no effect for these calculations. Amounts of kinetic reactions (KINETICS) are automatically saved during all batch-reaction, advection, transport, and RUN_CELLS calculations and cannot be saved with the **SAVE** keyword. The USE (or RUN_CELLS) keyword can be invoked to use the saved compositions in subsequent batch-reaction calculations.

Example problems

The keyword **SAVE** is used in example problems 3, 4, 7, 10, 14, and 20.

Related keywords

ADVECTION, EXCHANGE, EQUILIBRIUM_PHASES, GAS_PHASE, KINETICS, RUN_CELLS, SELECTED_OUTPUT, SOLID_SOLUTIONS, SOLUTION, SURFACE, TRANSPORT, and USE.

SELECTED_OUTPUT

This keyword data block is used to produce a file that is suitable for processing by spreadsheets and other data-management software. It is possible to print selected entities from the compositions of the solution, exchange assemblage, gas phase, pure-phase assemblage, solid-solution assemblage, and surface assemblage after the completion of each type of calculation. The selected-output file contains a column for each data item defined through the identifiers of **SELECTED_OUTPUT** and a row for each calculation. Print settings at program startup are shown in Lines 1–2, 4–18 and 30 of the following Example data block. The identifiers are listed in the order in which they are written to the selected-output file.

Example data block

```
Line  0: SELECTED_OUTPUT
Line  1:        -file                    selected.out
Line  2:        -high_precision          false
Line  3:        -reset                   true
Line  4:        -simulation              true
Line  5:        -state                   true
Line  6:        -solution                true
Line  7:        -distance                true
Line  8:        -time                    true
Line  9:        -step                    true
Line 10:        -pH                      true
Line 11:        -pe                      true
Line 12:        -reaction                false
Line 13:        -temperature             false
Line 14:        -alkalinity              false
Line 15:        -ionic_strength          false
Line 16:        -water                   false
Line 17:        -charge_balance          false
Line 18:        -percent_error           false
Line 19:        -totals                  Hfo_s  C   C(4)   C(-4)   N   N(0)
Line 20:                                 Fe  Fe(3)  Fe(2)  Ca  Mg  Na  Cl
Line 21:        -molalities              Fe+2  Hfo_sOZn+   ZnX2
Line 22:        -activities              H+  Ca+2  CO2  HCO3-   CO3-2
Line 23:        -equilibrium_phases      Calcite Dolomite   Sphalerite
Line 24:        -saturation_indices      CO2(g)   Siderite
Line 25:        -gases                   CO2(g)  N2(g)      O2(g)
Line 26:        -kinetic_reactants       CH2O       Pyrite
Line 27:        -solid_solutions         CaSO4    SrSO4
Line 28:        -isotopes                R(D) R(D)_H3O+ R(D)_H2O(g)
Line 29:        -calculate_values        R(D) R(D)_H3O+ R(D)_H2O(g)
Line 30:        -inverse_modeling        true
```

Explanation

Line 0: **SELECTED_OUTPUT**

> **SELECTED_OUTPUT** is the keyword for the data block. No additional data are read on this line. Optionally, **SELECTED_OUT**, **SELECT_OUTPUT**, or **SELECT_OUT**.

Line 1: **-file** *file name*

> **-file**—Identifier allows definition of the name of the file where the selected results are written. Optionally, **file** or **-f[ile]**.

> *file name*—File name where selected results are written. If the file exists, the contents will be overwritten. File names must conform to operating system conventions. Default is **selected.out**.

Line 2: **-high_precision** [(*True* or *False*)]

> **-high_precision**—Prints results to the selected-output file with extra numerical precision (12 decimal places, default is 3 or 4). In addition, the criterion for convergence of the calculations is set to 1×10^{-12} (default is 1×10^{-8}). The convergence criterion also may be set by **-convergence_tolerance** in KNOBS data block. Default is **false** at startup. Optionally, **high_precision** or **-h[igh_precision]**.

> (*True* or *False*)—If **true**, output is written to the selected-output file with extra decimal places; if **false**, output to the selected-output file is written with normal precision. If neither **true** nor **false** is entered on the line, **true** is assumed. Optionally, **t[rue]** or **f[alse]**.

Line 3: **-reset** [(*True* or *False*)]

> **-reset**—Resets all identifiers listed in Lines 4–18 to **true** or **false**. Optionally, **reset** or **-r[eset]**.

> (*True* or *False*)—If **true**, identifiers on Lines 4–18 are set to true; if **false**, identifiers on Lines 4–18 are set to false. If neither **true** nor **false** is entered on the line, **true** is assumed. Optionally, **t[rue]** or **f[alse]**.

Line 4: **-simulation** [(*True* or *False*)]

> **-simulation**—Prints simulation number, or for advective-dispersive transport calculations, the sequence number of the advective-dispersive transport simulation. Default is **true** at startup. Optionally, **simulation**, **sim**, or **-sim[ulation]**.

(*True* or *False*)—If **true**, simulation number is printed to the selected-output file; if **false**, simulation number is not printed. If neither **true** nor **false** is entered on the line, **true** is assumed. Optionally, **t**[**rue**] or **f**[**alse**].

Line 5: **-state** [(*True* or *False*)]

-state—Prints type of calculation to the selected-output file for each calculation. The following character strings are used to identify each calculation type: initial solution, "i_soln"; initial exchange composition, "i_exch"; initial surface composition, "i_surf"; initial gas-phase composition, "i_gas"; batch reaction (including RUN_CELLS), "react"; inverse, "inverse"; advection, "advect"; and transport, "transp". Default is **true** at startup. Optionally, **state** or **-st**[**ate**].

(*True* or *False*)—If **true**, state is printed to the selected-output file; if **false**, state is not printed. If neither **true** nor **false** is entered on the line, **true** is assumed. Optionally, **t**[**rue**] or **f**[**alse**].

Line 6: **-solution** [(*True* or *False*)]

-solution—Prints solution number used for the calculation for each calculation. Default is **true** at startup. Optionally, **soln**, **-solu**[**tion**], or **-soln**. Note the hyphen is required to avoid a conflict with the keyword SOLUTION.

(*True* or *False*)—If **true**, solution number is printed to the selected-output file; if **false**, solution number is not printed. If neither **true** nor **false** is entered on the line, **true** is assumed. Optionally, **t**[**rue**] or **f**[**alse**].

Line 7: **-distance** [(*True* or *False*)]

-distance—Prints to the selected-output file (1) the X-coordinate of the cell for advective-dispersive transport calculations (TRANSPORT), (2) the cell number for advection calculations (ADVECTION), or (3) -99 for other calculations. Default is **true** at startup. Optionally, **distance**, **dist**, or **-d**[**istance**].

(*True* or *False*)—If **true**, distance is printed to the selected-output file; if **false**, distance is not printed. If neither **true** nor **false** is entered on the line, **true** is assumed. Optionally, **t**[**rue**] or **f**[**alse**].

Line 8: **-time** [(*True* or *False*)]

-time—Prints to the selected-output file (1) the cumulative model time since the beginning of the simulation for batch-reaction calculations with kinetics, (2) the cumulative transport time since the beginning of the run (or since **-initial_time** identifier was last defined) for advective-dispersive transport calculations and advective-transport calculations for which **-time_step** is defined, (3) the advection shift number for advective-transport calculations for which **-time_step** is not defined, or (4) -99 for other calculations. Default is **true** at startup. Optionally, **time** or **-ti[me]**.

(*True* or *False*)—If **true**, time is printed to the selected-output file; if **false**, time is not printed. If neither **true** nor **false** is entered on the line, **true** is assumed. Optionally, **t[rue]** or **f[alse]**.

Line 9: **-step** [(*True* or *False*)]

-step—Prints to the selected-output file (1) advection shift number for transport calculations, (2) reaction step for batch-reaction calculations, or (3) -99 for other calculations. Default is **true** at startup. Optionally, **step** or **-ste[p]**.

(*True* or *False*)—If **true**, step is printed to the selected-output file; if **false**, step is not printed. If neither **true** nor **false** is entered on the line, **true** is assumed. Optionally, **t[rue]** or **f[alse]**.

Line 10: **-pH** [(*True* or *False*)]

(*True* or *False*)—Prints pH to the selected-output file for each calculation. Default is **true** at startup. Optionally, **pH** (as with all identifiers, case insensitive).

(*True* or *False*)—If **true**, pH is printed to the selected-output file; if **false**, pH is not printed. If neither **true** nor **false** is entered on the line, **true** is assumed. Optionally, **t[rue]** or **f[alse]**.

Line 11: **-pe** [(*True* or *False*)]

-pe—Prints pe to the selected-output file for each calculation. Default is **true** at startup. Optionally, **pe**.

(*True* or *False*)—If **true**, pe is printed to the selected-output file; if **false**, pe is not printed. If neither **true** nor **false** is entered on the line, **true** is assumed. Optionally, **t[rue]** or **f[alse]**.

Line 12: **-reaction** [(*True* or *False*)]

(*True* or *False*)—Prints (1) reaction increment to the selected-output file if REACTION is used in the calculation or (2) -99 for other calculations. Default is **false** at startup. Optionally, **rxn**, **-rea[ction]**, or **-rx[n]**. Note the hyphen is required to avoid a conflict with the keyword REACTION.

(*True* or *False*)—If **true**, reaction increment is printed to the selected-output file; if **false**, reaction increment is not printed. If neither **true** nor **false** is entered on the line, **true** is assumed. Optionally, **t[rue]** or **f[alse]**.

Line 13: **-temperature** [(*True* or *False*)]

> **-temperature**—Prints temperature (Celsius) to the selected-output file for each calculation. Default is **false** at startup. Optionally, **temp, temperature,** or **-te**[**mperature**].
>
> (*True* or *False*)—If **true**, temperature is printed to the selected-output file; if **false**, temperature is not printed. If neither **true** nor **false** is entered on the line, **true** is assumed. Optionally, **t**[**rue**] or **f**[**alse**].

Line 14: **-alkalinity** [(*True* or *False*)]

> **-alkalinity**—Prints alkalinity (eq/kgw, equivalent per kilogram water) to the selected-output file for each calculation. Default is **true** at startup. Initial value at start of program is **false**. Optionally, **alkalinity, alk,** or **-al**[**kalinity**].
>
> (*True* or *False*)—If **true**, alkalinity is printed to the selected-output file; if **false**, alkalinity is not printed. If neither **true** nor **false** is entered on the line, **true** is assumed. Optionally, **t**[**rue**] or **f**[**alse**].

Line 15: **-ionic_strength** [(*True* or *False*)]

> **-ionic_strength**—Prints ionic strength to the selected-output file. Default is **false** at startup. Optionally, **ionic_strength, mu, -io**[**nic_strength**], or **-mu**.
>
> (*True* or *False*)—If **true**, ionic strength is printed to the selected-output file; if **false**, ionic strength is not printed. If neither **true** nor **false** is entered on the line, **true** is assumed. Optionally, **t**[**rue**] or **f**[**alse**].

Line 16: **-water** [(*True* or *False*)]

> **-water**—Prints mass of water to the selected-output file for each calculation. Default is **false** at startup. Optionally, **water** or **-w**[**ater**].
>
> (*True* or *False*)—If **true**, mass of water is printed to the selected-output file; if **false**, mass of water is not printed. If neither **true** nor **false** is entered on the line, **true** is assumed. Optionally, **t**[**rue**] or **f**[**alse**].

Line 17: **-charge_balance** [(*True* or *False*)]

> **-charge_balance**—Prints charge balance of solution (eq, equivalent) to the selected-output file for each calculation. Default is **false** at startup. Optionally, **charge_balance** or **-c**[**harge_balance**].

(*True* or *False*)—If **true**, charge balance is printed to the selected-output file; if **false**, charge balance is not printed. If neither **true** nor **false** is entered on the line, **true** is assumed. Optionally, **t[rue]** or **f[alse]**.

Line 18: **-percent_error** [(*True* or *False*)]

-percent_error—Prints percent error in charge balance ($100 \dfrac{cations - |anions|}{cations + |anions|}$) to the selected-output file for each calculation. Default is **false** at startup. Optionally, **percent_error** or **-per[cent_error]**.

(*True* or *False*)—If **true**, percent error is printed to the selected-output file; if **false**, percent error is not printed. If neither **true** nor **false** is entered on the line, **true** is assumed. Optionally, **t[rue]** or **f[alse]**.

Line 19: **-totals** *element list*

-totals—Identifier allows definition of a list of total concentrations, in molality, that will be written to the selected-output file. Optionally, **totals** or **-t[otals]**.

element list—List of elements, element valence states, exchange sites, or surface sites for which total concentrations will be written. The list may continue on the subsequent line(s) (Line 2a). After each calculation, the concentration (mol/kgw) of each of the selected elements, element valence states, exchange sites, and surface sites will be written to the selected-output file. Elements, valence states, exchange sites, and surface sites are defined in the first column of SOLUTION_MASTER_SPECIES, EXCHANGE_MASTER_SPECIES, or SURFACE_MASTER_SPECIES input. If an element is not defined or is not present in the calculation, its concentration will be printed as 0.

Line 20: *element list*

element list—Continuation of a list for **-totals** of elements, element valence states, exchange sites, or surface sites.

Line 21: **-molalities** *species list*

-molalities—Identifier allows definition of a list of species for which concentrations will be written to the selected-output file. Optionally, **molalities**, **mol**, or **-m[olalities]**.

species list—List of aqueous, exchange, or surface species for which concentrations will be written to the selected-output file. The list may continue on the subsequent line(s). After each calculation, the concentration (mol/kgw) of each species in the list will be written to the selected-output file. Species are defined by SOLUTION_SPECIES, EXCHANGE_SPECIES, or SURFACE_SPECIES input. If a species is not defined or is not present in the calculation, its concentration will be printed as 0.

Line 22: **-activities** *species list*

    -activities—Identifier allows definition of a list of species for which log of activity will be written to the selected-output file. Optionally, **activities** or **-a[ctivities]**.

    species list—List of aqueous, exchange, or surface species for which log of activity will be written to the selected-output file. The list may continue on the subsequent line(s). After each calculation, the log (base 10) of the activity of each of the species will be written to the selected-output file. Species are defined by SOLUTION_SPECIES, EXCHANGE_SPECIES, or SURFACE_SPECIES input. If a species is not defined or is not present in the calculation, its log activity will be printed as -999.999.

Line 23: **-equilibrium_phases** *phase list*

    -equilibrium_phases—Identifier allows definition of a list of pure phases for which (1) total amounts in the pure-phase assemblage and (2) moles transferred will be written to the selected-output file. Optionally, **-e[quilibrium_phases]** or **-p[ure_phases]**. Note the hyphen is required to avoid a conflict with the keyword EQUILIBRIUM_PHASES and its synonyms.

    phase list—List of phases for which data will be written to the selected-output file. The list may continue on the subsequent line(s). After each calculation, two values are written to the selected-output file: (1) the moles of each of the phases (defined by EQUILIBRIUM_PHASES), and (2) the moles transferred. Phases are defined by PHASES input. If the phase is not defined or is not present in the pure-phase assemblage, the amounts will be printed as 0.

Line 24: **-saturation_indices** *phase list*

    -saturation_indices—Identifier allows definition of a list of phases for which saturation indices [or log (base 10) partial pressure for gases] will be written to the selected-output file. Optionally, **saturation_indices**, **si**, **-s[aturation_indices]**, or **-s[i]**.

    phase list—List of phases for which saturation indices [or log (base 10) partial pressure for gases] will be written to the selected-output file. The list may continue on the subsequent line(s).

After each calculation, the saturation index of each of the phases will be written to the selected-output file. Phases are defined by PHASES input. If the phase is not defined or if one or more of its constituent elements is not in solution, the saturation index will be printed as -999.999.

Line 25: **-gases** *gas-component list*

-gases—Identifier allows definition of a list of gas components for which the amount in the gas phase (mol) will be written to the selected-output file. Optionally, **gases** or **-g[ases]**.

gas-component list—List of gas components. The list may continue on the subsequent line(s). After each calculation, the moles of each of the selected gas components in the gas phase will be written to the selected-output file. Gas components are defined by PHASES input. If a gas component is not defined or is not present in the gas phase, the amount will be printed as 0. Before the columns for the gas components, the selected-output file will contain the total pressure, total moles of gas components, and the volume of the gas phase. Log partial pressures of any gas, including the components in the gas phase, can be obtained by use of the **-saturation_indices** identifier.

Line 26: **-kinetic_reactants** *reactant list*

-kinetic_reactants—Identifier allows definition of a list of kinetically controlled reactants for which two values are written to the selected-output file: (1) the current moles of the reactant, and (2) the moles transferred of the reactant. Optionally, **kin**, **-k[inetics]**, **kinetic_reactants**, or **-k[inetic_reactants]**. Note the hyphen is required to avoid a conflict with the keyword KINETICS.

reactant list—List of kinetically controlled reactants. The list may continue on the subsequent line(s). After each calculation, the moles and the moles transferred of each of the kinetically controlled reactants will be written to the selected-output file. Kinetic reactants are identified by the rate name in the KINETICS data block. (The rate name in turn refers to a rate expression defined with RATES data block.) If the kinetic reactant is not defined, the amounts will be printed as 0.

Line 27: **-solid_solutions** *component list*

-solid_solutions—Identifier allows definition of a list of solid-solution components for which the moles in a solid solution is written to the selected-output file. Optionally,

-**so**[**lid_solutions**]. Note the hyphen is required to avoid a conflict with the keyword SOLID_SOLUTIONS.

component list—List of solid-solution components. The list may continue on the subsequent line(s). After each calculation, the moles of each solid-solution component in the list will be written to the selected-output file. A solid-solution component is identified by the component name defined in the SOLID_SOLUTIONS data block. (The component names are also phase names that have been defined in the PHASES data block.) If the component is not defined in any of the solid solutions, the amount will be printed as 0.

Line 28: -**isotopes** *isotope ratio list*

-**isotopes**—Identifier selects isotopes for which values are written to the selected-output file. The units of the isotopic values printed to the selected output file are the units of the standard as entered with keyword ISOTOPES (for example, permil or percent modern carbon). Optionally, -**is**[**otopes**]. Note the hyphen is required to avoid a conflict with the keyword ISOTOPES.

isotope ratio list—List of ratios for isotopes and isotopic species to be written to the selected-output file. The list may continue on the subsequent line(s). After each calculation, the isotope ratios in the list will be written to the selected-output file. Isotope ratios are defined in the ISOTOPE_RATIOS data block and refer to Basic programs defined in CALCULATE_VALUES data blocks. If an isotope ratio is not defined or is absent from solution, the value printed is -9,999.999.

Line 29: -**calculate_values** *calculate values list*

-**calculate_values**—Identifier selects Basic functions for which function values will be written to the selected-output file. The list may continue on the subsequent line(s). After each calculation, the values for functions in the list will be written to the selected-output file. The Basic functions are defined in the CALCULATE_VALUES data block. Optionally, -**ca**[**lculate_values**]. Note the hyphen is required to avoid a conflict with the keyword CALCULATE_VALUES.

calculate values list—List of names of Basic functions; value calculated by function will be written to the selected-output file. The list may continue on the subsequent line(s).

Line 30: **-inverse_modeling** [(*True* or *False*)]

> **-inverse_modeling**—Prints results of inverse modeling to the selected-output file. For each inverse model, three values are printed for each solution and phase defined in the **INVERSE_MODELING** data block: the central value of the mixing fraction or mole transfer, and the minimum and maximum of the mixing fraction or mole transfer, which are zero unless **-range** is specified in the **INVERSE_MODELING** data block. Default is **true** at startup. Optionally, **inverse** or **-i**[**nverse_modeling**]. Note the hyphen is required to avoid a conflict with the keyword **INVERSE_MODELING**.

> (*True* or *False*)—If **true**, results of inverse modeling are printed to the selected-output file; if **false**, results of inverse modeling are not printed. If neither **true** nor **false** is entered on the line, **true** is assumed. Optionally, **t**[**rue**] or **f**[**alse**].

Notes

The selected-output file contains a row for each calculation and a column for each data item defined through the identifiers of **SELECTED_OUTPUT**. Additional columns may be defined through the USER_PUNCH data block. In the input for the **SELECTED_OUTPUT** data block, all element names, species names, and phase names must be spelled exactly, including the case and charge for the species names. One line containing an entry for each of the items will be written to the selected-output file after each calculation—that is, after any initial solution, initial exchange-composition, initial surface-composition, or initial gas-phase-composition calculation; after each step in a batch-reaction calculation; or for each cell (as defined by **-punch_cells**) after each shift (as selected by **-punch_frequency**) in transport calculations. In ADVECTION and TRANSPORT simulations, the cells selected for printing to the selected-output file are defined with the **-punch_cells** identifier and the frequency at which results are written to the selected-output file is controlled by the **-punch_frequency** identifier. The **-selected_output** identifier in the PRINT data block can be used to selectively suspend and resume writing results to the selected-output file.

Several data items are included by default at the beginning of each line in the selected-output file. These data are described in Lines 4 through 11. Data described in Lines 12 through 18 are not printed by default. All of the data described by Lines 4 through 18 simultaneously may be included or excluded from the selected-output file with the **-reset** identifier. Unlike most of PHREEQC input, the order in which the identifiers are entered is important when using the **-reset** identifier. Any identifier set before **-reset** in the data block will be reset when **-reset** is encountered. Thus, **-reset** should be specified before any of the identifiers described in Lines 4 through 18.

The first line of the selected-output file contains a description of each data column. The columns of data are written in the following order: items described by Lines 4 through 18, totals, molalities, log activities, pure phases (two columns for each phase—total amount of phase and mole transfer for current calculation), saturation indices, gas-phase data (multiple columns), kinetically controlled reactants (two columns for each reactant—total amount of reactant and mole transfer for current calculation), solid-solution components, and data defined by the USER_PUNCH data block. A data item within an input list (for example, an aqueous species within the **-molalities** list) is printed in the order of the list. If the selected-output file contains data for gases (**-gases** identifier), the total pressure, total moles in the gas phase, and the total volume of the gas phase precede the moles of each gas component specified by the identifier.

The **-isotopes** identifier allows isotopic values in the units of the isotope standard to be written to the selected output file. The isotopic values are identified by the names of CALCULATE_VALUES Basic functions that have been identified in the ISOTOPE_RATIOS data block. The isotopic values can also be obtained by using the Basic function ISO in any Basic program, where the argument of the function is the name of a ratio defined in the ISOTOPE_RATIOS data block. The units of the standard are available by the ISO_UNITS Basic function, which takes the same argument as the Basic function ISO.

The **-calculate_values** identifier allows the values of Basic functions defined in the CALCULATE_VALUES data block to be written to the selected-output file. The value of a calculate-values function also can be used in a Basic program by using the CALC_VALUE function, where the argument is the name of a function defined in a CALCULATE_VALUES data block.

Example problems

The keyword **SELECTED_OUTPUT** is used in example problems 2, 5, 6, 7, 8, 9, 10, 11, 12, 13, 14, 15, 20, and 21.

Related keywords

ADVECTION, CALCULATE_VALUES, EQUILIBRIUM_PHASES, EXCHANGE_MASTER_SPECIES, EXCHANGE_SPECIES, GAS_PHASE, INVERSE_MODELING, ISOTOPE_RATIOS, KINETICS, KNOBS, PHASES, PRINT, REACTION,

SOLUTION_MASTER_SPECIES, SOLID_SOLUTIONS, SOLUTION_SPECIES, SURFACE_MASTER_SPECIES, SURFACE_SPECIES, TRANSPORT, and USER_PUNCH.

SIT

This keyword data block is used to specify parameters for the SIT (Specific ion Interaction Theory) aqueous model. The **SIT** data block is used in the database file *sit.dat*, which is the primary context for its use.

Example data block

```
Line 0: SIT
Line 1: -epsilon
Line 2:     Mg+2    Cl-    0.19
Line 2a:    Mn+2    Cl-    0.13
Line 2b:    Na+     Cl-    0.03
```

Explanation

Line 0: **SIT**

SIT is the keyword for the data block. No other data are input on the keyword line.

Line 1: **-epsilon**

-epsilon—Identifier begins a block of data that define $\varepsilon(i, k)$ ion-ion interaction parameters for the SIT aqueous model (see Grenthe and others, 1997).

Line 2: *cation anion* $A_0, A_1, A_2, A_3, A_4, A_5$

cation anion—A cation-anion pair of aqueous species, defined in either order.

$A_0, A_1, A_2, A_3, A_4, A_5$—Coefficients for the temperature dependence of the **-epsilon** parameter. The expression for a SIT parameter is the same as for a Pitzer parameter:

$$P = A_0 + A_1\left(\frac{1}{T} - \frac{1}{T_r}\right) + A_2 \ln\left(\frac{T}{T_r}\right) + A_3(T - T_r) + A_4(T^2 - T_r^2) + A_5\left(\frac{1}{T^2} - \frac{1}{T_r^2}\right), \text{ where } P \text{ is the}$$

parameter, T is the temperature in kelvin, T_r is the reference temperature (298.15 K), and ln is the natural log. If fewer than six coefficients are entered, the undefined coefficients are assumed to be zero.

Notes

The implementation of the SIT aqueous model has been taken from Grenthe and others (1997). The *sit.dat* database (Dr. Lara Duro, Amphos 21, written commun., 2012) has been developed by Amphos 21, BRGM (French Bureau of Research for Geology and Mining), and HydrAsa for ANDRA (French Agency for the Management of Nuclear Waste). More details on the source of data for *sit.dat* can be found in the comments at the beginning of the file.

If a **SIT** data block is read in the database file or the input file, then the SIT aqueous model is used for the simulations. Only one aqueous model can be used in a PHREEQC run; it is an error to read both a PITZER data block and a **SIT** data block.

Example problems

The **SIT** keyword is used in the *sit.dat* database.

Related keywords

PITZER.

SOLID_SOLUTIONS

This keyword data block is used to define a solid-solution assemblage. Each solid solution may be nonideal with two components or ideal with any number of components. The initial amount of each component in each solid solution is defined in this keyword data block. Any calculation involving solid solutions assumes that all solid solutions dissolve entirely and reprecipitate in equilibrium with the solution. The formulation is sufficiently general that synthetic organic liquid solutions also can be simulated.

Example data block

```
Line 0:    SOLID_SOLUTIONS 1 Two solid solutions
Line 1a:   CaSrBaSO4                        # ideal
Line 2a:       -comp        Anhydrite       1.500
Line 2b:       -comp        Celestite       0.05
Line 2c:       -comp        Barite          0.05
Line 1b:   Ca(x)Mg(1-x)CO3                   # Binary, nonideal
Line 3:        -comp1       Calcite         0.097
Line 4:        -comp2       Ca.5Mg.5CO3     0.003
Line 5:        -temp        25.0
Line 6:        -tempk       298.15
Line 7:        -Gugg_nondim 5.08            1.90
```

Optional definitions of excess free-energy parameters for nonideal solid solutions:

```
Line 8:        -Gugg_kj                   12.593  4.70
Line 9:        -activity_coefficients     24.05   1075.  0.0001 0.9999
Line 10:       -distribution_coefficients 0.0483  1248.  0.0001 0.9999
Line 11:       -miscibility_gap           0.0428  0.9991
Line 12:       -spinodal_gap              0.2746  0.9483
Line 13:       -critical_point            0.6761  925.51
Line 14:       -alyotropic_point          0.5768  -8.363
Line 15:       -Thompson                  17.303  7.883
Line 16:       -Margules                  -0.62   7.6
```

Explanation

Line 0: **SOLID_SOLUTIONS** [*number*] [*description*]

SOLID_SOLUTIONS is the keyword for the data block. Optionally, **SOLID_SOLUTION**.

number—A positive number designates the following solid-solution assemblage and its composition. A range of numbers may also be given in the form *m-n*, where *m* and *n* are positive integers, *m* is less than *n*, and the two numbers are separated by a hyphen without intervening spaces. Default is 1.

description—Optional comment that describes the solid-solution assemblage.

Line 1: *solid-solution name*

solid-solution name—User-defined name of a solid solution.

Line 2: **-comp** *phase name, moles*

-comp—Identifier indicates a component of an ideal solid solution is defined. Component is part of the solid solution defined by the preceding Line 1. Optionally, **comp**, **component**, or **-c[omponent]**.

phase name—Name of the pure phase that is a component in the solid solution. A phase with this name must have been defined in a PHASES data block.

moles—Moles of the component in the solid solution.

Line 3: **-comp1** *phase name, moles*

-comp1—Identifier indicates the first component of a nonideal, binary solid solution is defined. The component is part of the solid solution defined by the preceding Line 1. Optionally, **comp1** or **-comp1**.

phase name—Name of the pure phase that is component 1 of the nonideal solid solution. A phase with this name must have been defined in a PHASES data block.

moles—Moles of the component in the solid solution.

Line 4: **-comp2** *phase name, moles*

-comp2—Identifier indicates the second component of a nonideal, binary solid solution is defined. The component is part of the solid solution defined by the preceding Line 1. Optionally, **comp2** or **-comp2**.

phase name—Name of the pure phase that is component 2 of the nonideal solid solution. A phase with this name must have been defined in a PHASES data block.

moles—Moles of the component in the solid solution.

Line 5: **-temp** *temperature in Celsius*

-temp—Temperature at which excess free-energy parameters are defined, in Celsius. Temperature, either **temp** or **tempk**, is used if excess free-energy parameters are input with any of the following identifiers: **-gugg_nondim**, **-activity_coefficients**,

-distribution_coefficients, -miscibility_gap, -spinodal_gap, -alyotropic_point, or -margules. Optionally, **temp**, **tempc**, or **-t[empc]**. Default is 25 °C.

Line 6: **-tempk** *temperature in kelvin*

-tempk—Temperature at which excess free-energy parameters are defined, in kelvin. Temperature, either **temp** or **tempk**, is used if excess free-energy parameters are input with any of the following options: **-gugg_nondim**, **-activity_coefficients**, **-distribution_coefficients**, **-miscibility_gap**, **-spinodal_gap**, **-alyotropic_point**, or **-margules**. Optionally, **tempk** or **-tempk**. Default is 298.15 K.

Line 7: **-Gugg_nondim** a_0, a_1

-Gugg_nondim—Nondimensional Guggenheim parameters are used to calculate dimensional Guggenheim parameters. Optionally, **gugg_nondimensional**, **parms**, **-g[ugg_nondimensional]**, or **-p[arms]**.

a_0—Guggenheim a_0 parameter, dimensionless. Default is 0.0.

a_1—Guggenheim a_1 parameter, dimensionless. Default is 0.0.

Line 8: **-Gugg_kJ** g_0, g_1

-Gugg_kJ—Guggenheim parameters with dimensions of kJ/mol define the excess free energy of the nonideal, binary solid solution. Optionally, **gugg_kJ** or **-gugg_k[J]**.

g_0—Guggenheim g_0 parameter, kJ/mol. Default is 0.0.

g_1—Guggenheim g_1 parameter, kJ/mol. Default is 0.0.

Line 9: **-activity_coefficients** a_{comp_1}, a_{comp_2}, x_1, x_2

-activity_coefficients—Activity coefficients for components 1 and 2 are used to calculate dimensional Guggenheim parameters. Optionally, **activity_coefficients** or **-a[ctivity_coefficients]**.

a_{comp_1}—Activity coefficient for component 1 in the solid solution. No default.

a_{comp_2}—Activity coefficient for component 2 in the solid solution. No default.

x_1—Mole fraction of component 2 for which a_{comp_1} applies. No default.

x_2—Mole fraction of component 2 for which a_{comp_2} applies. No default.

Line 10: **-distribution_coefficients** k_1, k_2, x_1, x_2

> **-distribution_coefficients**—Two distribution coefficients are used to calculate dimensional Guggenheim parameters. Optionally, **distribution_coefficients** or **-d[istribution_coefficients]**.
>
> k_1—Distribution coefficient of component 2 at mole fraction x_1 of component 2, expressed as $\dfrac{(\chi_2/\chi_1)}{(a_2/a_1)}$, where χ is the mole fraction in the solid and a is the aqueous activity. No default.
>
> k_2—Distribution coefficient of component 2 at mole fraction x_2 of component 2, expressed as above. No default.
>
> x_1—Mole fraction of component 2 for which k_1 applies. No default.
>
> x_2—Mole fraction of component 2 for which k_2 applies. No default.

Line 11: **-miscibility_gap** x_1, x_2

> **-miscibility_gap**—The mole fractions of component 2 that delimit the miscibility gap are used to calculate dimensional Guggenheim parameters. Optionally, **miscibility_gap** or **-m[iscibility_gap]**.
>
> x_1—Mole fraction of component 2 at one end of the miscibility gap. No default.
>
> x_2—Mole fraction of component 2 at the other end of the miscibility gap. No default.

Line 12: **-spinodal_gap** x_1, x_2

> **-spinodal_gap**—The mole fractions of component 2 that delimit the spinodal gap are used to calculate dimensional Guggenheim parameters. Optionally, **spinodal_gap** or **-s[pinodal_gap]**.
>
> x_1—Mole fraction of component 2 at one end of the spinodal gap. No default.
>
> x_2—Mole fraction of component 2 at the other end of the spinodal gap. No default.

Line 13: **-critical_point** x_{cp}, t_{cp}

> **-critical_point**—The mole fraction of component 2 at the critical point and the critical temperature (kelvin) are used to calculate dimensional Guggenheim parameters. Optionally, **critical_point** or **-cr[itical_point]**.
>
> x_{cp}—Mole fraction of component 2 at the critical point. No default.
>
> t_{cp}—Critical temperature, in kelvin. No default.

Line 14: **-alyotropic_point** x_{aly}, $\log_{10}(\Sigma\Pi)$

    -alyotropic_point—The mole fraction of component 2 at the alyotropic point and the total solubility product at that point are used to calculate dimensional Guggenheim parameters. Optionally, **alyotropic_point** or **-al[yotropic_point]**.

    x_{aly}—Mole fraction of component 2 at the alyotropic point. No default.

    $\log_{10}(\Sigma\Pi)$—Total solubility product at the alyotropic point, where $\Sigma\Pi = (a_1 + a_2)a_{\text{common ion}}$. No default.

Line 15: **-Thompson** wg_2, wg_1

    -Thompson—Thompson and Waldbaum parameters wg_2 and wg_1 are used to calculate dimensional Guggenheim parameters. Optionally, **thompson** or **-th[ompson]**.

    wg_2—Thompson and Waldbaum parameter wg_2, kJ/mol. No default.

    wg_1—Thompson and Waldbaum parameter wg_1, kJ/mol. No default.

Line 16: **-Margules** $alpha_2$, $alpha_3$

    -Margules—Margules parameters $alpha_2$ and $alpha_3$ are used to calculate dimensional Guggenheim parameters. Optionally, **Margules** or **-Ma[rgules]**.

    $alpha_2$—Margules parameter $alpha_2$, dimensionless. No default.

    $alpha_3$—Margules parameter $alpha_3$, dimensionless. No default.

Notes

Multiple solid solutions may be defined by multiple sets of Lines 1, 2, 3, and 4. Line 2 may be repeated as necessary to define all the components of an ideal solid solution. Nonideal solid solution components must be defined with Lines 3 and 4. Calculations with solid solutions assume that the entire solid recrystallizes to be in equilibrium with the aqueous phase. This assumption is usually unrealistic because it is likely that only the outer layer of a solid would re-equilibrate with the solution, even given long periods of time. In most cases, the use of ideal solid solutions is also unrealistic because nonideal effects are nearly always present in solids. Liquid solutions of synthetic organic liquids usually behave as ideal mixtures and can be modeled well with this keyword (Appelo and Postma, 2005, Chapter 10, Example 10.5).

Lines 7–16 provide alternative ways of defining the excess free energy of a nonideal, binary solid solution. Only one of these lines should be included in the definition of a single solid solution. The parameters in the Example data block are taken from Glynn (1991) and Glynn (1990) for "nondefective" calcite (log K -8.48) and dolomite (expressed as $Ca_{0.5}Mg_{0.5}CO_3$, log K -8.545; note that a phase for dolomite with the given name, composition, and log K would have to be defined in a PHASES data block because it differs from the standard stoichiometry for dolomite in the databases). In the Example data block, Lines 7 through 16, except Line 14 (alyotropic point), define the same dimensional Guggenheim parameters. Internally, the program converts any one of these forms of input into dimensional Guggenheim parameters. When a batch-reaction or transport calculation is performed, the temperature of the calculation (as defined by mixing of solutions, REACTION_TEMPERATURE data block, or heat transport in TRANSPORT simulations) is used to convert the dimensional Guggenheim parameters to nondimensional Guggenheim parameters, which are then used in the calculation.

The identifiers **-gugg_nondim**, **-activity_coefficients**, **-distribution_coefficients**, **-miscibility_gap**, **-spinodal_gap**, **-alyotropic_point**, or **-margules** define parameters for a particular temperature which are converted to dimensional Guggenheim parameters by using the default temperature of 25 °C or the temperature specified in Line 5 or 6. If more than one Line 5 and (or) 6 is defined, the last definition will take precedence. If **-alyotropic_point** or **-distribution_coefficients** identifiers are used to define excess free-energy parameters, the dimensional Guggenheim parameters are dependent on (1) the values included with these two identifiers, and (2) the equilibrium constants for the pure-phase components. The latter are defined by a PHASES data block in the input file or database file.

The parameters for excess free energy are dependent on which component is labeled "1" and which component is labeled "2". It is recommended that the component with the smaller value of log K be selected as component 1 and the component with the larger value of log K be selected as component 2. The excess free-energy parameters must be consistent with this numbering. A positive value of a_1 (nondimensional Guggenheim parameter) or g_1 (dimensional Guggenheim parameter) will result in skewing the excess free-energy function toward component 2 and, if a miscibility gap is present, it will not be symmetric about a mole fraction of 0.5, but instead will be shifted toward component 2. In the calcite-dolomite example, the positive value of a_1 (1.90) results in a miscibility gap extending almost to pure dolomite (mole fractions of miscibility gap are 0.0428 to 0.9991).

After a batch reaction with a solid-solution assemblage has been simulated, it is possible to save the resulting solid-solution compositions with the SAVE keyword. If the new compositions are not saved, the solid-solution compositions will remain the same as they were before the batch reaction. Use of RUN_CELLS for a batch reaction automatically saves the new compositions of all reactants. After it has been defined or saved, the solid-solution assemblage may be used in subsequent simulations through the

USE or RUN_CELLS keywords. Solid-solution compositions are automatically saved following each shift in advection and transport calculations.

Example problems

The keyword **SOLID_SOLUTIONS** is used in example problem 10 and 20.

Related keywords

PHASES, RUN_CELLS, SAVE **solid_solution**, and USE **solid_solution**.

SOLUTION

This keyword data block is used to define the temperature and chemical composition of an initial solution. Individual element concentrations can be adjusted to achieve charge balance or equilibrium with a pure phase. All input concentrations are converted internally to units of moles of elements and element valence states, including hydrogen and oxygen. From this information, mass of water and molality can be calculated. Speciation calculations are performed on each solution defined by a **SOLUTION** data block and each solution is then available for subsequent batch-reaction, transport, or inverse-modeling calculations. The density and specific conductance of the solution are listed in the output file when the appropriate parameters have been read from the database file.

Example data block

```
Line  0:   SOLUTION 25 Test solution number 25
Line  1:       temp          25.0
Line  2:       pressure      10
Line  3:       pH            7.0       charge
Line  4:       pe            4.5
Line  5:       redox         O(-2)/O(0)
Line  6:       units         ppm
Line  7:       density       1.02
Line 8a:       Ca            80.
Line 8b:       S(6)          96.       as SO4
Line 8c:       S(-2)         1.        as S
Line 8d:       N(5) N(3)     14.       as N
Line 8e:       O(0)          8.0
Line 8f:       C             61.0      as HCO3    CO2(g)      -3.5
Line 8g:       Fe            55.       ug/kgs as Fe  S(6)/S(-2)  Pyrite
Line 9a:       -isotope      13C       -12.       1.  # permil PDB
Line 9b:       -isotope      34S       15.        1.5 # permil CDT
Line 10:       -water        0.5       # kg
```

Explanation

Line 0: **SOLUTION** [*number*] [*description*]

> **SOLUTION** is the keyword for the data block.

> *number*—A positive number designates the solution composition. A range of numbers may also be given in the form *m-n*, where *m* and *n* are positive integers, *m* is less than *n*, and the two numbers are separated by a hyphen without intervening spaces. Default is 1.

description—Optional comment that describes the solution.

Line 1: **temp** *temperature*

> **temp**—Indicates temperature is entered on this line. Optionally, **temp**, **temperature**, or **-t[emperature]**.

> *temperature*—Temperature, °C. Default 25 °C.

Line 2: **pressure** *pressure*

> **pressure**—Indicates pressure is entered on this line. Optionally, **press**, **pressure**, or **-pr[essure]**.

> *pressure*—Pressure, atm. Default 1 atm.

Line 3: **pH** *pH* [(**charge** or *phase name* [*saturation index*])]

> **pH**—Indicates pH is entered on this line. Optionally, **-pH** (as with all identifiers, case insensitive).

> *pH*—pH value, negative log of the activity of hydrogen ion.

> **charge**—Indicates pH is to be adjusted to achieve charge balance. If **charge** is specified for pH, it may not be specified for any other element.

> *phase name*—pH will be adjusted to achieve specified saturation index with the specified phase.

> *saturation index*—pH will be adjusted to achieve this saturation index for the specified phase. Default is 0.0.

> If Line 2 is not entered, the default pH is 7.0. Specifying both **charge** and a phase name is not allowed. Be sure that specifying a phase is reasonable; it may not be possible to adjust the pH to achieve the specified saturation index.

Line 4: **pe** *pe* [(**charge** or *phase name* [*saturation index*])]

> **pe**—Indicates pe is entered on this line. Optionally, **-pe**.

> *pe*—pe value, conventional negative log of the activity of the electron.

> **charge**—(Not recommended) indicates pe is to be adjusted to achieve charge balance.

> *phase name*—pe will be adjusted to achieve specified saturation index with the specified phase.

> *saturation index*—pe will be adjusted to achieve this saturation index for the specified phase. Default is 0.0.

> If Line 4 is not entered, the default pe is 4.0. Specifying both **charge** and a phase name is not allowed. Adjusting pe for charge balance is not recommended. Care should also be used in adjusting pe to a fixed saturation index for a phase because frequently this is not possible.

Line 5: **redox** *redox couple*

> **redox**—Indicates the definition of a redox couple that is used to calculate a pe. This pe will be used for any redox element for which a pe is needed to determine the distribution of the element among its valence states. Optionally, **-r[edox]**.

> *redox couple*—Redox couple which defines pe. A redox couple is specified by two valence states of an element separated by a "/". No spaces are allowed.

> If Line 5 is not entered, the input pe value will be as specified by **pe** or the default of 4. The use of **-redox** does not change the input pe. The Example data block uses the dissolved oxygen concentration [defined by O(0) in Line 8e] and the redox half-reaction for formation of $O_{2(aq)}$ from water (defined in the SOLUTION_SPECIES data block of the default databases) to calculate a pe for calculation of the distribution of species of redox elements (C and Fe in this example).

Line 6: **units** *concentration units*

> **units**—Indicates default concentration units are entered on this line. Optionally, **-u[nits]**.

> *concentration units*—Default concentration units. Three groups of concentration units are allowed, concentration (1) per liter ("/L"), (2) per kilogram solution ("/kgs"), or (3) per kilogram water ("/kgw"). All concentration units for a solution must be within the same group. Within a group, either grams or moles may be used, and prefixes milli (m) and micro (u) are acceptable. The abbreviations for parts per thousand, "ppt"; parts per million, "ppm"; and parts per billion, "ppb", are acceptable in the "per kilogram solution" group. Default is mmol/kgw.

Line 7: **density** *density*

> **density**—Indicates density is entered on this line. Optionally, **dens** or **-d[ensity]**.

> *density*—Density of the solution, kg/L (kilogram per liter, which equals g/cm^3). Default is 1.0. The density is used only if the input concentration units are "per liter".

Line 8: *element list, concentration, [units], ([**as** formula] or [**gfw** gfw]), [redox couple], [(**charge** or phase name [saturation index])]*

> *element list*—An element name or a list of element valence states separated by white space. Line 8d demonstrates the use of a list of valence states and indicates that the sum of N(5) and N(3) valence states is 14 ppm as N. The element names and valence states must correspond to the items in the first column in SOLUTION_MASTER_SPECIES.

concentration—Concentration of element in solution or sum of concentrations of element valence states in solution.

units—Concentration unit for element (see Line 8g). If units are not specified, the default units (**units** value if Line 6 is present, or mmol/kgw if Line 6 is absent) are assumed.

as *formula*—Indicates a chemical formula, *formula*, will be given from which a gram formula weight will be calculated. A gram formula weight is needed only when the input concentration is in mass units. The calculated gram formula weight is used to convert mass units into mole units for this element and this solution; it is not stored for further use. If a gram formula weight is not specified, the default is the gram formula weight defined in **SOLUTION_MASTER_SPECIES**. For alkalinity, the formula should give the gram equivalent weight. For alkalinity reported as calcium carbonate, the formula for the gram equivalent weight is $Ca_{0.5}(CO3)_{0.5}$; this is the default in the *phreeqc.dat* and *wateq4f.dat* database files distributed with this program.

gfw *gfw*—Indicates a gram formula weight, *gfw*, will be entered. A gram formula weight (g/mol) is needed only when the input concentration is in mass units. The specified gram formula weight is used to convert mass units into mole units only for this element and this solution; it is not stored for further use. If a gram formula weight is not specified, the default is the gram formula weight defined in **SOLUTION_MASTER_SPECIES**. For alkalinity, the gram equivalent weight should be entered. For alkalinity reported as calcium carbonate, the gram equivalent weight is approximately 50.04 g/eq (gram per equivalent).

redox couple—Redox couple to use for the element or element valence states in *element list*. Definition of a redox couple is appropriate only when the element being defined is redox active and either (1) the total amount of the element is specified (no parentheses in the element name) or (2) two or more valence-states are specified (a valence state is defined in parentheses following element name); definition of a redox couple is not needed for non-redox-active elements or for individual valence states of an element. Initial solution calculations do not require redox equilibrium among all redox couples of all redox elements. Specifying a *redox couple* will force selective redox equilibrium; the redox element being defined will be in equilibrium with the specified *redox couple*. A redox couple is specified by two valence states of an element separated by a "/". No spaces are allowed. The specified redox couple overrides the default pe or default redox couple and is used to calculate a pe by which the element is distributed among valence states. If no redox couple is entered, the default redox couple defined by Line 5 will be used, or the pe if Line 5 is not entered.

charge—Indicates the concentration of this element will be adjusted to achieve charge balance. The element must have ionic species. If **charge** is specified for one element, it may not be specified for pH or any other element. (Note that it is possible to have a greater charge imbalance than can be adjusted by removing all of the specified element, in which case the problem is unsolvable.)

phase name—The concentration of the element will be adjusted to achieve a specified saturation index for the given pure phase. Be sure that specifying equilibrium with the phase is reasonable; the element should be a constituent in the phase. *Phase name* may not be used if **charge** has been specified for this element.

saturation index—The concentration of the element will be adjusted to achieve this saturation index for the given pure phase. Note that the entry for *concentration* will be used as an initial guess, but the final concentration for the element or valence state will differ from the initial guess. Default is 0.0.

Line 9: **-isotope** *name, value, [uncertainty limit]*

-isotope—Indicates isotopic composition for an element or element valence state is entered on this line. Isotope data are used only in inverse modeling (see table 4 for default isotopes). Optionally, **isotope** or **-i[sotope]**.

name—Name of the isotope. The name must begin with mass number followed by an element or element-valence-state name that is defined through SOLUTION_MASTER_SPECIES.

value—Isotopic composition of element or element valence state; units are a ratio, permil, or percent modern carbon, depending on the isotope (see table 4 for default units).

uncertainty limit—The uncertainty limit to be used in inverse modeling. This value is optional in the **SOLUTION** data block and alternatively a default uncertainty limit may be used (see INVERSE_MODELING, table 4) or an uncertainty limit may be defined with the **-isotopes** identifier of the INVERSE_MODELING data block.

Line 10: **-water** *mass*

-water—Indicates mass of water is entered on this line. Molalities of solutes are calculated from input concentrations and the moles of solutes are determined by the mass of water in solution. Optionally, **water** or **-w[ater]**.

SOLUTION

     mass—Mass of water in the solution (kg, kilogram). Default is 1 kg.

Notes

The SOLUTION_SPREAD data block is an alternative method for defining solution compositions, where data are entered in rows. Each row defines a solution composition. The capabilities for defining solutions are equivalent between **SOLUTION** and SOLUTION_SPREAD.

The order in which the lines of **SOLUTION** input are entered is not important. Specifying both "**as**" and "**gfw**" within a single line is not allowed. Specifying both "**charge**" and a phase name within a single line is not allowed. Specifying the concentration of a valence state or an element concentration twice is not allowed. For example, specifying concentrations for both total Fe and Fe(+2) is not allowed, because ferrous iron is implicitly defined twice.

Alkalinity or total carbon or both may be specified in solution input. If both alkalinity and total carbon are specified, the pH is adjusted to attain the specified alkalinity. If the units of alkalinity are reported as calcium carbonate, the correct formula to use is "**as** Ca0.5(CO3)0.5", because the gram equivalent weight is 50.04 g/eq, which corresponds to one half the formula $CaCO_3$. However, to avoid frequent errors, if "**as** CaCO3" is entered, the value of 50.04 g/eq will still be used as the equivalent weight.

All concentrations defined in the **SOLUTION** data block are converted into molality. The absolute number of moles is usually numerically equal to the molality because a kilogram of solvent water is assumed. It is possible to define a solution with a different mass of water by using the **-water** identifier. In that case, the moles of solutes are scaled to produce the molality as converted from the input data. A solution with 1 mol/kgw of NaCl and "**-water** 0.5" has 0.5 mol of Na and Cl and 0.5 kilograms of water. Batch-reaction calculations also may cause the mass of water in a solution to deviate from 1 kilogram.

Isotope values may be used in conjunction with the INVERSE_MODELING data block. Uncertainty limits for isotopes in mole-balance modeling may be defined in three ways: default uncertainty limits may be used, uncertainty limits may be defined in the **SOLUTION** data block, or uncertainty limits may be defined in the INVERSE_MODELING data block. Uncertainty limits defined in the INVERSE_MODELING data block take precedence over the **SOLUTION** data block, which in turn take precedence over the defaults given in table 4.

A **SOLUTION** data block causes an initial solution calculation to be performed. The composition of the solution is saved after the initial solution calculation, which includes the moles of solutes accounting for any adjustments related to charge balance or phase equilibria. After the initial solution calculation, the solution is available to be used in batch reactions within the same simulation. It is also available for use in subsequent simulations by using the USE or RUN_CELLS data block.

Example problems

The keyword **SOLUTION** is used in all example problems, 1 through 22.

Related keywords

INVERSE_MODELING, RUN_CELLS, SAVE **solution**, SOLUTION_MASTER_SPECIES, SOLUTION_SPECIES, SOLUTION_SPREAD, and USE **solution**.

SOLUTION_MASTER_SPECIES

This keyword is used to define the correspondence between element names and aqueous primary and secondary master species. The alkalinity contribution of the master species, the gram formula weight used to convert mass units, and the element gram formula weight also are defined in this data block. Normally, this data block is included in the database file and only additions and modifications are included in the input file.

Example data block

```
Line 0:    SOLUTION_MASTER_SPECIES
Line 1a:       H              H+         -1.0    1.008           1.008
Line 1b:       H(0)           H2          0.0    1.008
Line 1c:       S              SO4-2       0.0    SO4             32.06
Line 1d:       S(6)           SO4-2       0.0    SO4
Line 1e:       S(-2)          HS-         1.0    S
Line 1f:       Alkalinity     CO3-2       1.0    Ca0.5(CO3)0.5   50.04
Line 1g:       [18O]          H2[18O]     0      [18O]           18
```

Explanation

Line 0: **SOLUTION_MASTER_SPECIES**

Keyword for the data block. No other data are input on the keyword line.

Line 1: *element name, master species, alkalinity, (gram formula weight or formula), gram formula weight of element*

element name—An element name or an element name followed by a valence state in parentheses. Two forms for element names are available: (1) element names that begin with a capital letter followed by zero or more lower case letters and underscores ("_"), and no numbers; and (2) element names that are enclosed in square brackets (see Line 1g) and use any combination of alphanumeric characters and the characters plus (+), minus (-), equal (=), colon (:), decimal point (.), and underscore (_). In general, the element names using form 1 are simply the chemical symbols for elements, which have a capital letter and zero or one lower case letter. Element names using form 2 also are case dependent, but upper and lower case characters can be used in any position.

master species—Formula for the master species, including its charge. If the *element name* does not contain a valence state in parentheses, the corresponding master species is a primary master species. If the element name does contain a valence state in parentheses, the master species is a secondary master species. All *master species* must be defined in the SOLUTION_SPECIES data block.

alkalinity—Alkalinity contribution of the master species, eq. The alkalinity contribution of aqueous non-master species will be calculated from the alkalinities assigned to the master species.

gram formula weight—Default value used to convert input data in mass units to mole units for the element or element valence. For alkalinity, the gram equivalent weight is entered. Either *gram formula weight* or *formula* is required, but these items are mutually exclusive.

formula—Chemical formula used to calculate gram formula weight, which is used to convert input data from mass units to mole units for the element or element valence. For alkalinity, the formula for the gram equivalent weight is entered. Either *gram formula weight* or *formula* is required, but these items are mutually exclusive.

gram formula weight of element—This field is required for primary master species and must be the gram formula weight for the pure element, not for an aqueous species.

Notes

Line 1 must be repeated for each element and each element valence state to be used by the program. Each element must have a primary master species. If secondary master species are defined for an element, then the primary master species additionally must be defined as a secondary master species for one of the valence states. PHREEQC will reduce all chemical reaction equations to a form that contains only primary and secondary master species. Each primary master species must be defined by SOLUTION_SPECIES input to have an identity reaction with log K of 0.0. For example, the definition of the primary master species SO_4^{-2} in the SOLUTION_SPECIES data block of the database *phreeqc.dat* is SO4-2 = SO4-2, log K 0.0. Secondary master species that are not primary master species must be defined by SOLUTION_SPECIES input to have a reaction that contains electrons, and the log K in general will not be 0.0. For example, the definition of the secondary master species HS⁻ in the SOLUTION_SPECIES data block of the database *phreeqc.dat* is SO4-2 + 9 H+ + 8 e- = HS- + 4 H2O, log K 33.65. The treatment of alkalinity is a special case and "Alkalinity" is defined as an additional element. In most cases, the definitions in **SOLUTION_MASTER_SPECIES** for alkalinity and carbon in the default database files should be used without modification.

The *gram formula weight* and *formula* are defined for convenience in converting units from mass to moles. For example, if data for nitrate are consistently reported in mg/L (milligram per liter) of nitrate as NO_3^-, then *gram formula weight* should be set to 62.0 g/mol or *formula* should be set to "NO3" in the database file. Then it will not be necessary to use the **as** or **gfw** options in the SOLUTION or SOLUTION_SPREAD data block. If nitrate is reported as mg/L as N, then *gram formula weight* should be set to 14.0 g/mol or *formula* should be set to "N", as is the case in the default databases. These variables (*gram formula weight* and *formula*) are only used if the concentration units are in terms of mass; if the data are reported in moles, then the variables are not used. The value of *gram formula weight of element* is required for primary master species, and its value is used to calculate the gram formula weight when a *formula* is given either in a **SOLUTION_MASTER_SPECIES**, SOLUTION, or SOLUTION_SPREAD data block.

Example problems

The keyword **SOLUTION_MASTER_SPECIES** is used in example problems 1, 7, 9, 14, 15, and 21, and in all databases.

Related keywords

SOLUTION, SOLUTION_SPREAD, and SOLUTION_SPECIES.

SOLUTION_SPECIES

This keyword data block is used to define chemical reaction, log K, and activity-coefficient parameters for each aqueous species. In addition, parameters may be defined for each species that are used to calculate specific conductance, multicomponent diffusion, density, and enrichment in the diffuse layer of surfaces. Normally, this data block is included in the database file and only additions and modifications are included in the input file.

Example data block

```
Line  0:   SOLUTION_SPECIES
Line  1: CO3-2 + H+ = HCO3-
Line  2:        log_k          10.329
Line  3:        delta_h        -3.561          kcal
Line  4:        -analytic 107.8871 0.03252849 -5151.79 -38.92561 563713.9
Line  5:        -gamma         5.4000          0.0000
Line  6:        -dw            1.18e-9
Line  7:        -Vm     8.615   0  -12.21  0   1.667   0   0   264   0   1
Line  8:        -Millero 21.07 0.185 -0.002248 2.29 -0.006644 -3.667e-06
Line 1a: H2O = OH- + H+
Line 4a:        -a_e   -283.971 -0.05069842 13323.0  102.24447 -1119669.0
Line 5a:        -gamma         3.5000          0.0000
Line 1b: D2O = D2O
Line 2a:        log_k          0
Line  9:        -activity_water
Line 1c: OH-  + HDO = OD- + H2O
Line 2b:        log_k          -0.301029995663
Line 10:        -add_logk      Log_alpha_D_OH-/H2O(l)        1.0
Line 5b:        -gamma         3.5000          0.0000
Line 1d:  Cl- =  Cl-
Line 2c:        log_k          0
Line 11:        -llnl_gamma    3.0000
Line 1e:  2H2O =  O2 + 4H+ + 4e-
Line 2d:        log_k          -85.9951
Line 12:        -co2_llnl_gamma
Line 1f:  Cs+ = Cs+
Line 2d:        log_k          0
Line 13:        -erm_ddl       2.1
Line 1g:  HS-  = S2-2 + H+
Line 2e:        log_k          -14.528
Line 14:        -no_check
Line 15:        -mole_balance  S(-2)2
```

Explanation

Line 0: **SOLUTION_SPECIES**

Keyword for the data block. No other data are input on the keyword line.

Line 1: *Association reaction*

Association reaction for aqueous species. The defined species must be the first species to the right of the equal sign. The association reaction must precede any identifiers related to the aqueous species. The association reaction is an identity reaction for each primary master species.

Line 2: **log_k** *log K*

log_k—Identifier for log K at 25 °C. Optionally, **-log_k, logk, -l[og_k]**, or **-l[ogk]**.

log K—Log K at 25 °C for the reaction. *Log K* must be 0.0 for primary master species. Default is 0.0.

Line 3: **delta_h** *enthalpy,* [*units*]

delta_h—Identifier for enthalpy of reaction at 25 °C. Optionally, **-delta_h, deltah, -d[elta_h]**, or **-d[eltah]**.

enthalpy—Enthalpy of reaction at 25 °C for the reaction. Default is 0.0 kJ/mol.

units—Default units are kilojoules per mole. Units may be calories, kilocalories, joules, or kilojoules per mole. Only the energy unit is needed (per mole is assumed) and abbreviations of these units are acceptable. Default units are kJ/mol. Explicit definition of units for all enthalpy values is recommended. The enthalpy of reaction is used in the Van't Hoff equation to determine the temperature dependence of the equilibrium constant. Internally, all enthalpy calculations are performed with the units kJ/mol.

Line 4: **-analytic** $A_1, A_2, A_3, A_4, A_5, A_6$

-analytic—Identifier for coefficients for an analytical expression for the temperature dependence of log K. Optionally, **analytical_expression, a_e, ae, -a[nalytical_expression], -a[_e], -a[e]**.

$A_1, A_2, A_3, A_4, A_5, A_6$—Six values defining log K as a function of temperature in the expression

$$\log_{10}K = A_1 + A_2 T + \frac{A_3}{T} + A_4 \log_{10} T + \frac{A_5}{T^2} + A_6 T^2, \text{ where } T \text{ is in kelvin.}$$

Line 5: **-gamma** *Debye-Hückel a, Debye-Hückel b*

-gamma—Indicates activity-coefficient parameters are to be entered. If **-gamma** is entered, then

the equation from WATEQ (Truesdell and Jones, 1974) is used, $\log\gamma = \dfrac{-Az^2\sqrt{\mu}}{1 + B\overset{o}{a}\sqrt{\mu}} + b\mu$. In

this equations, γ is the activity coefficient, μ is ionic strength, and A and B are constants at

a given temperature. If **-gamma** is not input for a species, then for a charged species the

Davies equation is used to calculate the activity coefficient:

$\log\gamma = -Az^2\left(\dfrac{\sqrt{\mu}}{1+\sqrt{\mu}} - 0.3\mu\right)$; for an uncharged species the following equation is used:

$\log\gamma = 0.1\mu$. Optionally, **-g[amma]**.

Debye-Hückel a—Ion-size parameter $\overset{o}{a}$ in the WATEQ activity-coefficient equation.

Debye-Hückel b—Parameter b in the WATEQ activity-coefficient equation.

Line 6: **-dw** *diffusion coefficient*

-dw—Identifier for tracer diffusion coefficient. Tracer diffusion coefficients are used in the

multicomponent diffusion calculation in **TRANSPORT** and in calculating the specific

conductance of a solution (Basic function SC). Default is 0 m^2/s (square meter per second)

if **-dw** is not included. Optionally, **dw** or **-dw**.

diffusion coefficient—Tracer diffusion coefficient for the species at 25 °C, m^2/s.

Line 7: **-Vm** *a1, a2, a3, a4, W, $\overset{o}{a}$, i1, i2, i3, i4*

-Vm—Identifier for parameters used to calculate the specific volume (cm^3/mol) of aqueous

species with a Redlich-type equation (see Redlich and Meyer, 1964). As explained in the

following Notes section, the volume of species i is calculated, by convention relative to the

reference volume of H$^+$ of 0, as $\phi_i = \phi_{i,inf} + 0.5z_i^2 A_v\dfrac{\sqrt{\mu}}{1 + \overset{o}{a} B\sqrt{\mu}} + \beta_i\mu^{i4}$, where the first term

of the right-hand side, $\phi_{i,inf}$, is the specific volume at infinite dilution, and the second and

third terms are functions of the ionic strength μ and the ion-size parameter in the extended

Debye-Hückel equation, $\overset{o}{a}$, and *i1, i2, i3* and *i4*.

The specific volume at infinite dilution is parameterized with SUPCRT92 formulas

(Johnson and others, 1992):

$$\phi_{i,\,inf} = 41.84\left(a1 \times 0.1 + \frac{a2 \times 100}{(2600 + P_b)} + \frac{a3}{(T_K - 228)} + \frac{a4 \times 1e4}{(2600 + P_b)(T_K - 228)} - W \times Q_{Born}\right),$$

where 41.84 transforms cal mol^{-1} bar^{-1} (calorie per mole per bar) into cm^3/mol, P_b is

pressure in bar, T_K is temperature in kelvin, $W \times Q_{Born}$ is the Born volume, calculated from

W and the pressure dependence of the dielectric constant of water.

The second term contains A_v, the Debye-Hückel limiting slope, which is calculated as a

function of temperature and pressure, and the extended Debye-Hückel equation (see the

Notes).

The coefficient β_i is calculated as $\beta_i = i1 + \dfrac{i2}{(T_K - 228)} + i3(T_K - 228)$.

a1, a2, a3, a4, W, å, i1, i2, i3, i4—Numerical values for parameters *a1* to *a4* (cal mol^{-1} bar^{-1},

cal/mol (calorie per mole), cal K mol^{-1} bar^{-1}[calorie kelvin per mole per bar), cal K mol^{-1}

[calorie kelvin per mol], respectively), the Born coefficient *W* (cal/mol), the Debye-Hückel

ion-size parameter $\overset{o}{a}$ (10^{-10} m), and *i1* (cm^3/mol), *i2* (cm^{3}K mol^{-1}), *i2* (cm^{3}K^{-1} mol^{-1}) and

i4 (-), used in the equation for calculating the conventional specific volume of a solute

species.

Line 8: **-Millero** *a, b, c, d, e, f*

-Millero—Alternative formulation for calculating the specific volume for the aqueous species

(Millero, 2000) by convention relative to the volume of H$^+$ of 0 at ionic strength of 0. The

specific volume for species *i* is calculated according to the formula

$\phi_i = \phi_{i,\,inf} + 0.5z^2 A_v I^{0.5} + \beta_i I$, where $\phi_{i,\,inf}$ is the specific volume at infinite dilution; A_v is

the Debye-Hückel limiting slope, and *I* is the ionic strength. The volume at infinite dilution

is parameterized as $\phi_{i,\,inf} = a + bT + cT^2$ and the coefficient β_i is parameterized as

$\beta_i = d + eT + fT^2$, where *T* is °C. If both **-Vm** and **-Millero** are defined for a species, the

numbers from **-Vm** are used. *Warning*: the applicability of the Millero formulas is limited

to *T* < 50 °C, and the calculated densities may be incorrect at ionic strengths > 1.0 except

for NaCl solutions. Optionally, **Millero** or **-Mi[llero]**.

a, b, c, d, e, f—Numerical values for parameters *a* to *f* in the specific volume equation.

Line 9: **-activity_water**

    -activity_water—Identifier indicates that the species is an isotopic form of water. The activity coefficient for the species is such that its activity is equal to mole fraction in solution. Optionally, **activity_water** or **-ac[tivity_water]**.

Line 10: **-add_logk** *named log K, coefficient*

    -add_logk—Identifier defining an additional term for the equilibrium constant of the species. The identifier is used primarily in defining the equilibrium constant for isotopic species that require the addition of an isotopic fractionation factor. Optionally, **add_logk**, **add_log_k**, **-ad[d_logk]** or **-ad[d_log_k]**.

    named log K—Name of an expression defined in a NAMED_EXPRESSIONS data block.

    coefficient—Coefficient for the expression *named log K*; the value of the expression is multiplied by *coefficient* and added to the log K for the species.

Line 11: **-llnl_gamma** *diameter*

    -llnl_gamma—Identifier for the hard-core diameter in the expression for the activity coefficient in the Lawrence Livermore aqueous model; this identifier can be used only with the Lawrence Livermore National Laboratory aqueous model (*llnl.dat*). Optionally, **llnl_gamma** or **-ll[nl_gamma]**.

    diameter—Hard-core diameter for the species.

Line 12: **-co2_llnl_gamma**

    -co2_llnl_gamma—The activity coefficient for carbon dioxide is used as the activity coefficient for this uncharged species; this identifier can be used only with the Lawrence Livermore National Laboratory aqueous model (*llnl.dat*). Optionally, **co2_llnl_gamma** or **-co[2_llnl_gamma]**.

Line 13: **-erm_ddl** *factor*

    -erm_ddl—Identifier for the enrichment factor for a species in the diffuse double layer of surfaces calculated with the **-Donnan** identifier in the SURFACE data block. Optionally, **erm_ddl** or **-e[rm_ddl]**.

    factor—Enrichment factor. Default is 1.0 (unitless).

Line 14: **-no_check**

    -no_check—Indicates the reaction equation should not be checked for charge and elemental balance. Generally, equations should be checked for charge and elemental balance. The only exceptions might be polysulfide species that assume equilibrium with a solid phase; this assumption has the effect of removing solid sulfur from the mass-action equation. By default, all equations are checked. However, the identifier **-mole_balance** is needed to ensure that the proper number of atoms of each element are included in mole-balance equations (see **-mole_balance**). Optionally, **no_check** or **-n[o_check]**.

Line 15: **-mole_balance** *formula*

-mole_balance—Indicates the stoichiometry of the species will be defined explicitly. Optionally, **mole_balance, mass_balance, mb, -m[ole_balance], -mass_balance, -m[b]**.

formula—Chemical formula defining the stoichiometry of the species. Normally, both the stoichiometry and mass-action expression for the species are determined from the chemical equation that defines the species. Rarely, it may be necessary to define the stoichiometry of the species separately from the mass-action equation. The polysulfide species provide an example. These species are usually assumed to be in equilibrium with native sulfur. The activity of a pure solid is 1.0, and thus the term for native sulfur does not appear in the mass-action expression (Line 1g). The S_2^- species contains two atoms of sulfur, but the chemical equation indicates it is formed from species containing a total of one sulfur atom. The **-mole_balance** identifier is needed to give the correct stoichiometry. Note that unlike all other chemical formulas used in PHREEQC, the valence state of the element can and should be included in the formula of Line 15. The example indicates that the polysulfide species will be summed into the S(-2) mole-balance equation.

Notes

Line 1 must be entered first in the definition of a species. Additional sets of lines (Lines 1–7 as needed) may be added to define all of the aqueous species. A log K should be defined for each species with either **log_k** (Line 2) or **-analytical_expression** (Line 4); the default of 0.0 is not meaningful for most association reactions. In this Example data block, the following types of aqueous species are defined: (a) a primary master species, SO_4^{-2}, for which the reaction is an identity reaction and log K is 0.0; (b) a secondary master species, HS^-, for which the reaction contains electrons; (c) an aqueous species that is not a master species, OH^-; and (d) an aqueous species for which the chemical equation does not balance, S_2^{-2}. If an activity coefficient of 1 is needed for a species, use **-gamma** 1e5 0 in Line 5.

The tracer diffusion coefficient is for a trace concentration of the solute species in pure water. Usually, it is determined by measuring the specific conductance of solutions at various concentrations and extrapolating to zero concentration (Robinson and Stokes, 2002). The molar conductivity of a solute species and its diffusion coefficient are related by $\Lambda^0_{m,i} = \frac{z_i^2 F^2}{RT} D_{w,i}$, where $\Lambda^0_{m,i}$ is the molar conductivity (S/m / (mol/m^3) equals S m^2 mol^{-1}) (siemens square meter per mole), z_i is the charge number (unitless) of species i, F is Faraday's constant (C/mol, coulomb per mole), R is the gas constant (J K^{-1}mol^{-1}, joule per kelvin per mole), T is the absolute temperature (K), and $D_{w,i}$ is the diffusion coefficient (m^2/s). PHREEQC calculates the specific conductance of a solution by summing the product of the specific conductivity and the molal concentration of all the species in solution, while correcting the molal concentration with an electrochemical activity coefficient that is derived from a combination of Kohlrausch's law and the Debye-Hückel equation as explained in *http://www.hydrochemistry.eu/exmpls/sc.html* (accessed June 25, 2012). The tracer diffusion coefficient is corrected for the temperature T (K) of the solution by $D'_{w,i} = (D_{w,i})_{298} \times \frac{T}{298} \times \frac{\eta_{298}}{\eta_T}$, where η is the viscosity of water (Atkins and de Paula, 2002).

If **-Vm** is defined, the specific volume of species i is calculated, by convention relative to the reference volume of H$^+$ of 0, as

$$\phi_i = \phi_{i,inf} + 0.5 z_i^2 A_v \frac{\sqrt{\mu}}{1 + \overset{o}{a} B \sqrt{\mu}} + \beta_i \mu^{i4},$$

where the first term of the right-hand side, $\phi_{i,inf}$, is the volume at infinite dilution; and the second and third terms are functions of the ionic strength μ.

In the second term, z is the charge number of the species, A_v is the Debye-Hückel limiting slope,

$$A_v = B \frac{q_e^2}{\varepsilon_r k_B T} RT \left(\frac{\partial}{\partial P} \ln(\varepsilon_r) - \frac{\kappa_0}{3} \right) \text{(cm}^3\text{/mol) (mol/kg)}^{-0.5},$$

with the Debye length factor, $B = \left(\frac{8\pi N_A q_e^2 \rho_0}{\varepsilon_r k_B T} \right)^{0.5}$ (1/cm)(kg/mol)$^{0.5}$, Avogadro's number $N_A = 6.022 \times 10^{23}$ molecules per mole, the electron charge $q_e = 4.803 \times 10^{-10}$ esu (electrostatic unit of charge), the density of pure water ρ_0 (g/cm^3), the relative dielectric constant ε_r, the Boltzmann constant $k_B = 1.38 \times 10^{-16}$ erg/K (erg per kelvin), the temperature T (K), the pressure P (atm), and the compressibility of pure water κ_0 (atm^{-1}). PHREEQC calculates the relative dielectric constant as a function of temperature and pressure, as well as its pressure dependence, according to Bradley and Pitzer (1979), and the density of pure water along the saturation line with equation 2.6 of Wagner and Pruss (2002) and at higher pressures and temperatures with interpolation functions based on IAPWS (International Association for the Properties of Water and Steam) (*http://www.nist.gov/srd/upload/NISTIR5078-Tab3.pdf*) or with the IF97 (*http://www.iapws.org/release.htm*) polynomial for region 1 ($273 < T < 623$ °C, $P_{sat} < P < 100$ MPa, megapascal). The Bradley and Pitzer equations also are used to calculate $Q_{Born} = \left(\dfrac{1}{\varepsilon_r^2} \dfrac{\partial}{\partial P} \varepsilon_r \right)_T$, which is a part of $\phi_{i,\,inf}$. The specific volumes are used to derive the volume changes of reactions, and hence, the pressure dependency of reaction constants for species, and the pressure dependent solubilities of minerals and gases. The volumes also are used for calculating the density of solutions in PHREEQC as implemented by Vincent Post (Free University, Amsterdam, Netherlands, written commun., 2009) based on the work of Millero (2000). The parameters, entered with the identifier **-Vm** in the *phreeqc.dat* and *pitzer.dat* databases and commented with "# supcrt modified", were obtained by least squares fitting of the specific volumes of salts in aqueous solution, compiled by Laliberté (2009), supplemented with data at lower concentrations (omitted by Laliberté (2009)) and at higher temperatures. In the databases, the ion-size parameter for anions in the extended Debye-Hückel equation, $\mathring{a}$, is equal to 0, and for cations equal to the Debye-Hückel a parameter that is entered with -gamma a. The values defined with **-Millero** in some (now obsolete) databases are, in principle, for the temperature range from 0 to 50°C (Millero, 2000) and may be incorrect for high ionic strengths except for solutions containing predominantly alkali cations and chloride anions.

The Lawrence Livermore National Laboratory aqueous model (Daveler and Wolery, 1992) uses the following expression for the log (base 10) of an activity coefficient: $\log \gamma_i = \dfrac{A_\gamma z_i^2 \sqrt{I}}{1 + \mathring{a}_i B_\gamma \sqrt{I}} + \dot{B} I$, where A_γ, B_γ, and $\dot{B}$ are Debye-Hückel parameters that are functions of temperature as defined in the LLNL_AQUEOUS_MODEL_PARAMETERS data block; z_i is the charge number for species i, $\mathring{a}_i$ is the hard-core diameter, which is defined for an aqueous species in the **SOLUTION_SPECIES** data block with the **-llnl_gamma** identifier; and I is the ionic strength. The activity for an uncharged species in the Lawrence Livermore National Laboratory aqueous model can be set to a function of temperature by using the **-co2_llnl_gamma** identifier. The function of temperature is defined by the **-co2_coefs** identifier in the LLNL_AQUEOUS_MODEL_PARAMETERS data block.

The enrichment factor entered with **-erm_ddl** multiplies the concentration that is calculated with the Boltzmann equation for the Donnan space on a charged surface. With this factor, the concentrations in the Donnan space are calculated as:

$$c_{Donnan, i} = c_i \times erm\_DDL_i \times \exp\left(\frac{-z_i F \psi_D}{RT}\right),$$ (4)

where $c_{Donnan, i}$ is the concentration of species i in the Donnan pore space (mol/L), c_i is the concentration in the free (uncharged) solution, erm_DDL_i is an enrichment factor (unitless) that can be defined in keyword SOLUTION_SPECIES, z_i is charge number (unitless), F is the Faraday constant (96485 JV^{-1}eq^{-1}, joule per volt per equivalent), ψ_D is the potential of the Donnan volume (V, volt), R is the gas constant (8.314 JK^{-1}mol^{-1}), and T is the absolute temperature (K). The potential ψ_D is adapted to let the charge of the Donnan volume counterbalance the surface charge:

$$\sum_i z_i c_{Donnan, i} + \sigma_{surface} = 0,$$ (5)

where $\sigma_{surface}$ is the surface charge (eq/L, equivalent per liter). The enrichment factor is useful for modeling the relative enrichment or depletion of equally charged species in the electrostatic layer on a charged surface, which is related to enhanced complexation in a low dielectric permittivity medium (Appelo and others, 2010).

By default, equation checking for charge and elemental balance is in force for each equation that is processed. Checking can only be disabled by using **-no_check** for each equation that is to be excluded from the checking process.

Example problems

The keyword **SOLUTION_SPECIES** is used in example problems 1, 7, 9, 14, 15, and 21 and in all databases.

Related keywords

SOLUTION_MASTER_SPECIES.

SOLUTION_SPREAD

The **SOLUTION_SPREAD** data block is an alternative input format for SOLUTION; that is, compatible with many spreadsheet programs. Input for **SOLUTION_SPREAD** is transposed from the input for SOLUTION, that is, the rows of input for SOLUTION become the columns of input for **SOLUTION_SPREAD**. The data are entered one line per solution in columns that are tab-delimited ("\t" in the Example data block).

Example data block

```
Line 0:   SOLUTION_SPREAD    # "\t" indicates the tab character
Line 1:  -temp         25
Line 2:  -pressure     100
Line 3:  -pH           7.1
Line 4:  -pe           4
Line 5:  -redox        O(0)/O(-2)
Line 6:  -units        mmol/kgw
Line 7:  -density      1
Line 8:  -water        1.0
Line 9a: -isotope      34S       15.0    1.0
Line 9b: -isotope      13C       -12.0
Line 10:  -isotope_uncertainty   13C    1.0
Line 11: Number\t   13C\t   uncertainty\t  pH\t  Ca\t  Na\t    Alkalinity\t  Description
Line 12: \t          \t              \t    \t    \t    \t  mg/kgw as HCO3\t
Line 13a: 10-11\t -10.2\t          0.05\t 6.9\t 23\t   6\t            61\t   soln 10-11
Line 13b: 1\t      -12.1\t          0.1\t    \t 17\t   6\t            55\t   My well 1
Line 13c: 5\t      -14.1\t          0.2\t    \t 27\t   9\t            70\t   My well 5
```

Explanation

Line 0: **SOLUTION_SPREAD**

Keyword for the data block. No other data are input on the keyword line.

Line 1: **-temp** *temperature*

-temp—Identifier for temperature. The *temperature* will be used for all subsequent solutions in the data block if no column has the heading **temperature** (or **temp**) or if the entry for the **temperature** column is empty for a solution. Optionally, **temp**, **-t**[**emp**], **temperature**, or **-t**[**emperature**].

temperature—Temperature, °C. Default is 25 °C.

Line 2: **pressure** *pressure*

> **pressure**—Identifier for pressure. The *pressure* will be used for all subsequent solutions in the
> data block if no column has the heading **pressure** (or **press**) or if the entry for the **pressure**
> column is empty for a solution. Optionally, **press**, **pressure**, or -**pr**[**essure**].

> *pressure*—Pressure, atm. Default 1 atm.

Line 3: -**pH** *pH*

> -**pH**—Identifier for pH. The pH will be used for all subsequent solutions in the data block if no
> column has the heading **pH** or if the entry for the **pH** column is empty for a solution.
> Optionally, **pH** (as with all identifiers, case insensitive).

> *pH*—pH value, negative log of the activity of hydrogen ion. Default is 7.0.

Line 4: -**pe** *pe*

> -**pe**—Identifier for pe. The value *pe* will be used as the default for all subsequent solutions in the
> data block if no column has the heading **pe** or if the entry for the **pe** column is empty for a
> solution. Optionally, **pe**.

> *pe*—pe value, conventional negative log of the activity of the electron. Default is 4.0.

Line 5: -**redox** *redox couple*

> -**redox**—Identifier for the redox couple to be used to calculate pe. This pe will be used for any
> redox element for which a pe is needed to determine the distribution of the element among
> its valence states. The redox couple will be used for all subsequent solutions in the data
> block if no column has the heading **redox** or if the entry for the **redox** column is empty for
> a solution. If no redox couple is specified, the pe will be used. Optionally, **redox** or
> -**r**[**edox**].

> *redox couple*—Redox couple to use for pe calculations. A redox couple is specified by two
> valence states of an element separated by a "/". No spaces are allowed. Default is **pe**.

Line 6: -**units** *concentration units*

> -**units**—Identifier for concentration units. The concentration units will be used for all subsequent
> solutions in the data block if no column has the heading **units** (or **unit**) or if the entry for
> the **units** column is empty for a solution. Optionally, **unit**, **units**, or -**u**[**nits**].

> *concentration units*—Default concentration units. Three groups of concentration units are
> allowed, concentration (1) per liter ("/L"), (2) per kilogram solution ("/kgs"), or (3) per
> kilogram water ("/kgw"). All concentration units for a solution must be within the same

group. Within a group, either grams or moles may be used, and the prefixes milli (m) and micro (u) are acceptable. Parts per thousand, "ppt"; parts per million, "ppm"; and parts per billion, "ppb", are acceptable in the "per kilogram solution" group. Default is mmol/kgw.

Line 7: **-density** *density*

> **-density**—Identifier for solution density. The density will be used for all subsequent solutions in the data block if no column has the heading **density** (or **dens**) or if the entry for the **density** column is empty for a solution. Density is used only if concentration units are per liter. Optionally, **dens**, **density** or **-d[ensity]**.

> *density*—Density of solution, kg/L. Default is 1.0 kg/L.

Line 8: **-water** *mass*

> **-water**—Identifier for mass of water. The mass of water will be used for all subsequent solutions in the data block if no column has the heading **water** or if the entry for the **water** column is empty for a solution. Molalities of solutes are calculated from input concentrations and the moles of solutes are determined by the mass of water in solution. Optionally, **water** or **-w[ater]**.

> *mass*—Mass of water in the solution (kg). Default is 1.0 kg.

Line 9: **-isotope** *name, value, [uncertainty_limit]*

> **-isotope**—Indicates isotopic composition for an element, or element valence state is entered on this line. Isotope data are used only in inverse modeling (see table 4 for default isotopes). Optionally, **isotope** or **-i[sotope]**.

> *name*—Name of the isotope. The name must begin with a mass number followed by an element or element-valence-state name that is defined through SOLUTION_MASTER_SPECIES.

> *value*—Isotopic composition of an element or element valence state; units are a ratio, permil, or percent modern carbon, depending on the isotope (see table 4 for default units).

> *uncertainty limit*—The uncertainty limit to be used in inverse modeling. This value is optional in the **SOLUTION** data block and alternatively a default uncertainty limit may be used (see INVERSE_MODELING, table 4) or an uncertainty limit may be defined with the **-isotopes** identifier of the INVERSE_MODELING data block.

Line 10: **-isotope_uncertainty** *name, uncertainty_limit*

    -isotope_uncertainty—Identifier for uncertainty limit in the ratio for an isotope. The uncertainty limit for the isotope ratio will be used for all subsequent solutions in the data block if no column has the same *name* directly followed by a column headed **uncertainty** or if the entry for the **uncertainty** column is empty for a solution. Isotopes and isotope uncertainty limits are used only in inverse modeling. Optionally, **uncertainty**, **-unc[ertainty]**, **uncertainties**, **unc[ertainties]**, **isotope_uncertainty**, or **-isotope_[uncertainty]**.

    name—Name of the isotope, beginning with mass number.

    uncertainty_limit—Uncertainty limit for the isotope to be used in inverse modeling.

Line 11: *column headings*

    column headings—Column headings are element names, element valence-state names (element chemical symbol followed by valence state in parentheses), isotope names (element chemical symbol preceded by the mass number), one of the identifiers in Lines 1–7 (without the hyphen), **number**, **description**, or **uncertainty**. Most often the headings are equivalent to the first data item of Line 8 of the SOLUTION data block. A column heading "**number**" is used to specify solution numbers or range of solution numbers that are specified following the keyword in the SOLUTION data block. Similarly, a column heading "**description**" allows the entry of the descriptive information that is entered following the keyword and solution number in the SOLUTION data block. A column headed "**uncertainty**" may be entered adjacent to the right of any isotope column to define uncertainty limits for isotope data in inverse modeling. One and only one line of headings must be entered.

Line 12: [*subheadings*]

    subheadings—Subheadings are used to specify element-specific units, redox couples, and concentration-determining phases. Anything entered following the second data item of Line 8 of the SOLUTION data block can be entered on this line, including **as**, **gfw**, redox couple, or phase name and saturation index. Tabs, not spaces, must delimit the columns; data within a column must be space delimited. Subheadings are optional. At most one line of subheadings can be entered directly following the line of headings and it is identified as a line in which all fields begin with a character.

Line 13: *chemical data*

> *chemical data*—Analytical data, one line for each solution. For most columns, the data are equivalent to the second data item of Line 8 of the SOLUTION data block. Tabs, not spaces, must delimit the columns. Solution numbers or ranges of numbers are defined in a column with the heading **number**; default numbering begins sequentially from 1 or sequentially from the largest solution number that has been defined by any SOLUTION, **SOLUTION_SPREAD**, or SAVE data block in this or any previous simulation. Descriptive information can be entered in a column with the heading **description**. One Line 13 is needed for each solution.

Notes

SOLUTION_SPREAD is a complete equivalent to the SOLUTION data block that allows data entry in a tabular or spreadsheet format. In general, column headings are elements or element valence states and succeeding lines are the data values for each solution, with one solution defined on each line. Read the documentation for SOLUTION for detailed descriptions of input capabilities to convert mass units to mole units, to change default redox calculations, and to adjust concentrations to obtain equilibrium with a specified phase. This information is entered as a subheading in **SOLUTION_SPREAD**. The identifiers of SOLUTION are included in **SOLUTION_SPREAD**, but in **SOLUTION_SPREAD**, the values defined for the identifiers apply to all subsequently defined solutions. Identifiers can precede or follow data lines (Line 13), and will apply to any subsequently defined solutions until the end of the data block or until the identifier is redefined. In the Example data block, the pH of solutions 10-11 is defined to be 6.9 by an entry in the **pH** column; the pH for solutions 1 and 5 is the default defined by **-pH** identifier, 7.1. Empty entries in columns with headings that are not identifiers are interpreted as zero concentrations or missing values. If a column heading cannot be interpreted as part of the solution input, warnings are printed and the data for that column are ignored.

Example problems

The keyword **SOLUTION_SPREAD** is used in example problem 16.

Related keywords

SOLUTION.

SURFACE

This keyword data block is used to define the amount and composition of each surface in a surface assemblage. The composition of a surface assemblage can be defined in two ways: (1) implicitly, by specifying that the surface assemblage is in equilibrium with a solution of fixed composition, or (2) explicitly, by defining the amounts of the surfaces in their neutral form (for example, SurfbOH). A surface assemblage may have multiple surfaces and each surface may have multiple binding sites, which are identified by lowercase letters following an underscore. Three types of surfaces are available: DDL (diffuse-double layer) surfaces (Dzombak and Morel, 1990), CD-MUSIC (Charge Distribution MUltiSIte Complexation) surfaces (Hiemstra and Van Riemsdijk, 1996), and non electrostatic surfaces. For DDL and CD-MUSIC surfaces, the composition of the diffuse layer that balances the charged surface can be calculated explicitly (optional). For DDL, the diffuse-layer composition can be calculated by the method of Borkovec and Westall (1983) or by the Donnan approach. For CD-MUSIC, the diffuse-layer composition can be calculated only by the Donnan approach.

Example data block 1

```
Line 0:    SURFACE 1 Surface in equilibrium with solution 10
Line 1:        -equilibrate with solution 10
Line 2:        Surfa_w        1.0      1000.     0.33
Line 2a:       Surfa_s        0.01
Line 2b:       Surfb          0.5      1000.     0.33
Line 0a:   SURFACE 2 Explicit diffuse layer
Line 1a:       -equilibrate with solution 10
Line 3:        -sites_units   absolute
Line 2c:       Surfa_w        1.0      1000.   0.33
Line 2d:       Surfa_s        0.01
Line 2e:       Surfb          0.5      1000.   0.33
Line 4:        -diffuse_layer 2e-8
Line 0b:   SURFACE 3 CD_MUSIC surface with Donnan layer
Line 1b:       -equilibrate with solution 10
Line 3a:       -sites_units   density
Line 5:        -cd_music
Line 2f:       Goe_uni        3.45     96.8    16.52
Line 2g:       Goe_tri        2.7
Line 6:        -capacitances  0.98     0.73
Line 7:        -Donnan        1e-8
Line 8:        -only_counter_ions       true
```

```
Line 0c: SURFACE 4 Sites related to pure phase and kinetic reactant
Line 1c:    -equilibrate with solution 10
Line 9:     Surfc_wOH   Fe(OH)3(a)  equilibrium_phase 0.1      1e5
Line 9a:    Surfc_sOH   Fe(OH)3(a)  equilibrium_phase 0.001
Line 9b:    Surfd_sOH   Al(OH)3(a)  kinetic_reactant 0.001    2e4
Line 10:    -no_edl
Line 0d: SURFACE 5 Clay surface with diffusion through Donnan layer
Line 1d:    -equilibrate with solution 5
Line 2h:    Clay_planar    1.59     37.     1.407e4
Line 2i:    Clay_ii        0.01
Line 2j:    Clay_fes       0.85e-3
Line 7a:    -Donnan         9.6e-10 viscosity 1.0
Line 8a:    -only_counter_ions true
Line 0e: SURFACE 6 Clay surface with variable Donnan layer
Line 1e:    -equilibrate with solution 6
Line 2k:    Clay_planar    1.59     37.     1.407e4
Line 7b:    -Donnan         debye_lengths 3.4 limit_ddl 0.9 viscosity 1
Line 0f: SURFACE 7 Colloidal Ferrihydrite particles
Line 1f:    -equilibrate with solution 7
Line 2l:    Hfo_w          2.4e-3   600    1.06   Dw 1e-11
Line 2m:    Hfo_s          6e-5
Line 7c:    -Donnan         1e-12
```

Explanation 1

Line 0: **SURFACE** [*number*] [*description*]

SURFACE is the keyword for the data block.

number—A positive number designates the surface assemblage and its composition. A range of numbers may also be given in the form *m-n*, where *m* and *n* are positive integers, *m* is less than *n*, and the two numbers are separated by a hyphen without intervening spaces. Default is 1.

description—Optional comment that describes the surface assemblage.

Line 1: **-equilibrate** *number*

-equilibrate—Indicates that the surface assemblage is defined to be in equilibrium with a given solution composition. Optionally, **equil**, **equilibrate**, **-e[quilibrate]**, **equilibrium**, or **-e[quilibrium]**.

number—Solution number with which the surface assemblage is to be in equilibrium. Any alphabetic characters following the identifier and preceding an integer ("with solution" in Line 1) are ignored.

Line 2: *surface binding site, (sites* or *site density), specific_area_per_gram, grams,* [**Dw** *coefficient*]

> *surface binding site*—Name of a surface binding site.

> *sites*—Total number of sites for this binding site, in moles; applies when **-sites_units** is **absolute**.

> *site density*—Site density for this binding site, in sites per square nanometer; applies when **-sites_units** is **density**.

> *specific_area_per_gram*—Specific area of surface, in m^2/g (square meter per gram). Default is $600 \ m^2/g$.

> *grams*—Mass of solid for calculation of surface area, g (gram); surface area is *grams* times *specific_area_per_gram*. Default is 0 g.

> **Dw** *coefficient*—Optional diffusion coefficient for the surface, m^2/s; applies only when **-multi_D** is true in a TRANSPORT calculation. If *coefficient* > 0, the surface is transported as a colloid with advective, dispersive, and diffusive transport. Default is 0 m^2/s, which means the surface is immobile.

Line 3: **-sites_units** (**absolute** or **density**)

> **-sites_units**—Identifier specifies the units for the sites definition. **Absolute** indicates the number of surface sites is given in moles; **density** indicates that the site density is given in sites per square nanometer of surface area. The choice of units applies to all surfaces in the surface assemblage. Default is **absolute** if **-sites_units** is not included. Optionally, **sites_units**, **site_units**, **-s[ite_units]**, or **-s[ites_units]**.

> **absolute** or **density**—**Absolute** indicates the number of sites is given in moles; **density** indicates the site density is given and the number of sites is calculated from the site density and the surface area.

Line 4: **-diffuse_layer** [*thickness*]

> **-diffuse_layer**—Indicates that the composition of the diffuse layer will be calculated, such that the net surface charge plus the net charge in the diffuse layer will sum to zero. See Notes 1following this section. Either **-diffuse_layer** or **-Donnan** is necessary to calculate the explicit diffuse-layer composition that counterbalances the surface charge. The identifiers **-diffuse_layer**, **-Donnan**, and **-no_edl** are mutually exclusive and apply to all surfaces in the surface assemblage. The **-diffuse_layer** option is not available when using a CD-MUSIC surface (**-cd_music**). Optionally, **diffuse_layer** or **-d[iffuse_layer]**.

thickness—Thickness of the diffuse layer, m (meter). Default is 10^{-8} m (equals 100 angstrom).

Line 5: **-cd_music**

> **-cd_music**—Indicates that the surfaces in the surface assemblage are CD-MUSIC surfaces. See Notes 1 for using diffuse double layer and **-no_edl** surfaces in a CD-MUSIC surface assemblage. Optionally, **cd_music** or **-cd[_music]**.

Line 6: **-capacitances** c_1, c_2

> **-capacitances**—Identifier specifies the capacitances for the CD-MUSIC surface. Different surfaces within the surface assemblage may have different capacitances. This option has effect only when **-cd_music** is defined. Defaults are $c_1 = 1$ and $c_2 = 5$ F/m^2 (farad per square meter) if **-capacitances** is not included. Optionally, **capacitances** or **-ca[pacitances]**.

> c_1—Capacitance for the 0-1 plane in the CD-MUSIC formulation, F/m^2.

> c_2—Capacitance for the 1-2 plane in the CD-MUSIC formulation, F/m^2.

Line 7: **-Donnan** [(*thickness* or **debye_lengths** *lengths* [**limit_ddl** *limit*])] [**viscosity** *fraction*]

> **-Donnan**—Indicates that the Donnan approach will be used to calculate the composition of the diffuse layer. The identifiers **-diffuse_layer**, **-Donnan**, and **-no_edl** are mutually exclusive and apply to all surfaces in the surface assemblage. The **-Donnan** option is available when using diffuse-double-layer (default) or CD-MUSIC (**-cd_music**) surfaces. Optionally, **Donnan** or **-Do[nnan]** (as with all identifiers, case insensitive).

> *thickness*—Thickness of the diffuse layer in meters. Default is 10^{-8} m.

> **debye_lengths** *lengths*—Either *thickness* or **debye_lengths** may be used to define the thickness of the diffuse layer. If **debye_lengths** is used, the Debye length is calculated from the ionic strength of the solution. The thickness of the diffuse double layer is calculated by the product of *lengths* times the Debye length (Appelo and Wersin, 2007).

> **limit_ddl** *limit*—If **debye_lengths** is specified, then, optionally, the amount of water contained in the diffuse layer can be limited. *Limit* is the fraction of the total water (pore space plus diffuse double layer water) that can be in the diffuse double layer. Default for *limit* is 0.8.

> **viscosity** *fraction*—When considering multicomponent diffusion in a TRANSPORT calculation (**-multi_D** true), *fraction* affects the diffusion of ions in the diffuse layer. *Fraction* is the viscosity in the diffuse layer relative to the viscosity in the free pore space. Default is 1.0.

Line 8: **-only_counter_ions** [(*True* or *False*)]

> **-only_counter_ions**—Indicates that the surface charge will be counterbalanced in the diffuse layer with counter-ions only (the sign of charge of counter-ions is opposite to the surface charge). This option has effect only when **-diffuse_layer** or **-Donnan** is defined. When **-only_counter_ions** is true and **-diffuse_layer** is used, charge balance by co-ion exclusion is neglected (co-ions have the same sign of charge as the surface), meaning that co-ions have the same concentration in the diffuse layer as in the free pore space. When **-only_counter_ions** is true and **-Donnan** is used, co-ions are completely excluded from the diffuse layer. See Notes 1 following this section. Default is **false** if **-only_counter_ions** is not included. Optionally, **only_counter_ions** or **-o[nly_counter_ions]**.

> *(True* or *False)*—**True** indicates that the surface charge will be balanced by a surplus of counter-ions in the diffuse layer; **false** indicates that surface charge will be balanced by a counter-ion surplus and a co-ion deficit in the diffuse layer relative to the bulk solution. Optionally, **t[rue]** or **f[alse]**.

Line 9: *surface binding-site formula, name,* [(**equilibrium_phase** or **kinetic_reactant**)], *sites_per_mole, specific_area_per_mole*

> *surface binding-site formula*—Formula for surface species including stoichiometry of surface site and other surface-complexed elements connected with a pure phase or kinetic reactant. The formula must be charge balanced and is normally the OH-form of the surface binding site. If no elements other than the surface site are included in the formula, then the surface site must be uncharged. If elements are included in the formula and the surface is reacted with a solution, then these elements, in proportion to the mineral present, will be available to desorb and possibly be incorporated in other solids in the system. Further, if the mineral or kinetic reactant is dissolved, these elements will be removed from the solution and (or) other solids in the system in proportion to the mineral or kinetic reactant dissolution.

> *name*—Name of the pure phase or kinetic reactant that has this kind of surface site. If *name* is the name of a phase, the moles of the phase in the EQUILIBRIUM_PHASES data block with the same number as this surface number (4 for Lines 9 and 9a) will be used to determine the number of moles of surface sites (moles of phase times *sites_per_mole* equals moles of surface sites). If *name* is the rate name for a kinetic reactant, then the moles of the reactant in the KINETICS data block with the same number as this surface number (4 for line 9b)

will be used to determine the number of surface sites (moles of kinetic reactant times *sites_per_mole* equals moles of surface sites). Note that the stoichiometry of the phase or reactant must contain sufficient amounts of the elements in the surface complexes defined in Line 3. In the Example data block 1, there must be at least 0.101 mol of oxygen and hydrogen per mole of Fe(OH)3(a).

equilibrium_phase or **kinetic_reactant**—If **equilibrium_phase** is used, the *name* on the line is a phase defined in an EQUILIBRIUM_PHASES data block. If **kinetic_reactant** is used, the name on the line is the rate name for a kinetic reactant defined in a KINETICS data block. Default is **equilibrium_phase**. Optionally, **e** or **k**; only the first letter is checked.

sites_per_mole—Moles of surface sites per mole of phase or kinetic reactant, unitless (mol/mol).

specific_area_per_mole—Specific area of surface, in m^2/mol (square meter per mole) of equilibrium phase or kinetic reactant. Default is 0 m^2/mol.

Line 10: **-no_edl**

-no_edl—Indicates that no electrostatic terms will be used in the mass-action equations for surface species and no explicit calculation of the diffuse-layer composition is performed. The identifiers **-no_edl**, **-diffuse_layer**, and **-Donnan** are mutually exclusive and apply to all surfaces in the surface assemblage. Optionally, **no_edl**, **-n[o_edl]**, **no_electrostatic**, **-n[o_electrostatic]**.

Notes 1

The databases included with PHREEQC contain thermodynamic data for a diffuse-double-layer surface named "Hfo" (Hydrous ferric oxide) that are derived from Dzombak and Morel (1990). Two sites are defined for this surface: a strong binding site, Hfo_s, and a weak binding site, Hfo_w. Note that Dzombak and Morel (1990) used 0.2 mol weak sites and 0.005 mol strong sites per mol Fe, a surface area of 5.33×10^4 m^2/mol Fe, and a gram-formula weight of 89 g Hfo/mol Fe; to be consistent with their model, the relative number of strong and weak sites should remain constant as the total number of sites varies. To facilitate consistency, the identifier **-sites_units density** can be used, which calculates the number of sites from the site density (sites per square nanometer), the specific surface area (square meter per gram), and the mass (grams).

A surface assemblage can have multiple surfaces, each of which can have multiple site types. **SURFACE** 1 in Example data block 1 has two surfaces, Surfa and Surfb. Surfa has two binding sites, Surfa_w and Surfa_s; they share the same surface area and have the same electrostatic potential. The link between the two is indicated by the shared surface name, which is followed by an underscore, "_", and other letter(s) to distinguish the two types of sites. The surface area and mass for Surfa must be defined in the input data for at least one of the two binding sites. Surfb has only one kind of binding site and the area and mass must be defined as part of the input for the single binding site. **SURFACE** 2 is the same as **SURFACE** 1 except that an explicit calculation of the composition of the diffuse layer is specified. The identifier **-sites_units absolute** is equivalent to the default that is used in **SURFACE** 1.

SURFACE 3 defines a CD-MUSIC surface. The data are based on a description of binding on goethite (Hiemstra and Van Riemsdijk, 1996; Rahnemaie and others, 2006) that has two site types (Goe_uni and Goe_tri). The identifiers **-equilibrate**, **-diffuse_layer**, **-sites_units**, **-cd_music**, **-Donnan**, **-only_counter_ions**, and **-no_edl** apply to all surfaces and site types in the surface assemblage. The identifier **-capacitances** is defined for each individual CD-MUSIC surface and does not apply to the entire surface assemblage. A combination of CD-MUSIC, diffuse double layer, and **-no_edl** surfaces cannot be used directly in a **SURFACE** assemblage. However, a diffuse-double-layer surface, like Surfa in **SURFACE** 1, will keep its properties in a CD-MUSIC surface assemblage when defined with very high capacitances; for example, **-capacitances** 1e5 1e5. Similarly, a **-no_edl** surface will keep its properties in a diffuse-double-layer surface if the surface area is very large (1×10^{10} m^2 [square meter] for the product of *specific_area_per_gram* and *grams*). By extension, a **-no_edl** surface will keep its properties when defined in a CD_MUSIC surface assemblage if the surface area and capacitances are large. Thus, with these special definitions, **-cd_music** can be used to model simultaneously all of the available types of surfaces (**-no_edl**, diffuse double layer, and CD-MUSIC). If a **-no_edl** surface with a large surface area is included in an assemblage together with identifier **-Donnan**, the *thickness* may have to be set *to a small number* to avoid the situation where the volume of EDL water becomes unrealistically large.

SURFACE 4 has one surface, Surfc, which has two binding sites, Surfc_w and Surfc_s. The number of binding sites for these two kinds of sites is determined by the amount of Fe(OH)3(a) in EQUILIBRIUM_PHASES 4, where 4 is the same number as the surface number. If m represents the moles of Fe(OH)3(a) in EQUILIBRIUM_PHASES 4, then the number of sites of Surfc_w is $0.1m$ (mol) and of Surfc_s is $0.001m$ (mol). The surface area for Surfc is defined relative to the moles of Fe(OH)3(a), such that the surface area is $100,000m$ (m^2). During batch-reaction simulations the moles of Fe(OH)3(a) in EQUILIBRIUM_PHASES 4 may change, in which case the number of sites of Surfc will change as will the surface area associated with Surfc. Whenever Fe(OH)3(a) precipitates, the specified amounts of Surfc_wOH and Surfc_sOH are formed and all the species that are surface-complexed and in the electrical double layer will be taken from the elements in the system. These formulas are charge balanced and the OH groups are part of the formula for Fe(OH)3(a). The OH is not used in the initial surface-composition calculation, but is critical when amounts of Fe(OH)3(a) vary. Erroneous results will occur if the formula is not charge balanced, and a warning message will be printed if the elements in the surface complex (other than the surface site itself) are not contained in sufficient quantities in the equilibrium phase or kinetic formula.

The number of sites of Surfd in **SURFACE** 4 is determined by the amount of a kinetic reactant defined in **KINETICS** 4, where 4 is the same number as the surface number. Sites related to a kinetic reactant are exactly analogous to sites related to an equilibrium phase. The same restrictions apply—the formula must be charge balanced, and the elements in the surface complex (other than the surface site itself) should be included in the formula of the reactant.

SURFACE 5 is a template for modeling sorption and diffusion in clay rocks; in this case, the Opalinus Clay at Mont Terri in Switzerland (Appelo and others, 2010). A rock dry density of 2.7 kg/L, an overall porosity of 0.161, of which half is accessible for Cl^-, a specific surface area of 37 m^2/g, and an exchange capacity of 0.114 eq/kg (equivalent per kilogram) are translated to a surface that is defined relative to 1 L of pore water. Thus, 1 L of pore water is in contact with $2.7 \times (1 - 0.161) / 0.161 = 14.07$ kg rock. The surface has "clay_planar" sites that express the bulk exchange capacity of the rock (1.59 mol sites). The measured sorption isotherm for Cs^+ indicates the presence of two sites: "clay_ii" sites with an intermediate strength for binding Cs^+, and "clay_fes" sites on the frayed edges of illite with a very strong and very specific binding strength. The number of these sites and the complexation constants for Cs^+ and other cations are obtained by fitting the isotherm, while accounting for two sorption types: one type is for surface complexation, which is specific for the various ions; the other type is connected with charge compensation in the Donnan space, which is in principle, the same for all equal-charged ions (although it can be varied with **-erm_DDL** in keyword **SOLUTION_SPECIES** to match observed differences). In **SURFACE** 5, the exclusion of Cl^- is modeled with a Donnan pore space that contains only counter-ions (is Cl^- free). Thus, it holds 0.5 L water and has a thickness (derived from this volume and the total area) of $0.5 \times 10^{-3} / (37 \times 14.07 \times 10^3) =$ 9.6×10^{-10} m. This thickness equals 1.9 Debye lengths at the ionic strength of the pore water of 0.368, which is in good agreement with anion-exclusion theory (Schofield, 1947; Tournassat and Appelo, 2011).

An anion-free Donnan pore space is the simplest option and is in line with traditional calculations in soil science. Perhaps more realistically, the anion exclusion can be modeled with **-only_counter_ions** false and an increased thickness of the Donnan layer. The fraction of free (uncharged) pore water, f_{free}, follows from the Cl^- accessible pore space and the potential in the Donnan space, ψ_D (V):

$$\frac{\varepsilon_{a,Cl}}{\varepsilon_{tot}} = 0.5 = f_{free} + (1 - f_{free}) \times \exp\left(\frac{F\psi_D}{RT}\right), \tag{6}$$

where ε_a is the accessible porosity (unitless), ε_{tot} is the total porosity (unitless), F is the Faraday constant (96485 $JV^{-1}eq^{-1}$), R is the gas constant (8.314 $JK^{-1}mol^{-1}$), and T is the temperature (K). The potential in the Donnan space depends on the water composition and the surface charge, while the latter is also a function of the surface complexation constants: higher constants increase complexation and, usually, decrease the surface charge, provided the complexes are uncharged (charged complexes could increase the surface charge). By fixing the complexation constant for Na^+ to -0.7, and the constants for the other major cations by matching the measured distribution coefficients in Opalinus Clay, the surface charge—that is, the part of the exchange capacity that is compensated in the Donnan pore space—can be calculated to be 45 percent (Appelo and others, 2010). This results in $f_{free} = 0.117$ and thus, 0.883×10^{-3} m^3 (cubic meter) water in the Donnan pore space per m^3 pore water. Accordingly, the thickness of the Donnan space becomes $0.883 \times 10^{-3} / (37 \times 14.07 \times 10^3) = 1.7 \times 10^{-9}$ m, or 3.4 Debye lengths.

SURFACE 6 illustrates the option to set the thickness of the Donnan layer to be a number of Debye lengths, κ^{-1}, given by $\kappa^{-1} = \left(3.94 \cdot 10^{-24} \varepsilon_r \left(\frac{T}{I}\right)\right)^{0.5}$ (m), where ε_r is the relative dielectric constant of water, T is the temperature (kelvin), and I is the ionic strength. Thus, for **SURFACE** 5, with **-only_counter_ions** true, the thickness can be defined to be **-Donnan debye_lengths** 1.9, while with **-only_counter_ions false** (the default option) the thickness is defined to be **-Donnan debye_lengths** 3.4 as in **SURFACE** 6. The thickness will now vary with the ionic strength of the solution, and the fraction of free pore water will be adjusted to maintain the same total amount of water. For program convergence, the fraction of Donnan water is limited with **limit_ddl** 0.9 in **SURFACE** 6 (default is **limit_ddl** 0.8).

The chemical and physical properties of clay rocks can be measured precisely with diffusion experiments, and PHREEQC can model the experiments by calculating multicomponent diffusion with option **-multi_D** in keyword TRANSPORT. With this option, diffusion is calculated separately for the free (uncharged) pore water and the Donnan pore water, while each solute species has its own diffusion coefficient. It is probable that the electrostatic double layer, mimicked by the Donnan pore space in PHREEQC, has properties that are different from free pore water. The dielectric permittivity is lower in an electrostatic field, which will enhance the complexation of charged ions into neutral species. Such complexation will diminish anion exclusion, and will be different for different anions, depending on charge and hydration radius. This effect can be modeled by defining an enrichment factor in the Donnan pore space with **-erm_ddl** in keyword SOLUTION_SPECIES. Furthermore, the viscosity may be higher in the Donnan pore water than in ordinary water, and diffusion would be lower in proportion with the viscosity ratio. PHREEQC allows setting the viscosity ratio as illustrated in **SURFACE** 5 and **SURFACE** 6. (Default is 1.0)

SURFACE 7 defines a diffusion coefficient of 10^{-11} m^2/s for a surface consisting of ferrihydrite particles (Hfo in the database). When the diffusion coefficient is larger than 0, the surface will advect and disperse like a normal solute species in a column defined with keyword TRANSPORT and **-multi_D** true, and it will diffuse in accordance with the diffusion coefficient. The surface must be neutral, either charge-free by itself, or by adding a Donnan layer that neutralizes the surface charge. In **SURFACE** 7, the Donnan layer is given a small thickness of only 1 picometer to avoid significant variation in water contents in the cells during transport. The transported surfaces will carry the elements in the surface complexed species, as well as the solutes in the Donnan layer. The diffusion coefficient of the surface, either the whole, or part of it, can be modified with the special Basic function CHANGE_SURF. If changed to 0 m^2/s, the surface becomes immobile. Thus, **SURFACE** 7 is a colloidal particle that can transport heavy metals complexed on its surface while the diffusion coefficient is greater than zero, and it can coagulate with other particles and be deposited depending on chemical or physical conditions in the column by setting the diffusion coefficient to zero with CHANGE_SURF. An example is given at *http://www.hydrochemistry.eu/exmpls/colloid.html* (accessed June 25, 2012).

Line 1 requires the program to make a calculation to determine the composition of a surface assemblage, termed an "initial surface calculation". Before any batch-reaction or transport calculations, initial surface calculations are performed to determine the composition of the surface assemblages that would exist in equilibrium with the specified solution (solution 10 for **SURFACE** 1 in this Example data block). The composition of the solution will not change during these calculations. In contrast, during a batch-reaction calculation, when a surface assemblage (defined as in Example data block 1 or Example data block 2 of this section) is reacted with a solution with which it is not in equilibrium, both the surface composition and the solution composition will be adjusted to a new equilibrium.

When **-diffuse_layer** or **-Donnan** is not used (default), any charge that develops on the surface during a reaction step will be accompanied by an equal, but opposite, charge imbalance for the solution. Thus, charge imbalances accumulate in the solution and on the surface when surfaces and solutions are separated. This handling of charge imbalances for surfaces is physically incorrect. Consider the following example, where a charge-balanced surface is brought together with a charge-balanced solution. Assume a positive charge develops at the surface. Now remove the surface from the solution. With the default formulation, a positive charge imbalance is associated with the surface, Z_s, and a negative charge imbalance, Z_{soln}, is associated with the solution. In reality, the charged surface plus the diffuse layer surrounding it is electrically neutral and the combination should be removed. This would leave an electrically neutral solution as well. The default formulation is workable; its main defect is that the counter-ions that should be in the diffuse layer are retained in the solution. The model results are adequate, provided solutions and surfaces are not separated or the exact concentrations of aqueous counter-ions are not critical to the investigation.

The **-diffuse_layer** and **-Donnan** identifiers activate models to balance the accumulation of surface charge with an explicit calculation of the diffuse-layer composition. When these identifiers are used, the composition of the diffuse layer is calculated and an additional printout of the elemental composition of the diffuse layer is produced. The **-diffuse_layer** identifier calculates the moles of each aqueous species in the diffuse layer according to the method of Borkovec and Westall (1983) and the assumption that the diffuse layer is a constant thickness (optionally input with **-diffuse_layer**, default is 10^{-8} m). The variation of thickness of the diffuse layer with ionic strength is ignored. The net charge in the diffuse layer exactly balances the net surface charge. The **-diffuse_layer** calculation requires an integration that is slow and liable to failure under certain conditions. The **-Donnan** calculation is much faster and more robust, and gives results that are usually similar to the **-diffuse_layer** calculation.

In the **-Donnan** calculation, the concentrations in the diffuse layer are averaged and computed with:

$$c_{Donnan, i} = c_i \times erm\_DDL_i \times \exp\left(\frac{-z_i F \psi_D}{RT}\right), \tag{7}$$

where $c_{Donnan, i}$ is the concentration of species i in the Donnan pore space (mol/L), c_i is the concentration in the free (uncharged) solution, erm_DDL_i is an enrichment factor (unitless) that can be defined in keyword SOLUTION_SPECIES, and z_i is the charge number (unitless). The potential ψ_D is adjusted to let the charge of the Donnan volume counterbalance the surface charge:

$$\sum_i z_i c_{Donnan, i} + \sigma_{surface} = 0, \tag{8}$$

where $\sigma_{surface}$ is the surface charge (eq/L).

Conceptually, the results of using the explicit diffuse-layer calculations are correct—charge imbalances on the surface are balanced in the diffuse layer and the solution remains charge balanced. Great uncertainties exist in the true composition of the diffuse layer and the thickness of the diffuse layer. The ion complexation in the bulk solution is assumed to apply in the diffuse layer, which is unlikely because of changes in the dielectric constant of water near the charged surface. Identifier **-erm_ddl** in keyword SOLUTION_SPECIES can correct for such effects if needed, but it is a primitive and arbitrary correction. The explicit diffuse layer calculation is based on assumptions that allow the volume of water in the diffuse layer to remain small relative to the solution volume. It is possible, especially for solutions of low ionic strength, for the calculated concentration of an element to be negative in the integrated diffuse layer (calculated with identifier **-diffuse_layer**). In this case, the assumed thickness of the diffuse layer is too small (or perhaps the entire diffuse-layer approach is inappropriate) and the program stops with an error message. The identifier **-only_counter_ions** offers an option to let only the counter-ions increase in concentration in the diffuse layer, and to leave the co-ions at the same concentration in the diffuse layer as in the bulk solution. The counter-ions have a higher concentration in the diffuse layer than without this option, because co-ion exclusion is neglected. Alternatively, when using **-only_counter_ions** and **-Donnan**, the co-ion concentration is zero in the Donnan pore space. In this case, the counter-ions will have a smaller concentration in the Donnan layer with **-only_counter_ions** true, than with **-only_counter_ions** false.

A third alternative for modeling surface-complexation reactions, in addition to the default, **-diffuse_layer**, and **-cd_music**, is to ignore the surface potential entirely. The **-no_edl** identifier eliminates the potential term from mass-action expressions for surface species, eliminates any charge-balance equations for surfaces, and eliminates any charge-potential relationships. The charge on the surface is calculated and saved with the surface composition, and an equal and opposite charge is stored with the aqueous phase. All of the cautions about separation of charge, mentioned in the previous paragraphs, apply to the calculation using **-no_edl**. No explicit calculation of the diffuse-layer composition is available when using **-no_edl**.

For transport calculations, it is much faster in terms of CPU time to use either the default (no explicit diffuse layer calculation) or **-no_edl**. However, **-Donnan** and **-diffuse_layer** can be used to test the sensitivity of the results to diffuse-layer effects.

After a set of batch-reaction calculations has been simulated, it is possible to save the resulting surface composition with the SAVE keyword. If the new composition is not saved, the surface composition will remain the same as it was before the batch-reaction calculations. After it has been defined or saved, the surface composition may be used in subsequent simulations through the USE keyword. By using the RUN_CELLS data block, the results of the batch-reaction calculations, including the surface-assemblage composition, are automatically saved. In ADVECTION and TRANSPORT simulations, the surface assemblages in the column are automatically saved after each shift.

Example data block 2

```
Line 0d:     SURFACE 1 Neutral surface composition
Line 1:         Surf_wOH        0.3                 660.                0.25
Line 1a:        Surf_sOH        0.003
Line 2:         Surfc_sOH       Fe(OH)3(a)    equilibrium_phase        0.001
Line 2b:        Surfd_sOH       Al(OH)3(a)    kinetic_reactant         0.001
```

Explanation 2

Line 0d: **SURFACE** [*number*] [*description*]

Same as Example data block 1.

Line 1: *surface binding-site formula, (sites* or *site density), specific_area_per_gram, grams,* [**Dw** *coefficient*]

surface binding-site formula—Formula for a surface that is charge balanced.

sites—Total number of sites for this binding site, in moles; applies when **-sites_units** is **absolute**.

site density—Site density for this binding site, in sites per square nanometer; applies when **-sites_units** is **density**.

specific_area_per_gram—Specific area of surface, in m^2/g (square meter per gram). Default is $600\ m^2/g$.

grams—Mass of solid for calculation of surface area, g (gram); surface area is *grams* times *specific_area_per_gram*. Default is 0 g.

Dw *coefficient*—Optional diffusion coefficient for the surface, m^2/s; applies only when **-multi_D** is true in a TRANSPORT calculation. If *coefficient* > 0, the surface is transported as a colloid with advective, dispersive, and diffusive transport. Default is $0\ m^2/s$, which means the surface is immobile.

Line 2: *surface binding-site formula, name,* [(**equilibrium_phase** or **kinetic_reactant**)], *sites_per_mole, specific_area_per_mole*

Same as Line 9 in Example data block 1.

Notes 2

The difference between Example data block 2 and Example data block 1 is that no initial surface-composition calculation is performed in Example data block 2. The composition of the surface assemblage must be given precisely (from chemical analysis) and charge-balanced to avoid spurious pH and redox reactions. Additional surfaces and binding sites can be defined by repeating Lines 2 and 9 from Example data block 1. All other identifiers listed for Example data block 1 also can be included.

Example problems

The keyword **SURFACE** is used in example problems 8, 14, 19, and 21.

Related keywords

ADVECTION, COPY, DELETE, DUMP, RUN_CELLS, SURFACE_MASTER_SPECIES, SURFACE_SPECIES, SAVE **surface**, TRANSPORT, and USE **surface**.

SURFACE_MASTER_SPECIES

This keyword data block is used to define the correspondence between surface binding-site names and surface master species. Normally, this data block is included in the database file and only additions and modifications are included in the input file. The databases in the PHREEQC distribution contain master species for Hfo_s and Hfo_w, which represent the strong and weak binding sites of hydrous ferric oxides (Dzombak and Morel, 1990).

Example data block

```
Line  0:   SURFACE_MASTER_SPECIES
Line 1a:     Surf_s          Surf_sOH
Line 1b:     Surf_w          Surf_wOH
Line 1c:     [mySurf1]       [mySurf1]OH
```

Explanation

Line 0: **SURFACE_MASTER_SPECIES**

Keyword for the data block. No other data are input on the keyword line.

Line 1: *surface binding-site name, surface master species*

surface binding-site name—Name of a surface binding site. A binding site name is composed of a surface name and optionally an underscore and additional lower case letters to designate a specific binding site of the surface. Two forms for surface names are available: (1) surface names that begin with a capital letter followed by zero or more lower case letters, and no numbers; and (2) surface names that are enclosed in square brackets (see Line 1c) and use any combination of alphanumeric characters and the characters plus (+), minus (-), equal (=), colon (:), and decimal point (.). Surface names using form 2 are case dependent, and upper and lower case characters can be used in any position. Different binding sites for a surface name are designated with an underscore ("_") plus one or more lower case letters.

surface master species—Formula for the surface master species, usually the OH form of the binding site.

Notes

In this Example data block, a surface named "Surf" has a strong and a weak binding site and a surface named "[mySurf1]" has only one binding site. Association reactions must be defined with

SURFACE_SPECIES for the master species associated with each binding site and for any additional surface complexation species. Each surface master species (Surf_sOH, Surf_wOH, and [mySurf1]OH in this Example data block) must be defined by an identity reaction with log K of 0.0 in SURFACE_SPECIES input. Other surface species can be defined by association reactions with the surface master species. The number of sites, in moles, for each binding site is defined explicitly or implicitly (by defining site density and surface area) in the SURFACE data block. Information defining the surface area and capacitances (CD-MUSIC surfaces) also must be specified with the SURFACE data block.

Example problems

The keyword **SURFACE_MASTER_SPECIES** is used in example problems 14, 19, and 21. It is also found in the *Amm.dat*, *iso.dat, llnl.dat, minteq.dat, minteq.v4.dat, phreeqc.dat, pitzer.dat*, and *wateq4f.dat*, databases.

Related keywords

SURFACE and SURFACE_SPECIES.

SURFACE_SPECIES

This keyword data block is used to define a reaction and log K for each surface species, including surface master species. Normally, this data block is included in the database file and only additions and modifications are included in the input file. Surface species defined in Dzombak and Morel (1990) are defined in the standard set of databases; the master species are Hfo_w and Hfo_s for the weak and strong binding sites of hydrous ferric oxide.

Example data block 1

```
Line  0:  SURFACE_SPECIES
Line 1a: Surf_sOH = Surf_sOH
Line 2a:       log_k     0.0
Line 1b: Surf_sOH + H+ = Surf_sOH2+
Line 2b:       log_k     6.3
Line 1c: Surf_wOH = Surf_wOH
Line 2c        log_k     0.0
Line 1d: Surf_wOH + H+ = Surf_wOH2+
Line 2d:       log_k     4.3
```

Explanation 1

Line 0: **SURFACE_SPECIES**

Keyword for the data block. No other data are input on the keyword line.

Line 1: *Association reaction*

Association reaction for surface species. The defined species must be the first species to the right of the equal sign. The association reaction must precede all identifiers related to the surface species. Line 1a is a master-species identity reaction.

Line 2: **log_k** *log K*

log_k—Identifier for log K at 25 °C. Optionally, **-log_k**, **logk**, **-l[og_k]**, or **-l[ogk]**.

log K—Log K at 25 °C for the reaction. Log K for a master species is 0.0. Default is 0.0.

Notes 1

This Example data block assumes that Surf_w and Surf_s are defined in a SURFACE_MASTER_SPECIES data block. Lines 1 and 2 may be repeated as necessary to define all of the surface reactions. An identity reaction is needed to define each master surface species, Lines 1a and 1c in this Example data block. The log K for the identity reaction must be 0.0, Lines 2a and 2c in this Example data block.

An underscore plus one or more lowercase letters is used to define different binding sites for the same surface. In the Example data block, association reactions for a strong and a weak binding site are defined for the surface named "Surf". Multiple surfaces may be defined simply by defining multiple master surface species (for example, Surfa, Surfb, and Surfc). Multiple binding sites can be defined for each surface by using an underscore followed by one or more lower case letters. Binding sites on the same surface share the same surface area, and have the same electrostatic potential.

Temperature dependence of log K can be defined with enthalpy of reaction (identifier **delta_h**) and the Van't Hoff equation or with an analytical expression (**-analytical_expression**). It is also possible to define a temperature-dependent expression in NAMED_EXPRESSIONS and use **-add_logk** to add its value to the equilibrium constant for the surface species. See SOLUTION_SPECIES or PHASES for examples.

The identifier **-no_check** can be used to disable checking charge and elemental balances (see SOLUTION_SPECIES). The use of **-no_check** is not recommended. If **-no_check** is used, then the **-mole_balance** identifier is needed to ensure the correct stoichiometry for the surface species. In PHREEQC version 1, the **-no_check** option was included to permit the stoichiometry of a species to be defined separately from the mass-action equation. Specifically, the sorption of uranium on iron oxides as described by Waite and others (1994) provides an example, where they use different coefficients in the mass-action equation than in the mole-balance equations. However, activity of a surface species is defined as mole fraction of sites occupied by the species in PHREEQC versions 2 and 3, which is inconsistent with activity that is defined as molality by Waite and others (1994) and PHREEQC version 1. It is noted that formulas with coefficients of only 1 in the mass-action-equation will give identical results for all PHREEQC versions. The **-no_check** and **-mole_balance** identifiers have been retained in version 3, but their use should be restricted to special sorption formulas; for example, for modeling Freundlich isotherms (see example 19, in the Examples).

Example data block 2

```
Line 0:    SURFACE_SPECIES
Line 1a:      Goe_uniOH-0.5 = Goe_uniOH-0.5
Line 2a:           log_k 0
Line 3:            -Vm   8.51 1.15 -0.678 -2.83   2.38   0 2.0   14.5
Line 4a:           -cd_music  0 0 0
Line 1b:      Goe_uniOH-0.5 + H+ + AsO4-3 = Goe_uniOAsO3-2.5 + H2O
```

```
Line 2b:            log_k    20.1
Line 4b:            -cd_music 0.25  -2.25  0
Line 1c:     Goe_uniOH-0.5 + H+ + AsO4-3 = Goe_uniOAsO3-2.5 + H2O
Line 2c:            log_k    20.1
Line 5:             -cd_music  -1 -6 0 0.25 5
Line 1d:     Goe_triO-0.5 + Na+ = Goe_triONa+0.5
Line 2d:            log_k    -1
Line 4c:            -cd_music  0 0 1
```

Explanation 2

Line 0: **SURFACE_SPECIES**

Keyword for the data block. No other data are input on the keyword line.

Line 1: *Association reaction*

Association reaction for surface species. The defined species must be the first species to the right of the equal sign. The association reaction must precede all identifiers related to the surface species. Line 1a is a master-species identity reaction.

Line 2: **log_k** *log K*

log_k—Identifier for log *K* at 25 °C. Optionally, **-log_k**, **logk**, **-l[og_k]**, or **-l[ogk]**.

log K—Log *K* at 25 °C for the reaction. Log *K* for a master species is 0.0. Default is 0.0.

Line 3: Same as Example data block 1.

Line 4: **-cd_music** *deltaz$_0$*, *deltaz$_1$*, *deltaz$_2$*

-cd_music—Identifier for CD-MUSIC electrostatic parameters. More recent papers by Hiemstra and others (for example, Stachowicz and others, 2006) use this form to define the change in charge for the 0, 1, and 2 planes. Optionally, **cd_music** or **-cd[_music]**.

deltaz$_0$—The change in charge at the plane of specific adsorption, or 0 plane.

deltaz$_1$—The change in charge at the Stern layer, or 1 plane.

deltaz$_2$—The change in charge at the diffuse layer, or 2 plane.

Line 5: **-cd_music** *dz$_0$*, *dz$_1$*, *dz$_2$*, *fraction*, *Z$_{ion}$*

-cd_music—Identifier CD-MUSIC electrostatic parameters. Early papers by Hiemstra and others (for example, Hiemstra and van Riemsdijk, 1996) used this form to define the change in charge for the 0, 1, and 2 planes. Parameters defined by this method are converted to parameters as in Line 3 by the following equations: *deltaz$_0$* = *dz$_0$* + (*fraction*)*Z$_{ion}$* and

$deltaz_1 = dz_1 + (1 - fraction)Z_{ion}$. The meaning of $deltaz_2$ is the same in Lines 3 and 4.

Optionally, **cd_music** or **-cd[_music]**.

dz_0—The change in charge at the plane of specific adsorption, or 0 plane due to gain or loss of hydrogen and oxygen.

dz_1—The change in charge at the Stern layer, or 1 plane due to hydrogen and oxygen in the ligand.

$deltaz_2$—The change in charge at the diffuse layer, or 2 plane.

fraction—The fraction of the central ion charge that is associated with plane 0.

ion_z—The charge on the central ion.

Notes 2

This Example data block defines surface species for a CD-MUSIC surface. The definitions assume that Goe_uni has been defined as a site in a SURFACE_MASTER_SPECIES data block. Lines 1 through 3 in Example data block 2 are the same as for Example data block 1, but Lines 4 and 5 are specific to CD-MUSIC surfaces. Line 4b and Line 5 define the same electrostatic parameters for the same surface species in two alternative forms. Line 4b is now the more common form for definition of the distribution of surface charge for a species, where the change in charge at the three charge planes—the 0 or specific adsorption plane, the 1 or Stern layer plane, and the 2 or diffuse layer plane—is specified directly. Line 5 is an older form that separates the change in charge into that related to hydrogen and oxygen and that related to the distribution of charge from the central ion. Any parameters defined in the form of Line 5 are converted to the form of Line 4 for all surface calculations.

An underscore plus one or more lowercase letters is used to define different binding sites for the same CD-MUSIC surface. In the Example data block, Goe_uni and Goe_tri are two site types for a goethite surface.

Example problems

The keyword **SURFACE_SPECIES** is used in example problems 8, 14, 19, and 21. It is also found in the *Amm.dat*, *iso.dat*, *llnl.dat*, *minteq.dat*, *minteq.v4.dat*, *phreeqc.dat*, *pitzer.dat*, and *wateq4f.dat* databases.

Related keywords

PHASES, SOLUTION_SPECIES, SURFACE, and SURFACE_MASTER_SPECIES.

TITLE

This keyword data block is used to include a comment for a simulation in the output file. The comment will appear in the echo of the input data, and it will appear at the beginning of the simulation calculations.

Example data block

```
Line 0:  TITLE The title may begin on this line,
Line 1a: or on this line.
Line 1b: It continues until a keyword is found at the beginning of a line
Line 1c: or until the end of the file.
```

Explanation

Line 0: **TITLE** *comment*

TITLE is the keyword for the data block. Optionally, **COMMENT**.

comment—The first line of a title (or comment) may begin on the same line as the keyword.

Line 1: *comment*

comment—The title (or comment) may continue on as many lines as necessary. Lines are read and saved as part of the title until a keyword begins a line or until the end of the input file.

Notes

Be careful not to begin a line of the title with a keyword because that signals the end of the **TITLE** data block. The **TITLE** data block is intended to document a simulation in the output file. If more than one title keyword is entered for a simulation, each will appear in the output file as part of the echo of the input data, but only the last will also appear at the beginning of the simulation calculations. The characters "#" and ";" have special meanings in PHREEQC input files; in the **TITLE** data block, the "#" will cause the remainder of the line to be excluded from *comment* and ";" will have the same effect as a line break at that character position. Lines that are entirely white space (tabs and spaces) and comments (characters following "#") are eliminated.

Example problems

The keyword **TITLE** is used in all examples, 1 through 22.

TRANSPORT

This keyword data block is used to simulate one-dimensional (1D) transport of solutes, water, colloids, and heat due to the processes of advection and dispersion, diffusion, and diffusion into stagnant zones adjacent to the 1D flow system. Radial and three-dimensional (3D) diffusion can be modeled by using stagnant zones. Multicomponent diffusion allows individual tracer diffusion coefficients to be used in calculating the diffusion of ions; **TRANSPORT** has capabilities to model multicomponent diffusion in the aqueous solution and in the interlayers of swelling clay minerals. All the chemical processes modeled by PHREEQC, including kinetically controlled reactions, may be included in an advective-dispersive transport simulation. Purely advective transport plus reactions—without diffusion, dispersion, or stagnant zones—can be simulated with the ADVECTION data block.

Example data block

```
Line  0:  TRANSPORT
Line  1:        -cells              5
Line  2:        -shifts             25
Line  3:        -time_step          1 yr 2.0
Line  4:        -flow_direction     forward
Line  5:        -boundary_conditions  flux constant
Line  6:        -lengths            4*1.0 2.0
Line  7:        -dispersivities     4*0.1 0.2
Line  8:        -correct_disp       true
Line  9:        -diffusion_coefficient 1.0e-9
Line 10:        -stagnant           1  6.8e-6   0.3   0.1
Line 11:        -thermal_diffusion  3.0   0.5e-6
Line 12:        -initial_time       1000
Line 13:        -print_cells        1-3 5
Line 14:        -print_frequency    5
Line 15:        -punch_cells        2-5
Line 16:        -punch_frequency    5
Line 17:        -dump               dump.file
Line 18:        -dump_frequency     10
Line 19:        -dump_restart       20
Line 20:        -warnings           false
Line 21:        -multi_D            true   1e-9   0.3   0.05   1.0
Line 22:        -interlayer_D  true   0.09   0.01   250
```

Explanation

Line 0: **TRANSPORT**

TRANSPORT is the keyword for the data block. No other data are input on the keyword line.

Line 1: **-cells** *cells*

-cells—Indicates the number of cells in the 1D column for the advective-dispersive transport simulation. Optionally, **cells** or **-c[ells]**.

cells—Number of cells in a 1D column. Default is 1.

Line 2: **-shifts** *shifts*

-shifts—Indicates the number of shifts or diffusion periods in the advective-dispersive transport simulation. Optionally, **shifts** or **-s[hifts]**.

shifts—For advective-dispersive transport, *shifts* is the number of advective shifts or time steps, which is the number of times the solution in each cell will be shifted to the next higher or lower numbered cell; the total time simulated is $shifts \times time\ step$. For purely diffusive transport, *shifts* is the number of diffusion periods that are simulated; the total diffusion time is $shifts \times time\ step$. Default is 1.

Line 3: **-time_step** *time_step* [*unit*] [*substeps*]

-time_step—Defines time step associated with each advective shift or diffusion period. The number of shifts or diffusion periods is given by **-shifts**. Optionally, **timest**, **-t[imest]**, **time_step**, or **-t[ime_step]**.

time_step—Time (second) associated with each shift or diffusion period. Default is 0 s.

unit—Optional time unit may be **second**, **minute**, **hour**, **day**, **year**, or an abbreviation of one of these units. The *time_step* is converted to seconds after reading the data block; all internal calculations, Basic functions, and output times are in seconds. Default is second.

substeps— Subdivides the *time step* into *substeps* intervals. Used only in multicomponent diffusion calculations, where *time_step* reduction can help to avoid negative concentrations. The negative concentrations may occur when the time step is too large relative to the explicitly defined stagnant-cell mixing factors; too large a time step causes the Von Neumann criterion to be violated. Default is 1.

Line 4: **-flow_direction** (**forward**, **back**, or **diffusion_only**)

-flow_direction—Defines direction of flow. Default is **forward** at startup. Optionally, **direction**, **flow**, **flow_direction**, **-dir[ection]**, or **-f[low_direction]**.

forward, **back**, or **diffusion_only**—(1) **Forward**, advective flow direction is into higher numbered cells; optionally, **f[orward]**, (2) **Backward**, advective flow direction is into lower numbered cells; optionally **b[ackward]**, or (3) **Diffusion_only**, only diffusion occurs, there is no advective flow; optionally **d[iffusion_only]** or **n[o_flow]**.

Line 5: **-boundary_conditions** *first, last*

-boundary_conditions—Defines boundary conditions for the first and last cell. Optionally, **bc**, **bcond**, **-b[cond]**, **boundary_condition**, **-b[oundary_condition]**. Three types of boundary conditions are allowed at either end of the column (indicated by x_{end}):

constant—Concentration is constant $C(x_{end}, t) = C_0$, also known as first type or Dirichlet boundary condition. C_0 is the concentration outside the column (mol/kgw). Optionally, **co[nstant]** or **1**.

closed—No flux at boundary, $v = 0$ and $\dfrac{\partial C(x_{end}, t)}{\partial x} = 0$, also known as second type or Neumann boundary condition, where v is the flow velocity (m/s, meter per second). Optionally, **cl[osed]** or **2**.

flux—Flux boundary condition, $C(x_{end}, t) = C_0 + \dfrac{D_L}{v}\dfrac{\partial C(x_{end}, t)}{\partial x}$, also known as third type or Cauchy boundary condition, where D_L is the dispersion coefficient (m^2/s). Optionally, **f[lux]** or **3**.

first—Boundary condition at the first cell, **constant**, **closed**, or **flux**. Default is **flux**.

last—Boundary condition at the last cell, **constant**, **closed**, or **flux**. Default is **flux**.

Line 6: **-lengths** *list of lengths*

-lengths—Defines length of each cell for advective-dispersive transport simulations (m). Optionally, **length**, **lengths**, or **-l[engths]**.

list of lengths—Length of each cell (m). Any number of *lengths* up to the total number of cells (*cells*) may be entered. If *cells* is greater than the number of *lengths* entered, the final value read will be used for the remaining cells. Multiple lines may be used. Repeat factors can be used to input multiple data with the same value; in the Example data block, 4*1.0 is interpreted as 4 values of 1.0. Default is 1 m.

Line 7: **-dispersivities** *list of dispersivities*

-dispersivities—Defines dispersivity of each cell for advective-dispersive transport simulations (m). Optionally, **disp**, **dispersivity**, **dispersivities**, **-dis[persivity]**, or **-dis[persivities]**.

list of dispersivities—Dispersivity assigned to each cell (m). Any number of *dispersivities* up to the total number of cells (*cells*) may be entered. If *cells* is greater than the number of *dispersivities* entered, the final value read will be used for the remaining cells. Multiple lines may be used. Repeat factors can be used to input multiple data with the same value; in the Example data block, 4*0.1 is interpreted as 4 values of 0.1 m. Default is 0 m.

Line 8: **-correct_disp** [(*True* or *False*)]

-correct_disp—Dispersivity is multiplied by $(1 + 1/cells)$ for column ends with flux boundary conditions. This correction can improve modeling effluent composition from column experiments that are modeled with few cells. Default is **false** at startup. Optionally, **correct_disp** or **-co[rrect_disp]**.

(True or *False)*—**True** indicates that dispersivity is corrected for flux-boundary end cells; **false** indicates that no correction is made. If neither **true** nor **false** is entered on the line, **true** is assumed. Optionally, **t[rue]** or **f[alse]**.

Line 9: **-diffusion_coefficient** *diffusion coefficient*

-diffusion_coefficient—Defines diffusion coefficient for all aqueous species (m^2/s) when *not* using multicomponent diffusion (**-multi_D false**); this value of the diffusion coefficient is also used as the default thermal diffusion coefficient (see **-thermal_diffusion**). Default is 0.3×10^{-9} m^2/s at startup. Optionally, **diffusion_coefficient**, **diffc**, **-dif[fusion_coefficient]**, or **-dif[fc]**.

diffusion coefficient—Diffusion coefficient.

Line 10: **-stagnant** *stagnant_cells* [*exchange_factor* ε_m ε_{im}]

-stagnant—Defines the maximum number of stagnant (immobile) cells associated with each cell in which advection occurs (mobile cell). Each mobile cell may be connected with up to *stagnant_cells* immobile cells. The immobile cells associated with a mobile cell are usually conceived to be a 1D column of cells in which solutes from the mobile cell diffuse laterally.

However, the connections among the immobile cells can be defined freely with **MIX** data blocks, which allows calculation of multidimensional diffusion processes (Appelo and Wersin, 2007) and radial diffusion (Appelo and others, 2008, see example 21). The immobile cells associated with a mobile cell, *cell*, are numbered as follows: $n \times cells + 1 + cell$, where *cells* is the number of mobile cells and $1 \leq n \leq stagnant\_cells$. For each immobile cell, a solution (**SOLUTION**, **SOLUTION_SPREAD**, or **SAVE** data block) must be defined, and either a **MIX** data block or, for the first-order exchange model, the *exchange_factor* must be defined (only applicable if *stagnant_cells* equals 1). Mixing will be performed at each diffusion/dispersion time step. **EQUILIBRIUM_PHASES**, **EXCHANGE**, **GAS_PHASE**, **KINETICS**, **REACTION**, **REACTION_TEMPERATURE**, **SOLID_SOLUTIONS**, and **SURFACE** may be defined for an immobile cell. Thermal diffusion in excess of hydrodynamic diffusion can be calculated only for the first-order exchange model. Optionally, **stagnant** or **-st[agnant]**.

stagnant_cells—Maximum number of stagnant (immobile) cells associated with a mobile cell. Default is 0.

exchange_factor—Factor describing exchange between a mobile and its immobile cell (s^{-1}). The *exchange_factor* can be used only if *stagnant_cells* is 1, in which case all immobile cells have the same diffusion properties. WARNING: If *exchange_factor* is entered, all previously defined **MIX** structures will be deleted and **MIX** structures for the first-order exchange model for a dual porosity medium will be created. Default is 0 s^{-1}.

ε_m—Porosity in each mobile cell, expressed as a fraction of the total volume of mobile and immobile cells. The ε_m is used only if *stagnant_cells* is 1, in which case all mobile cells have the same porosity. Default is 0 (unitless).

ε_{im}—Porosity in each immobile cell, expressed as a fraction of the total volume of mobile and immobile cells. The ε_{im} is used only if *stagnant_cells* is 1, in which case all immobile cells have the same porosity. Default is 0 (unitless).

Line 11: **-thermal_diffusion** *temperature retardation factor, thermal diffusion coefficient*

-thermal_diffusion—Defines parameters for calculating the diffusive part of heat transport. Diffusive heat transport will be calculated as a separate process if the temperature in any of the solutions of the transport domain differs by more than 1 °C, and when the *thermal diffusion coefficient* is larger than the effective (aqueous) *diffusion coefficient*. Otherwise, diffusive heat transport is calculated as a part of aqueous diffusion. The *temperature retardation factor*, R_T, is defined as the ratio of the heat capacity of the total aquifer over the heat capacity of water in the pores, $R_T = 1 + \dfrac{(1-\varepsilon)\,\rho_s k_s}{\varepsilon\,\rho_w k_w}$, where ε is the water filled porosity, ρ is density (kg/m^3, kilogram per cubic meter), k is specific heat (kJ°C^{-1}kg^{-1}), and subscripts w and s indicate water and solid, respectively. The thermal diffusion coefficient, κ_e, can be estimated by using $\kappa_e = \dfrac{\kappa_t}{\varepsilon\rho_w k_w}$, where κ_t is the heat conductivity of the aquifer, including pore water and solid (kJ°C^{-1}m^{-1}s^{-1}, kilojoule per degree Celsius, per meter per second). The value of κ_e may be 100 to 1500 times larger than the aqueous diffusion coefficient, or about 1×10^{-6} m^2/s. A temperature change during transport is reduced by the temperature retardation factor (unitless) to account for the heat capacity of the matrix. Optionally, **-th[ermal_diffusion]**.

temperature retardation factor—Temperature retardation factor, unitless. Default is 2.0 (unitless).

thermal diffusion coefficient—Thermal diffusion coefficient. Default is the aqueous diffusion coefficient.

Line 12: **-initial_time** *initial_time*

-initial_time—Identifier to set the time at the beginning of a transport simulation. The identifier sets the initial value of the variable controlled by **-time** in the SELECTED_OUTPUT data block. Optionally, **initial_time** or **-i[nitial_time]**.

initial_time—Time (seconds) at the beginning of the transport simulation. Default is the cumulative time including all preceding ADVECTION simulations (for which **-time_step** has been defined) and all preceding TRANSPORT simulations.

Line 13: **-print_cells** *list of cell numbers*

-print_cells—Identifier to select cells for which results will be written to the output file. Optionally, **print**, **print_cells**, or **-pr[int_cells]**. Note that the hyphen is required to avoid a conflict with the keyword PRINT.

list of cell numbers—Printing to the output file will occur only for these cell numbers. The list of cell numbers may be continued on the succeeding line(s). A range of cell numbers may be included in the list in the form *m-n*, where *m* and *n* are positive integers, *m* is less than *n*, and the two numbers are separated by a hyphen without intervening spaces. Default is 1-*cells*.

Line 14: **-print_frequency** *print_modulus*

-print_frequency—Identifier to select shifts for which results will be written to the output file. Optionally, **print_frequency, -print_f[requency], output_frequency,** or **-o[utput_frequency]**.

print_modulus—Printing to the output file will occur for advection shifts or diffusion periods that are evenly divisible by *print_modulus*. Default is 1.

Line 15: **-punch_cells** *list of cell numbers*

-punch_cells—Identifier to select cells for which results will be written to the selected-output file. Optionally, **punch, punch_cells, -pu[nch_cells], selected_cells,** or **-selected_c[ells]**.

list of cell numbers—Printing to the selected-output file will occur only for these cell numbers. The list of cell numbers may be continued on the succeeding line(s). A range of cell numbers may be included in the list in the form *m-n*, where *m* and *n* are positive integers, *m* is less than *n*, and the two numbers are separated by a hyphen without intervening spaces. Default is 1-*cells*.

Line 16: **-punch_frequency** *punch_modulus*

-punch_frequency—Identifier to select shifts for which results will be written to the selected-output file. Optionally, **punch_frequency, -punch_f[requency], selected_output_frequency, -selected_o[utput_frequency]**.

punch_modulus—Printing to the selected-output file will occur for advection shifts or diffusion periods that are evenly divisible by *punch_modulus*. Default is 1.

Line 17: **-dump** *dump file*

-dump—Identifier to write a complete state of the advective-dispersive transport simulation to *dump file* after every *dump_modulus* advection shifts or diffusion periods. The file is formatted as an input file that can be used to restart calculations. Previous contents of the file are overwritten each time the file is written. Optionally, **dump** or **-du[mp]**.

dump file—Name of the file to which complete state of the advective-dispersive transport simulation will be written. Default is *phreeqc.dmp*.

Line 18: **-dump_frequency** *dump_modulus*

-dump_frequency—Complete state of the advective-dispersive transport simulation will be written to the dump file for advection shifts or diffusion periods that are evenly divisible by *dump_modulus*. Optionally, **dump_frequency** or **-dump_f**[**requency**].

dump_modulus—Number of advection shifts or diffusion periods. Default is *shifts*/2 or 1, whichever is larger.

Line 19: **-dump_restart** *shift number*

-dump_restart—If an advective-dispersive transport simulation is restarted from a dump file, the starting shift number is given on this line. Optionally, **dump_restart** or **-dump_r**[**estart**].

shift number—Starting shift number for the calculations, if restarting from a dump file. The shift number is written in the dump file by PHREEQC. It equals the shift number at which the dump file was created. Default is 1.

Line 20: **-warnings** [(*True* or *False*)]

-warnings—Identifier enables or disables printing of warning messages for transport calculations. In some cases, transport calculations could produce many warnings, which are not errors. Once it is determined that the warnings are not due to erroneous input, disabling the warning messages can avoid generating large output files. Default is **true** at startup. Optionally, **warnings**, **warning**, or **-w**[**arnings**].

(*True* or *False*)—If **true**, warning messages are printed to the screen and the output file; if **false**, warning messages are not printed to the screen nor the output file. The value set with **-warnings** is retained in all subsequent transport simulations until changed. If neither **true** nor **false** is entered on the line, **true** is assumed. Optionally, **t**[**rue**] or **f**[**alse**].

Line 21: **-multi_D** (*True* or *False*) *default_Dw porosity porosity_limit Archie_n*

-multi_D—Enables or disables the calculation of multicomponent diffusion. In multicomponent diffusion each solute can be given its own diffusion coefficient, allowing it to diffuse at its own rate, but with the constraint that overall charge balance is maintained (Vinograd and McBain, 1941; Appelo and Wersin, 2007), Optionally, **multi_D** or **-m**[**ulti_D**] (as with all identifiers, case insensitive). With **-multi_D true**, the diffusive flux is calculated by (see also Notes):

$$J_i = -(D_{w,i}' \times \varepsilon^n)\left(\frac{\partial \ln \gamma_i}{\partial \ln c_i} + 1\right) \mathrm{grad}(c_i) + CBt_i,$$

where i indicates the species; J_i is the flux (mol m^{-2}s^{-1}, mole per square meter per second); $D_{w,i}'$ is the (temperature corrected) tracer diffusion coefficient (m^2/s); ε is the water-filled (or accessible) porosity (unitless); n is an empirical exponent, known from Archie's law to be about 1; γ_i is the activity-coefficient (unitless); c_i is the concentration (mol/m^3, mole per cubic meter); $\mathrm{grad}(c_i)$ is the concentration gradient (mol/m^4, mole per meter to the fourth power), which may be different in free (uncharged) pore water and in the Donnan pore space on a surface (see Notes); and CBt_i is the charge balance term (Appelo and Wersin, 2007, see the Notes). The tracer diffusion coefficients are defined with keyword **SOLUTION_SPECIES** in *phreeqc.dat* for 25 °C, and corrected to temperature T (K) of the solution as follows:

$$D_w' = (D_w)_{298} \times \frac{T}{298} \times \frac{\eta_{298}}{\eta_T},$$

where η is the viscosity of water.

When **-multi_D** is **false**, the diffusive flux is calculated with $J_i = -D_p \times \mathrm{grad}(c_i)$, where D_p is the same for all species (defined with identifier **-diffusion_coefficient** as in Line 9) and not corrected for changes of temperature.

Note that PHREEQC assumes that, for diffusion, the cell contains water exclusively, and uses the pore water diffusion coefficient for calculating the flux. (The effective diffusion coefficient (D_e) is for a volume of grains and pores together, and is related to D_p as follows: $D_e = D_p \, \varepsilon$). The identifier **-stagnant** allows for nonuniform porosities, nonuniform tortuosities, and other variations in the diffusion domain, provided mixing factors among cells are defined explicitly in the input file with **MIX** data blocks.

(*True* or *False*)—If **true**, multicomponent diffusion is calculated; if **false**, diffusion is calculated with the diffusion coefficient given in Line 9.

default_Dw—The diffusion coefficient (m^2/s at 25 °C) given to solute species for which **-dw** is not defined in keyword **SOLUTION_SPECIES**. *T*he value must be used when calculating explicit mixing factors for stagnant cells. Default is 0 m^2/s.

porosity—The porosity filled with free and Donnan pore water in the cells; the *porosity* is a fraction of a representative volume of the porous medium (unitless). Initially all cells are defined with the same *porosity*. The porosity for a cell can be changed in any keyword data block that supports Basic programming (**RATES, USER_GRAPH, USER_PRINT**, and **USER_PUNCH**) by using the PHREEQC Basic function CHANGE_POR(*porosity, cell_no*). The porosity in a cell can be retrieved with the function GET_POR(*cell_no*). Default is 0 (unitless).

porosity_limit—The porosity limit, below which diffusion stops. Default is 0 (unitless).

Archie_n—The exponent *n* used for calculating the pore-water diffusion coefficient ($D_{p,i}$) from the tracer diffusion coefficient ($D_{w,i}$), $D_{p,i} = D_{w,i}\,\varepsilon^n$, where ε is the water filled (or the accessible) porosity (unitless), and *n* is an empirical exponent that varies from approximately 0.9 to 1.2 (Grathwohl, 1998; Van Loon and others, 2007) but may be higher for diffusion perpendicular to the bedding plane. The parameter ϑ^2 (approximately $1/\varepsilon_w^{\,n}$) is the tortuosity factor, which accounts for the longer diffusion path for a particle in a porous media than in pure water.

Line 22: **-interlayer_D** (*True* or *False*) *interlayer_porosity interlayer_porosity_limit interlayer_tortuosity_factor*

-interlayer_D—Enables or disables the calculation of interlayer diffusion in swelling clay minerals. If **-interlayer_**D is **true**, **-multi_D** also must be **true**, and the **-multi_D** parameters must be set as explained with Line 21. Optionally, **interlayer_D** or **-int[erlayer_D]** (as with all identifiers, case insensitive). The flux in the interlayers is calculated for the cations associated with X⁻ (as defined with keyword **EXCHANGE**):

$$J_i = -\frac{D_{w',i}}{\vartheta_{IL}^2} \times m_{CEC} \times \mathrm{grad}(\beta_l) + CBt_i,$$

where, *i* indicates an aqueous species, J_i is the flux (mol $m^{-2}s^{-1}$), $D_{w',i}$ is the temperature corrected diffusion coefficient, ϑ_{IL}^2 is the interlayer tortuosity factor (unitless), m_{CEC} is the concentration of total X^-, $mol(X^-)$ / $(m^3$ interlayer water), where $(m^3$ interlayer water) = $(m^3$ free pore water + m^3 Donnan water) $\times (\varepsilon_{IL} / \varepsilon)$, grad$(\beta_t)$ is the gradient of the equivalent fraction of species i on the exchange sites $(1/m)$, and CBt_i is the charge balance term (Appelo and Wersin, 2007). The tracer diffusion coefficients are defined with keyword **SOLUTION_SPECIES** in *phreeqc.dat* for 25 °C and are corrected to temperature T (K) of the solution with the equation $D_w' = (D_w)_{298} \times \frac{T}{298} \times \frac{\eta_{298}}{\eta_T}$, where η is the viscosity of water.

(True or *False)*—If **true**, interlayer diffusion is calculated; if **false**, interlayer diffusion is not calculated.

interlayer_porosity—The porosity of interlayer water, a fraction of the total volume. Default is 0 (unitless).

interlayer_porosity_limit—The porosity of interlayer water, below which interlayer diffusion stops. Default is 0 (unitless).

interlayer_tortuosity_factor—The tortuosity factor for interlayer diffusion, ϑ_{IL}^2 (unitless). Default is 100.0.

Notes

The advective-dispersive transport capabilities of PHREEQC are derived from a formulation of 1D, advective-dispersive transport presented by Appelo and Postma (2005). The 1D column is defined by a series of cells (number of cells is *cells*), each of which has the same pore volume. Lengths are defined for each cell and the time step (*time step*) gives the time necessary for a pore volume of water to move through each cell. Thus, the velocity of water in each cell is determined by the length of the cell divided by the time step. In the Example data block, a column of five cells (*cells*) is modeled and 5 pore volumes of filling solution are moved through the column (*shifts/cells* is 5). The total time of the simulation is 25 yr (year) (*shifts* × *time step*). The total length of the column is 6 m (four 1-m cells and one 2-m cell).

At each shift, advection is simulated by moving solution *cells* - 1 to cell *cells*, solution *cells* - 2 to cell *cells* - 1, and so on, until solution 0 is moved to cell 1 (upwind scheme). With flux-type boundary conditions, the dispersion steps follow the advective shift. With Dirichlet boundary conditions, the dispersion step and the advective shift are alternated. After each advective shift and dispersion step, kinetic reactions and chemical equilibria are calculated. The moles of pure phases and the compositions of the exchange assemblage, surface assemblage, gas phase, solid-solution assemblage, and kinetic reactants in each cell are updated after each chemical calculation.

For advective transport, the influent solution must be defined, otherwise the program stops with an error message. Solution 0 is the influent with **-flow_direction forward**; solution *cells* + 1 is the influent when the flow is **backward**. If the effluent (solution *cells* + 1, for direction **forward**) is not defined, the program copies it from the effluent boundary cell in the column. A closed boundary condition is not possible with advective flow, and the boundary condition will be changed to a flux boundary condition. Likewise, a flux-boundary condition is not possible when pure diffusion is modeled, and the boundary condition will be changed to a closed boundary condition.

The **-time_step** identifier defines the length of time associated with each advective shift or diffusion period. The program may subdivide this time step into smaller dispersion time steps if necessary to calculate dispersion accurately. Each dispersion time step may be further subdivided to integrate the kinetic reactions (**KINETICS** data block). Kinetic reactions are likely to slow the calculations by a factor of six or more compared to pure equilibrium calculations.

The numerical scheme is for cell-centered concentrations, which has consequences for data interpretation. Thus, the composition in a boundary cell is a half-cell distance away from the column outlet and needs a half time step to arrive at (or from) the column end. The half time step must be added to the total residence time in the column when effluent from a column is simulated [use (TOTAL_TIME + *time step* / 2) for time, see example 15, or ((STEP_NO + 0.5) / *cells*) for pore volumes, see example 11]. The kinetics time for advective transport into the boundary cell is the advective time step divided by 2. Also, the cell-centered scheme does not account for dispersion in the border half-cell in case of a flux boundary condition. The identifier **-correct_disp** provides an option to correct the ignored dispersion by increasing the dispersivity for all cells in the column by the appropriate amount. The correction will improve the comparison with analytical solutions for conservative elements when the number of cells is small.

A "dual porosity" model, in which part of the porosity allows advective flow and part of the porosity is accessible only by diffusion, can be developed with a first-order exchange model or with finite differences, and both approaches can be defined in terms of a mixing among cells (see "Transport in Dual Porosity Media" in Parkhurst and Appelo, 1999). With the **TRANSPORT** data block, one column of mobile cells is used to represent the part of the flow system in which advection occurs, and then additional immobile cells connected to the mobile cells are used to represent the stagnant zone that is accessible only by diffusion. The stagnant zone can be defined to be parallel or perpendicular to the column of mobile cells or to be a combination of the two by proper definition of mixing factors in MIX data blocks. A shortcut is available for the classical formulation of a dual porosity medium with a first-order rate of exchange. In this case, **-stagnant** is used to define one stagnant cell for each mobile cell (*stagnant_cells* = 1), an exchange factor (*exchange_factor*) for the exchange between immobile and mobile cells, and the porosities ϵ_m and ϵ_{im} for the mobile and immobile cells.

Thermal diffusion can be modeled for a stagnant zone with first-order exchange between mobile and immobile cells. Thermal exchange is calculated after subtracting the part of the exchange that is associated with hydrodynamic diffusion (see "Transport of Heat" in Parkhurst and Appelo, 1999). PHREEQC uses the value of the *diffusion coefficient* to find the correct heat exchange factor, and the value entered with identifier **-diffusion_coefficient** should be the same as has been used to calculate the exchange factor α (see equation 125 in Parkhurst and Appelo, 1999).

Most of the information for advective-dispersive transport calculations must be entered with other keyword data blocks. Advective-dispersive transport assumes that solutions with numbers 1 through *cells* have been defined by using SOLUTION, SOLUTION_SPREAD, or SAVE data blocks. In addition the infilling solution must be defined. If **-flow_direction** is **forward**, solution 0 is the infilling solution; if **-flow_direction** is **backward**, solution *cells + 1* is the infilling solution, if **-flow_direction** is **diffusion_only**, then infilling solutions at both column ends are optional. If stagnant zones are modeled, solution compositions for the stagnant-zone cells must be defined with SOLUTION, SOLUTION_SPREAD, or SAVE data blocks.

Pure-phase assemblages may be defined with EQUILIBRIUM_PHASES or SAVE, with the number of the assemblage corresponding to the cell number. Likewise, an exchange assemblage, a surface assemblage, a gas phase, or a solid-solution assemblage can be defined for each cell through EXCHANGE, SURFACE, GAS_PHASE, SOLID_SOLUTIONS, or SAVE keywords, with the identifying number corresponding to the cell number. Kinetically controlled reactions can be defined for each cell through the KINETICS data block. Note that ranges of numbers can be used to define multiple solutions, exchange assemblages, surface assemblages, gas phases, solid-solution assemblages, or kinetic reactions simultaneously and that SAVE allows definition of a range of numbers. Constant-rate reactions can be defined for mobile or immobile cells through REACTION data blocks, again with the identifying number of the REACTION data block corresponding to the cell number. REACTION_TEMPERATURE data blocks can be used to specify the initial temperatures of the cells in the transport simulation. Temperatures in the cells may change during the transport simulation depending on the temperature distribution and the *temperature retardation factor* defined by **-thermal_diffusion**. REACTION_PRESSURE data blocks can be used to set the pressure in each cell.

By default, the composition of the solution, pure-phase assemblage, exchange assemblage, surface assemblage, gas phase, solid-solution assemblage, and kinetic reactants are printed for each cell for each shift. Use of **-print_cells** and **-print_frequency** will limit the amount of data written to the output file. If **-print_cells** has been defined then only the specified cells will be written; otherwise, all cells will be written. The identifier **-print_frequency** will restrict writing to the output file to those shifts that are evenly divisible by *print_modulus*. In the Example data block, results for cells 1, 2, 3, and 5 are written to the output file after each integer pore volume (5 shifts) has passed through the column. Data written to the output file can be further limited with the keyword PRINT (see **-reset false**).

If a SELECTED_OUTPUT data block has been defined, then selected data are written to the selected-output file. Use of **-punch_cells** and **-punch_frequency** in the **TRANSPORT** data block will limit the data that are written to the selected-output file. If **-punch_cells** has been defined then only the specified cells will be written; otherwise, all cells will be written. The identifier **-punch_frequency** will restrict writing to the selected-output file to those shifts that are evenly divisible by *punch_modulus*. In the Example data block, results are written to the selected-output file for cells 2, 3, 4, and 5 after each integer pore volume (5 shifts) has passed through the column.

At the end of an advective-dispersive transport simulation, all the physical and chemical data (for example, compositions of solutions, equilibrium-phase assemblages, surfaces, exchangers, solid solutions, and kinetic reactants) are automatically saved and are identified by the cell number in which they reside. These data are available for subsequent simulations within a single run. Transient conditions can be simulated by including subsequent SOLUTION and **TRANSPORT** data blocks, which may define new chemical boundary and transport conditions. Only parameters that differ from the previous advective-dispersive transport simulation need to be redefined, such as new infilling solution (SOLUTION 0), a change from advection to diffusion only (**-flow_direction diffusion_only**), or a change in flow direction from forward to backward (**-flow_direction backward**). All parameters not specified in the new **TRANSPORT** data block remain the same as the previous advective-dispersive transport simulation. Normally, the diffusion coefficient, lengths of cells, dispersivities, and stagnant zone definitions remain the same through all advective-dispersive transport simulations and thus need not be redefined.

For long advective-dispersive transport calculations, it may be desirable to save intermediate states in the calculation, either because of hardware failure or because of nonconvergence of the numerical method.

The -**dump_frequency** identifier allows intermediate states to be saved at intervals during the calculation. The -**dump** identifier allows the definition of a file name in which to write these intermediate states. The dump file is formatted as an input file for PHREEQC, so calculations can be resumed from the point at which the dump file was made. The -**dump_restart** identifier allows a shift number to be specified from which to restart the calculations.

With the identifiers -**multi_D** and -**interlayer_D**, diffusion can be calculated as a multicomponent process in uncharged ("free") pore water, in electrostatic double layer (EDL, referred to as the Donnan pore space) water at charged surfaces, and in interlayer water of swelling clay minerals. Each solute species or surface defined with -**dw** greater than zero diffuses at its own speed, while overall, charge balance is maintained (Vinograd and McBain, 1941; Appelo and Wersin, 2007). Instead of concentration, as in Fick's laws, the thermodynamic potential forms the basis for calculating the multicomponent flux:

$$\mu_i = \mu_i^0 + RT \ln a_i + z_i F \psi, \tag{9}$$

where μ_i^0 is the standard thermodynamic potential of species i (J/mol, joule per mole), R is the gas constant (8.314 J K^{-1}mol^{-1}), T is the absolute temperature (K), a_i is the activity (unitless), z_i is charge number (unitless), F is the Faraday constant (96485 J V^{-1}eq^{-1}), and ψ is the electrical potential (V). The activity is related to concentration c_i (here, mol/m^3) by $a_i = \gamma_i c_i/c^0$, where γ_i is the activity coefficient (unitless) and c^0 is the standard state (here 1.0 mol/m^3). The diffusive flux of i as a result of chemical and electrical potential gradients is

$$J_i = -\frac{u_i c_i}{|z_i| F} \nabla(\mu_i) = -\frac{u_i RT}{|z_i| F} c_i \nabla(\ln a_i) - \frac{u_i}{|z_i|} c_i z_i \nabla(\psi), \tag{10}$$

where J_i is the flux of species i (mol m^{-2}s^{-1}), u_i is the mobility (m^2 s^{-1}V^{-1}, square meter per second per volt), c_i is the concentration (mol/m^3), z_i is charge number (unitless), $\nabla(\mu_i)$ is the gradient of the chemical potential (J mol^{-1}m^{-1}, joule per mole per meter), and similarly for the gradients of lna_i and ψ. If the electrical current is zero, $\sum_j z_j J_j = 0$, equation 10 can be rearranged to solve for the gradient of the electrical potential:

$$\nabla(\psi) = \frac{\sum_j -\frac{u_j \mathrm{R}T}{|z_j|\mathrm{F}} c_j z_j \nabla(\ln a_j)}{\sum_j \frac{u_j}{|z_j|} c_j z_j^2},$$ (11)

where the variables are linked with species j to indicate their origin from the zero-charge transfer condition. With $a_i = \gamma_i c_i/c^0$ and $c_i \mathrm{d}(\ln c_i) = \mathrm{d}(c_i)$, the gradient of the activity becomes:

$$c_i \nabla(\ln a_i) = \left(\frac{\mathrm{d}(\ln \gamma_i)}{\mathrm{d}(\ln c_i)} + 1 \right) \nabla c_i.$$ (12)

And, finally, using the identity

$$\frac{u_i \mathrm{R}T}{|z_i|\mathrm{F}} = D_{w,i},$$ (13)

equation 10 can be recast completely in known model variables:

$$J_i = -D_{w,i} \left(\frac{\mathrm{d}(\ln \gamma_i)}{\mathrm{d}(\ln c_i)} + 1 \right) \nabla c_i + D_{w,i} c_i z_i \frac{\sum_i D_{w,j} z_j \left(\frac{\mathrm{d}(\ln \gamma_j)}{\mathrm{d}(\ln c_j)} + 1 \right) \nabla c_j}{\sum_j D_{w,j} c_j z_j^2}.$$ (14)

Appelo and Wersin (2007) have shown that the thermodynamic potential gradients are the same in free pore water and in EDL water (the Donnan pore space); only the concentrations are different in the two water types. The concentration in the Donnan pore space is:

$$c_{Donnan,i} = c_i \times erm\_DDL_i \times \exp\left(\frac{-z_i \mathrm{F} \psi_D}{\mathrm{R}T} \right),$$ (15)

where erm_DDL_i is the enrichment factor in the Donnan pore space, defined with identifier **-erm_ddl** in keyword **SOLUTION_SPECIES**, and ψ_D is the potential in the Donnan pore space. Accordingly, for calculating the flux through the Donnan space, if **-only_counter_ions** is **false**, c_i in equation 14 is replaced by $c_{Donnan,i}$ for the fraction of Donnan water in the pore space. If **-only_counter_ions** is **true**, the concentration gradient in free pore water is used for counter-ions (c_i in equation 14 is *not* replaced for counter-ions), and for co-ions the flux is zero (their concentration is zero), see below.

Diffusion of exchangeable cations can be calculated when they are defined with exchange species X⁻, as in the databases *phreeqc.dat* and *wateq4f.dat*. The concentration of an exchangeable species is

$$c_i = \gamma_t \times \beta_t \times CEC / z_i, \tag{16}$$

where c_i is the concentration of species i (mol/m^3 interlayer water), γ_t is the activity coefficient, defined with -**gamma** in keyword **EXCHANGE_SPECIES**, β is the equivalent fraction of the exchangeable species, and CEC is the exchange capacity (eq/m^3, equivalent per cubic meter of interlayer water), defined as moles X$^-$ in keyword **EXCHANGE** (the volume of interlayer water is calculated from the interlayer porosity, as explained below). For calculating the flux through the interlayer space, ∇c_i in equation 14 is replaced by $\frac{CEC}{z_i}\nabla\beta_i$.

The actual mass transfer is found by multiplying the flux by the surface area, which PHREEQC calculates either from the amount of water in a cell and the cell length in a regular column, or from the mixing factor defined for stagnant cells, assuming a density of water of 1 kg/L. Thus, in a regular column, the surface area for diffusion between two cells i and j is

$$A_{ij} = \frac{w_{free}/1000 + V_{Donnan}}{\Delta x}, \tag{17}$$

where A_{ij} is the surface area (m^2), w_{free} is kgw defined with -**water** in keyword **SOLUTION**, V_{Donnan} is the volume of water in the Donnan pore space, equal to $A_{surf} \times t_{Donnan}$, the product of the area of the surface (m^2) and the thickness of the Donnan layer (m), both defined in keyword **SURFACE**, and Δx is the cell length (m) defined with -**lengths**.

For stagnant cells, the mixing factor is multiplied by the ratio of the diffusion coefficient of a species and the default diffusion coefficient that is used when calculating the mixing factor. The mixing factor is given by equation 128 of the PHREEQC version 2 manual (Parkhurst and Appelo, 1999) or equation S2 of Appelo and Wersin, 2007):

$$mixf_{ij} = \frac{D_p \Delta t A_{ij} f_{bc}}{h_{ij} V_j}, \tag{18}$$

where $mixf_{ij}$ is the mixing factor between cells i and j, defined with keyword **MIX**. D_p is the pore water diffusion coefficient, given by $D_p = D_w \varepsilon^n$, where D_w is the default diffusion coefficient; ε is the porosity; and n is Archie's factor, all defined with identifier -**multi_D**. Δt is the time step defined with -**time_step**. f_{bc} is a correction factor that equals 2 for constant concentration boundary cells and is 1 otherwise, h_{ij} is the distance between the midpoints of cells i and j, and V_j is the volume of pore water in cell j for which the mass transfer is calculated. The $mixf_{ij}$ that is defined in the input file must be calculated with a volume V_j of 0.001 m^3. PHREEQC will adapt the value of $mixf_{ij}$ by using the actual amount of water in the cell, which is given by (w_{free}/1000 + V_{Donnan}). Furthermore, (according to equation 18) the mixing factor for individual species is multiplied by $D_{w,\ i}/D_w$.

For interlayer diffusion, PHREEQC calculates the surface area by

$$A_{ij,\ IL} = \frac{\varepsilon_{IL}}{\varepsilon}A_{ij}, \tag{19}$$

where ε_{IL} is the porosity occupied by interlayer water, defined with identifier **-interlayer_D**, and ε is the porosity of free and Donnan pore water together, defined with **-multi_D**. For stagnant cells, the mixing factor is multiplied with the ratio of the interlayer porosity and the free and Donnan porosity, and with the ratio of the diffusion coefficients, and of the inverse tortuosity coefficients. If the surface areas and (or) the porosities are different for the two cells, the harmonic mean is used (see example 21, in Examples).

For the Donnan pore space, the enhanced concentration gradient for counter-ions and the decreased concentration gradient for co-ions is applied if identifier **-only_counter_ions** is **false** in keyword SURFACE. If identifier **-only_counter_ions** is **true**, the concentration of the co-ions is zero in the Donnan space, and also the flux of the co-ions is zero. In this case, the concentration gradient of counter-ions in free pore water is used for the Donnan pore space for two reasons. First, because with only counter ions in the Donnan pore space, the concentrations of the counter-ions will be smaller than in free pore water if the surface charge is smaller than the charge of the co-ions in solution. This could give unrealistic gradients when surface charges are different from cell to cell. The second reason is that it allows direct comparison of multicomponent results of PHREEQC with traditional calculations in which Fick's laws are used to model the behavior of individual tracers in clays and clay-containing rocks.

Example problems

The keyword **TRANSPORT** is used in example problems 11, 12, 13, 15, and 21. Examples of multicomponent diffusion are given in Appelo and Wersin (2007), supplementary information, Appelo and others (2008) and (2010), and in *http://www.hydrochemistry.eu/exmpls/index.html#new* (accessed June 25, 2012). An example of interlayer diffusion is available in *http://www.hydrochemistry.eu/exmpls/opa_col.html#new2* (accessed June 25, 2012).

Related keywords

ADVECTION, EQUILIBRIUM_PHASES, EXCHANGE, GAS_PHASE, KINETICS, MIX, PRINT, REACTION, REACTION_TEMPERATURE, SAVE, SELECTED_OUTPUT, SOLID_SOLUTIONS, SOLUTION, and SURFACE.

This keyword data block is used to specify explicitly which solution, exchange assemblage, gas phase, pure-phase assemblage, solid-solution assemblage, or surface assemblage is to be used in the batch-reaction calculation of a simulation. **USE** also can specify kinetically controlled reactions (KINETICS data block), reaction parameters (REACTION data block), reaction-pressure parameters (REACTION_PRESSURE data block), reaction-temperature parameters (REACTION_TEMPERATURE data block), and mixing parameters (MIX data block) to be used in a batch-reaction calculation.

Example data block

```
Line 0a: USE equilibrium_phases none
Line 0b: USE exchange 2
Line 0c: USE gas_phase 3
Line 0d: USE kinetics 1
Line 0e: USE mix 1
Line 0f: USE reaction 2
Line 0g: USE reaction_pressure 1
Line 0h: USE reaction_temperature 1
Line 0i: USE solid_solution 6
Line 0j: USE solution 1
Line 0k: USE surface 1
```

Explanation

Line 0: **USE** *keyword*, (*number* or **none**)

USE is the keyword for the data block.

keyword—One of 11 keywords, **equilibrium_phases**, **exchange**, **gas_phase**, **kinetics**, **mix**, **reaction**, **reaction_pressure**, **reaction_temperature**, **solid_solutions**, **solution**, or **surface**.

number—Positive integer associated with previously defined composition or reaction parameters.

none—No reactant of the type of the specified keyword will be used in the batch-reaction calculation.

Notes

Batch reactions are defined by allowing a solution or mixture of solutions to come to equilibrium with one or more of the following entities: an exchange assemblage, a pure-phase assemblage, a solid-solution assemblage, a surface assemblage, or a gas phase. In addition, kinetically controlled reactions, stoichiometric reactions, reaction pressures, and reaction temperatures can be specified for batch-reaction calculations.

Entities can be defined implicitly—a solution or mixture (SOLUTION or MIX keywords) must be defined within the simulation, then the first of each kind of entity defined in the simulation will be used to define the reaction system. Thus, the first solution (or mixture) will be brought together with the first of each of the following entities that is defined in the simulation: exchange assemblage (EXCHANGE), gas phase (GAS_PHASE), pure-phase assemblage (EQUILIBRIUM_PHASES), solid-solution assemblage (SOLID_SOLUTIONS), surface assemblage (SURFACE); equilibrium among these entities will be calculated and maintained. Irreversible reactions may also be added implicitly to the system, and again, the first of the following entities that is defined in the simulation is added: kinetically controlled reaction (KINETICS), stoichiometric reaction (REACTION), reaction pressure (REACTION_PRESSURE), and reaction temperature (REACTION_TEMPERATURE).

Entities to be included in the system can be defined explicitly with the USE keyword. Any combination of USE *keyword number* data blocks can be used to define a system. "USE *keyword* none" can be used to eliminate an entity that was implicitly defined to be in the system. For example, if only a solution and a surface are defined in a simulation and the surface is defined to be in equilibrium with the solution, then implicitly, an additional batch-reaction calculation will be made to equilibrate the solution with the surface. Though not incorrect, the batch-reaction calculation will produce the same compositions for the solution and surface as previously defined. By including "USE solution none", the batch-reaction calculation will be eliminated.

The composition of the solution, exchange assemblage, solid-solution assemblage, surface assemblage, pure-phase assemblage, or gas phase can be saved after a set of batch-reaction calculations with the SAVE keyword.

The RUN_CELLS data block can be used to define a specific batch-reaction calculation. With RUN_CELLS; -cells *n*, all reactants that are numbered *n* are put together and reacted. If MIX *n* has been defined, it will take precedence over SOLUTION *n*. If neither MIX *n* nor SOLUTION *n* have been defined, then no reaction will be calculated. USE data blocks have no effect on the selection of reactants for a

RUN_CELLS calculation. The compositions of reactants following a RUN_CELLS calculation are automatically saved with the same identifying number, n.

Example problems

The keyword **USE** is used in example problems 2, 3, 6, 7, 8, 10, 14, 20, and 22.

Related keywords

EQUILIBRIUM_PHASES, EXCHANGE, GAS_PHASE, KINETICS, MIX, REACTION, REACTION_PRESSURE, REACTION_TEMPERATURE, RUN_CELLS, SAVE, SOLID_SOLUTIONS, SOLUTION, and SURFACE.

USER_GRAPH

This keyword data block is used to create charts of simulation results. The data block defines the data to be charted and the parameters that control the appearance of the chart. Data to be plotted are defined with Basic programs. Observations or other data points can be added to a chart from user-specified files. Multiple charts may be defined for one or more simulations by using multiple **USER_GRAPH** data blocks with different identifying numbers. Different data may be added to a chart in a subsequent simulation by defining a data block with the same identifying number, but with a different Basic program to define the variables to be plotted. A new chart may be defined with the same identifying number if an intervening **USER_GRAPH** data block includes the identifier **-detach**. Each chart is a different program thread; once detached, the thread is still running, the chart is still viewable, and its data can be inspected or written to file, but no new data can be added. After a chart is detached, a new **USER_GRAPH** data block with the same identifying number will generate a new chart.

The keyword follows the syntax of the **USER_GRAPH** data block in PHREEQC for Windows (Post, 2012) and relies on John Champion's *zedgraph* software (*http://sourceforge.net/projects/zedgraph*, accessed May 18, 2012). **USER_GRAPH** is not available in the standard distribution of the Linux version of PHREEQC, but can be implemented on a Linux computer by installing *wine*; using *winetricks* to install *dotnet20*, *dotnet20sp2*, *vcrun2008*, and *gdiplus*; and running PHREEQC compiled for windows with **#define MULTICHART**.

Example data block 1

```
Line 0: USER_GRAPH 3 Plots F and pH against Ca concentration
Line 1:      -headings              F pH
Line 2:      -axis_titles   "Calcium, in milligrams per liter" \
                            "Fluoride, in milligrams per liter" "pH"
Line 3:      -chart_title   "Fluorite Equilibrium in Ca(OH)2 Solutions"
Line 4:      -axis_scale x_axis      0 350 50 25
Line 4a:     -axis_scale y_axis      0 7 1
Line 4b:     -axis_scale sy_axis     7 14 0 0
Line 5:      -initial_solutions      false
Line 6:      -connect_simulations    true
Line 7:      -plot_concentration_vs  x
Line 8:      -plot_tsv_file          filename
Line 9:      -batch                  filename.emf  false false
Line 10:     -start
```

```
Basic:  10   PLOT_XY  TOT("Ca")*40.08e3, TOT("F")*19e3, color = Red,\
                  symbol = Square, symbol_size = 6, y-axis = 1
Basic:  20   PLOT_XY  TOT("Ca")*40.08e3, -LA("H+"), color = Green, \
                  symbol = Diamond, symbol_size = 7, y-axis = 2, \
                  line_width = 1
Line 11:      -end
Line 0a: USER_GRAPH 4
Line 12:      -detach
Line 0b: USER_GRAPH 1
Line 13:      -active                    false
```

Explanation 1

Line 0: **USER_GRAPH** [*number*] [*description*]

USER_GRAPH is the keyword for the data block.

number—A positive number designates the user-graph definition. Default is 1.

description—Optional comment that describes the user-graph chart. The *description* will appear in the title of the chart window.

Line 1: **-headings** *labels*

-headings—Identifier provides labels for chart lines. The labels are separated by spaces and correspond with the order that Y and secondary Y curves are calculated with PLOT_XY Basic statements. Optionally, **heading**, **headings**, or **-h**[**headings**].

labels—List of labels, one for each of the curves. In Example data block 1 (see Basic lines 10–20), "F" corresponds to the first PLOT_XY curve, and "pH" corresponds to the second PLOT_XY curve. The labels can be changed in subsequent simulations for proper identification of the parameters graphed (without need of repeating the Basic statements that define the data to be plotted).

Line 2: **-axis_titles** *label$_1$ label$_2$ label$_3$*

-axis_titles—Identifier provides labels for the X, Y, and secondary Y axes. Optionally, **axis_titles** or **-a**[**xis_titles**].

label$_1$—Label printed below the chart along the X axis.

label$_2$—Label printed to the left of the chart along the Y axis.

label$_3$—Label printed to the right of the chart along the Y2 axis.

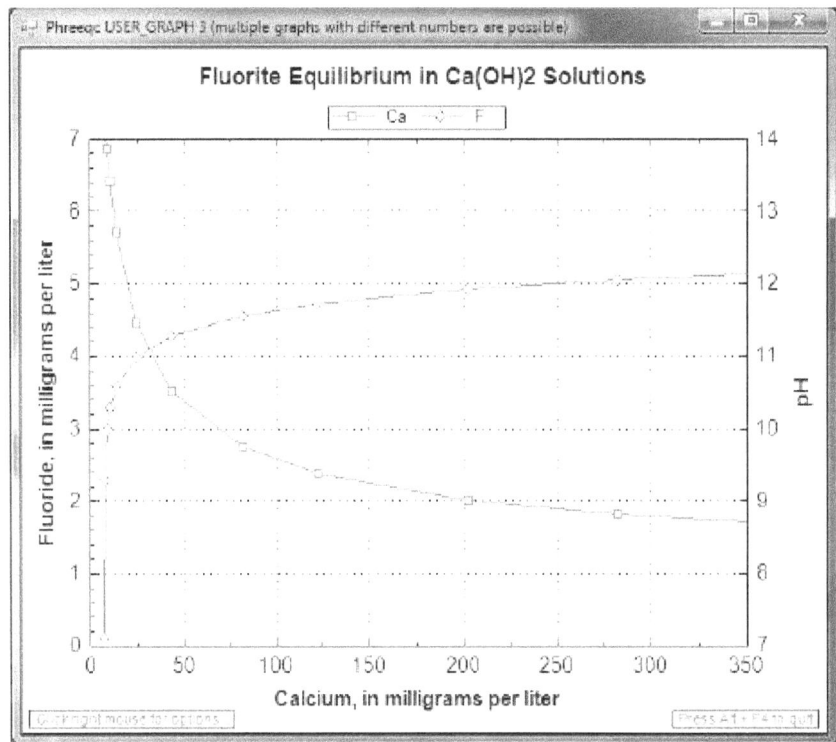

Figure 1. Chart from Example data block 1 plotting fluoride concentration and pH against calcium concentration for calcium hydroxide solutions in equilibrium with fluorite.

Line 3: **-chart_title** *title*

> **-chart_title**—Identifier provides a title that is printed at the top of the chart. Optionally, **chart_title** or **-c[hart_title]**.

> *title*—Title for the chart.

Line 4: **-axis_scale** (**x_axis**, **y_axis**, or **sy_axis**) [(*min* or **auto**) [(*max* or **auto**) [(*major* or **auto**) [(*minor* or **auto**) [**log**]]]]]

> **-axis_scale**—Identifier provides parameters for scaling the X, Y, or secondary Y axis. If less than five items are listed, the missing scaling parameters are determined by default algorithms. If **axis_scale** is not specified, the program will adjust the scale to a range that displays all the data points. Optionally, **axis_scale** or **-axis_s[cale]**.

> **x_axis**, **y_axis**, or **sy_axis**—Selects the axis for which scaling parameters are provided: X, Y, or secondary Y, respectively.

> *min* or **auto**—The minimum value for the axis, determined automatically if **auto** is specified.

> *max* or **auto**—The maximum value for the axis, determined automatically if **auto** is specified.

major or **auto**—The spacing of major tick marks for the axis, determined automatically if **auto** is specified.

minor or **auto**—The spacing of minor tick marks for the axis, determined automatically if **auto** is specified.

log—If specified, axis is scaled logarithmically.

Line 5: **-initial_solutions** [(*True* or *False*)]

-initial_solutions—Identifier selects whether to plot results from initial solution, initial exchange, initial surface, and initial gas-phase calculations. Default is **false** if **-initial_solutions** is not included. Optionally, **initial_solutions** or -i[nitial_solutions].

(*True* or *False*)—If **true**, results of initial calculations are plotted on the chart; if **false**, results of initial calculations are not plotted on the chart. If neither **true** nor **false** is entered on the line, **true** is assumed. Optionally, **t[rue]** or **f[alse]**.

Line 6: **-connect_simulations** [(*True* or *False*)]

-connect_simulations—Identifier selects whether to retain curve properties (colors, symbols, line widths) in subsequent simulations, or in subsequent shifts for transport and advection simulations. Default value is **true** if **-connect_simulations** is not included. Optionally, **connect_simulations** or -co[nnect_simulations].

(*True* or *False*)—If **true**, curve properties are retained for each additional simulation; if **false**, curve properties will differ with each simulation. If neither **true** nor **false** is entered on the line, **true** is assumed. Optionally, **t[rue]** or **f[alse]**.

Line 7: **-plot_concentration_vs** (**x** or **t**)

-plot_concentration_vs—Identifier selects whether to plot distance or time on the X axis in advection or transport simulations. Default is **x** if **-plot_concentration_vs** is not included. Optionally, **plot_concentration_vs** or -p[lot_concentration_vs].

x or **t**—**x** (or **d**) indicates distance, **t** indicates time.

Line 8: **-plot_tsv_file** *filename*

-plot_tsv_file—Identifier selects a file containing data to be plotted on the chart. The first line of the file is a set of headings, one for the X axis, followed by one for each curve to be plotted. All headings and data are tab delimited. It is possible to set curve properties by special values in the first column beginning at line 2 in the file. Up to five lines of special values may be defined; each special value is followed by settings for each curve (tab delimited).

The special values are "color", "symbol", "symbol_size", "line_width", and "y_axis". Legal values for these settings are described in the explanation for Basic line 10. The data lines follow the lines that define special values; each data line has an X value in the first column followed by Y values for each curve; missing values are indicated by consecutive tab characters. Data from multiple files can be added to the chart by using multiple instances of Line 8. Optionally, **plot_tsv_file** or **-plot_t[sv_file]**.

filename—Name of the file containing data to be plotted. The program stops with an error message if the file is not found. It is necessary to give the full path name for the file if the working directory is not the same as the directory that contains the file with data to be plotted.

Line 9: **-batch** [*filename.suffix* [(*True* or *False*) [(*True* or *False*)]]]

-batch—This identifier is used to close the chart automatically at the end of the run, and optionally, save the chart to file with or without the yellow background for the chart area. If **-batch** is defined, then the chart will be closed automatically at the end of the run. Optionally, **batch** or **-b[atch]**.

filename.suffix—If *filename.suffix* is entered following **-batch**, then a chart file of type *suffix* will be saved at the end of the run. *Suffix* may be any of the following: emf, bmp, jpeg, jpg, png, bmp, or tiff. If *filename.suffix* is not entered, then the chart will not be saved to a file at the end of the run. Note that if **-batch** is not defined, the chart can be saved by right clicking on the chart and choosing the "Save image as..." option.

(*True* or *False*)—If **true** is entered following *filename.suffix*, then the yellow background will be included in the saved chart; if **false** is entered following *filename.suffix*, then the saved chart will have no colored background. If neither **true** nor **false** is entered on the line, **true** is assumed. Optionally, **t[rue]** or **f[alse]**.

(*True* or *False*)—If **true** is entered, then the grid lines will be included in the saved chart; if **false** is entered then the saved chart will have not have grid lines. If neither **true** nor **false** is entered on the line, **true** is assumed. Optionally, **t[rue]** or **f[alse]**.

Line 10: **-start**

-start—Indicates the start of the Basic program. Optional.

Basic: *numbered Basic statement*

> *numbered Basic statement*—A valid Basic language statement that must be numbered. The statements are evaluated in the order of the line numbers. Statements and functions that are available through the Basic interpreter are listed in The Basic Interpreter, tables 7 and 8.

Basic: *number* PLOT_XY *expression$_1$, expression$_2$* [, **color** = *color*] [, **symbol** = *symbol*] [, **symbol_size** = *i*] [, **line_width** = *j*] [, **y-axis** = *k*]

> *number*—Number for a Basic statement.

> PLOT_XY—Basic command that sets the X and Y values of a data point for a curve in the chart.

> *expression$_1$*—Basic expression that evaluates to a number for the X value of a data point. Must be followed by a comma to separate it from the Y value.

> *expression$_2$*—Basic expression that evaluates to a number for the Y value of a data point.

> **color** = *color*—Option that sets the color for the data points. Legal values for *color* are Red, Green, Blue, Orange, Magenta, Yellow, Black, Cyan, Brown, Lime, and Gray (lower case is permitted). (Additional colors as defined in Microsoft .NET 2 also may be used, see *http://msdn.microsoft.com/en-us/library/aa358802%28v=VS.85%29.aspx*, accessed May 18, 2012). If the option is not specified, the colors for curves will cycle sequentially through the listed colors.

> **symbol** = *symbol*—Option that sets the symbol for the data points. Legal values for *symbol* are Square, Diamond, Triangle, Circle, XCross, Plus, Star, TriangleDown, HDash, VDash, and None (spelled exactly as noted). If the option is not specified, the symbols for curves will cycle sequentially through the listed symbols.

> **symbol_size** = *i*—Option that sets the symbol size for the data points. The *i* parameter is an integer greater than or equal to zero. Default is 6. The symbol is not plotted if $i = 0$ (similar to **symbol** = None).

> **line_width** = *j*—Option that sets the line width for the curve. The *j* parameter is an integer greater than or equal to zero. Default is 1. The line is not plotted if $j = 0$.

> **y_axis** = *k*—Option that sets the Y axis for the data points. If *k* is 1, the data are plotted relative to the primary Y axis; if *k* is 2, the data are plotted relative to the secondary Y axis.

Line 11: **-end**

-**end**—Indicates the end of the Basic program. Optional. Note the hyphen is required to avoid a conflict with the keyword **END**.

Line 12: **-detach**

-**detach**—Indicates that no more data will be added to the chart and that the identifying number is available for a new **USER_GRAPH** definition. The chart window remains visible and all mouse functions for the chart remain functional.

Line 13: **-active** [(*True* or *False*)]

-**active**—Allows plotting of data to the chart to be suspended or resumed. Default is **true** if -**active** is not included.

(*True* or *False*)—If **false**, plotting of data is suspended; if **true**, plotting is resumed. If neither **true** nor **false** is entered on the line, **true** is assumed. Optionally, **t[rue]** or **f[alse]**.

Notes 1

Example data block 1 defines a chart by using PLOT_XY Basic statements, whereas Example data block 2 (below) defines the same chart by using GRAPH_X, GRAPH_Y, and GRAPH_SY Basic statements. One, or a combination of these two types of Basic statements, is required to obtain a plot; all other identifiers are optional. PLOT_XY statements are useful for plotting curves when the X-axis values differ for the variables in the same reaction or transport step, and for controlling curve properties: colors, symbols, and line widths. The Basic program may contain any legitimate numbered Basic statements in addition to the lines containing PLOT_XY. Basic statements are evaluated in numerical order. Statements and functions that are available through the Basic interpreter are listed in The Basic Interpreter, tables 7 and 8.

The PLOT_XY command is followed by a pair of expressions that evaluate to the X and Y values for a data point of a curve. The point may be relative to the primary Y axis (**y-axis = 1**) or the secondary Y axis (**y-axis = 2**). The characteristics of the symbols and lines for a curve can be specified with the options **color**, **symbol**, **symbol_size**, and **line_width**. The options require an equals sign and may be separated by commas or spaces.

Identifier -**active** allows activation and deactivation of a chart for sequences of simulations during a run. Plotting of data in a specific chart is interrupted by setting -**active** *false*. Plotting can be resumed in later simulations by setting -**active** *true*. Identifier -**detach** will stop plotting of a specific chart during the

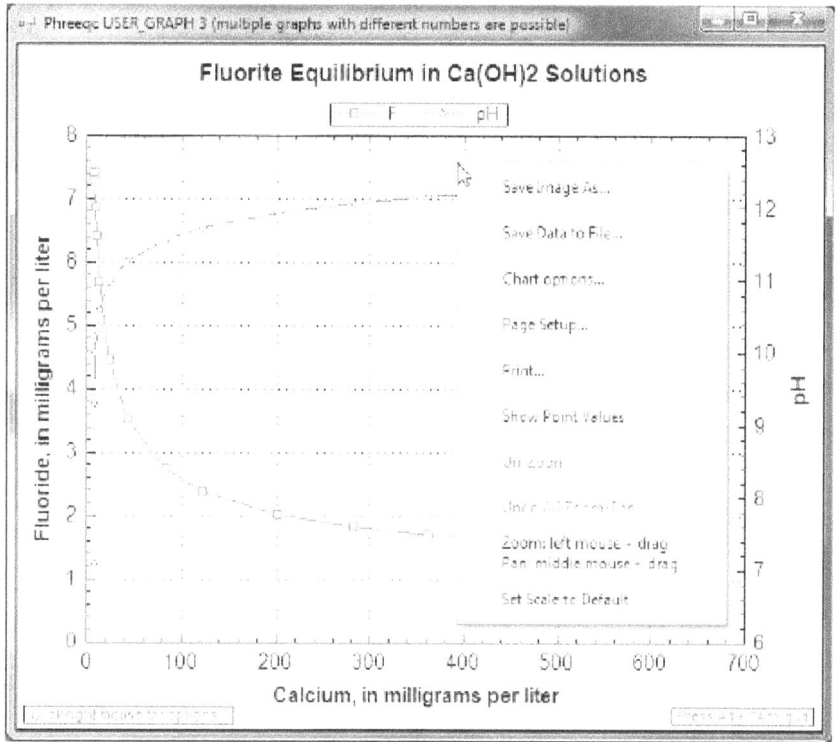

Figure 2. Chart created by Example data block 2 showing options available when right-clicking inside a chart window.

remaining simulations and permits redefinition of a chart with the same number. Plotting of all the charts can be interrupted and restarted with **PRINT; -user_graph** *false* / *true*.

The labels defined with **-headings** will be used for labeling curves in subsequent simulations until redefined. For example, the simulation in Example block 1 may be repeated for 50 °C in the same run. F and pH from that run may then be labeled with **USER_GRAPH**; -heading F(50C) pH; END.

Right-clicking on the window of a chart created by **USER_GRAPH** will display a number of options for manipulating the chart, as shown in figure 2.

Save Image As...—Allows the image to be saved in a variety of graphic formats.

Save Data to File...—Saves the X and Y values of the curves to a user-selected file.

Chart options...—Allows showing or hiding red hints boxes, colored chart background, and chart grid lines.

Page Setup...—Allows the page to be formatted for printing (alternatively, use Print...).

Print...—Prints the chart. The graphics quality is excellent if a PDF printer is selected.

Show Point Values—Displays the X–Y values of the point under the mouse cursor.

Zoom: left mouse + drag—Pressing and dragging the left mouse button will draw a rectangle; the rectangle will be enlarged to the plot window when the button is released.

Pan: middle mouse + drag—Rolling the middle mouse button up or down changes the scales of the axes. Pressing the middle mouse and dragging the mouse pans the viewable area of the chart.

Un-Zoom/Un-Pan—Undoes last zoom or pan. The option changes depending on whether the last operation was a zoom or a pan.

Undo All Zoom/Pan—Returns to the original scaling for the chart.

Set Scale to Default—Scales the axes to values that will display all the data points. This is the default scaling that appears when the identifier **-axis_scale** is not used.

Example data block 2

```
Line 0:  USER_GRAPH 5 Uses graph_x, graph_y, and graph_sy
Line 1:       -headings      Ca F pH
Line 2:       -axis_titles   "Calcium, in milligrams per liter" \
                             "Fluoride, in milligrams per liter" "pH"
Line 3:       -start
Basic:  10 GRAPH_X TOT("Ca") * 40.08e3
Basic:  20 GRAPH_y TOT("F") * 19e3
Basic:  30 GRAPH_SY -LA("H+")
Line 10:      -end
```

Explanation 2

Line 0: **USER_GRAPH** [*number*] [*description*]

USER_GRAPH is the keyword for the data block.

number—A positive number designates the user-graph definition. Default is 1.

description—Optional comment that describes the user-graph chart. The *description* will appear in the title of the chart window.

Line 1: **-headings** *labels*

-headings—Identifier provides labels for chart curves. The labels are separated by spaces and correspond with the order of the items calculated with Basic statements. Unlike Example data block 1, the headings include an entry for the item plotted on the X axis. Optionally, **heading**, **headings**, or **-h**[**eadings**].

labels—List of labels, one for the variable plotted on the X axis and one for each of the curves. In Example data block 2 (see Basic lines 10–30), "Ca" corresponds to the X-axis variable, "F" corresponds to the Y-axis variable, and "pH" corresponds to the secondary Y-axis variable. The labels can be changed in subsequent simulations for proper identification of the parameters graphed (without need of repeating the Basic statements that define the data to be plotted).

Line 2: **-axis_titles** *label₁ label₂ label₃*

Same as Line 2 in Example data block 1.

Line 3: **-start**

-start—Indicates the start of the Basic program. Optional.

Basic: *numbered Basic statement*

numbered Basic statement—A valid Basic language statement that must be numbered. The statements are evaluated in the order of the line numbers. Statements and functions that are available through the Basic interpreter are listed in The Basic Interpreter, tables 7 and 8.

Basic: *number GRAPH_X expression*

number—Number for a Basic statement.

GRAPH_X—Basic command that sets the X value of a data point for a curve of the chart. A point on a curve is defined by an X–Y pair, where the X value is defined by GRAPH_X and the Y value is defined by GRAPH_Y or GRAPH_SY.

expression—Basic expression that evaluates to a number.

Basic: *number GRAPH_Y expressions*

number—Number for a Basic statement.

GRAPH_Y—Basic command that sets the Y value of a data point for a curve on the primary Y axis of the chart. Multiple Y curves can be defined on the same line by defining multiple comma-separated expressions.

expressions—List of one or more Basic expressions that evaluate to numbers.

Basic: *number GRAPH_SY expressions*

number—Number for a Basic statement.

GRAPH_SY—Basic command that sets the Y value of a data point for a curve on the secondary Y axis of the chart. Multiple secondary Y curves can be defined on the same line by defining multiple comma-separated expressions.

expressions—List of one or more Basic expressions that evaluate to numbers.

Line 10: **-end**

> **-end**—Indicates the end of the Basic program. Optional. Note the hyphen is required to avoid a conflict with the keyword **END**.

Notes 2

Both PLOT_XY and GRAPH_X, GRAPH_Y, and GRAPH_SY statements can be used to define curves in the same Basic program. GRAPH_X, GRAPH_Y, and GRAPH_SY statements provide the more concise syntax, especially for multiple curves, which can be defined with a single GRAPH_Y or GRAPH_SY statement. However, the PLOT_XY allows X values to be defined individually for each curve and the colors, symbols, and lines can be set explicitly with command options.

When using both PLOT_XY and GRAPH_X, GRAPH_Y, and GRAPH_SY statements in the same Basic program, the order of the statements determines the order in which the headings must be defined. Headings are associated with each PLOT_XY statement, the GRAPH_X statement, and each expression of GRAPH_Y and GRAPH_SY statements in the order in which they occur in the Basic program.

Example problems

The keyword **USER_GRAPH** is used in example problems 2, 5, 6, 7, 8, 9, 10, 11, 12, 13, 14, 15, 17, 19, 20, 21, and 22.

Related keywords

USER_PRINT and USER_PUNCH.

USER_PRINT

This keyword data block is used to define a Basic program that prints user-defined quantities to the output file. Any Basic "PRINT" statement will write to the output file.

Example data block

```
Line 0: USER_PRINT
Line 1:     -start
Basic:  10 REM convert to ppm
Basic:  20 PRINT "Sodium:    ", MOL("Na+")* 22.99 * 1000
Basic:  30 PRINT "Magnesium: ", MOL("Mg+2")* 24.3 * 1000
Basic:  40 pairs = MOL("NaCO3-") + MOL("MgCO3")
Basic:  50 PRINT "Pairs (mol/kgw): ", pairs
Basic:  60 REM print reaction increment
Basic:  70 PRINT "Rxn incr:  ", RXN
Line 2:     -end
```

Explanation

Line 0: **USER_PRINT**

USER_PRINT is the keyword for the data block. No other data are input on the keyword line.

Line 1: **-start**

-start—Indicates the start of the Basic program. Optional.

Basic: *numbered Basic statement*

numbered Basic statement—A valid Basic language statement that must be numbered. The statements are evaluated in the order of the line numbers. Statements and functions that are available through the Basic interpreter are listed in The Basic Interpreter, tables 7 and 8.

Line 2: **-end**

-end—Indicates the end of the Basic program. Optional. Note the hyphen is required to avoid a conflict with the keyword **END**.

Notes

USER_PRINT allows the user to write Basic programs to make calculations and print selected results as the program is running. Results of PRINT Basic statements are written directly to the output file after each calculation. More information on the Basic interpreter is available in the section The Basic Interpreter. All of the functions defined in The Basic Interpreter, tables 7 and 8, are available in **USER_PRINT** Basic programs. Writing results of **USER_PRINT** can be enabled or suspended with the **-user_print** identifier in the PRINT data block. The USER_PUNCH data block is similar to **USER_PRINT**, except that PUNCH Basic statements are used to write results to the selected-output file.

Example problems

The keyword **USER_PRINT** is used in example problems 6, 10, 12, and 20.

Related keywords

PRINT, RATES, SELECTED_OUTPUT, and USER_PUNCH.

This keyword data block is used to define Basic programs that print user-defined quantities to the selected-output file. Any Basic "PUNCH" statement will write to the selected-output file.

Example data block

```
Line 0: USER_PUNCH
Line 1:     -headings Na+ Mg+2 Pairs Rxn_increment
Line 2:     -start
Basic:  10 REM convert to ppm
Basic:  20 PUNCH MOL("Na+")* 22.99 * 1000
Basic:  30 PUNCH MOL("Mg+2")* 24.3 * 1000
Basic:  40 pairs = MOL("NaCO3-") + MOL("MgCO3")
Basic:  50 PUNCH pairs
Basic:  60 REM punch reaction increment
Basic:  70 PUNCH RXN
Line 3:     -end
```

Explanation

Line 0: **USER_PUNCH**

USER_PUNCH is the keyword for the data block. No other data are input on the keyword line.

Line 1: **-headings** *list of column headings*

-headings—Headings will appear on the first line of the selected-output file. Optionally, **heading**, **headings**, or **-h[eadings]**.

list of column headings—White-space-delimited (any combination of spaces and tabs) list of column headings.

Line 2: **-start**

-start—Indicates the start of the Basic program. Optional.

Basic: *numbered Basic statement*

numbered Basic statement—A valid Basic language statement that must be numbered. The statements are evaluated in the order of the line numbers. Statements and functions that are available through the Basic interpreter are listed in The Basic Interpreter, tables 7 and 8.

Line 3: **-end**

-end—Indicates the end of the Basic program. Optional. Note the hyphen is required to avoid a conflict with the keyword **END**.

Notes

USER_PUNCH allows the user to write a Basic program to make calculations and print selected results to the selected-output file as PHREEQC is running. Results of PUNCH Basic statements are written directly to the selected-output file after each calculation. The Basic program is useful for writing results in the desired units or in a format that can be plotted directly. All of the functions defined in The Basic Interpreter (tables 7 and 8) are available in USER_PUNCH Basic programs. USER_PUNCH has no effect unless a SELECTED_OUTPUT data block has been defined. Writing results of USER_PUNCH can be enabled or suspended with the -selected_output identifier in the PRINT data block. If the -selected_output identifier in the PRINT data block is **false**, then all selected output, including USER_PUNCH, is disabled; if **true**, then all selected output, including USER_PUNCH, is enabled. The USER_PRINT data block is similar to USER_PUNCH, except that PRINT Basic statements are used to write results to the output file.

Example problems

The keyword USER_PUNCH is used in example problems 6, 8, 9, 10, 11, 12, 13, 14, 15, 20, and 21.

Related keywords

PRINT, RATES, SELECTED_OUTPUT, USER_GRAPH, and USER_PRINT.

PHREEQC has an embedded Basic interpreter (David Gillespie, Synaptics, Inc., San Jose, Calif., written commun., 1997; distributed with the Linux operating system, Free Software Foundation, Inc.). Basic is a computer language with statements on numbered lines. The statements are much like the formulas entered in a spreadsheet cell, but Basic allows, in addition, the conditional statements and looping operations of a programming language. Variables can be defined at will, given a value, and used in subsequent lines. Variable names must start with a letter, which can be followed by any number of letters and numbers, and the variable's name must be different from the general and PHREEQC Basic functions. Names ending with a "$" are for strings. Thus,

```
10 A = 1.246
20 A$ = 'A equals 1.246'
```

is perfect.

In Basic you can use the operators "+", "-", "*", "/", and "=", just as in written equations. A single variable is used on the left side of the equals sign. Expressions in parentheses, "(*expression*)", are evaluated first, and then used in the more general expression. Exponentiation is done with the ^ sign: $2^2 = 4$. The standard Basic and special PHREEQC Basic functions are listed in tables 7 and 8, respectively.

Basic programs are executed in line number order, regardless of the order used for writing the lines (but, for good programming, keep the number order intact). Basic variables, functions, and statements are case insensitive. Initially, a numeric variable is zero, and a string is empty, "". The scope of variables is limited to the program unit where they are defined (RATES, USER_GRAPH, USER_PRINT, USER_PUNCH, or CALCULATE_VALUES data block). Numeric data can be transferred between program units with the functions PUT and GET (see table 8). However, in multithreaded and multiprocessor applications (for instance, PHAST, Parkhurst and others, 2010), PUT and GET may not work correctly.

Basic in PHREEQC is quite powerful, and it could be used for other purposes than manipulating variables in PHREEQC. For example, the following input file (illustrated in figure 3) plots the sine function from 0 to 360 degrees:

```
SOLUTION 1
REACTION; H2O 0; 0 in 21
USER_GRAPH
-axis_titles 'ANGLE, IN DEGREES' 'SINE(ANGLE)'
-axis_scale x_axis 0 360 90
10 pi = 2 * arctan(1e20)
20 i = pi * (step_no - 1) / 10
30 graph_x i * 180 / pi
40 graph_y sin(i)
END
```

Originally, the Basic interpreter was a unique feature of PHREEQC version 2, aimed at calculating rates for kinetic geochemical processes. Rate expressions for kinetic reactions can have various forms, and they tend to be redefined or updated frequently as more data become available. In a PHREEQC input file, the rates can be adapted easily by the user as necessary. Because rate expressions often depend on conditions (for example, the rate expression may be different for mineral dissolution and precipitation), the conditional "if" statement of Basic can be necessary. Special Basic functions (table 8) have been written to retrieve geochemical quantities that frequently are used in kinetic rate expressions, such as molalities, activities, saturation indices, and moles of reactants.

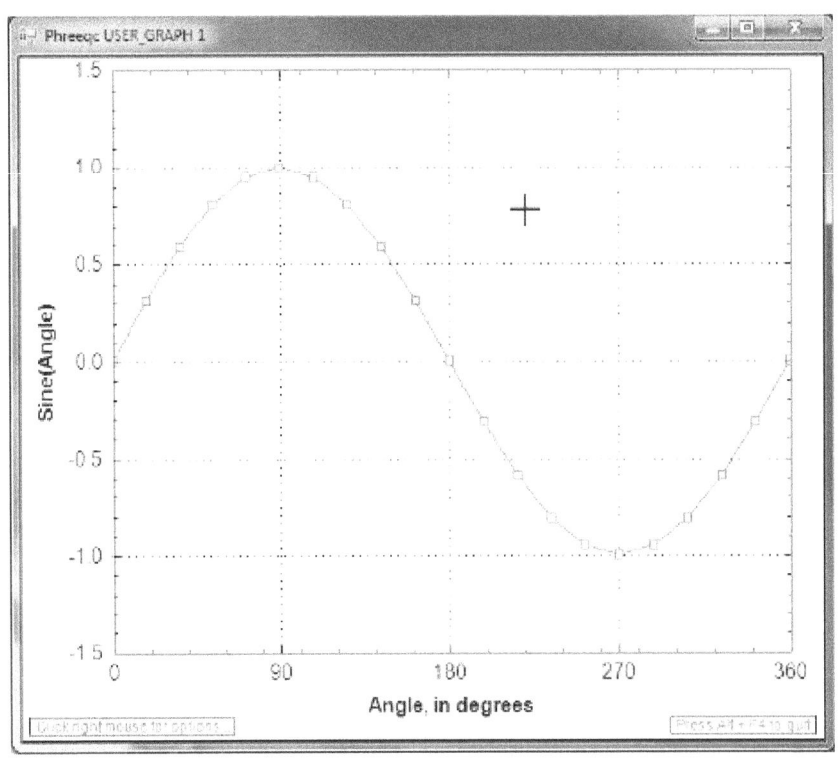

Figure 3. Sine function plotted with a USER_GRAPH data block and exported as a GIF (Graphics Interchange Format) file.

PHREEQC calculations generate a large number of geochemical quantities, possibly distributed in space and time as well. Rather than storing or writing all of these quantities for a run, small Basic programs in the data blocks USER_GRAPH, USER_PRINT and USER_PUNCH can be used to print selected items or to calculate and graph specific numbers such as sums of species or concentrations in milligrams per liter. Also, the implementation of isotopes as individual chemical components relies heavily on Basic programs in the CALCULATE_VALUES data block, where Basic is used to calculate specific isotopic ratios, such as of carbon-13 in various bicarbonate species. Functions defined in CALCULATE_VALUES data blocks can be used in any Basic program within PHREEQC.

Table 7. Standard Basic statements and functions.

Basic Statements and Functions	Explanation
+, -, *, /	Add, subtract, multiply, and divide.
string1 + *string2*	String concatenation.
$a \wedge b$	Exponentiation, a^b.
<, >, <=, >=, <>, =, AND, OR, XOR, NOT	Relational and Boolean operators.
ABS(*a*)	Absolute value.
ARCTAN(*a*)	Arctangent function.
ASC(*character*)	ASCII value for *character*.
CHR$(*number*)	Convert ASCII *number* to character.
CEIL(*a*)	Smallest integer not less than *a*.
COS(*a*)	Cosine function.
DATA *list*	List of data.
DIM *a*(*n*)	Define a dimensioned variable.
END	Ends the program
EOL$	End of line string that is appropriate for the operating system.
ERASE v	Revert the Basic variable to an undimensioned variable so that it can be used as a scalar or dimensioned with another DIM statement. Applies only to variables that have been dimensioned with a DIM statement.
EXP(*a*)	e^a.
FLOOR(*a*)	Largest integer not greater than *a*.

Table 7. Standard Basic statements and functions.—Continued

Basic Statements and Functions	Explanation
FOR $i = n$ TO m STEP k NEXT i	"For" loop.
GOSUB *line*	Go to subroutine at line number.
GOTO *line*	Go to line number.
IF (*expr*) THEN *statement* ELSE *statement*	If, then, else statement (on one line; a "\" may be used to concatenate lines).
INSTR(*a\$. b\$*)	The character position of the beginning of string *b\$* in *a\$*; 0 if not found.
LEN(*string*)	Number of characters in *string*.
LOG(*a*)	Natural logarithm.
LOG10(*a*)	Base 10 logarithm.
LTRIM(*a\$*)	Trims white space from the beginning of string *a\$*.
MID\$(*string, n*) MID\$(*string, n, m*)	Extract characters from position *n* to end of *string*. Extract *m* characters from *string* starting at position *n*.
a MOD *b*	Returns remainder of *a* / *b*.
ON *expr* GOTO *line1, line2, ...* ON *expr* GOSUB *line1, line2, ...*	If the value of the expression, rounded to an integer, is N, go to the Nth line number in the list. If N is less than one or greater than the number of line numbers listed, execution continues at the next statement after the ON statement.
PAD(*a\$, i*)	Pads *a\$* with spaces to a total of *i* characters; returns a copy of *a\$* if the length of *a\$* is more than *i* characters.
READ	Read from DATA statement.
REM	At beginning of line, line is a remark with no effect on the calculations.
RESTORE *line*	Set pointer to DATA statement at *line* for subsequent READ.
RETURN	Return from subroutine.
RTRIM(*a\$*)	Trims white space from the end of string *a\$*.
SGN(*a*)	Sign of *a*, +1 or -1.
SIN(*a*)	Sine function.
SQR(*a*)	a^2.
SQRT(*a*)	$\sqrt{a}$.
STR\$(*a*)	Convert number to a string.
TAN(*a*)	Tangent function.
TRIM(*a\$*)	Trims white space from the beginning and end of string *a\$*.
VAL(*string*)	Convert string to number.
WHILE (*expression*) WEND	"While" loop.

Table 8. Special Basic statements and functions for PHREEQC.

Special PHREEQC Statement or Function	Explanation
ACT("HCO3-")	Activity of an aqueous, exchange, or surface species.
ALK	Alkalinity of solution, equivalents per kilogram water.
CALC_VALUE("R(D)_OH-")	Value calculated by Basic function (here, "R(D)_OH-") defined in CALCULATE_VALUES data block.
CELL_NO	Cell number in TRANSPORT or ADVECTION calculations; otherwise solution or mix number.
CHANGE_POR(0.21, cell_no)	Modifies the porosity in a cell, used only in multicomponent diffusion calculations (keyword TRANSPORT). Here, porosity of cell cell_no is set to 0.21.
CHANGE_SURF("Hfo", 0.2, "Sorbedhfo", 0, cell_no)	Changes the diffusion coefficient of (part of) a surface (SURFACE), and renames the surface (if names are different). This function is for modeling transport, deposition, and remobilization of colloids. It is used in conjunction with multicomponent diffusion in a TRANSPORT data block. Here: take a fraction 0.2 of "Hfo" and rename it "Sorbedhfo" with a diffusion coefficient of 0, in cell cell_no. The diffusion coefficient of zero means that "Sorbedhfo" is not transported.
CHARGE_BALANCE	Charge balance of a solution, equivalents.
DESCRIPTION	Description associated with current solution or current mixture.
DH_A	Debye-Hückel A parameter in the activity coefficient equation, $(mol/kg)^{-0.5}$.
DH_Av	Debye-Hückel limiting slope of specific volume *vs.* ionic strength, $(cm^3/mol)(mol/kg)^{-0.5}$.
DH_B	Debye-Hückel B parameter in the activity coefficient equation, $angstrom^{-1}(mol/kg)^{-0.5}$.
DIST	Distance to midpoint of cell in TRANSPORT calculations, cell number in ADVECTION calculations, "-99" in all other calculations.
EDL("As", "Hfo")	Moles of element in the diffuse layer of a surface. The number of moles does not include the specifically sorbed species. The surface name should be used, not a surface site name (that is, no underscore). The first argument can have several special values, which return information for the surface: "charge", surface charge, in equivalents; "sigma", surface charge density, coulombs per square meter; "psi", potential, Volts; "water", mass of water in the diffuse layer, kg. For CD-MUSIC surfaces, charge, sigma and psi can be requested for the 0, 1 and 2 planes: EDL("Charge", "Goe") # Charge (eq) at the zero-plane of Goe (Goethite) EDL("Charge1", "Goe") # Charge (eq) at plane 1 of Goe EDL("Charge2", "Goe") # Charge (eq) at plane 2 of Goe and similar for "sigma" and "psi".
EOL$	End of line character, which is equivalent to "\n" in the C programming language.
EPS_R	Relative dielectric constant.
EQUI("Calcite")	Moles of a phase in the equilibrium-phase assemblage.
EQUI_DELTA("Calcite")	Moles of a phase in the equilibrium-phase assemblage that reacted during the current calculation.
EXISTS(i1[, i2, ...])	Determines if a value has been stored with a PUT statement for the list of one or more subscripts. The function equals 1 if a value has been stored and 0 if no value has been stored. Values are stored in global storage with PUT and are accessible by any Basic program. See description of PUT for more details.
GAMMA("H+")	Activity coefficient of a species.
GAS("CO2(g)")	Moles of a gas component in the gas phase.
GAS_P	Pressure of the GAS_PHASE (atm), either specified for a fixed-pressure gas phase, or calculated for a fixed-volume gas phase. Related functions are PR_P and PRESSURE.
GAS_VM	Molar volume (L/mol, liter per mole) of the GAS_PHASE (calculated with Peng-Robinson).

Table 8. Special Basic statements and functions for PHREEQC.—Continued

Special PHREEQC Statement or Function	Explanation
GET(*i1*[, *i2*, ...])	Retrieves the value that is identified by the list of one or more subscripts. Value is zero if PUT has not been used to store a value for the set of subscripts. Values stored in global storage with PUT are accessible by any Basic program. See description of PUT for more details.
GET_POR(10)	Porosity in a cell (here, cell 10), used in conjunction with Basic function CHANGE_POR in multicomponent diffusion.
GFW("CaCO3")	Returns the gram formula weight of the specified formula.
GRAPH_X tot("Ca") * 40.08e3	Used in USER_GRAPH data block to define the X values for points. Here, Ca in mg/L is the X value for points of the chart. See the description of the USER_GRAPH keyword for more details.
GRAPH_Y tot("F") * 19e3	Used in USER_GRAPH data block to define the Y values for points plotted on the primary Y axis. Here, F in mg/L is the Y value for points. See the description of the USER_GRAPH keyword for more details.
GRAPH_SY-la("H+")	Used in USER_GRAPH data block to define the Y values for points plotted on the secondary Y axis. Here, pH is the Y value for points plotted on the secondary Y axis. See the description of the USER_GRAPH keyword for more details.
ISO("[18O]"), ISO("R(D)_H3O+")	Isotopic composition in the input units (for example, permil) for an isotope (here, [18O]) or an isotope ratio defined in ISOTOPE_RATIOS (here, "R(D)_H3O+").
ISO_UNIT("[18O]"), ISO("R(D)_H3O+")	String value for the input units (for example, "permil") for an isotope or an isotope ratio defined in ISOTOPE_RATIOS.
KAPPA	Compressibility of pure water at current pressure and temperature.
KIN("CH2O")	Moles of a kinetic reactant.
KIN_DELTA("CH2O")	Moles of a kinetic reactant that reacted during the current calculation.
LA("HCO3-")	Log10 of activity of an aqueous, exchange, or surface species.
LG("H+")	Log10 of the activity coefficient for an aqueous species.
LIST_S_S("Carbonate_s_s", *count*, *comp$*, *moles*)	Returns the sum of the moles of components in a solid solution and the composition of the solid solution. The first argument is an input value specifying the name of the solid solution. *Count* is an output variable containing the number of components in the solid solution. *Comp$* is an output character array containing the names of each component in the solid solution. *Moles* is an output numeric array containing the number of moles of each component, in the order defined by *Comp$*. Arrays are in sort order by number of moles.
LK_NAMED("Log_alpha_D_OH-/ H2O(l)")	The value calculated by a named expression defined in the NAMED_EXPRESSIONS data block.
LK_PHASE("Calcite")	Log10 of the equilibrium constant for a phase defined in the PHASES data block.
LK_SPECIES("HCO3-")	Log10 of the equilibrium constant for an aqueous, exchange, or surface species.
LM("HCO3-")	Log10 of molality of an aqueous, exchange, or surface species.
M	Current moles of the kinetic reactant for which the rate is being calculated (see KINETICS).
M0	Initial moles of the kinetic reactant for which the rate is being calculated (see KINETICS).
MISC1("Ca(x)Sr(1-x)SO4")	Mole fraction of component 2 at the beginning of the miscibility gap, returns 1.0 if there is no miscibility gap (see SOLID_SOLUTIONS).
MISC2("Ca(x)Sr(1-x)SO4")	Mole fraction of component 2 at the end of the miscibility gap, returns 1.0 if there is no miscibility gap (see SOLID_SOLUTIONS).
MOL("HCO3-")	Molality of an aqueous, exchange, or surface species.
MU	Ionic strength of the solution.
OSMOTIC	Osmotic coefficient if using the Pitzer or SIT aqueous model, otherwise 0.0, unitless.
PARM(*i*)	The *i*th item in the parameter array defined in KINETICS data block.

Table 8. Special Basic statements and functions for PHREEQC.—Continued

Special PHREEQC Statement or Function	Explanation
PERCENT_ERROR	Percent charge-balance error [100(cations-|anions|)/(cations + |anions|)], unitless.
PHASE_FORMULA("Dolomite")	With a single argument, PHASE_FORMULA returns a string that contains the chemical formula for the phase; in this example, "CaMg(CO3)2".
PHASE_FORMULA("Dolomite", *count, elt$, coef*)	With four arguments, PHASE_FORMULA returns a string that contains the chemical formula for the phase, and, in addition, returns values for *count, elt$, coef. Count* is the dimension of the *elt$* and *coef* arrays. *Elt$* is a character array with the name of each element in the chemical formula for the phase. *Coef* is a numeric array containing the number of atoms of each element in the phase formula, in the order defined by *elt$*, which is alphabetical by element.
PLOT_XY tot("Ca") * 40.08e3, tot("F") * 19e3, color = Blue, symbol = Circle, symbol_size = 6, y-axis = 1, line_width = 0	Used in USER_GRAPH data block to define the points to chart; here, Ca in mg/L is the X value for points, F in mg/L is the Y value for points, the symbols are blue circles, the points are plotted relative to the Y axis, and no line connects the points. See the description of the USER_GRAPH keyword for more details.
PRINT	Write to output file.
PR_P("CO2(g)")	Pressure (atm) of a gas component in a Peng-Robinson GAS_PHASE.
PR_PHI("CO2(g)")	Fugacity coefficient of a gas component in a Peng-Robinson GAS_PHASE.
PRESSURE	Current pressure applied to the solution (atm). PRESSURE is a specified value except for fixed-volume GAS_PHASE calculations.
PUNCH	Write to selected-output file.
PUT(*x, i1*[, *i2, ...*])	Saves value of *x* in global storage that is identified by a sequence of one or more subscripts. Value of *x* can be retrieved with GET(*i1*,[, *i2, ...*]) and a set of subscripts can be tested to determine if a value has been stored with EXISTS(*i1*[, *i2, ...*]). PUT may be used in CALCULATE_VALUES, RATES, USER_GRAPH, USER_PRINT, or USER_PUNCH Basic programs to store a value. The value may be retrieved by any of these Basic programs. The value persists until overwritten by using a PUT statement with the same set of subscripts, or until the end of the run. For a KINETICS data block, the Basic programs for the rate expressions are evaluated in the order in which they are defined in the input file. Use of PUT and GET in parallel processing environments may be unreliable.
QBRN	The Born parameter for calculating the temperature dependence of the specific volume of an aqueous species at infinite dilution. This is the pressure derivative of the relative dielectric constant of water multiplied by 41.84 bar cm^3/cal (bar cubic centimeter per calorie): $41.84\left(\dfrac{1}{\varepsilon_r^2}\dfrac{\partial}{\partial P}\varepsilon_r\right)_T$, cm^3/mol
RHO	Density of solution, kilograms per liter.
RXN	Moles of reaction as defined in **-steps** in REACTION data block for a batch-reaction calculation; otherwise zero.
SAVE	Moles of kinetic reactant for a time step in a rates function or the value returned from a CALCULATE_VALUES function.
SC	Specific conductance, microsiemens per centimeter.
SI("Calcite")	Saturation index of a phase, $Log10\left(\dfrac{IAP}{K}\right)$ log 10 of the ion activity product divided by equilibrium constant.
SIM_NO	Simulation number, equals one more than the number of END statements before current simulation.
SIM_TIME	Time from the beginning of a kinetic batch-reaction or transport calculation, in seconds.
SOLN_VOL	Volume of the solution, in liters.
SR("Calcite")	Saturation ratio of a phase, $\dfrac{IAP}{K}$, ion activity product divided by equilibrium constant.

Table 8. Special Basic statements and functions for PHREEQC.—Continued

Special PHREEQC Statement or Function	Explanation
STEP_NO	Step number in batch-reaction calculations, or shift number in ADVECTION and TRANSPORT calculations.
SUM_GAS("template", "element")	Sums number of moles of the element in gases that match the template. The template selects a set of gases. For example, a template of "{C,[13C],[14C]}{O,[18O]}2" selects all the isotopic variants of $CO_2(g)$. Multiple elements at a stoichiometric position are separated by commas within braces; an asterisk (*) in the template matches any element. The number of moles of "element" is calculated by summing the stoichiometric coefficient of the element times the moles of the gas for all selected gases.
SUM_SPECIES("template", "element")	Sums number of moles of the element in aqueous, exchange, and surface species that match the template. The template selects a set of species. For example, a template of "*HCO3*" selects all bicarbonate species. Multiple elements at a stoichiometric position are separated by commas within braces; an asterisk (*) in the template matches any element. The number of moles of "element" is calculated by summing the stoichiometric coefficient of the element times the moles of the species for all selected species.
SUM_S_S("s_s_name", "element")	Sums number of moles of the element in the specified solid solution.
SURF("element", "surface")	Number of moles of the element sorbed on the surface. The second argument should be the surface name, not the surface-site name (that is, no underscore). A redox state may be specified; for example, "As" or "As(5)" is permitted.
SYS("element")	With a single argument, SYS calculates the number of moles of the element in all phases (solution, equilibrium phases, surfaces, exchangers, solid solutions, and gas phase) in the reaction calculation.
SYS("element", *count*, *name$*, *type$*, *moles*)	With five arguments, SYS returns the number of moles of the element in all phases in the reaction calculation (solution, equilibrium phases, surfaces, exchangers, solid solutions, and gas phase), and, in addition, returns values for *count_species*, *name$*, *type$*, *moles*. *Count* is the dimension of the *name$*, *type$*, and *moles* arrays. *Name$* is a character array with the name of each species that contains the element. *Type$*, is a character array with the type of the phase of each species: "aq", "equi", "surf", "ex", "s_s", "gas", or "diff"; where aq is aqueous, equi is equilibrium phase, surf is surface, ex is exchange, s_s is solid solution, gas is gas phase, and diff is surface diffuse layer. *Moles* is the number of moles of the element in the species (stoichiometry of element times moles of species). The sum of all items in the *moles* array is equal to the return value of the SYS function.

The five-argument form of SYS accepts the following arguments in place of "element":

"**elements**" returns the total number of moles of elements solution, exchangers, and surfaces in the calculation, other than H and O. *Count* is number of elements, valence states, exchangers, and surfaces. *Name$* contains the element name. *Type$* contains the type for each array item: "dis" for dissolved, "ex" for exchange, and "surf" for surface. *Moles* contains the number of moles of the element in each type of phase (stoichiometry of element times moles of species).

"**phases**" returns the maximum saturation index of all pure phases appropriate for the calculation. *Count* is number of pure phases. *Name$* contains the phase names as defined in the PHASES data block. *Type$* is "phase". *Moles* contains the saturation index for the phases.

"**aq**" returns the sum of moles of all aqueous species in the calculation. *Count* is number of aqueous species. *Name$* contains the aqueous species names. *Type$* is "aq". *Moles* contains the moles of species.

"**ex**" returns the sum of moles of all exchange species in the calculation. *Count* is number of exchange species. *Name$* contains the exchange species names. *Type$* is "ex". *Moles* contains the moles of species.

"**surf**" returns the sum of moles of all surface species in the calculation. *Count* is number of surface species. *Name$* contains the surface species names. *Type$* is "surf". *Moles* contains the moles of species. |

Table 8. Special Basic statements and functions for PHREEQC.—Continued

Special PHREEQC Statement or Function	Explanation
	"**s_s**" returns sum of moles of all solid-solution components in the calculation. *Count* is number of solid-solution components. *Name$* contains the names of the solid-solution components. *Type$* is "s_s". *Moles* contains the moles of components.
	"**gas**" returns sum of moles of all gas components in the calculation. *Count* is number of gas components. *Name$* contains names of the gas components. *Type$* is "gas". *Moles* contains the moles of gas components
S_S("Magnesite")	Current moles of a solid-solution component.
TC	Temperature in Celsius.
TK	Temperature in Kelvin.
TIME	Time interval for which moles of reaction are calculated in rate programs, automatically set in the time-step algorithm of the numerical integration method, in seconds.
TOT("Fe(2)")	Total molality of element or element redox state. TOT("water") is total mass of water, in kilograms.
TOTAL_TIME	Cumulative time (seconds) including all advective (for which -**time_step** is defined) and advective-dispersive transport simulations from the beginning of the run or from last -**initial_time** identifier.
TOTMOLE("Ca")	Moles of an element or element valence state in solution. TOTMOLE has two special values for the argument: "water", moles of water in solution; and "charge", equivalents of charge imbalance in solutions (same as Basic function CHARGE_BALANCE). Note the Basic function TOT returns moles per kilogram water, whereas TOTMOLE returns moles.
VM("Na+")	Returns the specific volume (cm^3/mol) of a SOLUTION_SPECIES, relative to VM("H+") = 0, a function of temperature, pressure, and ionic strength.

This page left blank intentionally.

Examples

In this section of the report example calculations are presented that demonstrate most of the capabilities of PHREEQC. The first 18 examples are derived from the version 2 manual but are updated with the new capabilities of version 3. Four new examples, 19 through 22, illustrate more capabilities of PHREEQC. Example 19 demonstrates the use of empirical sorption isotherms and compares measured data with a deterministic model for sorption of Cd^{+2} on iron oxyhydroxides, clay minerals, and organic matter for a soil. Example 20 calculates simultaneous multi-isotope fractionation between water and calcite. Example 21 uses the multicomponent diffusion transport capabilities, which allow calculation of diffusion processes with ion-specific tracer diffusion coefficients. Finally, example 22 shows capability of PHREEQC to calculate the solubilities of gases at high pressures. The input files for all examples are included in tables and can be used as templates for modeling other geochemical processes. The new keyword USER_GRAPH is used to display results for most examples, and selected output from each of the example runs is presented.

Example 1—Speciation Calculation

This example calculates the distribution of aqueous species in seawater and the saturation state of seawater relative to a set of minerals. To demonstrate how to expand the model to new elements, uranium is added to the aqueous model defined by *phreeqc.dat*. [Several of the database files distributed with the program (*wateq4f.dat*, *llnl.dat*, *minteq.dat*, *minteq.v4.dat*, and *sit.dat*) include the element uranium, and use of any one of these databases would make the uranium definitions in this example unnecessary.]

The essential data needed for a speciation calculation are the temperature, pH, and concentrations of elements and (or) element valence states. These data for seawater are given in table 9. The input file for this example calculation is shown in table 10. A comment about the calculations performed in this simulation is included with the TITLE keyword. The SOLUTION data block defines the composition of seawater. Note that valence states are identified by the chemical symbol for the element followed by the valence in parentheses [S(6), N(5), N(-3), and O(0)].

The pe to be used for distributing redox elements and for calculating saturation indices is specified by the **redox** identifier. In this example, a pe is to be calculated from the O(-2)/O(0) redox couple, which corresponds to the dissolved oxygen/water couple, and this calculated pe will be used for all calculations that require a pe. If **redox** were not specified, the default would be the input pe. The default redox identifier

Table 9. Seawater composition.

[Concentration is in parts per million (ppm) unless specified otherwise]

Analysis	PHREEQC notation	Concentration
Calcium	Ca	412.3
Magnesium	Mg	1291.8
Sodium	Na	10768.0
Potassium	K	399.1
Iron	Fe	0.002
Manganese	Mn	0.0002
Silica, as SiO_2	Si	4.28
Chloride	Cl	19353.0
Alkalinity, as HCO_3^-	Alkalinity	141.682
Sulfate, as SO_4^{2-}	S(6)	2712.0
Nitrate. as NO_3^-	N(5)	0.29
Ammonium, as NH_4^+	N(-3)	0.03
Uranium	U	0.0033
pH, standard units	pH	8.22
pe, unitless	pe	8.451
Temperature, °C	temperature	25.0
Density, kilograms per liter	density	1.023

can be overridden for any redox element, as demonstrated by the manganese input, where the input pe will be used to speciate manganese among its valence states, and the uranium input, where a pe calculated from the nitrate/ammonium couple will be used to speciate uranium among its valence states.

The default units are specified to be ppm in this file (**units** identifier). This default can be overridden for any concentration, as demonstrated by the uranium concentration, which is specified to be ppb instead of ppm. Because ppm is a mass unit, not a mole unit, the program must use a gram formula weight to convert each concentration into molal units. The default gram formula weights for each master species are specified in the **SOLUTION_MASTER_SPECIES** input (the formulas used to calculate gram formula weights for *phreeqc.dat* are listed in table 3). If the data are reported relative to a gram formula weight different from the default, it is necessary to specify the appropriate gram formula weight in the input file. This can be done with the **gfw** identifier, where the actual gram formula weight is input—the gram-formula weight by which to convert nitrate is specified to be 62.0 g/mol, or more simply with the **as** identifier, where the chemical formula for the reported units is input, as shown in the input for alkalinity and ammonium in this example.

Note finally that the concentration of O(0), dissolved oxygen, is given an initial estimate of 1 ppm, but that its concentration will be adjusted until a log partial pressure of oxygen gas of -0.7 is achieved. [O2(g) is defined under **PHASES** input in each database.] When using phase equilibria to specify initial concentrations [like O(0) in this example], only one concentration is adjusted. For example, if gypsum were used to adjust the calcium concentration, the concentration of calcium would vary, but the concentration of sulfate would remain fixed.

Table 10. Input file for example 1.

```
TITLE Example 1.--Add uranium and speciate seawater.
SOLUTION 1  SEAWATER FROM NORDSTROM AND OTHERS (1979)
        units     ppm
        pH        8.22
        pe        8.451
        density   1.023
        temp      25.0
        redox     O(0)/O(-2)
        Ca               412.3
        Mg               1291.8
        Na               10768.0
        K                399.1
        Fe               0.002
        Mn               0.0002   pe
        Si               4.28
        Cl               19353.0
        Alkalinity       141.682 as HCO3
        S(6)             2712.0
        N(5)             0.29     gfw   62.0
        N(-3)            0.03     as    NH4
        U                3.3      ppb   N(5)/N(-3)
        O(0)             1.0      O2(g) -0.7
SOLUTION_MASTER_SPECIES
        U        U+4       0.0       238.0290       238.0290
        U(4)     U+4       0.0       238.0290
        U(5)     UO2+      0.0       238.0290
        U(6)     UO2+2     0.0       238.0290
SOLUTION_SPECIES
        #primary master species for U
        #is also secondary master species for U(4)
        U+4 = U+4
               log_k           0.0
SOLUTION_SPECIES
        U+4 + 4 H2O = U(OH)4 + 4 H+
               log_k          -8.538
               delta_h        24.760 kcal
        U+4 + 5 H2O = U(OH)5- + 5 H+
               log_k          -13.147
               delta_h        27.580 kcal
        #secondary master species for U(5)
        U+4 + 2 H2O = UO2+ + 4 H+ + e-
```

Table 10. Input file for example 1.—Continued

```
                 log_k            -6.432
                 delta_h          31.130 kcal
       #secondary master species for U(6)
       U+4 + 2 H2O = UO2+2 + 4 H+ + 2 e-
                 log_k            -9.217
                 delta_h          34.430 kcal
       UO2+2 + H2O = UO2OH+ + H+
                 log_k            -5.782
                 delta_h          11.015 kcal
       2UO2+2 + 2H2O = (UO2)2(OH)2+2 + 2H+
                 log_k            -5.626
                 delta_h          -36.04 kcal
       3UO2+2 + 5H2O = (UO2)3(OH)5+ + 5H+
                 log_k           -15.641
                 delta_h          -44.27 kcal
       UO2+2 + CO3-2 = UO2CO3
                 log_k            10.064
                 delta_h           0.84 kcal
       UO2+2 + 2CO3-2 = UO2(CO3)2-2
                 log_k            16.977
                 delta_h           3.48 kcal
       UO2+2 + 3CO3-2 = UO2(CO3)3-4
                 log_k            21.397
                 delta_h          -8.78 kcal
PHASES
       Uraninite
       UO2 + 4 H+ = U+4 + 2 H2O
       log_k            -3.490
       delta_h         -18.630 kcal
END
```

Uranium is not included in *phreeqc.dat*, one of the database files that is distributed with the program. Thus, data to describe the thermodynamics and composition of aqueous uranium species must be included in the input data when using this database. Two keyword data blocks are needed to define the uranium species, SOLUTION_MASTER_SPECIES and SOLUTION_SPECIES. By adding these two data blocks to the input data file, aqueous uranium species will be defined for the duration of the run. To add uranium permanently to the list of elements, these data blocks should be added to the database file. The data for uranium shown here are intended to be illustrative and are not a complete description of uranium speciation.

It is necessary to define a primary master species for uranium with SOLUTION_MASTER_SPECIES input. Because uranium is a redox-active element, it is also necessary to define a secondary master species for each valence state of uranium. The data block SOLUTION_MASTER_SPECIES (table 10) defines U^{+4} as the primary master species for uranium and also as the secondary master species for the +4 valence state. UO_2^+ is the secondary master species for the +5 valence state, and UO_2^{+2} is the secondary master species for the +6 valence state. Equations defining these aqueous species plus any other complexes of uranium must be defined through SOLUTION_SPECIES input.

In the data block SOLUTION_SPECIES (table 10), the primary and secondary master species are noted with comments. A primary master species is always defined in the form of an identity reaction (U+4 = U+4). Secondary master species are the only aqueous species that contain electrons in their chemical reaction. Additional hydroxide and carbonate complexes are defined for the +4 and +6 valence states, but none for the +5 state.

Finally, a new phase, uraninite, is defined with PHASES input. This phase will be used in calculating saturation indices in speciation modeling, but could also be used, without redefinition, for batch-reaction, transport, or inverse calculations within the computer run.

The output from the model (table 11) contains several blocks of information delineated by headings. First, the names of the input, output, and database files for the run are listed. Next, all keywords encountered in reading the database file are listed under the heading "Reading data base". Then, the input data, excluding comments and empty lines, are echoed under the heading "Reading input data for simulation 1". The simulation is defined by all input data up to and including the END keyword.

Table 11. Output for example 1.

```
   Input file: ex1
  Output file: ex1.out
Database file: phreeqc.dat

------------------
Reading data base.
------------------

              SOLUTION_MASTER_SPECIES
              SOLUTION_SPECIES
              PHASES
              EXCHANGE_MASTER_SPECIES
              EXCHANGE_SPECIES
              SURFACE_MASTER_SPECIES
              SURFACE_SPECIES
              RATES
              END
-----------------------------------
Reading input data for simulation 1.
-----------------------------------
```

Table 11. Output for example 1.—Continued

```
TITLE Example 1.--Add uranium and speciate seawater.
SOLUTION 1   SEAWATER FROM NORDSTROM AND OTHERS (1979)
        units   ppm
        pH      8.22
        pe      8.451
        density 1.023
        temp    25.0
        redox   O(0)/O(-2)
        Ca              412.3
        Mg              1291.8
        Na              10768.0
        K               399.1
        Fe              0.002
        Mn              0.0002   pe
        Si              4.28
        Cl              19353.0
        Alkalinity      141.682 as HCO3
        S(6)            2712.0
        N(5)            0.29     gfw   62.0
        N(-3)           0.03     as    NH4
        U               3.3      ppb   N(5)/N(-3)
        O(0)            1.0      O2(g) -0.7
SOLUTION_MASTER_SPECIES
        U       U+4     0.0      238.0290      238.0290
        U(4)    U+4     0.0      238.0290
        U(5)    UO2+    0.0      238.0290
        U(6)    UO2+2   0.0      238.0290
SOLUTION_SPECIES
        U+4 = U+4
                log_k           0.0
        U+4 + 4 H2O = U(OH)4 + 4 H+
                log_k           -8.538
                delta_h         24.760 kcal
        U+4 + 5 H2O = U(OH)5- + 5 H+
                log_k           -13.147
                delta_h         27.580 kcal
        U+4 + 2 H2O = UO2+ + 4 H+ + e-
                log_k           -6.432
                delta_h         31.130 kcal
        U+4 + 2 H2O = UO2+2 + 4 H+ + 2 e-
                log_k           -9.217
                delta_h         34.430 kcal
        UO2+2 + H2O = UO2OH+ + H+
                log_k           -5.782
                delta_h         11.015 kcal
        2UO2+2 + 2H2O = (UO2)2(OH)2+2 + 2H+
                log_k           -5.626
                delta_h         -36.04 kcal
        3UO2+2 + 5H2O = (UO2)3(OH)5+ + 5H+
                log_k           -15.641
                delta_h         -44.27 kcal
```

Table 11. Output for example 1.—Continued

```
                       UO2+2 + CO3-2 = UO2CO3
                               log_k          10.064
                               delta_h         0.84 kcal
                       UO2+2 + 2CO3-2 = UO2(CO3)2-2
                               log_k          16.977
                               delta_h         3.48 kcal
                       UO2+2 + 3CO3-2 = UO2(CO3)3-4
                               log_k          21.397
                               delta_h        -8.78 kcal
               PHASES
               Uraninite
               UO2 + 4 H+ = U+4 + 2 H2O
               log_k          -3.490
               delta_h       -18.630 kcal
               END
-----
TITLE
-----


 Example 1.--Add uranium and speciate seawater.


 ------------------------------------------
 Beginning of initial solution calculations.
 ------------------------------------------


 Initial solution 1.SEAWATER FROM NORDSTROM AND OTHERS (1979)

 ---------------------------Solution composition----------------------------

           Elements         Molality        Moles

           Alkalinity       2.406e-003    2.406e-003
           Ca               1.066e-002    1.066e-002
           Cl               5.657e-001    5.657e-001
           Fe               3.711e-008    3.711e-008
           K                1.058e-002    1.058e-002
           Mg               5.507e-002    5.507e-002
           Mn               3.773e-009    3.773e-009
           N(-3)            1.724e-006    1.724e-006
           N(5)             4.847e-006    4.847e-006
           Na               4.854e-001    4.854e-001
           O(0)             4.377e-004    4.377e-004   Equilibrium with O2(g)
           S(6)             2.926e-002    2.926e-002
           Si               7.382e-005    7.382e-005
           U                1.437e-008    1.437e-008

 ---------------------------Description of solution--------------------------

                            pH  =    8.220
                            pe  =    8.451
           Specific Conductance (uS/cm, 25 oC) = 53257
                       Density (g/cm3)  =    1.02327
```

Table 11. Output for example 1.—Continued

```
                          Volume (L)   =   1.01473
                   Activity of water   =   0.981
                      Ionic strength   =   6.745e-001
                 Mass of water (kg)    =   1.000e+000
                 Total carbon (mol/kg) =   2.257e-003
                    Total CO2 (mol/kg) =   2.257e-003
                   Temperature (deg C) =   25.00
                Electrical balance (eq) =  7.936e-004
Percent error, 100*(Cat-|An|)/(Cat+|An|) =   0.07
                          Iterations   =   7
                            Total H  = 1.110149e+002
                            Total O  = 5.563077e+001
```

```
-------------------------------Redox couples--------------------------------

          Redox couple            pe   Eh (volts)

          N(-3)/N(5)             4.6750     0.2766
          O(-2)/O(0)            12.4062     0.7339


---------------------------Distribution of species--------------------------
```

| | | | Log | Log | Log | mole V |
Species	Molality	Activity	Molality	Activity	Gamma	cm3/mol
OH-	2.705e-006	1.647e-006	-5.568	-5.783	-0.215	-2.63
H+	7.983e-009	6.026e-009	-8.098	-8.220	-0.122	0.00
H2O	5.551e+001	9.806e-001	1.744	-0.009	0.000	18.07
C(4)	2.257e-003					
HCO3-	1.238e-003	8.359e-004	-2.907	-3.078	-0.170	27.87
NaHCO3	6.168e-004	7.205e-004	-3.210	-3.142	0.067	19.41
MgHCO3+	2.136e-004	1.343e-004	-3.670	-3.872	-0.201	5.82
MgCO3	7.301e-005	8.527e-005	-4.137	-4.069	0.067	-17.09
CaHCO3+	3.717e-005	2.572e-005	-4.430	-4.590	-0.160	9.96
CO3-2	3.128e-005	6.506e-006	-4.505	-5.187	-0.682	-0.34
CaCO3	2.256e-005	2.636e-005	-4.647	-4.579	0.067	-14.60
NaCO3-	1.477e-005	9.972e-006	-4.831	-5.001	-0.170	1.77
CO2	9.887e-006	1.155e-005	-5.005	-4.937	0.067	30.26
UO2(CO3)3-4	1.221e-008	1.143e-010	-7.913	-9.942	-2.029	(0)
UO2(CO3)2-2	2.148e-009	6.681e-010	-8.668	-9.175	-0.507	(0)
MnCO3	2.157e-010	2.519e-010	-9.666	-9.599	0.067	(0)
MnHCO3+	5.475e-011	3.631e-011	-10.262	-10.440	-0.178	(0)
UO2CO3	1.074e-011	1.255e-011	-10.969	-10.901	0.067	(0)
FeCO3	1.498e-020	1.749e-020	-19.825	-19.757	0.067	(0)
FeHCO3+	1.255e-020	9.369e-021	-19.902	-20.028	-0.127	(0)
Ca	1.066e-002					
Ca+2	9.645e-003	2.412e-003	-2.016	-2.618	-0.602	-16.70
CaSO4	9.560e-004	1.117e-003	-3.020	-2.952	0.067	7.50
CaHCO3+	3.717e-005	2.572e-005	-4.430	-4.590	-0.160	9.96
CaCO3	2.256e-005	2.636e-005	-4.647	-4.579	0.067	-14.60
CaOH+	8.721e-008	6.513e-008	-7.059	-7.186	-0.127	(0)
CaHSO4+	5.922e-011	4.422e-011	-10.228	-10.354	-0.127	(0)

Table 11. Output for example 1.—Continued

Cl	5.657e-001					
Cl-	5.657e-001	3.568e-001	-0.247	-0.448	-0.200	18.79
MnCl+	1.068e-009	7.086e-010	-8.971	-9.150	-0.178	7.01
MnCl2	9.449e-011	1.104e-010	-10.025	-9.957	0.067	(0)
MnCl3-	1.635e-011	1.085e-011	-10.786	-10.965	-0.178	(0)
FeCl+2	1.519e-018	2.939e-019	-17.819	-18.532	-0.713	(0)
FeCl2+	7.062e-019	4.684e-019	-18.151	-18.329	-0.178	(0)
FeCl+	7.393e-020	5.521e-020	-19.131	-19.258	-0.127	(0)
FeCl3	1.431e-020	1.671e-020	-19.844	-19.777	0.067	(0)
Fe(2)	6.437e-019					
Fe+2	4.891e-019	1.121e-019	-18.311	-18.950	-0.640	-20.66
FeCl+	7.393e-020	5.521e-020	-19.131	-19.258	-0.127	(0)
FeSO4	4.443e-020	5.190e-020	-19.352	-19.285	0.067	(0)
FeCO3	1.498e-020	1.749e-020	-19.825	-19.757	0.067	(0)
FeHCO3+	1.255e-020	9.369e-021	-19.902	-20.028	-0.127	(0)
FeOH+	8.697e-021	5.768e-021	-20.061	-20.239	-0.178	(0)
Fe(OH)2	6.840e-024	7.989e-024	-23.165	-23.097	0.067	(0)
Fe(OH)3-	7.283e-026	4.830e-026	-25.138	-25.316	-0.178	(0)
FeHSO4+	2.752e-027	2.056e-027	-26.560	-26.687	-0.127	(0)
Fe(3)	3.711e-008					
Fe(OH)3	2.771e-008	3.237e-008	-7.557	-7.490	0.067	(0)
Fe(OH)4-	7.114e-009	4.804e-009	-8.148	-8.318	-0.170	(0)
Fe(OH)2+	2.286e-009	1.544e-009	-8.641	-8.811	-0.170	(0)
FeOH+2	1.481e-013	2.865e-014	-12.830	-13.543	-0.713	(0)
FeCl+2	1.519e-018	2.939e-019	-17.819	-18.532	-0.713	(0)
FeSO4+	1.174e-018	7.786e-019	-17.930	-18.109	-0.178	(0)
FeCl2+	7.062e-019	4.684e-019	-18.151	-18.329	-0.178	(0)
Fe+3	3.431e-019	2.727e-020	-18.465	-19.564	-1.100	(0)
Fe(SO4)2-	5.939e-020	4.435e-020	-19.226	-19.353	-0.127	(0)
FeCl3	1.431e-020	1.671e-020	-19.844	-19.777	0.067	(0)
Fe2(OH)2+4	2.360e-024	2.210e-026	-23.627	-25.656	-2.029	(0)
FeHSO4+2	4.039e-026	1.256e-026	-25.394	-25.901	-0.507	(0)
Fe3(OH)4+5	1.054e-029	7.129e-033	-28.977	-32.147	-3.170	(0)
H(0)	0.000e+000					
H2	0.000e+000	0.000e+000	-44.470	-44.402	0.067	28.61
K	1.058e-002					
K+	1.040e-002	6.483e-003	-1.983	-2.188	-0.205	9.66
KSO4-	1.756e-004	1.186e-004	-3.755	-3.926	-0.170	(0)
Mg	5.507e-002					
Mg+2	4.759e-002	1.374e-002	-1.322	-1.862	-0.540	-20.41
MgSO4	7.178e-003	8.384e-003	-2.144	-2.077	0.067	5.84
MgHCO3+	2.136e-004	1.343e-004	-3.670	-3.872	-0.201	5.82
MgCO3	7.301e-005	8.527e-005	-4.137	-4.069	0.067	-17.09
MgOH+	1.152e-005	8.116e-006	-4.939	-5.091	-0.152	(0)
Mn(2)	3.773e-009					
Mn+2	2.127e-009	4.875e-010	-8.672	-9.312	-0.640	-15.99
MnCl+	1.068e-009	7.086e-010	-8.971	-9.150	-0.178	7.01
MnCO3	2.157e-010	2.519e-010	-9.666	-9.599	0.067	(0)
MnSO4	1.932e-010	2.257e-010	-9.714	-9.646	0.067	4.99
MnCl2	9.449e-011	1.104e-010	-10.025	-9.957	0.067	(0)
MnHCO3+	5.475e-011	3.631e-011	-10.262	-10.440	-0.178	(0)
MnCl3-	1.635e-011	1.085e-011	-10.786	-10.965	-0.178	(0)

Table 11. Output for example 1.—Continued

MnOH+	3.074e-012	2.039e-012	-11.512	-11.691	-0.178	(0)
Mn(OH)3-	5.020e-020	3.329e-020	-19.299	-19.478	-0.178	(0)
Mn(NO3)2	1.344e-020	1.570e-020	-19.871	-19.804	0.067	(0)
Mn(3)	5.354e-026					
Mn+3	5.354e-026	4.255e-027	-25.271	-26.371	-1.100	(0)
N(-3)	1.724e-006					
NH4+	1.610e-006	9.049e-007	-5.793	-6.043	-0.250	18.44
NH3	7.327e-008	8.558e-008	-7.135	-7.068	0.067	24.46
NH4SO4-	4.064e-008	3.035e-008	-7.391	-7.518	-0.127	(0)
N(5)	4.847e-006					
NO3-	4.847e-006	2.845e-006	-5.314	-5.546	-0.232	30.32
Mn(NO3)2	1.344e-020	1.570e-020	-19.871	-19.804	0.067	(0)
Na	4.854e-001					
Na+	4.781e-001	3.431e-001	-0.320	-0.465	-0.144	-0.58
NaSO4-	6.631e-003	4.478e-003	-2.178	-2.349	-0.170	22.62
NaHCO3	6.168e-004	7.205e-004	-3.210	-3.142	0.067	19.41
NaCO3-	1.477e-005	9.972e-006	-4.831	-5.001	-0.170	1.77
NaOH	4.839e-017	5.652e-017	-16.315	-16.248	0.067	(0)
O(0)	4.377e-004					
O2	2.188e-004	2.556e-004	-3.660	-3.592	0.067	30.40
S(6)	2.926e-002					
SO4-2	1.432e-002	2.604e-003	-1.844	-2.584	-0.740	16.99
MgSO4	7.178e-003	8.384e-003	-2.144	-2.077	0.067	5.84
NaSO4-	6.631e-003	4.478e-003	-2.178	-2.349	-0.170	22.62
CaSO4	9.560e-004	1.117e-003	-3.020	-2.952	0.067	7.50
KSO4-	1.756e-004	1.186e-004	-3.755	-3.926	-0.170	(0)
NH4SO4-	4.064e-008	3.035e-008	-7.391	-7.518	-0.127	(0)
HSO4-	2.042e-009	1.525e-009	-8.690	-8.817	-0.127	40.96
MnSO4	1.932e-010	2.257e-010	-9.714	-9.646	0.067	4.99
CaHSO4+	5.922e-011	4.422e-011	-10.228	-10.354	-0.127	(0)
FeSO4+	1.174e-018	7.786e-019	-17.930	-18.109	-0.178	(0)
Fe(SO4)2-	5.939e-020	4.435e-020	-19.226	-19.353	-0.127	(0)
FeSO4	4.443e-020	5.190e-020	-19.352	-19.285	0.067	(0)
FeHSO4+2	4.039e-026	1.256e-026	-25.394	-25.901	-0.507	(0)
FeHSO4+	2.752e-027	2.056e-027	-26.560	-26.687	-0.127	(0)
Si	7.382e-005					
H4SiO4	7.061e-005	8.248e-005	-4.151	-4.084	0.067	52.08
H3SiO4-	3.210e-006	2.018e-006	-5.494	-5.695	-0.201	28.72
H2SiO4-2	1.095e-010	2.278e-011	-9.960	-10.642	-0.682	(0)
U(4)	1.830e-021					
U(OH)5-	1.830e-021	1.367e-021	-20.738	-20.864	-0.127	(0)
U(OH)4	2.922e-025	3.413e-025	-24.534	-24.467	0.067	(0)
U+4	0.000e+000	0.000e+000	-46.746	-48.775	-2.029	(0)
U(5)	2.871e-018					
UO2+	2.871e-018	2.144e-018	-17.542	-17.669	-0.127	(0)
U(6)	1.437e-008					
UO2(CO3)3-4	1.221e-008	1.143e-010	-7.913	-9.942	-2.029	(0)
UO2(CO3)2-2	2.148e-009	6.681e-010	-8.668	-9.175	-0.507	(0)
UO2CO3	1.074e-011	1.255e-011	-10.969	-10.901	0.067	(0)
UO2OH+	5.991e-014	4.474e-014	-13.222	-13.349	-0.127	(0)
UO2+2	5.350e-016	1.664e-016	-15.272	-15.779	-0.507	(0)
(UO2)2(OH)2+2	5.579e-021	1.736e-021	-20.253	-20.761	-0.507	(0)

Table 11. Output for example 1.—Continued

```
   (UO2)3(OH)5+   1.610e-022 1.203e-022    -21.793    -21.920      -0.127      (0)

----------------------------------Saturation indices----------------------------------

              Phase           SI   log IAP   log K(298 K,   1 atm)

              Anhydrite      -0.92    -5.20    -4.28  CaSO4
              Aragonite       0.53    -7.80    -8.34  CaCO3
              Calcite         0.68    -7.80    -8.48  CaCO3
              Chalcedony     -0.52    -4.07    -3.55  SiO2
              Chrysotile      3.36    35.56    32.20  Mg3Si2O5(OH)4
              CO2(g)         -3.48    -4.94    -1.46  CO2
              Dolomite        2.24   -14.85   -17.09  CaMg(CO3)2
              Fe(OH)3(a)      0.18     5.07     4.89  Fe(OH)3
              Goethite        6.08     5.08    -1.00  FeOOH
              Gypsum         -0.64    -5.22    -4.58  CaSO4:2H2O
              H2(g)         -41.30   -44.40    -3.10  H2
              H2O(g)         -1.51    -0.01     1.50  H2O
              Halite         -2.48    -0.91     1.57  NaCl
              Hausmannite     1.57    62.60    61.03  Mn3O4
              Hematite       14.17    10.17    -4.01  Fe2O3
              Jarosite-K     -7.57   -16.78    -9.21  KFe3(SO4)2(OH)6
              Manganite       2.40    27.74    25.34  MnOOH
              Melanterite   -19.39   -21.59    -2.21  FeSO4:7H2O
              NH3(g)         -8.86    -7.07     1.80  NH3
              O2(g)          -0.70    -3.59    -2.89  O2 Pressure   0.2 atm, phi
1.000.
              Pyrochroite    -8.09     7.11    15.20  Mn(OH)2
              Pyrolusite      6.97    48.35    41.38  MnO2:H2O
              Quartz         -0.09    -4.07    -3.98  SiO2
              Rhodochrosite  -3.37   -14.50   -11.13  MnCO3
              Sepiolite       1.15    16.91    15.76  Mg2Si3O7.5OH:3H2O
              Sepiolite(d)   -1.75    16.91    18.66  Mg2Si3O7.5OH:3H2O
              Siderite      -13.25   -24.14   -10.89  FeCO3
              SiO2(a)        -1.35    -4.07    -2.71  SiO2
              Sylvite        -3.54    -2.64     0.90  KCl
              Talc            6.03    27.43    21.40  Mg3Si4O10(OH)2
              Uraninite     -12.42   -15.91    -3.49  UO2

------------------
End of simulation.
------------------


------------------------------------
Reading input data for simulation 2.
------------------------------------


------------------------------
End of Run after 0.64 Seconds.
------------------------------
```

Any comment entered within the simulation with the TITLE keyword is printed next. The title is followed by the heading "Beginning of initial solution calculations", below which are the results of the speciation calculation for seawater. The concentration data, converted to molality, are given under the subheading "Solution composition". For initial solution calculations, the number of moles in solution is numerically equal to molality because 1 kg of water is assumed. The **-water** identifier can be used to define a different mass of water for a solution. During batch-reaction calculations, the mass of water may change and the moles in the aqueous phase will not exactly equal the molality of a constituent. Note that the molality of dissolved oxygen that produces a log partial pressure of -0.7 has been calculated and is annotated in the output.

After the subheading "Description of solution", some of the properties listed in the first block of output are equal to their input values and some are calculated. In this example, pH, pe, and temperature are equal to the input values. The specific conductance, density, activity of water, ionic strength, total carbon (alkalinity was the input datum), total inorganic carbon ("Total CO2"), electrical balance, percent error, total hydrogen, and total oxygen have all been calculated by the model.

Under the subheading "Redox couples" the pe and Eh are printed for each redox couple for which data were available; in this case, ammonium/nitrate and water/dissolved oxygen.

Under the subheading "Distribution of species", the molalities, activities, activity coefficients, and specific volumes of all species of each element and element valence state are listed. The lists are alphabetical by element name and are descending in terms of molality within each element or element valence state. Beside the name of each element or element valence state, the total molality is given. If **-Vm** parameters are defined in **SOLUTION_SPECIES**, specific volumes are calculated relative to the volume of H^+ (which is zero by convention at all pressures, temperatures and ionic strengths); otherwise, specific volumes are listed as (0).

Finally, under the subheading "Saturation indices", saturation indices for all minerals that are appropriate for the given analytical data are listed alphabetically by phase name near the end of the output. The saturation index is given in the column headed "SI", followed by the columns for the log of the ion activity product ("log IAP") and the log of the solubility constant ("log KT"). The chemical formulas for each of the phases is printed in the right-hand column. Note, for example, that no aluminum-bearing minerals are included because aluminum was not included in the analytical data. Also note that mackinawite (FeS) and other sulfide minerals are not included in the output because no analytical data were specified for S(-2). If a concentration for S [instead of S(6)] or S(-2) had been entered, then a concentration of S(-2) would have been calculated and a saturation index for mackinawite and other sulfide minerals would have been calculated.

Example 2—Equilibration With Pure Phases

This example shows how to calculate the solubility and relative thermodynamic stability of two minerals, gypsum and anhydrite. First, as a function of temperature at 1 atm, and second, as a function of temperature and pressure, while comparing the calculations with experimental solubility data.

Conceptually, the two models define a beaker with pure water to which the minerals gypsum and (or) anhydrite are added. Step-wise, the beaker is heated, the minerals dissolve to equilibrium, and the concentrations and saturation indexes are calculated and plotted. If, at a given temperature, gypsum is less soluble than anhydrite, anhydrite dissolves completely while gypsum precipitates; similarly, if gypsum is more soluble than anhydrite, anhydrite dissolves completely and gypsum precipitates. Adding a single mineral allows the (possibly metastable) solubility at any temperature and pressure to be calculated.

The input file for the first model is given in table 12. It defines a single simulation in which various keywords define the actions that will be processed together. The water is defined with keyword SOLUTION. It is given a pH of 7 and a temperature of 25 °C, but these are equal to default and could be omitted. Also, by default, the pe is 4 and the density is 1 kg/L, and by omitting these parameters the default values will be used. The two minerals are defined with keyword EQUILIBRIUM_PHASES. The mineral name is followed by the target saturation index and the amount in moles (defaults are 0 and 10, respectively). If a phase is not present initially, it can be given 0 mol. Of course, the mineral names must have been defined before through a PHASES data block in the database or the input file. The REACTION_TEMPERATURE data block lets the temperature change from 25 °C to 75 °C in 51 steps (25, 26, ..., 75 °C).

Table 12. Input file for example 2.

```
TITLE Example 2.--Temperature dependence of solubility
              of gypsum and anhydrite
SOLUTION 1 Pure water
      pH      7.0
      temp    25.0
EQUILIBRIUM_PHASES 1
      Gypsum          0.0     1.0
      Anhydrite       0.0     1.0
REACTION_TEMPERATURE 1
      25.0 75.0 in 51 steps
SELECTED_OUTPUT
      -file    ex2.sel
      -temperature
      -si      anhydrite  gypsum
USER_GRAPH 1 Example 2
      -headings Temperature Gypsum Anhydrite
```

Table 12. Input file for example 2.—Continued

```
        -chart_title "Gypsum-Anhydrite Stability"
        -axis_scale x_axis 25 75 5 0
        -axis_scale y_axis auto 0.05 0.1
        -axis_titles "Temperature, in degrees celsius" "Saturation index"
        -initial_solutions false
  -start
  10 graph_x TC
  20 graph_y SI("Gypsum") SI("Anhydrite")
  -end
END
TITLE Example 2.--Temperature dependence of solubility
               of gypsum and anhydrite
SOLUTION 1 Pure water
        pH      7.0
        temp    25.0
EQUILIBRIUM_PHASES 1
        Gypsum            0.0      1.0
        Anhydrite         0.0      1.0
REACTION_TEMPERATURE 1
        25.0 75.0 in 51 steps
SELECTED_OUTPUT
        -file   ex2.sel
        -temperature
        -si     anhydrite  gypsum
USER_GRAPH 1 Example 2
        -headings Temperature Gypsum Anhydrite
        -chart_title "Gypsum-Anhydrite Stability"
        -axis_scale x_axis 25 75 5 0
        -axis_scale y_axis auto 0.05 0.1
        -axis_titles "Temperature, in degrees celsius" "Saturation index"
        -initial_solutions false
  -start
  10 graph_x TC
  20 graph_y SI("Gypsum") SI("Anhydrite")
  -end
END
```

At each step, the temperature and the saturation indices for gypsum and anhydrite are written to the file *ex2.sel* as defined by **SELECTED_OUTPUT**, and plotted by **USER_GRAPH** as shown in figure 4. The figure shows that below 58 °C, the solution is in equilibrium with gypsum, but subsaturated with respect to anhydrite. Above that temperature, anhydrite is the more stable phase.

PHREEQC starts the simulation by calculating the solution composition if a new **SOLUTION** (or **SOLUTION_SPREAD**) data block is defined, continues with the reactions defined with the keywords, and prints the results of the calculations. The printout of the initial solution and the first batch-reaction step is listed in table 13. Headings define the various parts and are self-explanatory.

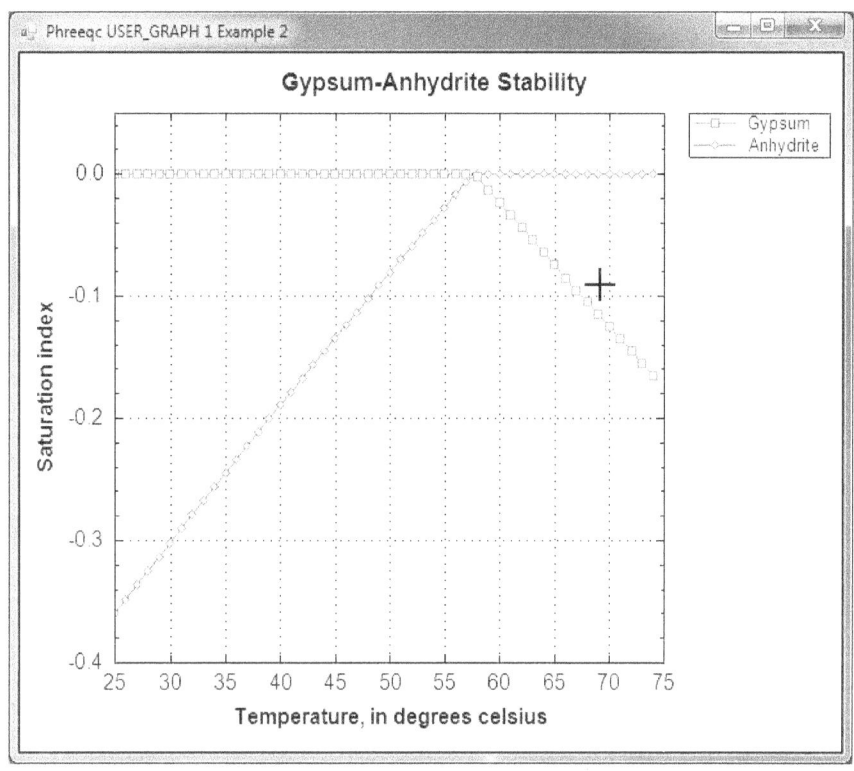

Figure 4. Saturation indices of gypsum and anhydrite in water that has equilibrated with the more stable of the two phases over the temperature range 25 to 75 °C.

Table 13. Selected output for example 2.

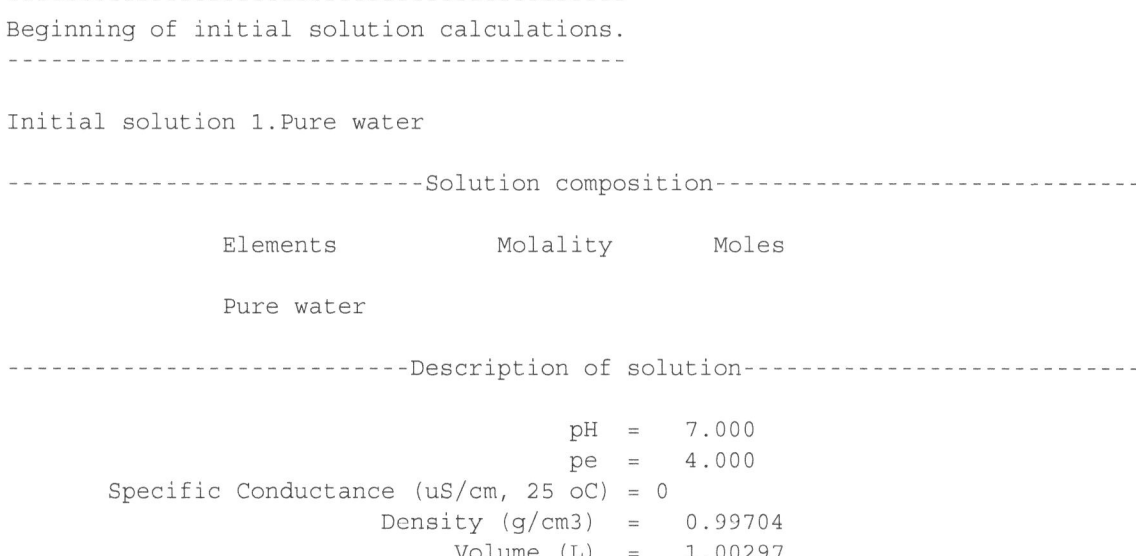

```
--------------------------------------------
Beginning of initial solution calculations.
--------------------------------------------

Initial solution 1.Pure water

----------------------------Solution composition----------------------------

            Elements          Molality       Moles

            Pure water

----------------------------Description of solution----------------------------

                            pH  =    7.000
                            pe  =    4.000
     Specific Conductance (uS/cm, 25 oC) = 0
                   Density (g/cm3)  =    0.99704
                      Volume (L)  =    1.00297
```

Table 13. Selected output for example 2.—Continued

```
               Activity of water  =   1.000
                   Ionic strength  =   1.007e-007
               Mass of water (kg)  =   1.000e+000
          Total alkalinity (eq/kg)  =   1.217e-009
            Total carbon (mol/kg)  =   0.000e+000
               Total CO2 (mol/kg)  =   0.000e+000
               Temperature (deg C)  =   25.00
            Electrical balance (eq)  =  -1.217e-009
Percent error, 100*(Cat-|An|)/(Cat+|An|)  =   -0.60
                       Iterations  =   0
                          Total H  = 1.110124e+002
                          Total O  = 5.550622e+001
```

---------------------------Distribution of species---------------------------

```
                                          Log      Log      Log     mole V
    Species        Molality   Activity  Molality Activity  Gamma   cm3/mol

    OH-          1.013e-007  1.012e-007  -6.995   -6.995  -0.000   -4.14
    H+           1.001e-007  1.000e-007  -7.000   -7.000  -0.000    0.00
    H2O          5.551e+001  1.000e+000   1.744    0.000   0.000   18.07
H(0)          1.416e-025
    H2           7.079e-026  7.079e-026 -25.150  -25.150   0.000   28.61
O(0)          0.000e+000
    O2           0.000e+000  0.000e+000 -42.080  -42.080   0.000   30.40
```

----------------------------Saturation indices----------------------------

```
    Phase              SI    log IAP   log K(298 K,   1 atm)

    H2(g)           -22.05   -25.15   -3.10  H2
    H2O(g)           -1.50     0.00    1.50  H2O
    O2(g)           -39.19   -42.08   -2.89  O2
```

Beginning of batch-reaction calculations.

Reaction step 1.

Using solution 1.Pure water
Using pure phase assemblage 1.
Using temperature 1.

----------------------------Phase assemblage----------------------------

```
                                              Moles in assemblage
Phase            SI   log IAP  log K(T, P)   Initial       Final        Delta

Anhydrite      -0.30   -4.58     -4.28     1.000e+000          0 -1.000e+000
Gypsum          0.00   -4.58     -4.58     1.000e+000  1.985e+000  9.855e-001
```

Table 13. Selected output for example 2.—Continued

```
--------------------------Solution composition----------------------------

            Elements         Molality      Moles

            Ca             1.508e-002  1.455e-002
            S              1.508e-002  1.455e-002

-------------------------Description of solution---------------------------

                              pH  =    7.066      Charge balance
                              pe  =   10.745      Adjusted to redox equilibrium
      Specific Conductance (uS/cm, 25 oC) = 2161
                   Density (g/cm3)  =    0.99909
                      Volume (L)  =    0.96829
              Activity of water  =    1.000
                 Ionic strength  =    4.183e-002
                Mass of water (kg)  =    9.645e-001
          Total alkalinity (eq/kg)  =    1.261e-009
             Total carbon (mol/kg)  =    0.000e+000
                Total CO2 (mol/kg)  =    0.000e+000
               Temperature (deg C)  =   25.00
          Electrical balance (eq)  =   -1.217e-009
  Percent error, 100*(Cat-|An|)/(Cat+|An|)  =   -0.00
                      Iterations  =   19
                         Total H  = 1.070706e+002
                         Total O  = 5.359351e+001

-------------------------Distribution of species---------------------------

                                          Log     Log      Log    mole V
       Species        Molality   Activity Molality Activity Gamma  cm3/mol

       OH-          1.431e-007 1.178e-007  -6.844   -6.929  -0.084  -3.90
       H+           9.974e-008 8.587e-008  -7.001   -7.066  -0.065   0.00
       H2O          5.551e+001 9.996e-001   1.744   -0.000   0.000  18.07
Ca          1.508e-002
    Ca+2            1.046e-002 5.176e-003  -1.981   -2.286  -0.305 -17.66
    CaSO4           4.627e-003 4.672e-003  -2.335   -2.331   0.004   7.50
    CaOH+           1.203e-008 1.000e-008  -7.920   -8.000  -0.080    (0)
    CaHSO4+         3.172e-009 2.637e-009  -8.499   -8.579  -0.080    (0)
H(0)        3.354e-039
    H2              1.677e-039 1.693e-039 -38.776  -38.771   0.004  28.61
O(0)        2.878e-015
    O2              1.439e-015 1.453e-015 -14.842  -14.838   0.004  30.40
S(-2)       0.000e+000
    HS-             0.000e+000 0.000e+000 -118.111 -118.195  -0.084  20.77
    H2S             0.000e+000 0.000e+000 -118.324 -118.320   0.004  37.16
    S-2             0.000e+000 0.000e+000 -123.735 -124.047  -0.312    (0)
S(6)        1.508e-002
    SO4-2           1.046e-002 5.075e-003  -1.981   -2.295  -0.314  14.66
    CaSO4           4.627e-003 4.672e-003  -2.335   -2.331   0.004   7.50
```

Table 13. Selected output for example 2.—Continued

```
HSO4-           5.096e-008  4.237e-008    -7.293    -7.373    -0.080    40.44
CaHSO4+         3.172e-009  2.637e-009    -8.499    -8.579    -0.080    (0)

-----------------------------Saturation indices----------------------------

          Phase              SI   log IAP   log K(298 K,   1 atm)

          Anhydrite        -0.30    -4.58    -4.28  CaSO4
          Gypsum            0.00    -4.58    -4.58  CaSO4:2H2O
          H2(g)           -35.67   -38.77    -3.10  H2
          H2O(g)           -1.50    -0.00     1.50  H2O
          H2S(g)         -117.27  -125.26    -7.99  H2S
          O2(g)           -11.95   -14.84    -2.89  O2
          Sulfur          -87.58   -82.70     4.88  S
```

The heading "Phase assemblage" records the saturation indices and amounts of each of the phases defined by **EQUILIBRIUM_PHASES**. In the first batch-reaction step, the solution is undersaturated with respect to anhydrite (saturation index is -0.30), and in equilibrium with gypsum (saturation index is 0.0). Consequently, all of the anhydrite has dissolved and most of the calcium and sulfate have precipitated as gypsum. The "Solution composition" shows that 15.1 mmol/kgw of calcium and sulfate are in solution, which is the solubility of gypsum in pure water at 25 °C. However, the total moles of the two constituents in the aqueous phase is only 14.6 because the mass of water has decreased to 0.964 kg by precipitating gypsum ($CaSO_4 \cdot 2H_2O$), as printed below "Description of solution". Accordingly, the mass of solvent water is not constant in batch-reaction calculations because reactions and waters of hydration in dissolving and precipitating phases may increase or decrease the mass of solvent water. Also listed under "Description of solution" are the calculated specific conductance (2161 μS/cm, microsiemens per centimeter), the density (0.999 g/cm^3), the cation-anion balance (0), and more.

To illustrate that the temperature where gypsum transforms into anhydrite is a function of pressure, the calculation is repeated with input file *ex2b* at pressures of 1, 500, and 1,000 bars. The calculated results are compared with experimental data summarized by Blount and Dickson (1973, figure 2). Part of the input file is listed in table 14.

Table 14. Input file for the first and second simulation in example 2B.

```
TITLE Calculate gypsum/anhydrite transitions, 30 - 170 oC, 1 - 1000 atm
     Data in ex2b.tsv from Blount and Dickson, 1973, Am. Mineral. 58, 323, fig. 2.
PRINT; -reset false
SOLUTION 1
EQUILIBRIUM_PHASES
Gypsum
REACTION_TEMPERATURE
```

Table 14. Input file for the first and second simulation in example 2B.—Continued

```
 30 90 in 10
USER_GRAPH 1 Example 2B, (P, T)-dependent solubilities of Gypsum and Anhydrite
 -plot_tsv_file ex2b.tsv
 -axis_titles "Temperature, in degrees celsius" "Solubility, in moles per \
     kilogram water"
 -axis_scale x_axis 30 170
 -axis_scale y_axis 1e-3 0.05 auto auto log
 10 plot_xy tc, tot("Ca"), color = Red, symbol = None
 -end
END # 1st simulation

USE solution 1
USE equilibrium_phases 1
USE reaction_temperature 1
REACTION_PRESSURE 2
 493
USER_GRAPH
 10 plot_xy tc, tot("Ca"), color = Red, symbol = None
END

USE solution 1
USE equilibrium_phases 1
USE reaction_temperature 1
REACTION_PRESSURE 3
 987
USER_GRAPH
 20 plot_xy tc, tot("Ca"), color = Red, symbol = None
END # 2nd simulation
```

Like before in table 13 (example file *ex2* in the PHREEQC distribution), the first simulation of table 14 (example file *ex2b* in the PHREEQC distribution) defines the **SOLUTION**, the **EQUILIBRIUM_PHASES**, the **REACTION_TEMPERATURE**s, and **USER_GRAPH** for plotting the experimental solubilities from file *ex2b.t*sv and the calculated concentrations. To accelerate the calculations, the output is reduced by using **PRINT**; **-reset false**. The second simulation of table 14 uses the same definitions through the **USE** data blocks and defines the reaction pressure as 493 atmospheres (= 500 bar). The third simulation in *ex2b* (not listed in table 14) does the same for 1,000 bars, and further simulations repeat the calculations with anhydrite as the equilibrium phase.

Figure 5 shows the solubility (note the logarithmic scale) as a function of temperature. The temperature of the gypsum to anhydrite transition increases from 58 °C at 1 atm, to 63 °C at 493 atm (= 500 bar), and 70 °C at 987 atm (= 1,000 bar). Thus, the stability of gypsum relative to anhydrite increases with pressure, which is because water in the gypsum crystal has a smaller volume than water in solution:

$$CaSO_4 \cdot 2H_2O = CaSO_4 + 2H_2O.$$

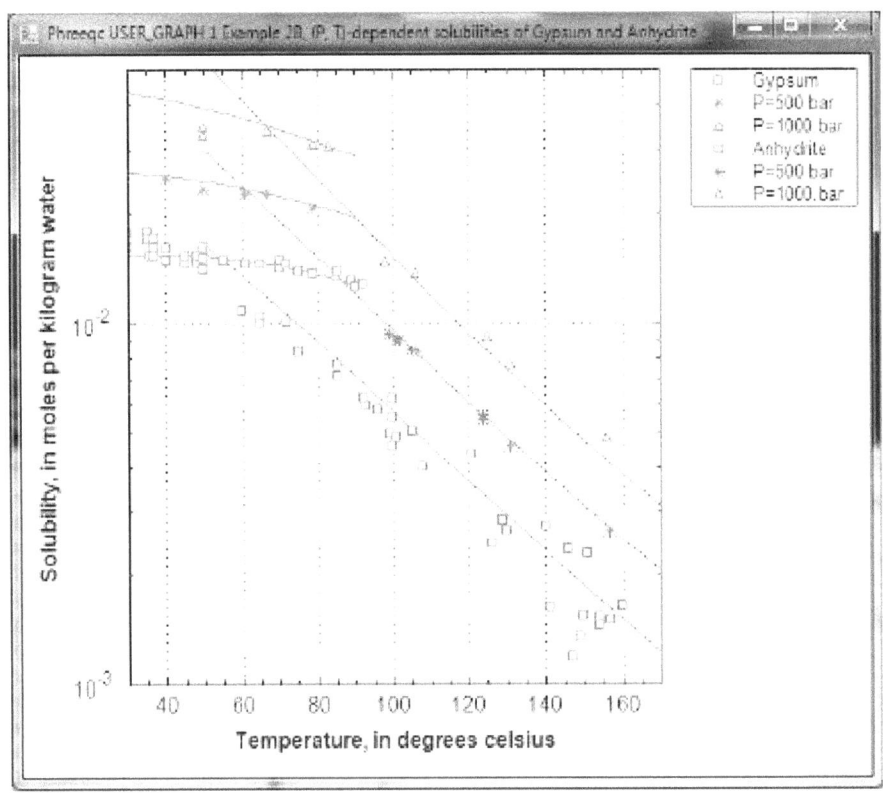

Figure 5. The solubility of gypsum and anhydrite as a function of temperature at 1, 500, and 1,000 bars. Data points from Blount and Dickson (1973); lines calculated by PHREEQC.

However, as illustrated in figure 5, the solubility of gypsum increases with pressure because the sum of the aqueous molar volumes of the solute species together is smaller than the molar volume of gypsum. Another point to note is that the experimental data for the gypsum solubility extend into the stability field of anhydrite. Apparently, the precipitation of anhydrite is too slow to reduce the concentrations in the experiments.

Example 3—Mixing

This example demonstrates the capabilities of PHREEQC to perform a series of geochemical simulations, with the final simulations relying on results from previous simulations within the same run. The example investigates diagenetic reactions that may occur in zones where seawater mixes with carbonate groundwater. The example is divided into five simulations, labeled part A through part E in table 15.

(A) Carbonate groundwater is defined by equilibrating pure water with calcite at a P_{CO_2} of $10^{-2.0}$ atm.

(B) Seawater is defined by using the major-ion data given in table 9.

(C) The two solutions are mixed together in the proportions 70 percent groundwater and 30 percent seawater.

(D) The mixture is equilibrated with calcite and dolomite.

(E) The mixture is equilibrated with calcite only to investigate the chemical evolution if dolomite precipitation is assumed to be negligible.

The input for part A (table 15) consists of (a) the definition of pure water with SOLUTION input, and (b) the definition of a pure-phase assemblage with EQUILIBRIUM_PHASES input. In the definition of the phases, only a saturation index was given for each phase. Because it was not entered, the amount of each phase defaults to 10.0 mol, which is essentially an unlimited supply for most phases. The batch reaction is implicitly defined to be the equilibration of the first solution defined in this simulation with the first pure-phase assemblage defined in the simulation. (Explicit definition of batch-reaction entities is done with the USE keyword.) The SAVE keyword instructs the program to save the batch-reaction solution composition from the final batch-reaction step as solution number 1. Thus, when the simulation begins, solution number 1 is pure water. After the batch-reaction calculations for the simulation are completed, the batch-reaction solution—water in equilibrium with calcite and CO_2—is stored as solution 1.

Table 15. Input file for example 3.

```
TITLE Example 3, part A.--Calcite equilibrium at log Pco2 = -2.0 and 25C.
SOLUTION 1   Pure water
        pH      7.0
        temp    25.0
EQUILIBRIUM_PHASES
        CO2(g)          -2.0
        Calcite         0.0
SAVE solution 1
END
TITLE Example 3, part B.--Definition of seawater.
SOLUTION 2   Seawater
        units   ppm
        pH      8.22
        pe      8.451
        density 1.023
        temp    25.0
        Ca              412.3
        Mg              1291.8
        Na              10768.0
        K               399.1
        Si              4.28
```

Table 15. Input file for example 3.—Continued

```
        Cl                  19353.0
        Alkalinity          141.682 as HCO3
        S(6)                2712.0
END
TITLE Example 3, part C.--Mix 70% groundwater, 30% seawater.
MIX 1
        1       0.7
        2       0.3
SAVE solution   3
END
TITLE Example 3, part D.--Equilibrate mixture with calcite and dolomite.
EQUILIBRIUM_PHASES 1
        Calcite             0.0
        Dolomite            0.0
USE solution 3
END
TITLE Example 3, part E.--Equilibrate mixture with calcite only.
EQUILIBRIUM_PHASES 2
        Calcite             0.0
USE solution 3
END
```

Part B defines the composition of seawater, which is stored as solution number 2. Part C mixes groundwater, (solution 1) with seawater (solution 2) in a closed system in which P_{CO_2} is calculated, not specified. The **MIX** keyword is used to define the mixing fractions (approximate mixing volumes) of each solution in the mixture. The **SAVE** keyword causes the mixture to be saved as solution number 3. The **MIX** keyword allows the mixing of an unlimited number of solutions in whatever fractions are specified. The fractions (volumes) need not sum to 1.0. If the fractions were 7.0 and 3.0 instead of 0.7 and 0.3, the number of moles of each element in solution 1 (including hydrogen and oxygen) would be multiplied by 7.0, the number of moles of each element in solution 2 would be multiplied by 3.0, and the resulting moles of elements would be added together. The mass of water in the mixture would be approximately 10 kg (7.0 from solution 1 and 3.0 from solution 2) instead of approximately 1 kg when the fractions are 0.7 and 0.3. The concentrations in the mixture would be the same for both sets of mixing fractions because the relative proportions of solution 1 and solution 2 are the same. However, during subsequent reactions it would take 10 times more mole transfer for mixing fractions 7.0 and 3.0 than shown in table 16 because the mass of water would be 10 times greater in that system.

Part D equilibrates the mixture with calcite and dolomite. The **USE** keyword specifies that solution number 3, which is the mixture from part C, is to be the solution with which the phases will equilibrate. By defining the phase assemblage with "**EQUILIBRIUM_PHASES** 1", the phase assemblage replaces the

Table 16. Selected results for example 3.

[Simulation A generates carbonate groundwater; B defines seawater; C mixes a fraction of 0.7 of A with 0.3 of B, with no other reaction; D equilibrates the mixture with calcite and dolomite; and E equilibrates the mixture with calcite only. Mole transfer is relative to the moles in the phase assemblage; positive numbers indicate an increase in the amount of the phase present, that is, precipitation; negative numbers indicate a decrease in the amount of the phase present, or dissolution. Saturation index: "--" indicates saturation index calculation not possible because one of the constituent elements was not in solution. Mole transfer: "--" indicates no mole transfer of this mineral was allowed in the simulation]

Simulation	pH	log P_{CO_2}	Saturation index		Mole transfer, millimoles		
			Calcite	Dolomite	CO_2	Calcite	Dolomite
A	7.292	-2.00	0.00	--	-1.993	-1.657	--
B	8.220	-3.48	0.68	2.24	--	--	--
C	7.263	-2.20	-0.24	0.25	--	--	--
D	7.083	-2.05	0.00	0.00	--	-15.61	7.853
E	7.472	-2.39	0.00	0.72	--	-0.085	--

previous assemblage number 1 that was defined in part A. Part E performs a similar calculation to part D, but uses phase assemblage 2, which does not contain dolomite as a reactant.

Selected results from the output for example 3 are presented in table 16. The groundwater produced by part A is in equilibrium with calcite and has a log P_{CO_2} of -2.0, as specified by the input. The moles of CO_2 in the phase assemblage decreased by about 2.0 mmol, which means that about 2.0 mmol dissolved into solution. Likewise, about 1.7 mmol of calcite dissolved. Part B defined seawater, which is calculated to have slightly greater than atmospheric carbon dioxide (-3.48 compared to about -3.5), and is supersaturated with calcite (saturation index 0.68) and dolomite (2.24). No mole transfers of minerals were allowed for part B. Part C performed the mixing and calculated the equilibrium distribution of species in the mixture, again with no mole transfers of the minerals allowed. The resulting log P_{CO_2} is -2.20, calcite is undersaturated, and dolomite is supersaturated. The saturation indices indicate that thermodynamically, dolomitization should occur; that is, calcite dissolves and dolomite precipitates. Part D calculates the amounts of calcite and dolomite that should react. To produce equilibrium, 15.61 mmol of calcite should dissolve and 7.853 mmol of dolomite should precipitate. Dolomitization is not observed to occur in present-day mixing zone environments, even though dolomite is the thermodynamically stable phase. The lack of significant dolomitization is due to the slow reaction kinetics of dolomite formation. Therefore, part E simulates what would happen if dolomite does not precipitate; in that case, only a very small amount of calcite dissolves (0.085 mmol) for this mixing ratio.

Example 4—Evaporation and Homogeneous Redox Reactions

Evaporation is accomplished by removing water from the chemical system. Water can be removed by several methods: (1) water can be specified as an irreversible reactant with a negative reaction coefficient in the **REACTION** keyword input, (2) the solution can be mixed with pure water which is given a negative mixing fraction in **MIX**, or (3) "H2O" can be specified as the alternative reaction in **EQUILIBRIUM_PHASES** keyword input, in which case water is removed or added to the aqueous phase to attain equilibrium with a specified phase. This example uses the first method; the **REACTION** data block is used to simulate concentration of rainwater by approximately 20-fold by removing 95 percent of the water. The resulting solution contains only about 0.05 kg of water. In a subsequent simulation, the **MIX** keyword is used to generate a solution that has the same concentrations as the evaporated solution, but has a total mass of water of approximately 1 kg.

The first simulation input file (table 17) contains four keywords: (1) **TITLE** is used to specify a description of the simulation to be included in the output file, (2) **SOLUTION** is used to define the composition of rainwater from central Oklahoma, (3) **REACTION** is used to specify the amount of water, in moles, to be removed from the aqueous phase, and (4) **SAVE** is used to store the result of the batch-reaction calculation as solution number 2.

Table 17. Input file for example 4.

```
TITLE Example 4a.--Rainwater evaporation
SOLUTION 1  Precipitation from Central Oklahoma
        units           mg/L
        pH              4.5   # estimated
        temp            25.0
        Ca              .384
        Mg              .043
        Na              .141
        K               .036
        Cl              .236
        C(4)            .1        CO2(g)   -3.5
        S(6)            1.3
        N(-3)           .208
        N(5)            .237
REACTION 1
        H2O     -1.0
        52.73 moles
SAVE solution 2
END
TITLE Example 4b.--Factor of 20 more solution
MIX
        2        20.
SAVE solution 3
END
```

All solutions defined by SOLUTION input are scaled to have exactly 1 kg (approximately 55.5 mol) of water, unless -water identifier is used. To concentrate the solution by 20-fold, it is necessary to remove 52.73 mol of water (55.506×0.95).

The second simulation uses MIX to multiply by 20 the moles of all the elements in the solution, including hydrogen and oxygen. This procedure effectively increases the total mass (or volume) of the aqueous phase but maintains the same concentrations. For identification in table 18, the solution that results from the MIX simulation is stored as solution 3 with the SAVE keyword. Solution 3 will have the same concentrations as solution 2 (from the previous simulation) but will have a mass of water of approximately 1 kg.

Selected results of the simulation are presented in table 18. The concentration factor of 20 is reasonable in terms of a water balance for the process of evapotranspiration in central Oklahoma (Parkhurst and others, 1996). The PHREEQC modeling assumes that evaporation and evapotranspiration have the same effect and that evapotranspiration has no effect on the ion ratios. These assumptions have not been verified and may not be correct. After evaporation, the simulated solution composition is still undersaturated with respect to calcite, dolomite, and gypsum. As expected, the mass of water decreases from 1 kg in rainwater (solution 1) to approximately 0.05 kg in solution 2 after water was removed by the reaction. In general, the amount of water remaining after the reaction varies because water may be consumed or produced by homogeneous hydrolysis reactions, surface complexation reactions, and dissolution and precipitation of pure phases. The number of moles of chloride (μmol, micromole) was unaffected by the removal of water; however, the concentration of chloride (μmol/kgw, micromole per kilogram water) increased because the amount of water decreased. The second mixing simulation increased the mass of water and the moles of chloride by a factor of 20. Thus, the moles of chloride increased, but the chloride concentration is the same before (solution 2) and after (solution 3) in the mixing simulation because the mass of water increased proportionately.

An important point about homogeneous redox reactions is illustrated in the results of these simulations (table 18). Batch-reaction calculations (and transport calculations) always produce aqueous equilibrium among all redox elements. The rainwater analysis contained data for both ammonium and nitrate, but none for dissolved nitrogen. The pe of the rainwater has no effect on the distribution of species in the initial solution because concentrations of the individual redox states of redox elements (C, N, and S) are specified. Although nitrate and ammonium should not coexist at thermodynamic equilibrium, the speciation calculation allows redox disequilibria and accepts the concentrations of the two redox states of

Table 18. Selected results for example 4.

[kg, kilogram; Cl, chloride; μmol, micromole; μmol/kgw, micromole per kilogram water]

Constituent	Solution 1 Rainwater	Solution 2 Concentrated 20-fold	Solution 3 Mixed with factor 20
Mass of water, kg	1.000	0.05002	1.000
Cl, μmol	6.657	6.657	133.1
Cl, μmol/kgw	6.657	133.1	133.1
Nitrate [N(5)], μmol/kgw	16.9	160.1	160.1
Dissolved nitrogen [N(0)], μmol/kgw	0	475.1	475.1
Ammonium [N(-3)], μmol/kgw	14.8	0	0
Calcite saturation index	-9.20	-9.36	-9.36
Dolomite saturation index	-19.00	-19.33	-19.33
Gypsum saturation index	-5.35	-2.91	-2.91

nitrogen that are defined by the input data, regardless of thermodynamic equilibrium. During the batch-reaction (evaporation) step, redox equilibrium is attained for the aqueous phase, which causes ammonium to be oxidized and nitrate to be reduced, generating dissolved nitrogen [$N_{2(aq)}$, or N(0) in PHREEQC notation]. The first batch-reaction solution (solution 2) contains the equilibrium distribution of nitrogen, which consists of nitrate and dissolved nitrogen, but no ammonium (table 18). The oxidation of ammonium and reduction of nitrate occur in the batch-reaction calculation to produce redox equilibrium from the inherent redox disequilibrium in the definition of the rainwater composition. Nitrogen redox reactions would have occurred in the simulation even if the REACTION keyword had specified that no water was to be removed. Solution 3 (table 18) also is the result of a batch-reaction calculation and has the same redox equilibrium as solution 2. The only way to prevent complete equilibration of the nitrogen redox states would be to define the individual redox states as separate SOLUTION_MASTER_SPECIES and SOLUTION_SPECIES; for example, by defining a new element in SOLUTION_MASTER_SPECIES called "Amm" and defining NH_3 and other N(-3) species in terms of Amm (Amm, $AmmH^+$, and others). In this case, equilibrium would be attained among all species of N and all species of Amm, but no equilibria would exist between N and Amm species. This option has been implemented in the database *Amm.dat*.

Example 5—Irreversible Reactions

This example demonstrates the irreversible reaction capabilities of PHREEQC in modeling the oxidation of pyrite. Oxygen (O_2) and NaCl are added irreversibly to pure water in six amounts (0.0, 0.001, 0.005, 0.01, 0.03, and 0.05 mol); the relative proportion of O_2 to NaCl in the irreversible reaction is 1.0 to 0.5. Pyrite, calcite, and goethite are allowed to dissolve to equilibrium and the carbon dioxide partial pressure is maintained at $10^{-3.5}$ (atmospheric partial pressure). In addition, gypsum is allowed to precipitate if it becomes supersaturated.

Pure water is defined with **SOLUTION** input (table 19), and the pure-phase assemblage is defined with **EQUILIBRIUM_PHASES** input. By default, 10 mol of pyrite, goethite, calcite, and carbon dioxide are present in the pure-phase assemblage, but gypsum is defined to have 0.0 mol in the pure-phase

Table 19. Input file for example 5.

```
TITLE Example 5.--Add oxygen, equilibrate with pyrite, calcite, and goethite.
SOLUTION 1  PURE WATER
        pH       7.0
        temp     25.0
EQUILIBRIUM_PHASES 1
        Pyrite          0.0
        Goethite        0.0
        Calcite         0.0
        CO2(g)          -3.5
        Gypsum          0.0       0.0
REACTION 1
        O2      1.0
        NaCl    0.5
        0.0     0.001   0.005   0.01    0.03    0.05
SELECTED_OUTPUT
        -file    ex5.sel
        -total   Cl
        -si      Gypsum
        -equilibrium_phases   pyrite goethite calcite CO2(g) gypsum
USER_GRAPH Example 5
        -headings Pyrite Goethite Calcite CO2(g) Gypsum SI_Gypsum
        -chart_title "Pyrite Oxidation"
        -axis_titles "O2 added, in millimoles" "Millimoles dissolved" \
            "Saturation index"
 10 x = RXN * 1e3
 20 PLOT_XY x, 1e3 * (10 - EQUI("Pyrite")), symbol = Plus
 30 PLOT_XY x, 1e3 * (10 - EQUI("Goethite")), symbol = Plus
 40 PLOT_XY x, 1e3 * (10 - EQUI("Calcite")), symbol = Plus
 50 PLOT_XY x, 1e3 * (10 - EQUI("CO2(g)")), symbol = Plus
 60 PLOT_XY x, 1e3 * (-EQUI("Gypsum")), symbol = Plus, color = Magenta
 70 PLOT_XY x, SI("Gypsum"), y-axis = 2, line_width = 2, symbol = Circle, \
            symbol_size = 8, color = Magenta
END
```

assemblage. Gypsum can precipitate if it becomes supersaturated; it cannot dissolve initially because no moles are present. The **REACTION** data block defines the irreversible reaction that is to be modeled. In this example, oxygen ("O2") will be added with a relative coefficient of 1.0 and NaCl will be added with a relative coefficient of 0.5. The steps of the reaction are defined to be 0.0, 0.001, 0.005, 0.01, 0.03, and 0.05 mol. The reactants can be defined by a chemical formula, as in this case ("O2" and "NaCl") or by a phase name that has been defined with **PHASES** input. Thus, the phase names "O2(g)" or "Halite" from the default database file could have been used in place of "O2" or "NaCl" to achieve the same result. The number of moles of oxygen atoms added is equal to the stoichiometric coefficient of oxygen in the formula "O2" (2.0) times the relative coefficient (1.0) times the moles of reaction defined by the reaction step (0.0, 0.001, 0.005, 0.01, 0.03 or 0.05). Thus, in the last step, $2.0 \times 1.0 \times 0.05 = 0.1$ mol of O atoms are added. Similarly, the number of moles of chloride added at each step is the stoichiometric coefficient of chlorine in the formula "NaCl" (1.0) times the relative coefficient (0.5) times the moles in the reaction step. **SELECTED_OUTPUT** and **USER_GRAPH** are used to write the total concentration of chloride, the saturation index of gypsum, and the total amounts and mole transfers of pyrite, goethite, calcite, carbon dioxide, and gypsum to the file *ex5.sel* and to plot the reactions in the chart after each equilibrium calculation.

The results for example 5 are summarized in table 20 and displayed in figure 6. When no oxygen or sodium chloride is added to the system, a small amount of calcite and carbon dioxide dissolves, and trace amounts of pyrite and goethite react; the pH is 8.27, the pe is low (-4.94) because of equilibrium with pyrite, and gypsum is six orders of magnitude undersaturated (saturation index -6.13). As oxygen and sodium

Table 20. Selected results for example 5.
[Mole transfer is relative to the moles in the phase assemblage. Positive numbers indicate an increase in the amount of the phase present; that is, precipitation. Negative numbers indicate a decrease in the amount of the phase; that is, dissolution]

Reactants added, millimoles		pH	pe	Mole transfer, millimoles					Saturation index of gypsum
O_2	NaCl			Pyrite	Goethite	Calcite	$CO_{2(g)}$	Gypsum	
0.0	0.0	8.27	-4.94	-0.00003	0.00001	-0.50	-0.49	0.0	-6.13
1.0	0.5	8.17	-4.28	-0.27	0.27	-0.93	0.14	0.0	-2.01
5.0	2.5	7.98	-3.96	-1.33	1.33	-2.94	2.39	0.0	-1.05
10.0	5.0	7.88	-3.81	-2.67	2.67	-5.56	5.10	0.0	-0.64
30.0	15.0	7.72	-3.57	-8.00	8.00	-16.18	15.82	0.0	-0.01
50.0	25.0	7.72	-3.56	-13.33	13.33	-26.84	26.49	9.55	0.00

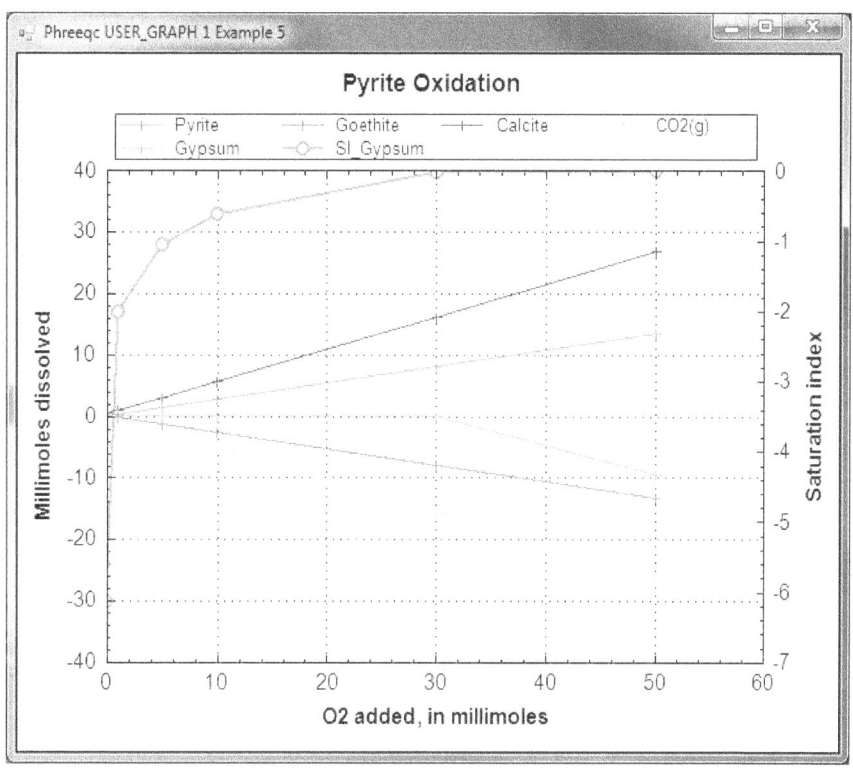

Figure 6. Reaction of pyrite, calcite, and goethite in a beaker filled with 1 kilogram water to which oxygen is added in fixed steps. Pyrite and calcite dissolve, goethite precipitates, and gypsum starts to precipitate when more than 30 millimoles of O_2 is added to the beaker. Carbon dioxide, originating from calcite, is lost from the acidifying solution, which remains in equilibrium with the atmospheric partial pressure.

chloride are added, pyrite oxidizes, and goethite, being relatively insoluble, precipitates. This reaction generates sulfuric acid, decreases the pH, slightly increases the pe, and causes calcite to dissolve and carbon dioxide to be released. When slightly more than 30 mmol of oxygen is added, gypsum reaches saturation and begins to precipitate. When 50 mmol of oxygen and 25 mmol of sodium chloride have been added, a total of 9.55 mmol of gypsum has precipitated.

Example 6—Reaction-Path Calculations

In this example, the precipitation of phases as a result of incongruent dissolution of K-feldspar (microcline) is investigated. Only the four phases originally addressed by Helgeson and others (1969)—K-feldspar, gibbsite, kaolinite, and K-mica (muscovite)—are considered. The thermodynamic data for the phases (table 21, **PHASES** keyword) are derived from Robie and others (1978) and are the same as for test problem 5 in the PHREEQE manual (Parkhurst and others, 1980).

PHREEQC can be used to solve this problem in three ways: the individual intersections of the reaction path and the phase boundaries on a phase diagram can be calculated (simulation 6A, table 21), the reaction path can be calculated incrementally (simulation 6B, table 21), or the reaction path can be calculated as a kinetic process (simulation 6C, table 21). In the first approach, no knowledge of the amounts of reaction is needed, but a number of simulations are necessary to find the appropriate phase-boundary intersections. In the second approach, only one simulation is sufficient, but the appropriate amounts of reaction must be known beforehand. In the third approach, a kinetic rate expression is used to calculate the reaction path by using a step-size adjusting algorithm, which takes care of phase boundary transitions by automatically decreasing the time interval when necessary. Only the total time to arrive at the point of K-feldspar equilibrium is required. All three approaches are demonstrated here. PHREEQC implicitly contains all the logic of a complete reaction-path program (for example, Helgeson and others, 1970; Wolery, 1979; Wolery and others, 1990). Moreover, the capability to calculate directly the phase boundary intersections provides an efficient way to outline reaction paths on phase diagrams, and the option to add the reaction incrementally and automatically find the stable phase assemblage allows points on the reaction path between phase boundaries to be obtained easily. The kinetic approach and the Basic interpreter that is embedded in PHREEQC can be used to save and print the arrival time and the aqueous composition at each phase transition.

Conceptually, example 6 considers the reactions that would occur if K-feldspar were placed in a beaker and allowed to react slowly. As K-feldspar dissolves, other phases may begin to precipitate. It is assumed that only gibbsite, kaolinite, or K-mica can form, and that these phases will precipitate reversibly if they reach saturation. That is, the phases that precipitated at the beginning of the reaction are allowed to redissolve, if necessary, to maintain equilibrium as the reaction proceeds.

Table 21. Input file for example 6.

```
TITLE Simulation 6A.--React to phase boundaries.
SOLUTION 1  PURE WATER
        pH    7.0 charge
      temp    25.0
PHASES
     Gibbsite
            Al(OH)3 + 3 H+ = Al+3 + 3 H2O
            log_k        8.049
            delta_h      -22.792 kcal
     Kaolinite
            Al2Si2O5(OH)4 + 6 H+ = H2O + 2 H4SiO4 + 2 Al+3
            log_k        5.708
            delta_h      -35.306 kcal
```

Table 21. Input file for example 6.—Continued

```
            K-mica
                    KAl3Si3O10(OH)2 + 10 H+ = 3 Al+3 + 3 H4SiO4 + K+
                    log_k          12.970
                    delta_h        -59.377 kcal
            K-feldspar
                    KAlSi3O8 + 4 H2O + 4 H+ = Al+3 + 3 H4SiO4 + K+
                    log_k          0.875
                    delta_h        -12.467 kcal
SELECTED_OUTPUT
            -file           ex6A-B.sel
            -activities     K+ H+ H4SiO4
            -si             Gibbsite Kaolinite K-mica K-feldspar
            -equilibrium    Gibbsite Kaolinite K-mica K-feldspar
END
TITLE Simulation 6A1.--Find amount of K-feldspar dissolved to
                    reach gibbsite saturation.
USE solution 1
EQUILIBRIUM_PHASES 1
            Gibbsite        0.0     KAlSi3O8        10.0
            Kaolinite       0.0     0.0
            K-mica          0.0     0.0
            K-feldspar      0.0     0.0
USER_GRAPH 1 Simulation 6
            -headings 6A--Intersections
            -chart_title "K-Feldspar Reaction Path"
            -axis_titles "Log[H4SiO4]" "Log([K+] / [H+])"
            -axis_scale x_axis -8.0 0.0 1 1
            -axis_scale y_axis -1.0 8.0 1 1
    10 PLOT_XY LA("H4SiO4"),(LA("K+")-LA("H+")), color = Red, line_w = 0, \
               symbol = Circle, symbol_size = 10
END
TITLE Simulation 6A2.--Find amount of K-feldspar dissolved to
                    reach kaolinite saturation.
USE solution 1
EQUILIBRIUM_PHASES 1
            Gibbsite        0.0     0.0
            Kaolinite       0.0     KAlSi3O8        10.0
            K-mica          0.0     0.0
            K-feldspar      0.0     0.0
END
TITLE Simulation 6A3.--Find amount of K-feldspar dissolved to
                    reach K-mica saturation.
USE solution 1
EQUILIBRIUM_PHASES 1
            Gibbsite        0.0     0.0
            Kaolinite       0.0     0.0
            K-mica          0.0     KAlSi3O8        10.0
            K-feldspar      0.0     0.0
END
TITLE Simulation 6A4.--Find amount of K-feldspar dissolved to
                    reach K-feldspar saturation.
USE solution 1
```

Table 21. Input file for example 6.—Continued

```
EQUILIBRIUM_PHASES 1
        Gibbsite          0.0       0.0
        Kaolinite         0.0       0.0
        K-mica            0.0       0.0
        K-feldspar        0.0       KAlSi3O8        10.0
END
TITLE Simulation 6A5.--Find point with kaolinite present,
                   but no gibbsite.
USE solution 1
EQUILIBRIUM_PHASES 1
        Gibbsite          0.0       KAlSi3O8        10.0
        Kaolinite         0.0       1.0
END
TITLE Simulation 6A6.--Find point with K-mica present,
                   but no kaolinite
USE solution 1
EQUILIBRIUM_PHASES 1
        Kaolinite         0.0       KAlSi3O8        10.0
        K-mica            0.0       1.0
END
TITLE Simulation 6B.--Path between phase boundaries.
USE solution 1
EQUILIBRIUM_PHASES 1
        Kaolinite         0.0       0.0
        Gibbsite          0.0       0.0
        K-mica            0.0       0.0
        K-feldspar        0.0       0.0
REACTION 1
        K-feldspar        1.0
        0.04 0.08 0.16 0.32 0.64 1.0 2.0 4.0
        8.0 16.0 32.0 64.0 100 200 umol
USER_GRAPH
        -headings 6B--Increments
  10 PLOT_XY LA("H4SiO4"),(LA("K+")-LA("H+")), color = Blue, line_w = 0, \
                     symbol = XCross, symbol_size = 7
END
TITLE Simulation 6C.--kinetic calculation
SOLUTION 1
        -units mol/kgw
        Al        1.e-13
        K         1.e-13
        Si        3.e-13
EQUILIBRIUM_PHASES 1
        Gibbsite   0.0  0.0
        Kaolinite  0.0  0.0
        K-mica     0.0  0.0
KINETICS 1
K-feldspar
# k0 * A/V = 1e-16 mol/cm2/s * (10% fsp, 0.1mm cubes) 136/cm = 136.e-13 mol/dm3/s
        -parms 1.36e-11
        -m0 2.16
        -m  1.94
```

Table 21. Input file for example 6.—Continued

```
          -step_divide 1e-6
          -steps     1e2 1e3 1e4 1e5 1e6 1e7 1e8
#         -steps     1e2 1e3 1e4 1e5 63240.0 64950.0 1347610.0 1010300.0 45242800.0
INCREMENTAL_REACTIONS true
RATES
K-feldspar
-start
  10  REM store the initial amount of K-feldspar
  20  IF EXISTS(1) = 0 THEN PUT(M, 1)
  30  REM calculate moles of reaction
  40  SR_kfld = SR("K-feldspar")
  50  moles = PARM(1) * (M/M0)^0.67 * (1 - SR_kfld) * TIME
  60  REM The following is for printout of phase transitions
  80  REM      Start Gibbsite
  90  if ABS(SI("Gibbsite")) > 1e-3 THEN GOTO 150
 100   i = 2
 110    GOSUB 1500
 150 REM       Start Gibbsite -> Kaolinite
 160 if ABS(SI("Kaolinite")) > 1e-3 THEN GOTO 200
 170   i = 3
 180    GOSUB 1500
 200 REM       End Gibbsite -> Kaolinite
 210 if ABS(SI("Kaolinite")) > 1e-3 OR EQUI("Gibbsite") > 0 THEN GOTO 250
 220   i = 4
 230    GOSUB 1500
 250 REM       Start Kaolinite -> K-mica
 260 if ABS(SI("K-mica")) > 1e-3 THEN GOTO 300
 270   i = 5
 280    GOSUB 1500
 300 REM       End Kaolinite -> K-mica
 310 if ABS(SI("K-mica")) > 1e-3 OR EQUI("Kaolinite") > 0 THEN GOTO 350
 320   i = 6
 330    GOSUB 1500
 350 REM       Start K-mica -> K-feldspar
 360 if ABS(SI("K-feldspar")) > 1e-3 THEN GOTO 1000
 370   i = 7
 380    GOSUB 1500
1000 SAVE moles
1010 END
1500 REM subroutine to store data
1510 if GET(i) >= M THEN RETURN
1520     PUT(M, i)
1530     PUT(TOTAL_TIME, i, 1)
1540     PUT(LA("K+")-LA("H+"), i, 2)
1550     PUT(LA("H4SiO4"), i, 3)
1560 RETURN
-end
USER_PRINT
  10 DATA "A: Gibbsite              ", "B: Gibbsite  -> Kaolinite ", \
          "C: Gibbsite  -> Kaolinite ", "D: Kaolinite -> K-mica    ", \
          "E: Kaolinite -> K-mica    ", "F: K-mica    -> K-feldspar"
  20 PRINT \
```

Table 21. Input file for example 6.—Continued

```
    "       Transition                 Time   K-feldspar     LA(K/H)    LA(H4SiO4)"
 30 PRINT "                                             transfer"
 40 PRINT "                                             (umoles)"
 50 FOR i = 2 TO 7
 60    READ s$
 70    IF EXISTS(i) THEN PRINT s$, GET(i,1), (GET(1) - GET(i))*1e6, GET(i,2), GET(i,3)
 80 NEXT i
SELECTED_OUTPUT
        -file ex6C.sel
        -reset false
USER_PUNCH
  -headings pH+log[K]   log[H4SiO4]
  10 PUNCH LA("K+")-LA("H+") LA("H4SiO4")
USER_GRAPH
  -headings 6C--Kinetics
  10 PLOT_XY LA("H4SiO4"),(LA("K+")-LA("H+")), color = Blue, line_w = 2, symbol = None
END
PRINT; -user_print false
# --Plot the phase boundaries with USER_GRAPH..
PHASES
 K_H; KH = K+ - H+; -no_check
USER_GRAPH
-initial_solutions true
 10 PLOT_XY LA("H4SiO4"), SI("K_H"), color = Black, symbol = None
SOLUTION 1
 pH 11; K 1 K_H 8; Al 1 Gibbsite; Si 1 K-mica
SOLUTION 2
 pH 7; K 1 K-mica; Al 1 Gibbsite; Si 1 Kaolinite
SOLUTION 3
 pH 7; K 1 K-mica; Al 1 K-feldspar; Si 1 Kaolinite
SOLUTION 4
 pH 7; K 1 K_H -1; Al 1 Kaolinite; Si 1 K-feldspar
END
USER_GRAPH
 10 PLOT_XY LA("H4SiO4"), SI("K_H"), color = Black, symbol = None
SOLUTION 1
 pH 11; K 1 K_H 8; Al 1 K-feldspar; Si 1 K-mica
SOLUTION 2
 pH 7; K 1 K-mica; Al 1 K-feldspar; Si 1 Kaolinite
SOLUTION 3
 pH 7; K 1 K-mica; Al 1 Gibbsite; Si 1 Kaolinite
SOLUTION 4
 pH 7; K 1 K_H -1; Al 1 Gibbsite; Si 1 Kaolinite
END
```

The input file (table 21) first defines pure water with **SOLUTION** input and the thermodynamic data for the phases with **PHASES** input. Some of the minerals are defined in the database file (*phreeqc.dat*), but inclusion in the input file replaces any previous definitions for the duration of the run (the database file is not altered). **SELECTED_OUTPUT** is used to write a file of data that was used to produce table 22, and

Table 22. Selected results for simulation 6A.

[Simulation refers to labels in the input file for simulation 6A. Negative mole transfers indicate dissolution, positive mole transfers indicate precipitation. Simulations 6A1–6A6 refer to label points A–F on figure 7 and table 24. H^+, hydrogen ion; K^+, potassium ion; H_4SiO_4, silicic acid]

Simulation	K-feldspar mole transfer micromoles	Log activity			Mole transfer, micromoles			Saturation index				Point on graph
		H^+	$\dfrac{K^+}{H^+}$	H_4SiO_4	Gibbsite	Kaolinite	K-mica	Gibbsite	Kaolinite	K-mica	K-feldspar	
6A1	-0.03	-7.00	-0.57	-7.10	0.00	0.00	0.00	0.0	-3.8	-10.7	-14.7	A
6A2	-2.18	-8.21	2.55	-5.20	1.78	0.00	0.00	0.0	0.0	-1.9	-5.9	B
6A3	-20.10	-9.11	4.41	-4.47	0.00	9.76	0.00	-0.7	0.0	0.0	-2.5	D
6A4	-190.9	-9.39	5.49	-3.55	0.00	0.00	63.62	-2.0	-0.7	0.0	0.0	F
6A5	-3.02	-8.35	2.83	-5.20	0.00	1.24	0.00	0.0	0.0	-1.6	-5.6	C
6A6	-32.81	-9.07	4.41	-4.25	0.00	0.00	10.83	-0.9	0.0	0.0	-2.1	E

USER_GRAPH produces the chart shown in figure 7. SELECTED_OUTPUT specifies that the log of the activities of the potassium ion, hydrogen ion, and silicic acid; the saturation indices for gibbsite, kaolinite, K-mica, and K-feldspar; and the total amounts in the phase assemblage and mole transfers for gibbsite, kaolinite, K-mica, and K-feldspar will be written to the file *ex6A-B.sel* after each calculation. The definitions for SELECTED_OUTPUT remain in effect for all simulations in the run until a new SELECTED_OUTPUT data block is read or until writing to the file is suspended with the identifier **-selected_output** in the PRINT data block.

Simulation 6A1 allows K-feldspar to react until equilibrium with gibbsite is reached. This is set up in the EQUILIBRIUM_PHASES input by specifying equilibrium for gibbsite (saturation index equals 0.0) and an alternative reaction to reach equilibrium, KAlSi3O8 (the formula for K-feldspar). A large amount of K-feldspar (10.0 mol) is present to ensure that equilibrium with gibbsite can be obtained. Kaolinite, K-mica, and K-feldspar are allowed to precipitate if they become saturated (which does not occur in this part of the simulation), but they cannot dissolve because they were given zero initial moles in the phase assemblage. The amount of reaction is the dissolution of precisely enough K-feldspar to reach equilibrium with gibbsite. No gibbsite will dissolve or precipitate; the alternative reactant (KAlSi3O8) will dissolve or precipitate in its place. Simulations 6A2–6A4 perform the same calculations for kaolinite, K-mica, and K-feldspar. At other temperatures or using other minerals, a target phase may remain undersaturated regardless of the amount of the alternative reaction that is added because the phase is unstable relative to the

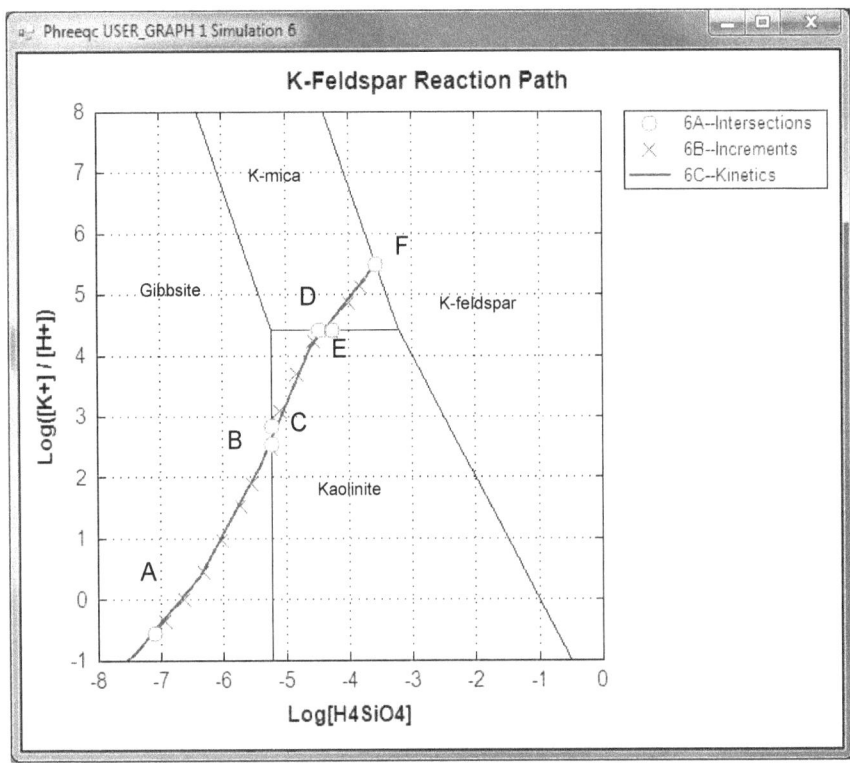

Figure 7. Phase diagram for the dissolution of K-feldspar (microcline) in pure water at 25 °C showing stable phase-boundary intersections (example 6A) and reaction paths across stability fields (simulations 6B and 6C). Phase boundaries are plotted by using thermodynamic data from Robie and others (1978). (Names of the stability fields and letter designations of points have been added to clarify the **USER_GRAPH** chart.)

other defined phases. If this were the case, the numerical method would find the amount of the alternative reaction that produces the maximum saturation index for the target phase.

Selected results for simulations 6A1–6A4 are presented in table 22 and are plotted on figure 7 as points A, B, D, and F. The stability fields for the phases, which are based on the thermodynamic data, are calculated and plotted with **USER_GRAPH** as explained later. At point B, gibbsite starts to be transformed into kaolinite, a reaction that consumes Si. From this point, the reaction path follows the gibbsite-kaolinite phase boundary until all gibbsite is converted (point C), and then crosses the kaolinite field to point D. Similarly, there is a point E on the kaolinite-K-mica phase boundary, where the reaction path starts crossing the K-mica field to point F. Simulations 6A5 and 6A6 (table 21) solve for the two points C and E. In simulation 6A5, point C is calculated by specifying that kaolinite is present at equilibrium, while K-feldspar dissolves until gibbsite is at saturation, but with zero concentration in the phase assemblage. Likewise, simulation 6A6 solves for the point where K-mica is at saturation and present in the phase assemblage, while kaolinite is at saturation, but with zero concentration (point E). Assigning an initial amount of 1 mol to kaolinite in 6A5 and to K-mica in 6A6 is arbitrary, but the amount must be sufficient to reach equilibrium with the mineral.

A simpler approach to determining the reaction path is to react K-feldspar incrementally, allowing the stable phase assemblage among gibbsite, kaolinite, K-mica, and K-feldspar to form at each point along the path. The only difficulty in this approach is to know the appropriate amounts of reaction to add. From points A to F in figure 7, K-feldspar dissolution ranges from 0.03 (simulation 6A1, table 22) to 190.9 mmol (simulation 6A4, table 22). In simulation 6B (table 21) a logarithmic range of reaction increments is used to define the path (solid line) across the phase diagram (fig. 7) from its beginning at gibbsite equilibrium (point A) to equilibrium with K-feldspar (point F). However, the exact locations of points A through F will not be determined with the arbitrary set of reaction increments that are used in simulation 6B (table 21). The reaction path calculated by simulation 6B is plotted on the phase diagram in figure 7 with points A through F from simulation 6A (table 21) included in the set of points.

Finally, in the kinetic approach in simulation 6C (table 21), kinetic dissolution of K-feldspar is followed with time, while the phases gibbsite, kaolinite, and K-mica are allowed to precipitate and redissolve as the kinetic reaction proceeds. SOLUTION 1 is defined to have a small amount of dissolved K-feldspar (1×10^{-13} mol). The solution then contains all elements related to phases in EQUILIBRIUM_PHASES, which, although not required for the program to run successfully, eliminates some warning messages.

During the integration of the reaction rates, a simple dissolution rate law was assumed based on transition-state theory:

$$R_{K-feldspar} = k_1 \frac{A}{V}\left(\frac{m}{m_0}\right)^{0.67}\left(1 - \left(\frac{IAP}{K}\right)_{K-feldspar}\right) , \qquad (20)$$

where R is the rate, mol cm^{-2}s^{-1} (mole per square centimeter per second); k_1 is the rate constant, equal to 1×10^{-16} mol cm^{-2}s^{-1}; m is the amount of kinetic reactant, mol; m_0 is the initial amount of kinetic reactant, mol; IAP is the ion activity product; and K is the equilibrium constant.

The KINETICS data block is used to enter specific data for the kinetic simulation. The stoichiometry of the kinetic reaction is the chemical formula of K-feldspar; by default, the name of the rate is assumed to be a phase defined in the PHASES data block and the formula of the phase is used as the stoichiometry of the reaction. It was assumed that the pristine soil contained 10 percent K-feldspar in the form of 0.1 mm cubes, and had $\rho_b / \varepsilon = 6$ g/cm^3, so that $A/V = 136$/cm. The value of $k_1 \frac{A}{V} = 1.36 \times 10^{-11}$ mol L^{-1}s^{-1} (mole per liter per second) is entered in the **KINETICS** data block with the identifier **-parms** (assuming that 1 kgw = 1 liter), and can be recalled as "PARM(1)" in the Basic rate definition in the **RATES** data block. It was assumed that the soil had already been weathered to some extent, and that only 90 percent of the initial K-feldspar was left [**-m0** 2.16 and **-m** 1.94, where **m0** indicates the initial mass (1 kg soil × 0.1 = 100 g / 278.3 g/mol = 0.359 mol/kg × 6 kg/L = 2.16 mol/L), and **m** is the remaining mass (90 percent of 2.16 is 1.94 mol/L)]. The maximum amount of reaction for any time interval is restricted to 1×10^{-6} mol (**-step_divide** 1e-6). Time steps (s) are defined with the identifier **-steps**. INCREMENTAL_REACTIONS **true** causes each new time step to start at the end of the previous time step, so that the total time is the sum of all the time steps (= 1.111111×10^8 s).

The rate for K-feldspar dissolution is defined in the form of Basic statements in the **RATES** data block. To demonstrate some of the features of the Basic interpreter, the Basic program also identifies and saves information at phase transitions, which is printed at the end of the run by using **USER_PRINT**. The accuracy of locating a phase transition is determined by the user-definable accuracy of the integration. A small tolerance (**-tol**), a large **-step_divide** that is greater than 1 (initial time interval will be divided by this number), or a small **-step_divide** that is less than 1 (specifies maximum moles of reaction) will force smaller time intervals and more accurate identification of phase transitions. In this simulation (6C), **-step_divide** is set to 1×10^{-6}, which limits the maximum amount of reaction for any time interval to be less than 1 micromole. Thus, the amount of reaction to reach a phase transition will be identified with an accuracy of 1 micromole. However, limiting the amount of reaction requires smaller time intervals during the integration and, consequently, more time intervals to complete the integration, which increases the CPU time of the run.

The purpose of each part of the Basic program is described in table 23. The functions PUT, GET, and EXISTS are used to manipulate data in static, global storage. The arguments used in the PUT function identify a number and the storage location. For example, PUT(M,i) will place the value of "M" in the location identified by the variable *i*. EXISTS can be used to determine if a storage location contains a number, and GET is used to retrieve data that have been stored. For example, IF EXISTS(i) then m2 = GET(i) will assign the number in location *i* to *m2* if *i* contains a number, and do nothing if it does not. Once a number has been stored with PUT, it exists for the remainder of the run, unless it is overwritten with another PUT statement in the same location (that is, with the same set of subscripts). Data stored with PUT

Table 23. Description of Basic program for K-feldspar dissolution kinetics and identification of phase transitions.
[mol, mole]

Line number	Purpose
20	Save initial amount of K-feldspar (1.94 mol)
40–50	Integrate K-feldspar dissolution rate over time interval given by TIME.
90–110	Identify greatest amount of K-feldspar present (least amount of reaction) at which gibbsite is saturated.
160–180	Identify greatest amount of K-feldspar present at which kaolinite is saturated.
200–230	Identify greatest amount of K-feldspar present at which kaolinite is saturated, but gibbsite is absent.
250–280	Identify greatest amount of K-feldspar present at which K-mica is saturated.
300–330	Identify greatest amount of K-feldspar present at which K-mica is saturated, but kaolinite is absent.
350–380	Identify greatest amount of K-feldspar present at which K-feldspar is saturated.
1000	Save integrated reaction.
1010	End of Basic program
1500–1560	Subroutine for saving values for phase transitions. If amount of K-feldspar is greater than current saved value for the index i, save amount of K-feldspar, cumulative time, log activity ratio of potassium ion divided by hydrogen ion, and log activity of silicic acid.

can be retrieved by any Basic program defined in RATES, USER_GRAPH, USER_PRINT, and USER_PUNCH. In this simulation (6C), data are stored by the RATES Basic program, and the USER_PRINT Basic program retrieves the data and prints a summary of the phase transitions. Whereas the RATES program is run many times during the kinetic integration of a time step, the USER_PRINT program is run only once at the end of each integration time step.

Table 24 gives the phase transitions encountered by the end of the last time step of simulation 6C. For each phase transition, the time at which the phase transition occurred, the total amount of K-feldspar that has reacted, and the coordinates of the transition on figure 7 are given. Although the values in table 24 are.

Table 24. Phase transitions identified by the **RATES** Basic program and printed to the output file by the **USER_PRINT** Basic program in simulation 6C, which simulates the kinetic dissolution of K-feldspar.

```
-------------------------------User print---------------------------------

        Transition              Time    K-feldspar     LA(K/H)    LA(H4SiO4)
                                        transfer
                                        (umoles)
A: Gibbsite                      1100   1.4048e-001  3.5642e-001 -6.3763e+000
B: Gibbsite   -> Kaolinite  1.7434e+005  2.2064e+000  2.5575e+000 -5.1950e+000
C: Gibbsite   -> Kaolinite  2.3929e+005  3.0284e+000  2.8317e+000 -5.1945e+000
D: Kaolinite  -> K-mica     1.5967e+006  2.0194e+001  4.4080e+000 -4.4630e+000
E: Kaolinite  -> K-mica     2.6017e+006  3.2848e+001  4.4087e+000 -4.2499e+000
F: K-mica     -> K-feldspar 4.7638e+007  1.9074e+002  5.4868e+000 -3.5536e+000
```

approximate, the amount of K-feldspar and the coordinates of the transition can be compared with table 22 As expected, the approximate mole transfers to reach the phase transitions in table 24 (column 3) are within 1 micromole of the values in table 22 (column 2).

The **SELECTED_OUTPUT** data block specifies that a new selected-output file will be used for this simulation, *ex6C.sel*, and all printing to the selected-output file is eliminated (**-reset** false). The **USER_PUNCH** data block causes two columns to be written to each line of the selected-output file, the log of the ratio of the activities of potassium ion to hydrogen ion and the log activity of silicic acid. The data are written after each time step has been simulated (**-steps**, **KINETICS** data block). Table 25 shows the results written to *ex6C.sel*.

The phase boundaries in figure 7 are plotted with **USER_GRAPH** at the end of the file. First, a dummy phase "K_H" is defined that will provide the activity ratio of K^+ and H^+ in the solution, which is plotted on the y-axis. The **-log_k** of zero for the reaction is the default value in PHREEQC and may be omitted. The saturation index of a solution for this phase is given by $\log([K^+] / [H^+])$, and can be obtained

Table 25. Results written to the selected-output file by the **USER_PUNCH** Basic program in simulation 6C, which simulates the kinetic dissolution of K-feldspar.
["\t" indicates a tab character]

```
  pH+log[K]\t      log[H4SiO4]\t
-6.0002e+000\t    -1.2524e+001\t
-1.8975e+000\t    -8.4212e+000\t
-8.5318e-001\t    -7.3798e+000\t
 3.5742e-001\t    -6.3763e+000\t
 2.1790e+000\t    -5.3817e+000\t
 4.1326e+000\t    -4.6001e+000\t
 5.2599e+000\t    -3.7183e+000\t
 5.4872e+000\t    -3.5533e+000\t
 8.0000e+000\t    -6.3924e+000\t
 4.4080e+000\t    -5.1950e+000\t
 4.4080e+000\t    -3.1935e+000\t
-1.0000e+000\t    -4.9484e-001\t
 8.0000e+000\t    -4.3909e+000\t
 4.4080e+000\t    -3.1935e+000\t
 4.4080e+000\t    -5.1950e+000\t
-1.0000e+000\t    -5.1950e+000\t
```

with SI("K_H") in a Basic program. Next, four solutions are defined for the points that delimit the phase boundaries, going from top to bottom and from left to right. **USER_GRAPH** connects the points with a black line, without symbols, and without entry in the legend (no **-headings**). After the **END**, another four solutions are defined for the phase boundaries from top to bottom and from right to left, and another curve is drawn by redefining line 10 in USER_GRAPH (in this way avoiding connecting the last bottom-right point with the new top-right point). Note that the text is compacted in this part of the input file by concatenating lines with semicolons, ";".

Example 7—Gas-Phase Calculations

This example demonstrates the capabilities of PHREEQC to model the evolution of gas compositions in equilibrium with a solution with a fixed (total) pressure or a fixed volume of the gas phase. In the case of a fixed-pressure gas phase, a gas bubble forms as soon as the sum of the partial pressures of the component gases exceeds the specified pressure of the gas phase. Once the bubble forms, its volume and composition will vary with the extent of reactions. This case applies to gas bubbles forming in surface water or groundwater at a given depth, where the total pressure is constant. With a fixed-volume gas phase, the aqueous solution is in contact with a head space of a fixed volume, which is typical for a laboratory experiment with a closed bottle. The gas phase always exists in this head space, but its pressure and composition will vary with the reactions. Another way to model gas-liquid reactions in PHREEQC is to maintain a fixed partial pressure by using the EQUILIBRIUM_PHASES data block. This fixed-partial-pressure approach is illustrated in this example by fixing the CO_2 pressure for a SOLUTION.

Conceptually, an infinite gas reservoir is assumed for the fixed-partial-pressure approach, for example, water in contact with ambient air. In this case, the partial pressure of a gas component remains constant regardless of the extent of reactions. If the gas reservoir is finite and the total pressure is constant, as in gas bubbles in estuarine and lake sediments and in groundwater, then a fixed-pressure gas phase should be used. If the gas reservoir is finite and its volume is constant, as in a bottle with a fixed head-space, then the fixed-volume gas phase is appropriate.

In this example, the fixed-partial-pressure approach is used to define a groundwater in equilibrium with a given CO_2 pressure and with calcite. The GAS_PHASE data block is used to model the decomposition of organic matter under fixed-pressure and fixed-volume conditions, with the assumption that carbon, nitrogen, hydrogen, and oxygen are released in the stoichiometry $CH_2O(NH_3)_{0.07}$ by the decomposition reaction. Without electron acceptors, the organic matter decomposes to CH_4 and CO_2, and NH_3 and N_2. The carbon and nitrogen will react to redox and gas-solution equilibrium in the model, but it should be noted that these redox reactions require bacterial mediation and are almost always in disequilibrium in groundwater systems. Aqueous carbon species are defined in SOLUTION_MASTER_SPECIES and SOLUTION_SPECIES of the default databases for two valence states, carbon(+4) and carbon(-4) (methane); no intermediate valence states of carbon are defined. Aqueous nitrogen may occur in the +5, +3, 0, and -3 valence states, depending on the database. The gas components considered here are water vapor (H_2O), carbon dioxide (CO_2), methane (CH_4), nitrogen (N_2), and ammonia (NH_3).

In the first simulation, the initial water is groundwater in equilibrium with calcite at a partial pressure of carbon dioxide of $10^{-1.5}$ [log $P(CO_2)$ = -1.5]. Pure water is defined with the **SOLUTION** data block with default values for pH (7.0), pe (4.0), and temperature (25 °C); calcite and carbon dioxide, which dissolve to equilibrium, are defined with **EQUILIBRIUM_PHASES**. **SAVE** is used to save the equilibrated solution so that it can be recalled later in the run (table 26). **USER_GRAPH** data blocks specify data to be plotted, and **SELECTED_OUTPUT** defines a file (*ex7.sel*) to which data (similar to the plotted data) are written for each calculation. As an alternative to writing the selected output file, the data plotted in the charts can be saved to a file by using the pop-up menu that appears when right-clicking the mouse while the cursor is inside a chart.

Table 26. Input file for example 7.

```
TITLE Example 7.--Organic decomposition with fixed-pressure and
                  fixed-volume gas phases
SOLUTION_MASTER_SPECIES
N(-3)    NH4+              0.0      N
SOLUTION_SPECIES
NH4+ = NH3 + H+
        log_k           -9.252
        delta_h 12.48   kcal
        -analytic    0.6322    -0.001225    -2835.76

NO3- + 10 H+ + 8 e- = NH4+ + 3 H2O
        log_k           119.077
        delta_h -187.055        kcal
        -gamma    2.5000    0.0000
PHASES
NH3(g)
        NH3 = NH3
        log_k           1.770
        delta_h -8.170  kcal
SOLUTION 1
EQUILIBRIUM_PHASES 1
        Calcite
        CO2(g)   -1.5
SAVE solution 1
SELECTED_OUTPUT
        -reset false
        -file ex7.sel
        -simulation      true
        -state           true
```

Table 26. Input file for example 7.—Continued

```
            -reaction        true
            -si CO2(g) CH4(g) N2(g) NH3(g)
            -gas CO2(g) CH4(g) N2(g) NH3(g)
END
#   Simulation 2: Decomposition of organic matter, CH2O(NH3).07,
#   at fixed pressure of 1.1 atm
USE solution 1
GAS_PHASE 1 Fixed-pressure gas phase
            -fixed_pressure
            -pressure        1.1
            CO2(g)           0.0
            CH4(g)           0.0
            N2(g)            0.0
            H2O(g)           0.0
REACTION 1
            CH2O(NH3)0.07     1.0
            1. 2. 3. 4. 8. 16. 32 64. 125. 250. 500. 1000. mmol
USER_GRAPH 1 Example 7
            -headings Fixed_Pressure: CH4 CO2 N2 H2O #Volume
            -chart_title "Gas Composition"
            -axis_titles "Organic matter reacted, in millimoles" \
                       "Log(Partial pressure, in atmospheres)" "Volume, in liters"
            -axis_scale x_axis 1 1e3 auto auto log
            -axis_scale y_axis -5.0 1.0 1 1
            -connect_simulations false
    -start
    10 IF GAS("CH4(g)") < 1e-10 THEN GOTO 100
    20 mM_OM = RXN * 1e3
    30 PLOT_XY -10, -10, line_width = 0, symbol_size = 0
    40 PLOT_XY mM_OM, SI("CH4(g)"), color = Black, symbol = XCross
    50 PLOT_XY mM_OM, SI("CO2(g)"), color = Red, symbol = XCross
    60 PLOT_XY mM_OM, SI("N2(g)"), color = Teal, symbol = XCross
    70 PLOT_XY mM_OM, SI("H2O(g)"), color = Blue, symbol = XCross
    100 REM end of program
    -end
USER_GRAPH 2 Example 7
            -headings  Fixed_P:...Pressure Fixed_P:...Volume
            -chart_title \
                "Total Gas Pressure and Volume"
            -axis_titles "Organic matter reacted, in millimoles" \
                       "Log(Pressure, in atmospheres)" "Volume, in liters"
            -axis_scale x_axis 1 1e3 auto auto log
            -axis_scale y_axis -5.0 1.0 1 1
            -axis_scale y2_axis 1e-3 1e5 auto auto log
            -connect_simulations false
    -start
    10 IF GAS("CH4(g)") < 1e-10 THEN GOTO 100
    20 mM_OM = RXN * 1e3
    30 moles = (GAS("CH4(g)") + GAS("CO2(g)") + GAS("N2(g)") + GAS("H2O(g)"))
    40 vol = moles * 0.08207 * TK / 1.1
    50 PLOT_XY mM_OM, LOG10(1.1), color = Magenta, symbol = XCross
    60 PLOT_XY mM_OM, vol, color = Cyan, symbol = XCross, y_axis = 2
```

Table 26. Input file for example 7.—Continued

```
   100 REM end of program
   -end
END
#  Simulation 3: Decomposition of organic matter, CH2O(NH3).07,
#  at fixed volume of 23.19 L
USE solution 1
USE reaction 1
GAS_PHASE 1 Fixed volume gas phase
         -fixed_volume
         -volume          23.19
         CO2(g)           0.0
         CH4(g)           0.0
         N2(g)            0.0
         H2O(g)           0.0
         -equilibrate 1
USER_GRAPH 1
         -headings Fixed_Volume: CH4 CO2 N2 H2O
  -start
  10 mM_OM = RXN * 1e3
  20 PLOT_XY -10, -10, line_width = 0, symbol_size = 0
  30 PLOT_XY mM_OM, SI("CH4(g)"), color = Black, symbol = Circle
  40 PLOT_XY mM_OM, SI("CO2(g)"), color = Red, symbol = Circle
  50 PLOT_XY mM_OM, SI("N2(g)"), color = Teal, symbol = Circle
  60 PLOT_XY mM_OM, SI("H2O(g)"), color = Blue, symbol = Circle, symbol_size = 5
  -end
USER_GRAPH 2
         -headings Fixed_V:...Pressure Fixed_V:...Volume
  -start
  10 mM_OM = RXN * 1e3
  20 tot_p = SR("CH4(g)") + SR("CO2(g)") + SR("N2(g)") + SR("H2O(g)")
  30 PLOT_XY mM_OM, LOG10(tot_p), color = Magenta, symbol = Circle
  40 PLOT_XY mM_OM, 23.19, color = Cyan, line_width = 1 symbol = Circle, y_axis = 2
  -end
END
```

In the second simulation, organic matter decomposes with a carbon to nitrogen ratio of 1:0.07 in reaction steps ranging from 1 to 1,000 mmol (**REACTION** keyword). A fixed-pressure gas phase will form when the sum of the partial pressures exceeds 1.1 atm; only H_2O, CO_2, CH_4, N_2, and NH_3 enter the gas phase, as defined by the **GAS_PHASE** data block. The third simulation uses the same initial solution and reaction. However, the gas phase starts with H_2O and CO_2 in equilibrium with the initial solution and has a fixed volume of 23.19 L, which is the final volume of the fixed-pressure gas phase in the previous simulation. After 1,000 mmol of reaction, the fixed-pressure and fixed-volume gas phases have (very nearly) the same pressure, volume, and composition, with slightly higher concentrations in the fixed-volume simulations because $H_2O(g)$ and $CO_2(g)$ entered the volume in the equilibration stage. At the other reaction increments, the pressure, volume, and composition are different for the two gas phases, except for the pressure of water vapor (fig. 8).

Figure 8 illustrates the two different approaches of GAS_PHASE. For the fixed-pressure gas phase, a bubble forms when nearly 3 mmol of reaction have been added. Initially, the composition reflects the solubility of the gases—more than 90 percent CH_4 and less than 10 percent CO_2—even though CH_4 and CO_2 are produced in equal proportion by the reaction. N_2 and NH_3 are minor components (NH_3 partial pressures are less than 10^{-7} atm throughout the batch-reaction calculation and are not plotted). The solubility effect lessens as the reaction progresses; CO_2 becomes the major carbonate species in the solution as pH decreases. From 200 mmol of reaction onwards, the partial pressures remain nearly constant and the gases reflect the stoichiometry of the organic matter decomposition. The volume of gas produced by the reactions ranges from less than 1 mL at 3 mmol of reaction to 23.19 L after 1,000 mmol of the stoichiometric reaction has been added. The pressure of H_2O remains the same throughout because the salinity of the solution remains the same.

For the fixed-volume gas phase, the gas phase exists from the beginning of the reaction (fig. 8). Initially, the gas is H_2O and CO_2, but as the reaction proceeds, CH_4 and N_2 in the ratio 0.5:0.03 enter the gas phase. This ratio occurs because one-half of the C released becomes CH_4, and slightly less than one-half of the NH_3 becomes N_2. The solution becomes acidic because of the CO_2 produced; hence, partitioning of CO_2 to the gas phase increases (relative to CH_4) as the reaction proceeds (fig. 8). In the final stage, the CO_2 and CH_4 partial pressures become nearly equal. All the partial pressures of the fixed-volume gas phases are smaller than the fixed-pressure gas phase up to 1,000 mmol of reaction (except H_2O, which remains the same). If the reaction continued beyond 1,000 mmol, the pressure of the fixed-volume gas phase would become greater and greater. Conversely, the volume of the fixed-pressure gas phase is less than the volume of the fixed-volume gas phase until 1,000 mmol of reaction, but would expand further if the reaction continued.

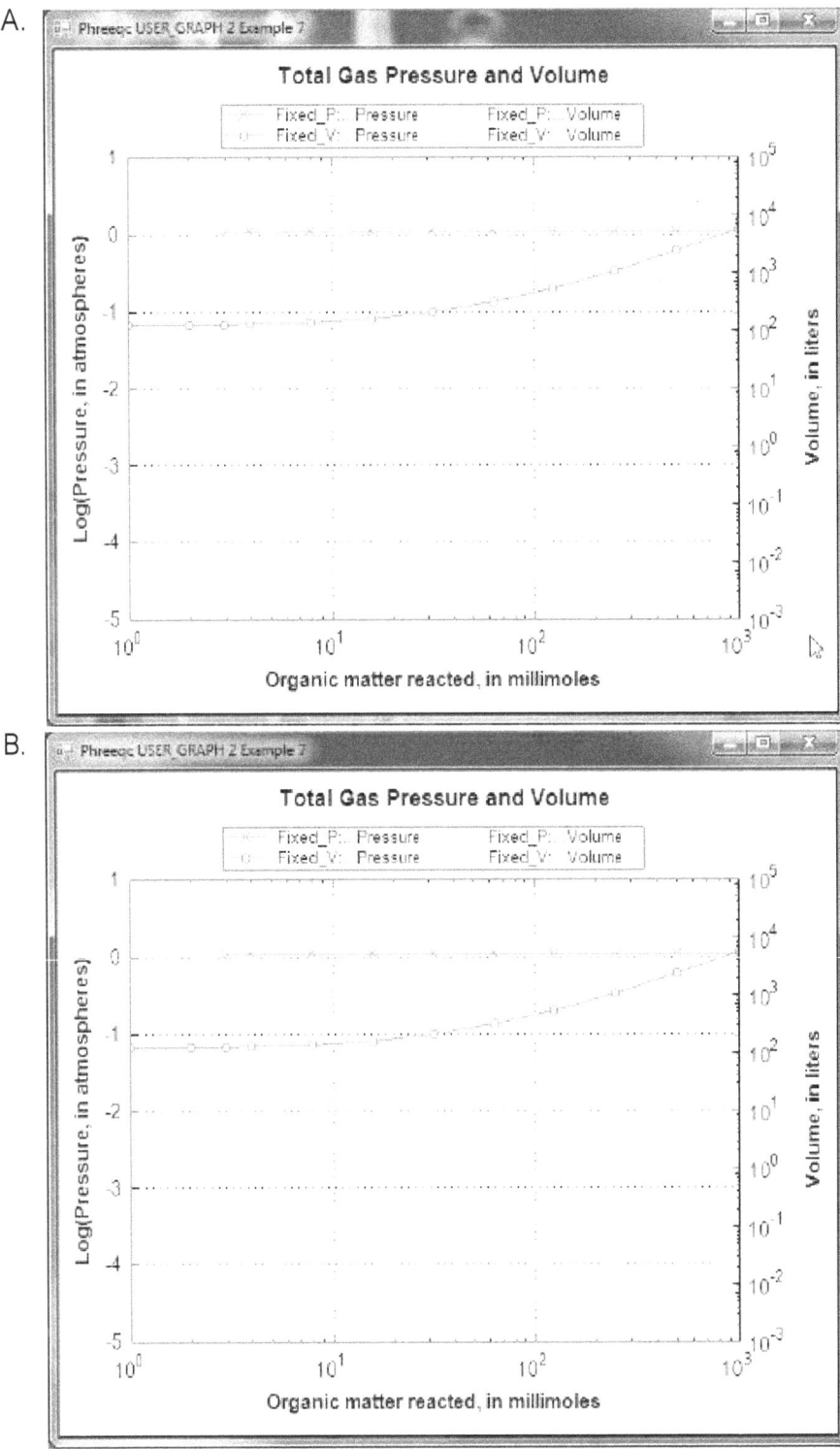

Figure 8. (A) Compositions and (B) total gas pressures and volumes of a fixed-pressure and a fixed-volume gas phase during decomposition of $CH_2O(NH_3)_{0.07}$ (organic matter) in water. (The partial pressure of ammonia gas is less than 10^{-7} atmospheres throughout and is not shown.)

Example 8—Surface Complexation

In all surface complexation models, sorption is a function of both chemical and electrostatic energy as described by the free energy relationship:

$$\Delta G_{tot} = \Delta G_{ads} + zF\psi, \tag{21}$$

where ΔG is the Gibbs energy (J/mol), z is the charge number (unitless) of the sorbed species, F is the Faraday constant (96,485 C/mol), ψ is the potential (V), and subscripts $_{tot}$ and $_{ads}$ indicate total and chemical adsorption energy, respectively. Sorption is stronger when the Gibbs energy decreases. Thus, a counter-ion that carries a charge opposite to the surface charge tends to be sorbed electrostatically, while a co-ion that carries a charge with the same sign as the surface tends to be rejected.

PHREEQC has two models for surface complexation. One is based on the Dzombak and Morel (1990) database for complexation of heavy metal ions on hydrous ferric oxide (Hfo), or ferrihydrite. Ferrihydrite, like many other oxy-hydroxides, binds metals and protons on strong and weak sites and develops a charge depending on the ions sorbed. The model uses the Gouy-Chapman equation to relate surface charge and potential.

The other model is CD-MUSIC, which also can accommodate multiple surface sites. In addition, the charge, the potential, and even the sorbed species can be distributed over two additional planes in the double layer that extends from the surface into the free (electrically neutral) solution. The CD-MUSIC model has more options to fit experimental data and initially was developed for sorption on goethite, but has been applied to many metal-oxide surfaces. An example is given in the PHREEQC Help file, which is available by installing *http://www.hydrochemistry.eu/phreeqc.chm.exe* (accessed June 25, 2012).

Neither of the models considers that a charged surface, when centrifuged and separated from a solution, must have a shell of co- and counter-ions that compensates the surface charge in an electrical double layer (EDL). However, PHREEQC can integrate the concentrations in the diffuse layer (**-diffuse_layer**), or calculate the average concentrations with the Donnan option (**-Donnan**). It also is possible to ignore the electrostatic contribution by use of the identifier **-no_edl**. This non-electrostatic model does not consider the effects of the development of surface charge on the formation of surface complexes, with the result that surface complexes are treated mathematically, much like aqueous complexes.

The following example of the Gouy-Chapman model is taken from Dzombak and Morel (1990, chapter 8). Sorption of zinc on hydrous ferric oxide is simulated by using weak and strong sites on the oxide surface. Protons and zinc ions compete for the two types of binding sites, which is described by mass-action equations. The equations take into account the dependence of the activities of surface species on the potential at the surface; in turn, the potential at the surface is related to the surface charge by the Gouy-Chapman relation. The example considers the sorption of zinc on hydrous ferric oxides as a function of pH for low (10^{-7} mol/kgw) and high (10^{-4} mol/kgw) zinc concentration in 0.1 mol/kgw sodium nitrate electrolyte.

Three keyword data blocks are required to define surface-complexation data for a simulation: SURFACE_MASTER_SPECIES, SURFACE_SPECIES, and SURFACE. The SURFACE_MASTER_SPECIES data block in the default database files defines a surface named "Hfo" (hydrous ferric oxides) with two binding sites. The name of a binding site is composed of a name for the surface, "Hfo", optionally followed by an underscore and a lowercase binding site designation, here "Hfo_w" and "Hfo_s" for "weak" and "strong". The underscore notation is necessary only for surfaces with two or more binding sites. The notation allows a mole-balance equation to be derived for each of the binding sites (Hfo_w and Hfo_s, in this example). The charges that develop on the two binding sites are summed, and the total is used to calculate the potential at the surface.

Surface-complexation reactions derived from the summary of Dzombak and Morel (1990) are defined by the SURFACE_SPECIES in the default database files for PHREEQC. However, the intrinsic stability constants used in this example of Dzombak and Morel (1990, chapter 8) differ from these summary values, and are therefore specified explicitly with a SURFACE_SPECIES data block in the input file (table 27). The reactions are taken from Dzombak and Morel (1990, p. 259) and entered in the input file (table 27). Note that the activity effect of the potential term is not included in the mass-action expression but is added internally by PHREEQC.

Table 27. Input file for example 8.

```
TITLE Example 8.--Sorption of zinc on hydrous iron oxides.
SURFACE_SPECIES
    Hfo_sOH  + H+ = Hfo_sOH2+
    log_k  7.18
    Hfo_sOH = Hfo_sO- + H+
    log_k  -8.82
    Hfo_sOH + Zn+2 = Hfo_sOZn+ + H+
    log_k  0.66
    Hfo_wOH  + H+ = Hfo_wOH2+
    log_k  7.18
    Hfo_wOH = Hfo_wO- + H+
    log_k  -8.82
    Hfo_wOH + Zn+2 = Hfo_wOZn+ + H+
```

Table 27. Input file for example 8.—Continued

```
      log_k  -2.32
SURFACE 1
      Hfo_sOH          5e-6      600.      0.09
      Hfo_wOH          2e-4
#       -Donnan
END
SOLUTION 1
      -units  mmol/kgw
      pH      8.0
      Zn      0.0001
      Na      100.     charge
      N(5)    100.
SELECTED_OUTPUT
      -file Zn1e_7
      -reset false
USER_PUNCH
  10 FOR i = 5.0 to 8 STEP 0.25
  20 a$ = EOL$ + "USE solution 1" + CHR$(59) + " USE surface 1" + EOL$
  30 a$ = a$ + "EQUILIBRIUM_PHASES 1" + EOL$
  40 a$ = a$ + "    Fix_H+ " + STR$(-i) + " NaOH 10.0" + EOL$
  50 a$ = a$ + "END" + EOL$
  60 PUNCH a$
  70 NEXT i
END
SOLUTION 2
      -units  mmol/kgw
      pH      8.0
      Zn      0.1
      Na      100.     charge
      N(5)    100.
SELECTED_OUTPUT
      -file Zn1e_4
      -reset false
USER_PUNCH
  10 FOR i = 5 to 8 STEP 0.25
  20 a$ = EOL$ + "USE solution 2" + CHR$(59) + " USE surface 1" + EOL$
  30 a$ = a$ + "EQUILIBRIUM_PHASES 1" + EOL$
  40 a$ = a$ + "    Fix_H+ " + STR$(-i) + " NaOH 10.0" + EOL$
  50 a$ = a$ + "END" + EOL$
  60 PUNCH a$
  70 NEXT i
END
#
# Model definitions
#
PHASES
      Fix_H+
      H+ = H+
      log_k  0.0
END
#
#   Zn = 1e-7
```

Table 27. Input file for example 8.—Continued

```
SELECTED_OUTPUT
     -file ex8.sel
     -reset true
     -molalities     Zn+2     Hfo_wOZn+      Hfo_sOZn+
USER_PUNCH
 10
USER_GRAPH 1 Example 8
     -headings pH Zn_solute Zn_weak_sites Zn_strong_sites Charge_balance
     -chart_title "Total Zn = 1e-7 molal"
     -axis_titles pH "Moles per kilogram water" "Charge balance, in milliequivalents"
     -axis_scale x_axis 5.0 8.0 1 0.25
     -axis_scale y_axis 1e-11 1e-6 1 1 log
     -axis_scale sy_axis -0.15 0 0.03
  -start
  10 GRAPH_X -LA("H+")
  20 GRAPH_Y MOL("Zn+2"), MOL("Hfo_wOZn+"), MOL("Hfo_sOZn+")
  30 GRAPH_SY CHARGE_BALANCE * 1e3
  -end
INCLUDE$ Zn1e_7
END
USER_GRAPH 1
     -detach
END
#
#    Zn = 1e-4
USER_GRAPH 2 Example 8
     -chart_title "Total Zn = 1e-4 molal"
     -headings pH Zn_solute Zn_weak_sites Zn_strong_sites Charge_balance
     -axis_titles pH "Moles per kilogram water" "Charge balance, in milliequivalents"
     -axis_scale x_axis 5.0 8.0 1 0.25
     -axis_scale y_axis 1e-8 1e-3 1 1 log
     -axis_scale sy_axis -0.15 0 0.03
  -start
  10 GRAPH_X -LA("H+")
  20 GRAPH_Y MOL("Zn+2"), MOL("Hfo_wOZn+"), MOL("Hfo_sOZn+")
  30 GRAPH_SY CHARGE_BALANCE * 1e3
  -end
INCLUDE$ Zn1e_4
END
```

The composition and other characteristics of an assemblage of surfaces are defined with the **SURFACE** data block. For each surface, the moles of each type of site and the surface area must be defined. In the input file, all the surface sites initially are in the uncharged, protonated form. Alternatively, the surface can be initialized to be in equilibrium with a solution with **-equilibrate** *solution_number*. In both cases, the composition of the surfaces will vary with the extent of subsequent reactions.

The number of binding sites and surface areas may remain fixed or may vary if the surface is related to an equilibrium phase or a kinetic reaction. In this example, the number of strong binding sites (Hfo_s,

5×10^{-6} mol) and of weak binding sites (Hfo_w, 2×10^{-4} mol) remain fixed. With **-sites_units** density, the number of sites per nm^2 (square nanometer) may be entered, instead of moles. The surface area must be defined with two numbers, the area per mass of surface material (here, 600 m^2/g) and the total mass of surface material (here, 0.09 g). The use of these two numbers is traditional, but only the surface area obtained from the product of the numbers is used to determine the specific charge and the surface potential. The surface area may be entered with any of the binding sites for a surface; in table 27, the surface area is entered with Hfo_s.

Two sodium nitrate solutions are defined with different concentrations of zinc (SOLUTION 1 and 2 data blocks), which can be recalled later in the run by USE **solution** 1 or 2. Together with the definitions of these solutions, USER_PUNCH is used to generate a PHREEQC input file, as explained in the next paragraph. A pseudo-phase, "Fix_H+" is defined with the PHASES data block. This phase is not real, but is used in the batch-reaction simulations to fix the pH at specified values.

The remaining simulations in the input file equilibrate the surface with either solution 1 or solution 2 for pH values that range from 5 to 8. It is possible to use the REACTION data block to add or remove varying amounts of NaOH from the solution in a single simulation, but the reaction increments will not produce evenly spaced pH values and the size of the reaction increments is not known beforehand. Alternatively, evenly spaced pH values can be obtained by using the phase "Fix_H+" in an EQUILIBRIUM_PHASES data block; the saturation indices correspond to the desired pH. A separate simulation is needed for each pH, illustrated here for pH = 5.

```
USE solution 1
USE surface 1
EQUILIBRIUM_PHASES 1
    Fix_H+          -5 NaOH 10.0
END
```

Writing a similar set of data blocks for each pH is easy, but tedious. However, a shortcut is available. The simulations can be generated with a small Basic program in keyword USER_PUNCH and punched to the files *zn1e_7* and *zn1e_4*. The Basic program punches a string, *a$*, that contains the PHREEQC keywords and instructions for each pH, which is set by the FOR-loop variable *i*. EOL$ is a function that returns a new-line character. During the run, the files are inserted in the input file with INCLUDE$ zn1e_7 and INCLUDE$ zn1e_4.

NaOH is added or removed from each solution to produce the specified saturation index for "Fix_H+" or log activity of H^+ (which is the negative of pH). However, a very low pH may not be attainable by removing all of the sodium from the solution. In this case HNO$_3$ should be used as reactant instead of NaOH.[1]

The results of the simulation are plotted in figure 9 and are consistent with the results shown in Dzombak and Morel (1990, figure 8.9). Zinc is more strongly sorbed at high pH values than at low pH values. In addition, at low concentrations of zinc, the strong binding sites outcompete the weak binding sites for zinc over the entire pH range, and at high pH, most of the zinc resides at the strong binding sites. At larger zinc concentrations, the strong binding sites predominate only at low pH. Because all the strong binding sites become filled at higher pH, most of the zinc resides at the more numerous weak binding sites at high pH and large zinc concentrations.

There is one more point to be noted in figure 9. The charge balance of the solution, plotted on the secondary Y axis, becomes negative with decreasing pH because the surface sorbs protons from the solution. If the solution and surface are separated (for example by **TRANSPORT** or by a **SAVE** and subsequent **USE**), the surface will keep its positive charge, and the solution its negative counter-charge. Such a separation of electrical charge is physically impossible. In reality, an electrical double layer exists on the surface that counterbalances the surface charge, and which remains with the surface when surface and solution are separated, thus keeping both of them electrically neutral. If the identifier **-Donnan** is added in keyword **SURFACE**, the composition of the electrical double layer is calculated and stored with each of the surfaces in the assemblage; this explicitly calculated electrical double layer makes each surface a neutral entity. In the input file this option (**-Donnan**) can be uncommented, resulting in a zero charge balance when the file is run again.

---

[1]It is possible to let the program determine whether NaOH or HNO$_3$ should be added to attain a pH if an additional pseudo phase is defined. The following example attains a pH of 2 and 10 without specifying different alternate reactions for the phase Fix_H+.

```
PHASES
NaNO3
            NaNO3 = Na+ + NO3-
            log_K  -20

Fix_H+
            H+ = H+
            log_K  0
END
SOLUTION 1
EQUILIBRIUM_PHASES
            NaNO3 0 10
            Fix_H+ -2 HNO3 10
END
SOLUTION 1
EQUILIBRIUM_PHASES
            NaNO3 0 10
            Fix_H+ -12 HNO3 10
END
```

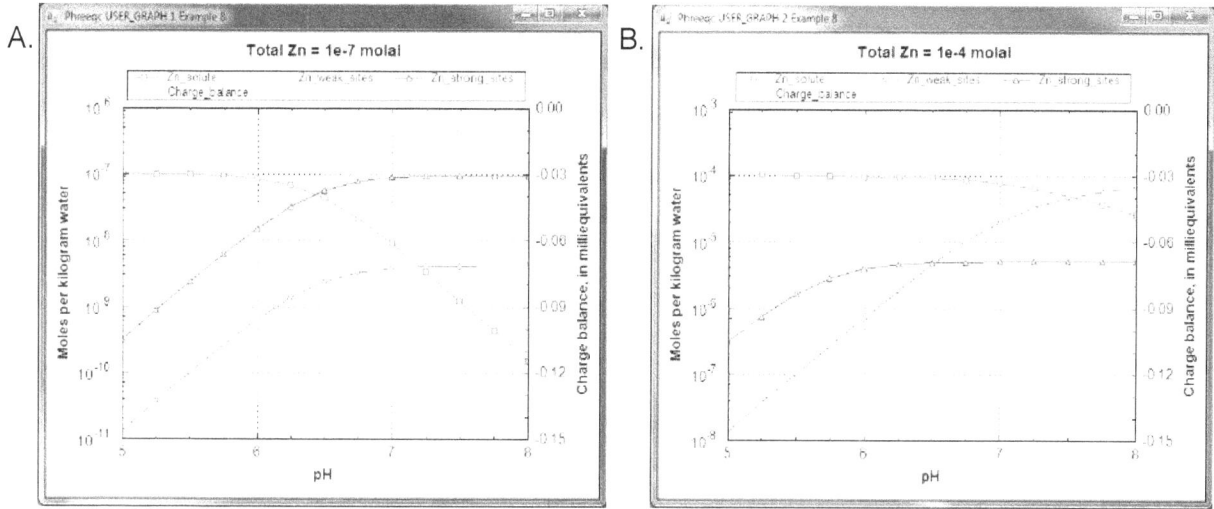

Figure 9. Distribution of zinc among the aqueous phase and strong and weak surface sites of hydrous iron oxide as a function of pH for total zinc concentrations of (A) 10^{-7} and (B) 10^{-4} molal. The secondary Y axis shows the charge balance of the solution, which (in the model) becomes negative at decreasing pH because protons are sorbed on the surface, or, at the higher Zn concentration and high pH, because Zn^{+2} is sorbed.

Example 9—Kinetic Oxidation of Dissolved Ferrous Iron With Oxygen

Kinetic rate expressions can be defined in a completely general way in PHREEQC by using Basic statements in the **RATES** data block. The rate expressions can be used in batch-reaction and transport calculations with the **KINETICS** data block. For transport calculations (**ADVECTION** or **TRANSPORT**), kinetic reactions can be defined cell by cell by the number range following the **KINETICS** keyword (**KINETICS** *m-n*). The rate expressions are integrated with an embedded (up to) 5th-order Runge-Kutta-Fehlberg algorithm, or with a stiff, variable-order, variable-step multistep solver (Cohen and Hindmarsh, 1996). Equilibrium is calculated before a kinetic calculation is initiated and again when a kinetic reaction increment is added. Equilibrium includes solution species equilibrium; exchange-, equilibrium-phase-, solid-solution-, and surface-assemblage equilibrium; and gas-phase equilibrium. A check is performed to ensure that the difference between estimates of the integrated rate over a time interval is smaller than a user-defined tolerance. If the tolerance is not satisfied, then the integration over the time interval is automatically restarted with a smaller time interval.

Kinetic reactions between solids and the aqueous phase can be calculated without modification of the database. PHREEQC also can calculate kinetic reactions among aqueous species that are normally assumed to be in equilibrium, but this requires that the database be redefined with separate

SOLUTION_MASTER_SPECIES for the aqueous species that react kinetically. Example 9 illustrates the procedure for decoupling two valence states of an element (iron) and shows how PHREEQC can be used to calculate the kinetic oxidation of Fe^{+2} to Fe^{+3} in water.

The rate of oxidation of Fe^{+2} by O_2 in water is given by (Singer and Stumm, 1970):

$$\frac{dm_{Fe^{+2}}}{dt} = -\left(2.91e\text{-}9 + 1.33e12\ a^2_{OH^-}P_{O_2}\right)m_{Fe^{+2}},$$ (22)

where t is time in seconds, a_{OH^-} is the activity of the hydroxyl ion, $m_{Fe^{+2}}$ is the total molality of ferrous iron in solution, and P_{O_2} is the oxygen partial pressure (atm).

The time for complete oxidation of ferrous iron is a matter of minutes in an aerated solution when pH is above 7.0. However, Fe^{+3} forms solute complexes with OH^- and it may also precipitate as iron oxyhydroxides, so that pH decreases during oxidation. Because the rate has quadratic dependence on the activity of OH^-, the oxidation rate rapidly diminishes as pH decreases. The rate equation is highly non-linear in an unbuffered solution and must be integrated numerically. This example models a reaction vessel with 10 mmol/kgw NaCl and 0.1 mmol/kgw $FeCl_2$ at pH = 7.0 through which air is bubbled; the change in solution composition over time is calculated.

The calculation requires the uncoupling of equilibrium among the Fe(2) and Fe(3) species. Two new "elements" are defined in SOLUTION_MASTER_SPECIES—"Fe_di", which corresponds to Fe(2), and "Fe_tri", which corresponds to Fe(3). The master species for these elements are defined to be Fe_di+2 and Fe_tri+3, and all solution species, phases, exchange species, and surface species must be rewritten using these new elements and master species. A few of the transcriptions are shown in table 28, which gives the partial input file for this example.

Table 28. Partial input file for example 9.

```
TITLE Example 9.--Kinetically controlled oxidation of ferrous
                  iron. Decoupled valence states of iron.
SOLUTION_MASTER_SPECIES
Fe_di              Fe_di+2    0.0      Fe_di           55.847
Fe_tri             Fe_tri+3   0.0      Fe_tri          55.847
SOLUTION_SPECIES
Fe_di+2 = Fe_di+2
        log_k   0.0
Fe_tri+3 = Fe_tri+3
        log_k   0.0
#
# Fe+2 species
#
Fe_di+2 + H2O = Fe_diOH+ + H+
```

Table 28. Partial input file for example 9.—Continued

```
          log_k   -9.5
          delta_h 13.20    kcal
#
#... and also other Fe+2 species
#
#
# Fe+3 species
#
Fe_tri+3 + H2O = Fe_triOH+2 + H+
          log_k   -2.19
          delta_h 10.4     kcal
#
#... and also other Fe+3 species
#
PHASES
Goethite
          Fe_triOOH + 3 H+ = Fe_tri+3 + 2 H2O
          log_k   -1.0
END
SOLUTION 1
          pH   7.0
          pe 10.0   O2(g) -0.67
          Fe_di   0.1
          Na   10.
          Cl   10.   charge
EQUILIBRIUM_PHASES 1
          O2(g)               -0.67
RATES
Fe_di_ox
  -start
  10  Fe_di = TOT("Fe_di")
  20  if (Fe_di <= 0) then goto 200
  30  p_o2 = SR("O2(g)")
  40  moles = (2.91e-9 + 1.33e12 * (ACT("OH-"))^2 * p_o2) * Fe_di * TIME
  200 SAVE moles
  -end
KINETICS 1
Fe_di_ox
          -formula  Fe_di  -1.0  Fe_tri  1.0
          -steps 100 400 3100 10800 21600 5.04e4 8.64e4 1.728e5 1.728e5 1.728e5 1.728e5
          -step_divide 1e-4
INCREMENTAL_REACTIONS true
SELECTED_OUTPUT
          -file ex9.sel
          -reset false
USER_PUNCH
          -headings Days  Fe(2)  Fe(3)  pH  si_goethite
  10 PUNCH SIM_TIME / 3600 / 24, TOT("Fe_di")*1e6, TOT("Fe_tri")*1e6, -LA("H+"),\
              SI("Goethite")
USER_GRAPH Example 9
          -headings _time_ Fe(2) Fe(3) pH
          -chart_title "Oxidation of Ferrous Iron"
```

Table 28. Partial input file for example 9.—Continued

```
        -axis_titles "Time, in days" "Micromole per kilogram water" "pH"
        -axis_scale secondary_y_axis 4.0 7.0 1.0 0.5
  -start
10 GRAPH_X TOTAL_TIME / 3600 / 24
20 GRAPH_Y TOT("Fe_tri")*1e6, TOT("Fe_di")*1e6
30 GRAPH_SY -LA("H+")
  -end
END
```

The **SOLUTION** data block defines a sodium chloride solution that has 0.1 mmol/kgw ferrous iron (Fe_di) and is in equilibrium with atmospheric oxygen. The **EQUILIBRIUM_PHASES** data block specifies that all batch-reaction solutions will also be in equilibrium with atmospheric oxygen; thus, there is a continuous supply of oxygen for oxidation of ferrous iron.

In the **RATES** data block, the rate expression is designated with the name "Fe_di_ox" and defined according to equation 22. Note the use of the special Basic function "TOT" to obtain the total concentration (molality) of ferrous iron (line 10), "SR" to obtain the saturation ratio, or, in the case of a gas, the partial pressure, here of oxygen (line 30), and "ACT" to obtain the activity of OH⁻ (line 40). Line 40 defines the moles of reaction. Notice also that the variable *moles* is calculated by multiplying the rate times the current time interval (TIME) and that the rate definition ends with a SAVE statement. The SAVE and TIME statements must be included in a rate definition; they specify the moles that reacted over the time (sub-)interval. The interval given by TIME is an internal PHREEQC variable that is adapted automatically by the code to obtain the required accuracy of the integration.

In the **KINETICS** data block, the rate expression named "Fe_di_ox" is invoked and parameters are defined. When the rate name in the **KINETICS** data block is identical to a mineral name that is defined under **PHASES**, the stoichiometry of that mineral will be used in the reaction. However, because no mineral is associated with the rate name of this example, the identifier **-formula** must be used to specify the reaction stoichiometry. The reaction involves loss of Fe_di [equivalent to Fe(2)] from solution as indicated by the stoichiometric coefficient of -1.0. The loss is balanced by a gain in solution of Fe_tri [equivalent to Fe(3)] with a stoichiometric coefficient of +1.0. Note that the formula contains only the elements for which the mass changes in the system. Thus, the overall kinetic reaction of the example is

$Fe^{2+} + H^+ + 0.25O_2 = Fe^{3+} + 0.5H_2O$, but the reaction of protons and oxygen to form water does not change the total mass of hydrogen or oxygen in the system. Hydrogen and oxygen are therefore not included in the formula. In table 28, the phase O2(g) in **EQUILIBRIUM_PHASES** allows dissolved oxygen to be maintained in equilibrium with atmospheric oxygen gas. In a system closed to oxygen, the dissolved oxygen would be partly consumed.

The identifier **-steps** in the KINETICS data block gives the time step(s) over which the kinetic reactions must be integrated. When INCREMENTAL_REACTIONS **true** is used, each time step increments the total time to be simulated, and the results from the previous time step are used as the starting point for the current time step.

The SELECTED_OUTPUT data block specifies the file name of the selected-output file and eliminates all default printing to that file (**-reset** false). The only output to the selected-output file in this example is defined with the USER_PUNCH data block. The Basic program in USER_PUNCH specifies that the following be printed after each kinetic time step (**-steps** defines 11 kinetic time steps): the cumulative time of the simulation, in days; the total ferrous and ferric iron, in μmol/kgw; the pH; and the saturation index of goethite. The results also are plotted with USER_GRAPH, and the points can be saved to a file by right-clicking the mouse when the cursor is inside the chart area.

When the input file is run, two warning messages are generated during the integration. If the integration time interval is too large, it is possible that the initial estimates of kinetic reaction increments produce negative solution concentrations. When this happens, the program prints a warning message, decreases the size of the time interval, and restarts the integration. The messages are warnings, not errors, and the program successfully completes the calculation. It is possible to eliminate the warning messages by reducing the initial integration interval. No warning messages are printed if the identifier **-step_divide** 100 is used (KINETICS), which divides the initial (overall) time step by 100. Likewise, no warning messages are printed if the identifier **-step_divide** 1e-7 is used, which causes the reaction increment to be less than 1×10^{-7} mol. The former approach, with **-step_divide** 100, is usually preferable because, although initial reaction increments are compelled to be small, later in the integration, large reaction increments are possible. Using **-step_divide** 1e-7 forces reaction increments to remain small throughout the entire integration, and in this example, the run time is about 5 times longer than using **-step_divide** 100, and about 10 times longer than not using **-step_divide** at all. Figure 10 shows the concentration of total Fe(2), total Fe(3), and pH in the reaction vessel over the 10 days of the simulation. It can be seen that the pH rapidly decreases at the beginning of the reaction. The slope of Fe(2) against time is initially steep, but lessens as the reaction progresses, which is consistent with equation 22. When the experiment is performed in reality in an unbuffered solution, it is noted that the pH initially rises. This rise in pH is consistent with slowly forming hydroxy-complexes of Fe(3). Because the oxidation reaction by itself consumes protons, the pH

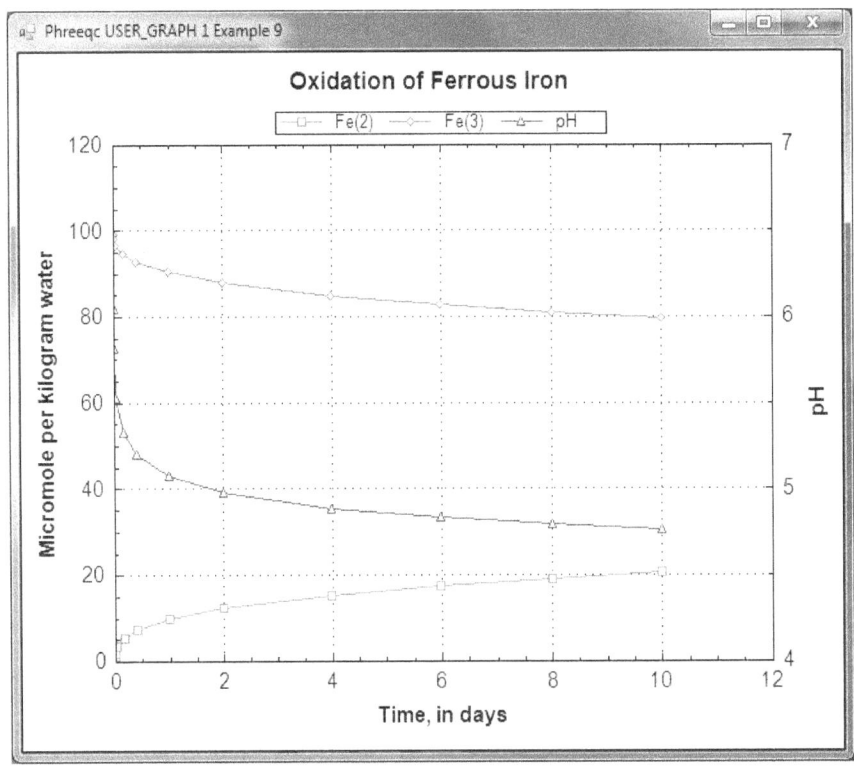

Figure 10. Concentration of total Fe(2), total Fe(3), and pH as dissolved ferrous iron [Fe(2)] is kinetically oxidized to ferric iron [Fe(3)] by oxygen.

would initially rise if the hydroxy-complexes that lower the pH form slowly. Such kinetic formation of aqueous complexes also could be included in PHREEQC simulations, but it would require that the hydroxy-complexes of Fe(3) also be defined by using a separate **SOLUTION_MASTER_SPECIES** and that a rate expression be defined for the kinetic formation of the complexes.

Example 10—Aragonite-Strontianite Solid Solution

PHREEQC has the capability to model multicomponent ideal and binary nonideal solid solutions. For ideal solid solutions, the activity of each end member solid is equal to its mole fraction. For nonideal solid solutions, the activity of each end member is the product of the mole fraction and an activity coefficient, which is determined from the mole fraction and Guggenheim excess free-energy parameters. Example 10 considers an aragonite ($CaCO_3$)-strontianite ($SrCO_3$) solid solution and demonstrates how the composition of the solid solution and the aqueous phase change as strontium carbonate is added to an initially pure calcium carbonate system.

The example is derived from a diagram presented in Glynn and Parkhurst (1992). The equilibrium constants at 25 °C, $K_{SrCO_3} = 10^{-9.271}$ and $K_{CaCO_3} = 10^{-8.336}$, and the Guggenheim parameters, $a_0 = 3.43$ and $a_1 = -1.82$, are derived from Plummer and Busenberg (1987). The input file is shown in table 29. The **PHASES** data block defines the log Ks for aragonite and strontianite and overrides any data for these minerals that might be present in the database file. The **SOLID_SOLUTIONS** data block defines the unitless Guggenheim excess free-energy parameters and the initial composition of the solid solution, which is zero moles of aragonite and strontianite. Initial solution 1 is defined to be a calcium bicarbonate solution. The solution is then equilibrated with aragonite at nearly 1 atm partial pressure of carbon dioxide and saved as the new composition of solution 1.

Table 29. Input file for example 10.

```
TITLE Example 10.--Solid solution of strontianite and aragonite.
PHASES
        Strontianite
                SrCO3 = CO3-2 + Sr+2
                log_k           -9.271
        Aragonite
                CaCO3 = CO3-2 + Ca+2
                log_k           -8.336
END
SOLID_SOLUTIONS 1
        Ca(x)Sr(1-x)CO3
                -comp1    Aragonite         0
                -comp2    Strontianite      0
                -Gugg_nondim    3.43    -1.82
END
SOLUTION 1
        -units mmol/kgw
        pH 5.93 charge
        Ca      3.932
        C       7.864
EQUILIBRIUM_PHASES 1
        CO2(g) -0.01265 10
        Aragonite
SAVE solution 1
END
#
#   Total of 0.00001 to 0.005 moles of SrCO3 added
#
USE solution 1
USE solid_solution 1
REACTION 1
        SrCO3    1.0
        .005 in 500 steps
PRINT
        -reset false
        -echo true
```

Table 29. Input file for example 10.—Continued

```
        -user_print true
USER_PRINT
-start
 10 sum = (S_S("Strontianite") + S_S("Aragonite"))
 20 if sum = 0 THEN GOTO 110
 30 xb = S_S("Strontianite")/sum
 40 xc = S_S("Aragonite")/sum
 50 PRINT "Simulation number:     ", SIM_NO
 60 PRINT "Reaction step number: ", STEP_NO
 70 PRINT "SrCO3 added:          ", RXN
 80 PRINT "Log Sigma pi:         ", LOG10 (ACT("CO3-2") * (ACT("Ca+2") + ACT("Sr+2")))
 90 PRINT "XAragonite:           ", xc
100 PRINT "XStrontianite:        ", xb
110 PRINT "XCa:                  ", TOT("Ca")/(TOT("Ca") + TOT("Sr"))
120 PRINT "XSr:                  ", TOT("Sr")/(TOT("Ca") + TOT("Sr"))
130 PRINT "Misc 1:               ", MISC1("Ca(x)Sr(1-x)CO3")
140 PRINT "Misc 2:               ", MISC2("Ca(x)Sr(1-x)CO3")
-end
SELECTED_OUTPUT
        -file ex10.sel
        -reset false
        -reaction true
USER_PUNCH
-head   lg_SigmaPi X_Arag X_Stront X_Ca_aq X_Sr_aq mol_Misc1 mol_Misc2 \
    mol_Arag mol_Stront
-start
 10 sum = (S_S("Strontianite") + S_S("Aragonite"))
 20 if sum = 0 THEN GOTO 60
 30 xb = S_S("Strontianite")/(S_S("Strontianite") + S_S("Aragonite"))
 40 xc = S_S("Aragonite")/(S_S("Strontianite") + S_S("Aragonite"))
 50 REM Sigma Pi
 60 PUNCH LOG10(ACT("CO3-2") * (ACT("Ca+2") + ACT("Sr+2")))
 70 PUNCH xc                              # Mole fraction aragonite
 80 PUNCH xb                              # Mole fraction strontianite
 90 PUNCH TOT("Ca")/(TOT("Ca") + TOT("Sr"))  # Mole aqueous calcium
100 PUNCH TOT("Sr")/(TOT("Ca") + TOT("Sr")) # Mole aqueous strontium
110 x1 = MISC1("Ca(x)Sr(1-x)CO3")
120 x2 = MISC2("Ca(x)Sr(1-x)CO3")
130 if (xb < x1 OR xb > x2) THEN GOTO 250
140    nc = S_S("Aragonite")
150    nb = S_S("Strontianite")
160    mol2 = ((x1 - 1)/x1)*nb + nc
170    mol2 = mol2 / ( ((x1 -1)/x1)*x2 + (1 - x2))
180    mol1 = (nb - mol2*x2)/x1
190    REM                               # Moles of misc. end members if in gap
200    PUNCH mol1
210    PUNCH mol2
220    GOTO 300
250    REM                               # Moles of misc. end members if not in gap
260    PUNCH 1e-10
270    PUNCH 1e-10
300 PUNCH S_S("Aragonite")               # Moles aragonite
```

Table 29. Input file for example 10.—Continued

```
  310 PUNCH S_S("Strontianite")              # Moles Strontianite
-end
USER_GRAPH Example 10
        -headings x_Aragonite  x_Srontianite
        -chart_title "Aragonite-Strontianite Solid Solution"
        -axis_titles "Log(SrCO3 added, in moles)" "Log(Mole fraction of component)"
        -axis_scale x_axis -5 1 1 1
        -axis_scale y_axis -5 0.1 1 1
        -connect_simulations true
        -start
  10 sum = (S_S("Strontianite") + S_S("Aragonite"))
  20 IF sum = 0 THEN GOTO 70
  30 xb = S_S("Strontianite")/ sum
  40 xc = S_S("Aragonite")/ sum
  50 PLOT_XY LOG10(RXN), LOG10(xc), line_w = 2, symbol_size = 0
  60 PLOT_XY LOG10(RXN), LOG10(xb), line_w = 2, symbol_size = 0
  70 rem
  -end
END
#
#  Total of 0.005 to 0.1 moles of SrCO3 added
#
USE solution 1
USE solid_solution 1
REACTION 1
        SrCO3   1.0
        .1 in 20 steps
END
#
#  Total of 0.1 to 10 moles of SrCO3 added
#
USE solution 1
USE solid_solution 1
REACTION 1
        SrCO3   1.0
        10.0 in 100 steps
END
```

In the next simulation, solution 1 is brought together with the solid solution (USE keywords) and 5 millimoles of strontium carbonate are added in 500 steps (REACTION data block). The PRINT keyword data block excludes all default printing to the output file and includes only the printing defined in the USER_PRINT data block. The USER_PRINT data block specifies that the following information about the solid solution be printed to the output file after each reaction step: the simulation number, reaction-step number, amount of strontium carbonate added, $\log\left(\sum\Pi\right)$ (log of the sum of the ion activity products), mole fractions of strontianite and aragonite, aqueous mole fractions of calcium and strontium, and the composition of the two solids that exist within the miscibility gap. The SELECTED_OUTPUT data block defines the selected-output file to be *ex10.sel*, cancels any default printing to the selected-output file (**-reset** false), and requests that the amount of reaction added at each step (as defined in the REACTION data block) be written to the selected-output file (**-reaction** true). The USER_PUNCH data block prints additional columns of information to the selected-output file, including all of the information needed to make figure 11. Two additional simulations add successively larger amounts of strontium carbonate to the system up to a total addition of 10 mol.

The excess free-energy parameters describe a nonideal solid solution that has a miscibility gap. For compositions that fall within the miscibility gap, the activities of calcium and strontium within the aqueous phase remain fixed and are in equilibrium with solids of two compositions, one solid with a strontium mole fraction of 0.0048 and one solid with a strontium mole fraction of 0.8579. For the simulations of example 10, each incremental addition of strontium carbonate increases the mole fraction of strontium carbonate in the solid until about 0.001 mol of strontium carbonate have been added (fig. 11A). That point is the beginning of the miscibility gap (fig. 11) and the composition of the solid is 0.0048 strontium mole fraction. The next increments of strontium carbonate (up to 0.005 mol strontium carbonate added) produce constant mole fractions of calcium and strontium in the solution (fig. 11B) and equilibrium with both the miscibility-gap end members. However, the amounts of calcium carbonate and strontium carbonate in the solid phases (fig. 11C) and the amounts of each of the miscibility gap end members (fig. 11D) vary with the amount of strontium carbonate added. Finally, the end of the miscibility gap is reached after about 0.005 mol of strontium carbonate have been added. At this point, the solution is in equilibrium with a single solid with a strontium mole fraction of 0.8579. Addition of more strontium carbonate increases the mole fractions of strontium in the aqueous phase and in the solid solution until both mole fractions are nearly 1.0 after the addition of 10 mol of strontium carbonate.

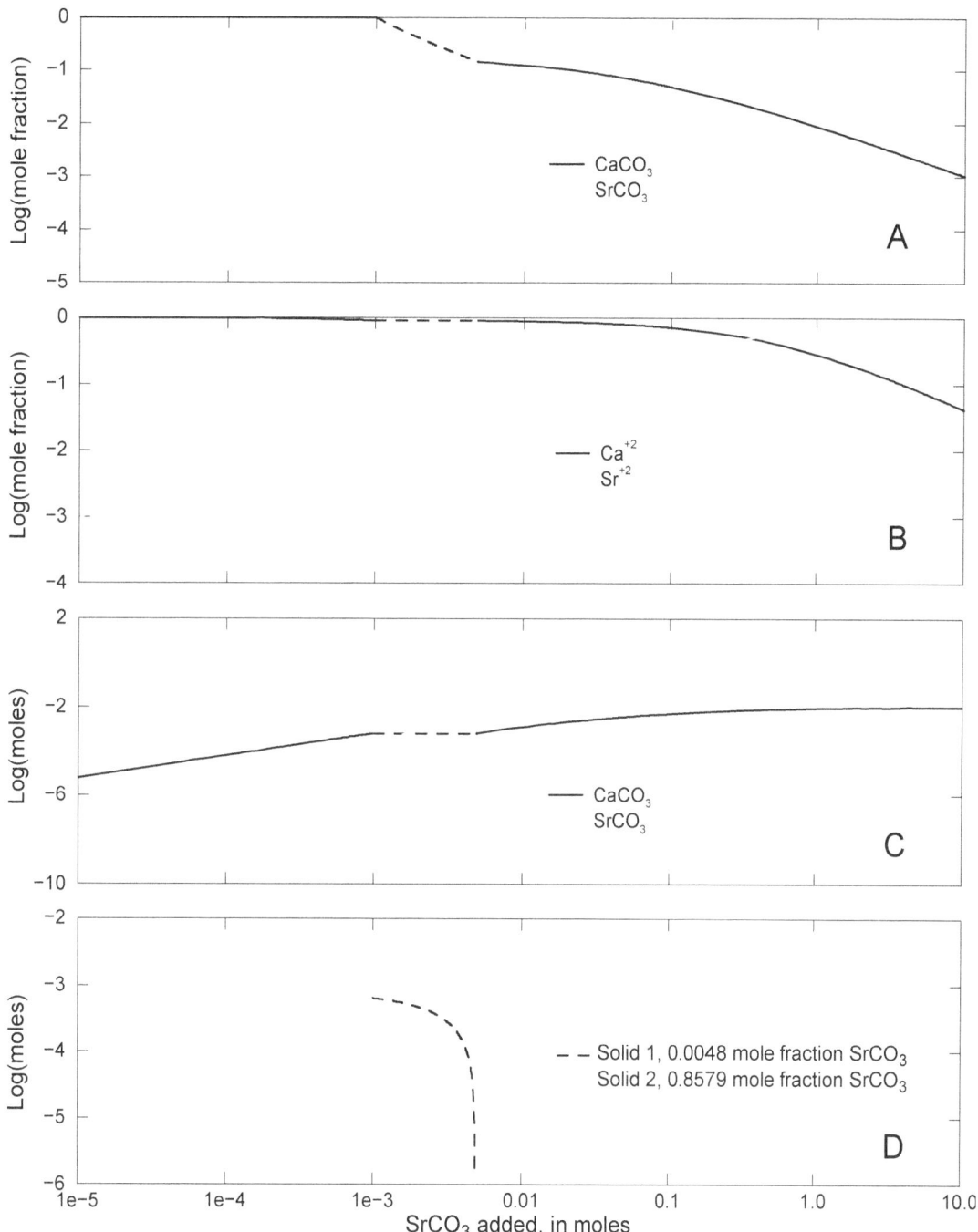

Figure 11. (A) Mole fraction of strontianite and aragonite in solid solution, (B) mole fraction of calcium and strontium in aqueous phase, (C) moles of strontianite and aragonite in solid solution, and (D) moles of miscibility-gap end members in solid solution, as a function of the amount of strontium carbonate added. Dashed lines indicate compositions within the miscibility gap.

Example 11—Transport and Cation Exchange

The following example simulates the chemical composition of the effluent from a column containing a cation exchanger (Appelo and Postma, 2005). Initially, the column contains a sodium-potassium-nitrate solution in equilibrium with the exchanger. The column is flushed with three pore volumes of calcium chloride solution. Calcium, potassium, and sodium react to equilibrium with the exchanger at all times. The problem is run two ways—by using the **ADVECTION** data block, which models only advection, and by using the **TRANSPORT** data block, which simulates advection and dispersive mixing.

The input file is listed in table 30. It starts with defining **SOLUTION** 0, which is the influent at cell 1 of the column. The column has 40 cells, and 40 solutions must be defined and numbered 1 through 40. The definition of a solution for each cell is mandatory. In this example, all cells contain the same solution, but this is not required. Solutions can be defined differently for each cell and could be changed by reactions in the current or preceding simulations (by using the **SAVE** keyword).

Table 30. Input file for example 11.

```
TITLE Example 11.--Transport and cation exchange.
SOLUTION 0  CaCl2
        units           mmol/kgw
        temp            25.0
        pH              7.0     charge
        pe              12.5    O2(g)   -0.68
        Ca              0.6
        Cl              1.2
SOLUTION 1-40   Initial solution for column
        units           mmol/kgw
        temp            25.0
        pH              7.0     charge
        pe              12.5    O2(g)   -0.68
        Na              1.0
        K               0.2
        N(5)            1.2
END
EXCHANGE 1-40
        -equilibrate 1
        X               0.0011
COPY cell 1 101
END
ADVECTION
        -cells          40
        -shifts         100
        -punch_cells    40
        -punch_frequency 1
        -print_cells    40
        -print_frequency 20
```

344 PHREEQC Version 3

Table 30. Input file for example 11.—Continued

```
PRINT; -reset false; -status false
SELECTED_OUTPUT
        -file            ex11adv.sel
        -reset           false
        -step
        -totals          Na Cl K Ca
USER_PUNCH
  -heading  Pore_vol
  10 PUNCH (STEP_NO + .5) / 40.
USER_GRAPH 1 Example 11
  -chart_title "Using ADVECTION Data Block"
  -headings Cl Na K Ca
  -axis_titles "Pore volumes" "Millimoles per kilogram water"
  -axis_scale x_axis 0 2.5
  -axis_scale y_axis 0 1.4
  -plot_concentration_vs time
  -start
  10 x = (STEP_NO + 0.5) / cell_no
  20 PLOT_XY x, TOT("Cl")*1000, symbol = None
  30 PLOT_XY x, TOT("Na")*1000, symbol = None
  40 PLOT_XY x, TOT("K")*1000, symbol = None
  50 PLOT_XY x, TOT("Ca")*1000, symbol = None
  60 PUT(1, 1)
  -end
COPY cell 101 1-40
END
USER_GRAPH 1
        -detach
END
TRANSPORT
        -cells           40
        -lengths         0.002
        -shifts          100
        -time_step       720.0
        -flow_direction  forward
        -boundary_conditions  flux  flux
        -diffusion_coefficient 0.0
        -dispersivities  0.002
        -correct_disp    true
        -punch_cells     40
        -punch_frequency 1
        -print_cells     40
        -print_frequency 20
SELECTED_OUTPUT
        -file            ex11trn.sel
        -reset           false
        -step
        -totals          Na Cl K Ca
        -high_precision true
USER_GRAPH 2 Example 11
  -chart_title "Using TRANSPORT Data Block"
  -headings Cl Na K Ca Cl_analytical
```

Table 30. Input file for example 11.—Continued

```
   -axis_titles "Pore volumes" "Millimoles per kilogram water"
   -axis_scale x_axis 0 2.5
   -axis_scale y_axis 0 1.4
#  -batch \temp\11.gif false # After saving, the chart on the monitor is closed.
   -plot_concentration_vs time
   10 x = (STEP_NO + 0.5) / cell_no
   20 PLOT_XY x, TOT("Cl")*1000, symbol = Plus, symbol_size = 2
   30 PLOT_XY x, TOT("Na")*1000, symbol = Plus, symbol_size = 2
   40 PLOT_XY x, TOT("K") *1000, symbol = Plus, symbol_size = 2
   50 PLOT_XY x, TOT("Ca")*1000, symbol = Plus, symbol_size = 2
# calculate Cl_analytical...
   60 DATA 0.254829592, -0.284496736, 1.421413741, -1.453152027, 1.061405429, 0.3275911
   70 READ a1, a2, a3, a4, a5, a6
   80 Peclet = 0.08 / 0.002
   90  z = (1 - x) / SQRT(4 * x / Peclet)
   100 PA = 0
   110 GOSUB 2000 # calculate e_erfc = exp(PA) * erfc(z)
   120 e_erfc1 = e_erfc
   130 z = (1 + x) / SQRT(4 * x / Peclet)
   140 PA = Peclet
   150 GOSUB 2000 # calculate exp(PA) * erfc(z)
   160 y = 0.6 * (e_erfc1 + e_erfc)
   170 PLOT_XY x, y, line_width = 0, symbol = Circle, color = Red
   180 d = (y - TOT("Cl")*1000)^2
   190 IF EXISTS(10) THEN PUT(d + GET(10), 10) ELSE PUT(d, 10)
   200 IF STEP_NO = 2 * CELL_NO THEN print 'SSQD for Cl after 2 Pore Volumes: ',
GET(10), '(mmol/L)^2'
   210 END
   2000 REM calculate e_erfc = exp(PA) * erfc(z)...
   2010 sgz = SGN(z)
   2020 z = ABS(z)
   2050 b = 1 / (1 + a6 * z)
   2060 e_erfc = b * (a1 + b * (a2 + b * (a3 + b * (a4 + b * a5)))) * EXP(PA - (z * z))
   2070 IF sgz = -1 THEN e_erfc = 2 * EXP(PA) - e_erfc
   2080 RETURN
END
```

The amount and composition of the exchanger in each of the 40 cells is defined by the **EXCHANGE** 1-40 data block in the second simulation. The **EXCHANGE** data block could have been written in the first simulation where the initial solutions are defined, but generally, it is preferable to define additional reactants in separate simulations. The intervening **END** (between the first and second simulation) avoids a redundant reaction calculation in which the first solution defined (here **SOLUTION** 0) is reacted with the first exchanger that is defined (cell 1). The definition of an exchanger for cells in the column is optional. The number of the exchanger corresponds to the number of the cell in a column, and, if an exchanger exists for a cell number, it is used automatically in the transport calculations for that cell. In this column, each cell has 1.1 mmol exchange sites that are initially in equilibrium with solution 1. Note that the initial exchange composition is calculated by an initial exchange calculation that does not affect the composition of solution 1. At the end of this simulation, cell 1 is copied to cell 101, which includes the data blocks SOLUTION 1 and EXCHANGE 1. By copying cell 101 back to cell 1-40 at the end of the advection simulation, the column is reinitialized for the next transport calculation.

The ADVECTION data block need only include the number of cells and the number of shifts for the simulation. The calculation accounts for the numbers of pore volumes that flow through the cells; cell lengths are not used. The identifiers **-punch_cells** and **-punch_frequency** specify that data will be written to the selected-output file for cell 40 at each shift. The identifiers **-print_cells** and **-print_frequency** indicate that data will be written to the output file for cell 40 at every 20 shifts.

The SELECTED_OUTPUT data block specifies that the shift (or advection step number) and the total dissolved concentrations of sodium, chloride, potassium, and calcium will be written to the file *ex11adv.sel*. Pore volumes can be calculated from the shift number; one shift moves a solution to the next cell, and the last solution out of the column. PHREEQC calculates cell-centered concentrations, so that the concentrations in the last cell arrive *one-half* shift later at the column end. In this example, one shift represents 1/40 of the column pore volume. The number of pore volumes (*PV*) that have been flushed from the column is therefore PV = (number of shifts + 0.5) / 40. The number of pore volumes is calculated and printed to the selected-output file using the USER_PUNCH data block. Similarly, USER_GRAPH defines the chart with the same data plotted on-screen.

At the end of the advection calculation (ADVECTION), the initial conditions are reset for the advection and dispersion calculation (TRANSPORT) by copying solution and exchange 101 back to solution and exchange 1-40. SOLUTION 0 is unchanged by the ADVECTION simulation and need not be redefined. The TRANSPORT data block includes a much more explicit description of the transport process than the ADVECTION data block. The length of each cell (**-lengths**), the boundary conditions at the column ends (**-boundary_conditions**), the direction of flow (**-flow_direction**), the dispersivity (**-dispersivities**), and the diffusion coefficient (**-diffusion_coefficient**) can all be specified. The identifier **-correct_disp** should be set to true when modeling outflow from a column with flux boundary conditions. The identifiers **-punch_cells**, **-punch_frequency**, **-print**, and **-print_frequency** serve the same function as in the ADVECTION data block. The second SELECTED_OUTPUT data block specifies that the transport step (shift) number and total dissolved concentrations of sodium, chloride, potassium, and calcium will be written to the file *ex11trn.sel*. The USER_PUNCH data block from the advection simulation is still in effect and the pore volume at each transport step is calculated and written to the selected-output file.

USER_GRAPH 1 is detached, which prevents adding any more data points. USER_GRAPH 2 is defined to make the curves distinct from the advection simulation [a plus ("+") symbol is used for the calculated points] and to plot the analytical solution for chloride breakthrough with dispersion, which is

$$C_L = \frac{C_i}{2}\left(erfc\left(\frac{1-x}{2\sqrt{x\alpha/L}}\right) + \exp\left(\frac{L}{\alpha}\right)\left(erfc\left(\frac{1+x}{2\sqrt{x\alpha/L}}\right)\right)\right),$$

where C_L is the concentration at the column-outlet (mol/kg H_2O), C_i is the concentration at the inlet (mol/kgw), x is the number of pore volumes (-), α is the dispersivity (m) and L is the length of the column (m).

The results for example 11 using the **ADVECTION** and **TRANSPORT** keywords are shown in figure 12. The concentrations in cell 40, which is the end cell, are plotted against pore volumes. The main features of the calculations are the same between the two transport simulations. Chloride is a conservative solute and arrives in the effluent after one pore volume. The sodium initially present in the column exchanges with the incoming calcium and is eluted as long as the exchanger contains sodium. The midpoint of the breakthrough curve for sodium occurs at about 1.5 pore volumes. Because potassium exchanges more strongly than sodium (because of the larger log K in the exchange reaction), potassium is released after sodium. Finally, when all of the potassium has been released, the concentration of calcium increases to the steady-state concentration in the influent.

The concentration changes of sodium and potassium in the effluent form a chromatographic pattern, which often can be calculated by simple means (Appelo, 1994b; Appelo and Postma, 2005). The number of pore volumes needed for the arrival of the sodium-decrease front can be calculated with the formula $P_v = 1 + V^s$, where $V^s = \Delta q/\Delta c$, Δq indicates the change in sorbed concentration (mol/kgw), and Δc indicates the change in solute concentration over the front. The sodium concentration in the solution that initially fills the column is 1.0 mmol/kgw, and the initial sorbed concentration of sodium is 0.55; the concentration of sodium in the infilling solution is zero, which must eventually result in 0 sorbed sodium. Thus, $(V^s)_{Na} = \Delta q/\Delta c = (0.55 - 0)/(1 - 0) = 0.55$ and $P_v = 1.55$, which indicates that the midpoint of the sodium front should arrive at the end of the column after 1.55 pore volumes.

Next, potassium is displaced from the exchanger. The concentration in solution increases to 1.2 mmol/kgw to balance the Cl⁻ concentration, and then falls to 0 when the exchanger is exhausted. When potassium is the only cation in solution, it will also be the only cation on the exchanger. For potassium,

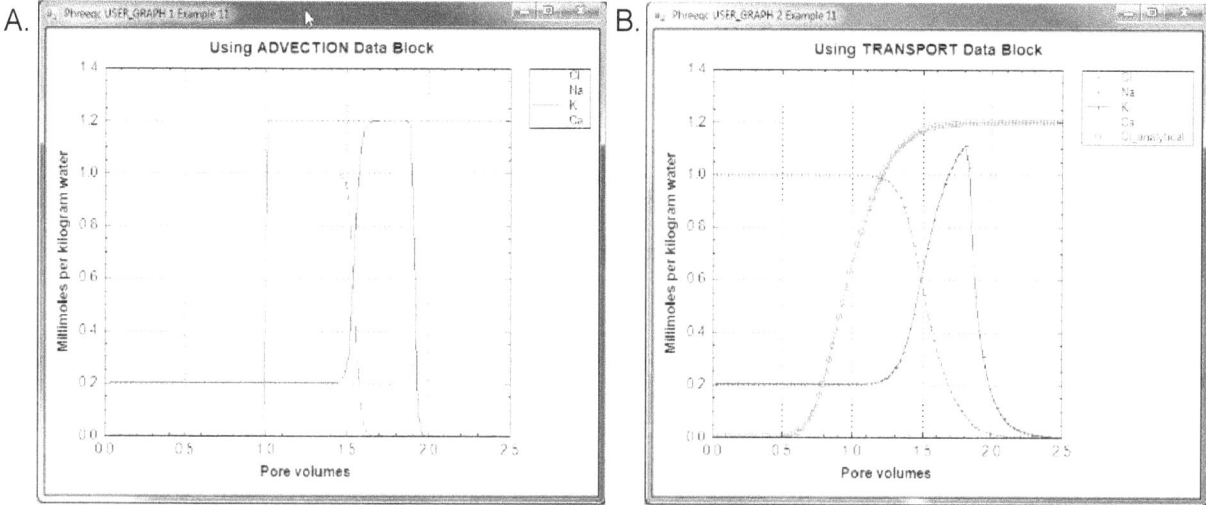

Figure 12. Results of (A) advective and (B) advective-dispersive transport simulations of the replacement of sodium and potassium on a cation exchanger by infilling calcium chloride solution. Lines display concentrations at the outlet of the column as calculated with PHREEQC with advection only (**ADVECTION** keyword) and with advection and dispersion (**TRANSPORT** keyword).

$(V^s)_K = \Delta q / \Delta c = (1.1 - 0)/(1.2 - 0) = 0.917$ and $P_v = 1.917$ pore volumes. It can be seen that the front locations for $(V^s)_{Na}$ and $(V^s)_K$ are closely matched by the midpoints of the concentration changes shown in figure 12.

The differences between the two simulations are due entirely to the inclusion of dispersion in the **TRANSPORT** calculation. The breakthrough curve for chloride in the **TRANSPORT** calculation coincides with an analytical solution to the advection dispersion equation for a conservative solute (Appelo and Postma, 2005), which is plotted in figure 12 with **USER_GRAPH**. The sum of the squared deviations for the first two pore volumes (transport step 80) is printed in the output file:

```
Transport step  80. Mixrun   4.

SSQD for Cl after 2 Pore Volumes:   7.985703292556e-005 (mmol/L)^2
```

The result shows that the calculations are in astonishingly good agreement. Without dispersion, the **ADVECTION** calculation produces a square-wave breakthrough curve for chloride. The characteristic smearing effects of dispersion are absent in the fronts calculated for the other elements as well, although some curvature exists because of the exchange reactions. The peak potassium concentration is larger in the **ADVECTION** calculation because dispersion is neglected.

Example 12—Advective and Diffusive Flux of Heat and Solutes

The following example demonstrates the capability of PHREEQC to calculate transient transport of heat and solutes in a column or along a 1D flowline. A column is initially filled with a dilute KCl solution at 25 °C in equilibrium with a cation exchanger. A KNO_3 solution then advects into the column and establishes a new temperature of 0 °C. Subsequently, a sodium chloride solution at 24 °C is allowed to diffuse from both ends of the column, assuming no heat is lost through the column walls. At one end, a constant boundary condition is imposed, and at the other end, the final cell is filled with the sodium chloride solution and a closed boundary condition is prescribed. For the column end with a constant boundary condition, an analytical solution is compared with PHREEQC results, for unretarded Cl^- ($R = 1.0$) and retarded Na^+ and temperature ($R = 3.0$). Finally, the second-order accuracy of the numerical method is verified by increasing the number of cells by a factor of three and demonstrating a decrease in the error of the numerical solution by approximately one order of magnitude relative to the analytical solution.

The input file for example 12 is shown in table 31. The **EXCHANGE_SPECIES** data block is used (1) to make the exchange constant for KX equal to NaX (log_k = 0.0), (2) to effectively remove the possibility of hydrogen ion exchange, and (3) to set activity coefficients for exchange species equal to their aqueous counterparts (**-gamma** identifier), so that the exchange between Na^+ and K^+ is linear and the retardation is constant. The influent, **SOLUTION** 0, is defined with temperature 0 °C and 24 mmol/kgw KNO_3. All solutions are defined to be in equilibrium with atmospheric oxygen partial pressure. The column is discretized in 60 cells, which are filled initially with a 1-µmol/kgw KCl solution (**SOLUTION** 1-60). Each cell has 48 mmol of exchange sites, which are defined to contain only potassium by the data block **EXCHANGE** 1-60.

Table 31. Input file for example 12.

```
TITLE Example 12.--Advective and diffusive transport of heat and solutes.
     Two different boundary conditions at column ends.
     After diffusion temperature should equal Na-conc in mmol/l.
SOLUTION 0   24.0 mM KNO3
    units mol/kgw
    temp  0      # Incoming solution 0C
    pH    7.0
    pe    12.0    O2(g) -0.67
    K     24.e-3
    N(5) 24.e-3
SOLUTION 1-60   0.001 mM KCl
    units mol/kgw
    temp 25     # Column is at 25C
    pH    7.0
```

Table 31. Input file for example 12.—Continued

```
    pe  12.0    O2(g) -0.67
    K    1e-6
    Cl  1e-6
EXCHANGE_SPECIES
    Na+ + X- = NaX
    log_k        0.0
    -gamma       4.0       0.075

    H+ + X- = HX
    log_k        -99.
    -gamma       9.0       0.0

    K+ + X- = KX
    log_k        0.0
    gamma        3.5       0.015
EXCHANGE 1-60
    KX    0.048
PRINT
    -reset    false
    -selected_output false
    -status false
SELECTED_OUTPUT
    -file     ex12.sel
    -reset    false
    -dist     true
    -high_precision  true
    -temp     true
USER_PUNCH
        -head Na_mmol K_mmol Cl_mmol
  10 PUNCH TOT("Na")*1000, TOT("K")*1000, TOT("Cl")*1000
TRANSPORT                # Make column temperature 0C, displace Cl
    -cells    60
    -shifts   60
    -flow_direction  forward
    -boundary_conditions flux  flux
    -lengths 0.333333
    -dispersivities          0.0     # No dispersion
    -diffusion_coefficient 0.0       # No diffusion
    -thermal_diffusion       1.0     # No retardation for heat
END

SOLUTION 0   Fixed temp 24C, and NaCl conc (first type boundary cond) at inlet
    units   mol/kgw
    temp 24
    pH  7.0
    pe  12.0    O2(g) -0.67
    Na  24.e-3
    Cl  24.e-3
SOLUTION 58-60  Same as soln 0 in cell 20 at closed column end (second type boundary
cond)
    units  mol/kgw
    temp 24
    pH  7.0
```

Table 31. Input file for example 12.—Continued

```
    pe  12.0    O2(g) -0.67
    Na  24.e-3
    Cl  24.e-3

EXCHANGE 58-60
    NaX  0.048
PRINT
   -selected_output true
TRANSPORT               # Diffuse 24C, NaCl solution from column end
   -shifts 1
   -flow_direction          diffusion
   -boundary_conditions     constant  closed
   -thermal_diffusion       3.0       # heat is retarded equal to Na
   -diffusion_coefficient   0.3e-9    # m^2/s
   -time_step               1.0e+10   # 317 years give 19 mixes

USER_GRAPH 1 Example 12
   -headings Na Cl Temp Analytical
   -chart_title "Diffusion of Solutes and Heat"
   -axis_titles "Distance, in meters" "Millimoles per kilogram water", "Degrees cel-
sius"
   -axis_scale x_axis 0 20
   -axis_scale y_axis 0 25
   -axis_scale sy_axis 0 25
   -initial_solutions false
   -plot_concentration_vs x
   -start
 10 x = DIST
 20 PLOT_XY x, TOT("Na")*1000, symbol = Plus
 30 PLOT_XY x, TOT("Cl")*1000, symbol = Plus
 40 PLOT_XY x, TC, symbol = XCross, color = Magenta, symbol_size = 8, y-axis 2
 50 if (x > 10 OR SIM_TIME <= 0) THEN END
 60 DATA 0.254829592, -0.284496736, 1.421413741, -1.453152027, 1.061405429, 0.3275911
 70 READ a1, a2, a3, a4, a5, a6
# Calculate and plot Cl analytical...
 80 z = x / (2 * SQRT(3e-10 * SIM_TIME / 1.0))
 90 GOSUB 2000
100 PLOT_XY x, 24 * erfc, color = Blue, symbol = Circle, symbol_size = 10,\
                         line_width = 0
# Calculate and plot 3 times retarded Na and temperature analytical...
110 z = z * SQRT(3.0)
120 GOSUB 2000
130 PLOT_XY x, 24 * erfc, color = Blue, symbol = Circle, symbol_size = 10,\
                         line_width = 0
140 END
2000 REM calculate erfc...
2050 b = 1 / (1 + a6 * z)
2060 erfc = b * (a1 + b * (a2 + b * (a3 + b * (a4 + b * a5)))) * EXP(-(z * z))
2080 RETURN
   -end
END
```

The TRANSPORT data block defines cell lengths of 1 m (**-lengths** 1), no dispersion (**-dispersivities** 0), no diffusion (**-diffusion_coefficient** 0), and no retardation for temperature (**-thermal_diffusion** 1). SOLUTION 0 is shifted 60 times into the column (**-shifts** 60, **-flow_direction** forward), and arrives in cell 60 at the last shift of the advective-dispersive transport simulation. The boundary conditions at the column ends are of flux type (**-boundary_conditions** flux flux). In this initial advective-dispersive transport simulation, no exchange occurs because the exchange sites are already completely filled with potassium, and the concentrations and temperatures in the 60 cells evolve to 24 mmol/kgw KNO_3 and 0 °C. This result could be more easily achieved with SOLUTION data blocks directly, but the simulation demonstrates how transient boundary and flow conditions can be represented. The dissolved and solid compositions and the temperature of each cell of the column is automatically saved after each shift in the advective-dispersive transport simulation. The keyword PRINT is used to exclude all printing to the output file (**-reset** false).

In the next advective-dispersive transport simulation, diffusion is calculated from the column ends. The column composition and temperatures are initially the conditions produced by the first advective-dispersive transport calculation, except that an NaCl solution is now defined as solution 0, which diffuses into the top of the column, and as solution 60, which diffuses from the bottom of the column. The new SOLUTION 0 is defined with a temperature of 24 °C and with 24 mmol/kgw NaCl. The last cell (SOLUTION 60) also is defined to have this solution composition and temperature. The exchanger in cell 60 is defined to be in equilibrium with the new solution composition in cell 60 (EXCHANGE 60).

The TRANSPORT data block defines one diffusive transport period (**-shifts** 1; **-flow_direction** diffusion). The boundary condition at the first cell is constant concentration, and at the last cell the column is closed (**-boundary_conditions** constant closed). The effective diffusion coefficient (**-diffusion_coefficient**) is set to 0.3×10^{-9} m^2/s, and the time step (**-time_step**) is defined to be 1.0×10^{10} s. Because only one diffusive time period is defined (**-shifts** 1), the total time modeled is equal to the time step, 1.0×10^{10} s. However, the time step will automatically be divided into a number of time substeps to satisfy stability criteria for the numerical method. The heat retardation factor is set to 3.0 (**-thermal_diffusion** 3.0). For Na^+ the ratio of exchangeable concentration (maximum NaX is 48 mmol/kgw) to solute concentration (maximum Na^+ = 24 mmol/kgw) is 2.0 for all concentrations, and the retardation is therefore $R = 1 + d$NaX$/d$Na$^+$ = 3.0, which is numerically equal to the temperature retardation.

The SELECTED_OUTPUT data block specifies the name of the selected-output file to be *ex12.sel*. The identifier "**-high_precision** true" is used to obtain an increased number of digits in the printing, and the identifiers **-dist** and **-temp** specify that the distance and temperature of each cell will be printed to the file. The USER_PUNCH data block is used to print concentrations of sodium, potassium, and chloride to the selected-output file in units of mmol/kgw, and USER_GRAPH plots the data.

In the model, the temperature is calculated with the (linear) retardation formula; however, the Na^+ concentration is calculated by the cation-exchange reactions. Even though the Na^+ concentration and the temperature are calculated by different methods, the numerical values should be the same because the initial and the transient conditions are numerically equal and the retardation factors are the same. The temperature and the Na^+ concentration are equal to at least 6 digits in the PHREEQC selected-output file, which indicates that the algorithm for the chemical transport calculations is correct for the simplified chemistry considered in this example. A further check on the accuracy is obtained by comparing simulation results with an analytical solution. For an infinite column with $C_x = 0$ for $t = 0$, and diffusion from $x = 0$ with $C_{x=0} = C_0$ for $t > 0$ the analytical solution is

$$C_{x,t} = C_0 \, erfc\left(\frac{x}{2\sqrt{D_e t / R}}\right),$$
(23)

where D_e is the effective diffusion coefficient and R is the retardation factor.

The PHREEQC results are compared with the analytical solution for Cl^- and for temperature and Na^+ in figure 13 and show excellent agreement. Notice that diffusion of Cl^- from the column ends has not yet "touched" in the mid-section, so that the column is still effectively infinite and the analytical solution is appropriate. Although both ends of the column started with the same temperature and concentration, at $x = 0$ m, the same temperature and concentrations are maintained because of the constant boundary condition. At $x = 20$ m (plotted at the midpoint of cell 60, $x = 19.83$ m), the temperature and concentrations decrease because of the closed boundary condition; no flux of heat or mass through this boundary is allowed, and the temperature and concentrations are diminishing because of diffusion into the column. The sodium concentration is not dissipating as rapidly as the chloride concentration because exchange sites must be filled with sodium along the diffusive reach.

Because this example has an analytical solution, it is possible to verify the second-order accuracy of the numerical algorithms used in PHREEQC. For a second-order method, decreasing the cell size by a factor of three should improve the results by about a factor of nine. The deviations from the analytical solution at the end of the time step are calculated at distances from 0.5 to 8.5 m in 0.5 m increments. The results are shown in table 32. As expected for a second-order method, the deviations from the analytical

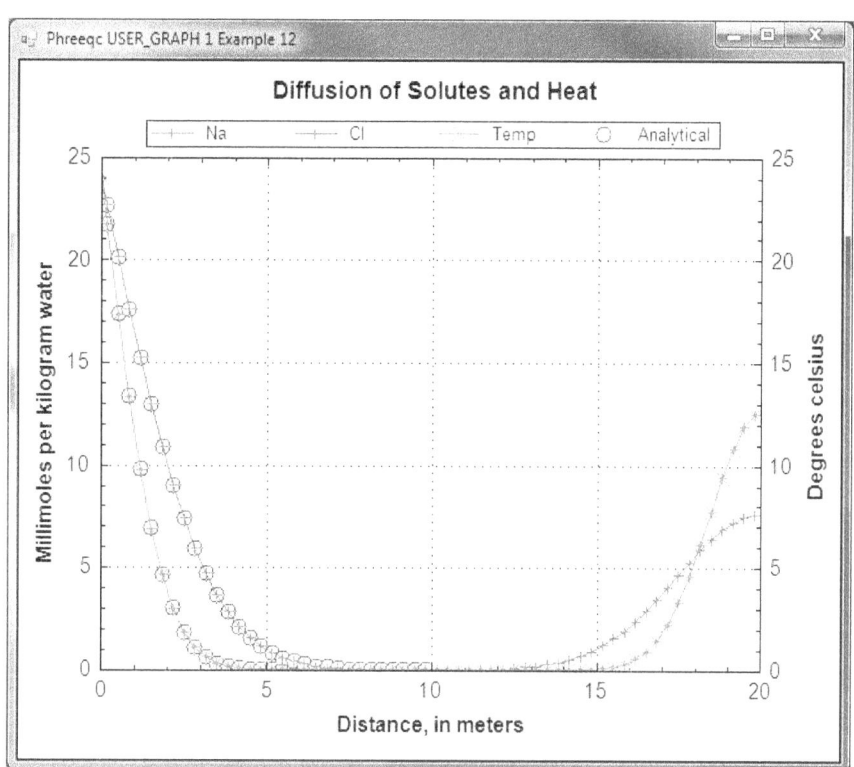

Figure 13. Simulation results for diffusion from column ends of heat and Na$^+$ (retardation $R = 3$) and Cl$^-$ ($R = 1$) compared with constant-boundary-condition analytical solution.

solution decreased by approximately an order of magnitude when the cell size was decreased by a factor of three.

Table 32. Numerical errors relative to the analytical solution for example 12 for a 20-cell and a 60-cell model.

Distance	Error in Cl concentration		Error in Na concentration	
	20-cell model	60-cell model	20-cell model	60-Cell model
0.5	3.32e-05	3.03e-06	5.75e-04	4.42e-05
1.5	8.17e-05	7.66e-06	5.54e-04	6.08e-05
2.5	9.18e-05	9.09e-06	8.29e-05	1.43e-05
3.5	7.15e-05	7.65e-06	-5.07e-05	-5.64e-06
4.5	4.24e-05	4.98e-06	-2.54e-05	-3.26e-06
5.5	2.00e-05	2.61e-06	-5.44e-06	-6.27e-07
6.5	7.81e-06	1.12e-06	-7.20e-07	-6.15e-08
7.5	2.55e-06	3.97e-07	-6.77e-08	-3.48e-09
8.5	5.58e-07	7.65e-08	-4.90e-09	-1.21e-10

Example 13—1D Transport in a Dual Porosity Column With Cation Exchange

This example demonstrates the capabilities of PHREEQC to calculate flow in a dual-porosity medium with diffusive exchange among the mobile and immobile pores. The flexible input format and the modular definition of additional solutions and reactants in PHREEQC allow inclusion of heterogeneities and various complexities within a 1D column. This example considers a column filled with small clay beads of 2 cm diameter, which act as dual porosity or stagnant zones. Both the first-order exchange approximation and finite differences are applied in this example, and transport of both a conservative and a retarded (by ion exchange) chemical is considered. It is furthermore shown how a heterogeneous column can be modeled by prescribing mixing factors to account for diffusion between mobile and immobile cells with the keyword **MIX**.

Stagnant Zone Calculation Using the First-Order Exchange Approximation With Implicit Mixing Factors

The first example input file, example 13A (table 33), is for a column with a uniform distribution of the stagnant porosity along the column. The 20 mobile cells are numbered 1 through 20. Each mobile cell, n, exchanges with one immobile cell, which is numbered $20 + 1 + n$ (cells 22 through 41 are immobile cells). All cells are given an identical initial solution and exchange complex, but these could be defined individually for each cell. It also is possible to distribute the immobile cells over the column non-uniformly, simply by omitting solutions for the stagnant cells that are not present. The connections between the mobile-zone and the stagnant-zone cells and among stagnant-zone cells can be varied along the column as well, but this requires that mixing factors among the mobile and immobile cells are prescribed by using the keyword **MIX**.

As defined in table 33, the column initially contains a 1 mmol/L KNO_3 solution in both the mobile and the stagnant zones (**SOLUTION 1-41**). An $NaCl-NO_3$ solution flows into the column (**SOLUTION 0**). An exchange complex with 1 mmol of sites is defined for each cell (**EXCHANGE 1-41**), and exchange coefficients are specified to give linear retardation $R = 2$ for Na^+ (**EXCHANGE_SPECIES**). The first **TRANSPORT** data block is used to define the physical and flow characteristics of the first transport simulation. The column is 2 m in length and is discretized in 20 cells (**-cells**) of 0.1 m (**-lengths**). A pulse of five shifts (**-shifts**) of the infilling solution (**SOLUTION 0**) is introduced into the column. The length of time for each shift is 3,600 s (**-time_step**), which results in a velocity in the mobile pores $v_m = 0.1 / 3,600 =$

2.78×10^{-5} m/s. The dispersivity is set to 0.015 m for all cells (**-dispersivities**). The diffusion coefficient is set to 0.0 (**-diffusion_coefficient**).

The stagnant/mobile interchange is defined by using the first-order exchange approximation. The stagnant zone consists of spheres with radius $r = 0.01$ m, diffusion coefficient $D_e = 3. \times 10^{-10}$ m^2/s, and shape factor $f_{s \to 1} = 0.21$ (Parkhurst and Appelo, 1999, "Transport in Dual Porosity Media", table 1). These variables give an exchange factor $\alpha = 6.8 \times 10^{-6}$ s^{-1}. Mobile porosity is $\varepsilon_m = 0.3$ (**-stagnant**) and immobile porosity $\varepsilon_{im} = 0.1$. For the first-order exchange approximation in PHREEQC, a single cell immobile zone and the parameters α, ε_m and ε_{im} are specified with **-stagnant**. This stagnant zone definition causes each cell in the mobile zone (numbered 1 through 20) to have an associated cell in the immobile zone (numbered 22 through 41). The **PRINT** data block is used to eliminate all printing to the output file.

Following the pulse of NaCl solution, 10 shifts of 1 mmol KNO$_3$/L (second **SOLUTION** 0) are introduced into the column. The second **TRANSPORT** data block does not redefine any of the column or flow characteristics, but specifies that results for cells 1 through 20 (**-punch_cells**) be written to the selected output file after 10 shifts (**-punch_frequency**). The data blocks **SELECTED_OUTPUT** and **USER_PUNCH** specify the data to be written to the selected-output file, and **USER_GRAPH** plots the same data.

Table 33. Input file for example 13A: Stagnant zone with implicitly defined mixing factors.

```
TITLE Example 13A.--1 mmol/L NaCl/NO3 enters column with stagnant zones.
                  Implicit definition of first-order exchange model.
SOLUTION 0    # 1 mmol/L NaCl
   units    mmol/l
   pH       7.0
   pe       13.0    O2(g)    -0.7
   Na       1.0     # Na has Retardation = 2
   Cl       1.0     # Cl has Retardation = 1, stagnant exchange
   N(5)     1.0     # NO3 is conservative
#        charge imbalance is no problem ...
END
SOLUTION 1-41  # Column with KNO3
   units    mmol/l
   pH       7.0
   pe       13.0    O2(g)    -0.7
   K        1.0
   N(5)     1.0
EXCHANGE_SPECIES # For linear exchange, make KX exch. coeff. equal to NaX
   K+ + X- = KX
   log_k   0.0
   -gamma  3.5      0.015
EXCHANGE 1-41
```

```
   -equil  1
   X         1.e-3
END
PRINT
   -reset false
   -echo_input true
   -status false
TRANSPORT
   -cells  20
   -shifts 5
   -flow_direction  forward
   -time_step       3600
   -boundary_conditions   flux   flux
   -diffusion_coefficient 0.0
   -lengths         0.1
   -dispersivities 0.015
   -stagnant  1  6.8e-6  0.3        0.1
#  1 stagnant layer^, ^alpha, ^epsil(m), ^epsil(im)
END
SOLUTION 0  # Original solution with KNO3 reenters
   units   mmol/l
   pH       7.0
   pe       13.0   O2(g)     -0.7
   K        1.0
   N(5)     1.0
END
SELECTED_OUTPUT
   -file    ex13a.sel
   -reset   false
   -solution
   -distance         true
USER_PUNCH
   -headings Cl_mmol Na_mmol
  10 PUNCH TOT("Cl")*1000, TOT("Na")*1000
TRANSPORT
   -shifts 10
   -punch_cells        1-20
   -punch_frequency    10
USER_GRAPH 1 Example 13A
   -headings Distance Na Cl
   -chart_title "Dual Porosity, First-Order Exchange with Implicit Mixing Factors"
   -axis_titles "Distance, in meters" "Millimoles per kilogram water"
   -axis_scale x_axis 0 2
   -axis_scale y_axis 0 0.8
   -plot_concentration_vs x
   -start
  10 GRAPH_X DIST
  20 GRAPH_Y TOT("Na")*1000 TOT("Cl")*1000
   -end
END
```

The mixing factors $mixf_m$ and $mixf_{im}$ for the first-order exchange approximation for this example are as follows (Parkhurst and Appelo, 1999, equations 121 and 123):

$$mixf_{im} = \frac{\varepsilon_m}{\varepsilon_m + \varepsilon_{im}} \times \left(1 - \exp\left(- \frac{\alpha\, t(\varepsilon_m + \varepsilon_{im})}{\varepsilon_m \varepsilon_{im}} \right) \right) \tag{24}$$

where t is the time step, and

$$mixf_m = mixf_{im} \frac{\varepsilon_{im}}{\varepsilon_m}. \tag{25}$$

The retardation factors R_m and R_{im} do not appear in the formulas for $mixf_{im}$ and $mixf_m$ because in PHREEQC, the retardation is a consequence of chemical reactions. According to equations 24 and 25, for this example, the mixing factors are calculated to be $mixf_{im} = 0.20886$ and $mixf_m = 0.06962$. The mixing factors differ for the mobile cell and the immobile cell to account for the different volumes of mobile and immobile water.

In PHREEQC, a mixing of mobile and stagnant water is done after each diffusion/dispersion step. This means that the time step decreases when the cells are made smaller and when more diffusive steps ("mixruns") are performed. A 20-cell model as in table 33 has one mixrun. A 100-cell model would have three mixruns, and the time step for calculating $mixf_{im}$ would be $(3,600/5) / 3 = 240$ s. A time step $t = 240$ s leads to $mixf_{im} = 0.01614$ in the 100-cell model.

Stagnant Zone Calculation Using the First-Order Exchange Approximation With Explicit Mixing Factors

The input file with explicit mixing factors for a uniform distribution of the stagnant zones is given in table 34. The **SOLUTION** data blocks are identical to the previous input file (table 33). One stagnant layer without further information is defined (**-stagnant** 1, in the **TRANSPORT** data block). The mobile/immobile exchange is set by the mix fraction given in the **MIX** data blocks. The results of this input file are identical with the results from the previous input file in which the shortcut notation was used. However, the explicit definition of mix factors illustrates that a non-uniform distribution of the stagnant zones, or other physical properties of the stagnant zone, can be included in PHREEQC simulations by varying the mixing fractions that define the exchange among mobile and immobile cells.

Table 34. Input file for example 13B: Stagnant zone with explicitly defined mixing factors.

```
TITLE Example 13B.--1 mmol/l NaCl/NO3 enters column with stagnant zones.
                     Explicit definition of first-order exchange factors.
SOLUTION 0    # 1 mmol/l NaCl
   units    mmol/l
   pH       7.0
   pe       13.0    O2(g)    -0.7
   Na       1.0     # Na has Retardation = 2
   Cl       1.0     # Cl has Retardation = 1, stagnant exchange
   N(5)     1.0     # NO3 is conservative
#        charge imbalance is no problem ...
END
SOLUTION 1-41  # Column with KNO3
   units    mmol/l
   pH       7.0
   pe       13.0    O2(g)     -0.7
   K        1.0
   N(5)     1.0
EXCHANGE_SPECIES # For linear exchange, make KX exch. coeff. equal to NaX
   K+ + X- = KX
   log_k   0.0
   -gamma  3.5      0.015
EXCHANGE 1-41
   -equil  1
   X       1.e-3
END
PRINT
        -reset false
        -echo_input true
        -status false
MIX  1;  1 .93038;      22 .06962      ;MIX  2;      2 .93038;    23 .06962;
MIX  3;  3 .93038;      24 .06962      ;MIX  4;      4 .93038;    25 .06962;
MIX  5;  5 .93038;      26 .06962      ;MIX  6;      6 .93038;    27 .06962;
MIX  7;  7 .93038;      28 .06962      ;MIX  8;      8 .93038;    29 .06962;
MIX  9;  9 .93038;      30 .06962      ;MIX 10;     10 .93038;    31 .06962;
MIX 11; 11 .93038;      32 .06962      ;MIX 12;     12 .93038;    33 .06962;
MIX 13; 13 .93038;      34 .06962      ;MIX 14;     14 .93038;    35 .06962;
MIX 15; 15 .93038;      36 .06962      ;MIX 16;     16 .93038;    37 .06962;
MIX 17; 17 .93038;      38 .06962      ;MIX 18;     18 .93038;    39 .06962;
MIX 19; 19 .93038;      40 .06962      ;MIX 20;     20 .93038;    41 .06962;
#
MIX 22;  1 .20886;      22 .79114      ;MIX 23;      2 .20886;    23 .79114;
MIX 24;  3 .20886;      24 .79114      ;MIX 25;      4 .20886;    25 .79114;
MIX 26;  5 .20886;      26 .79114      ;MIX 27;      6 .20886;    27 .79114;
MIX 28;  7 .20886;      28 .79114      ;MIX 29;      8 .20886;    29 .79114;
MIX 30;  9 .20886;      30 .79114      ;MIX 31;     10 .20886;    31 .79114;
MIX 32; 11 .20886;      32 .79114      ;MIX 33;     12 .20886;    33 .79114;
MIX 34; 13 .20886;      34 .79114      ;MIX 35;     14 .20886;    35 .79114;
MIX 36; 15 .20886;      36 .79114      ;MIX 37;     16 .20886;    37 .79114;
MIX 38; 17 .20886;      38 .79114      ;MIX 39;     18 .20886;    39 .79114;
MIX 40; 19 .20886;      40 .79114      ;MIX 41;     20 .20886;    41 .79114;
TRANSPORT
   -cells  20
   -shifts 5
```

```
      -flow_direction   forward
      -time_step        3600
      -boundary_conditions    flux   flux
      -diffusion_coefficient 0.0
      -lengths          0.1
      -dispersivities   0.015
      -stagnant         1
END
SOLUTION 0  # Original solution reenters
   units    mmol/l
   pH       7.0
   pe       13.0    O2(g)     -0.7
   K        1.0
   N(5)     1.0
END
SELECTED_OUTPUT
   -file    ex13b.sel
   -reset   false
   -distance        true
   -solution
USER_PUNCH
   -headings Cl_mmol Na_mmol
  10 PUNCH TOT("Cl")*1000, TOT("Na")*1000
TRANSPORT
   -shifts  10
   -punch_cells         1-20
   -punch_frequency     10
USER_GRAPH 1 Example 13B
   -headings Distance Na Cl
   -chart_title "Dual Porosity, First-Order Exchange with Explicit Mixing Factors"
   -axis_titles "Distance, in meters" "Millimoles per kilogram water"
   -axis_scale x_axis 0 2
   -axis_scale y_axis 0 0.8
   -plot_concentration_vs x
   -start
  10 GRAPH_X DIST
  20 GRAPH_Y TOT("Na")*1000 TOT("Cl")*1000
   -end
END
```

Stagnant Zone Calculation Using a Finite Difference Approximation

The stagnant zone consists of spheres with radius $r = 0.01$ m. Diffusion into the spheres induces radially symmetric concentration changes according to the differential equation:

$$\frac{\partial C}{\partial t} = D_e\left(\frac{\partial^2 C}{\partial r^2} + \frac{2}{r}\frac{\partial C}{\partial r}\right). \tag{26}$$

The calculation in finite differences can therefore be simplified to one (radial) dimension. The calculation follows the theory outlined in Parkhurst and Appelo (1999, section "Transport in Dual Porosity Media"). The stagnant zone is divided into a number of layers that mix by diffusion. In this example, the sphere is cut in five equidistant layers with $\Delta r = 0.002$ m. Five stagnant layers are defined under keyword **TRANSPORT** with **-stagnant** 5 (table 35). Mixing is specified among adjacent cells in the stagnant layers with **MIX** data blocks; the mixing factors are calculated by equations 27 and 28. For a neighbor cell, the mixing factor is

$$mixf_{ij} = \frac{D_e \Delta t A_{ij} f_{bc}}{h_{ij} V_j},$$
(27)

and for the central cell, it is

$$mixf_{jj} = 1 - D_e \Delta t \sum_{i \neq j}^{n} \frac{A_{ij} f_{bc}}{h_{ij} V_j},$$
(28)

where, D_e is the effective diffusion coefficient, Δt is the time interval, V_j is the volume of cell j (m^3), A_{ij} is the shared surface area of cell i and j (m^2), h_{ij} is the distance between midpoints of cells i and j (m), and f_{bc} is the correction factor for boundary cells (dimensionless). The values for mobile cell 1 and associated.

Table 35. Input file for example 13C: Stagnant zone with diffusion calculated by finite differences (partial listing).

```
TITLE Example 13C.--1 mmol/1 NaCl/NO3 enters column with stagnant zones.
                    5 layer stagnant zone with finite differences.
SOLUTION 0    # 1 mmol/1 NaCl
        units   mmol/1
        pH      7.0

        pe      13.0    O2(g)   -0.7
        Na      1.0     # Na has Retardation = 2
        Cl      1.0     # Cl has Retardation = 1, stagnant exchange
        N(5)    1.0     # NO3 is conservative
#       charge imbalance is no problem ...
END
SOLUTION 1-121
        units   mmol/1
        pH      7.0
        pe      13.0    O2(g)    -0.7
        K       1.0
        N(5)    1.0
EXCHANGE_SPECIES # For linear exchange, make KX exch. coeff. equal to NaX
        K+ + X- = KX
        log_k   0.0
        -gamma  3.5     0.015
EXCHANGE 1-121
        -equilibrate  1
```

```
        X               1.e-3
END
PRINT
        -reset false
        -echo_input true
MIX    1;     1  0.90712;   22  0.09288
MIX   22;     1  0.57098;   22  0.21656;   42  0.21246
MIX   42;    22  0.35027;   42  0.45270;   62  0.19703
MIX   62;    42  0.38368;   62  0.44579;   82  0.17053
MIX   82;    62  0.46286;   82  0.42143;  102  0.11571
MIX  102;    82  0.81000;  102  0.19000

#
#   MIX definitions omitted for mobile cells
#   2-19 and associated immobile cells

#
MIX   20;    20  0.90712;   41  0.09288
MIX   41;    20  0.57098;   41  0.21656;   61  0.21246
MIX   61;    41  0.35027;   61  0.45270;   81  0.19703
MIX   81;    61  0.38368;   81  0.44579;  101  0.17053
MIX  101;    81  0.46286;  101  0.42143;  121  0.11571
MIX  121;   101  0.81000;  121  0.19000
TRANSPORT
   -cells  20
   -shifts 5
   -flow_direction  forward
   -time_step       3600
   -boundary_conditions   flux   flux
   -diffusion_coefficient 0.0
   -lengths         0.1
   -dispersivities  0.015
   -stagnant        5
END
SOLUTION 0  # Original solution reenters
   units    mmol/l
   pH       7.0
   pe       13.0   O2(g)    -0.7
   K        1.0
   N(5)     1.0
END
SELECTED_OUTPUT
   -file    ex13c.sel
   -reset   false
   -distance         true
   -solution
USER_PUNCH
   -headings Cl_mmol Na_mmol
  10 PUNCH TOT("Cl")*1000, TOT("Na")*1000
TRANSPORT
   -shifts  10
```

```
    -punch_cells          1-20
    -punch_frequency      10
USER_GRAPH 1 Example 13C
        -headings Distance Na Cl
    -chart_title "Dual Porosity, Finite-Difference Approximation"
    -axis_titles "Distance, in meters" "Millimoles per kilogram water"
    -axis_scale x_axis 0 2
    -axis_scale y_axis 0 0.8
    -plot_concentration_vs x
    -start
 10 GRAPH_X DIST
 20 GRAPH_Y TOT("Na")*1000 TOT("Cl")*1000
    -end
END
```

immobile cells are given in table 36. The cells in the immobile layer are numbered as $n + (l \times cells) + 1$, where n is the number of a mobile cell, l is the number of the stagnant layer, and *cells* is the total number of mobile cells. In this example, the boundary cell in the stagnant zone for cell number 1 is cell number 22 and the other four stagnant layers are cell numbers 42, 62, 82, and 102, with number 102 being the innermost cell of the sphere, which is connected only to one other cell (cell 82). The volume of the mobile cell (cell 1) is expressed relative to the volume of a sphere of radius 0.01 m, by multiplying this volume by

$\varepsilon_m / \varepsilon_{im}$ ($4.19 \times 10^{-6} \times 0.3 / 0.1 = 1.26 \times 10^{-5}$). In table 36 the value for f_{bc} is 1.72, as calculated with the following equation from Parkhurst and Appelo (1999, equation 127):

$$f_{bc} = 2\frac{V_m}{V_m + V_{bc}} , \qquad (29)$$

where, V_m is the volume of the mobile zone and V_{bc} is the volume of the boundary cell that contacts the mobile zone. It is noted that using $f_{bc} = 1.81$ slightly improves the fit to an analytical solution given in

Table 36. Mixing factors for finite difference calculation of diffusion in spheres.
[-, undefined value for cell]

Cell	r, m	V_j, m^3	A_{ij}, m^2	h_{ij}, m	f_{bc}	$mixf_{ij}$	$mixf_{jj}$	$mixf_{jk}$
102	0.001	3.35e-8	5.03e-5	0.002	1	0.81	0.19	-
82	0.003	2.35e-7	2.01e-4	0.002	1	0.463	0.421	0.116
62	0.005	6.37e-7	4.52e-4	0.002	1	0.384	0.446	0.170
42	0.007	1.24e-6	8.04e-4	0.002	1	0.350	0.453	0.197
22	0.009	2.04e-6	1.26e-3	0.002	1.72	0.571	0.217	0.212
1	-	1.26e-5	-	-	-	-	0.907	0.093

Crank (1975) for diffusion into spheres in a closed vessel (a beaker with solution and clay beads). However, changing f_{bc} to 1.81 has little effect on the concentration profiles for the columns that are shown in figure 14, and various calculations have shown that $f_{bc} = 2$ provides good results in general.

Note in table 35 that 121 solutions are defined, 1 through 20 for the mobile cells, and the rest for the immobile cells. The input file is identical with the previous one, except for **-stagnant** 5 and the mixing factors among the cells. Although not all the mixing factors are shown in table 35, the remaining mixing factors are identical for subsequent cells and their neighboring stagnant cells. In this example with clay beads, only radial (1D) diffusion is considered, and only mixing among cells in different layers is defined; however, it is possible to include mixing among the immobile cells of adjacent mobile cells.

Figure 14 compares the concentration profiles in the mobile cells obtained with examples 13A and 13B with example 13C. The basic features of the two simulations are the same. The positions of the peaks, as calculated by the two simulations, are similar. The Cl⁻ peak is near 1.2 m, but would be at about 1.45 m in the absence of stagnant zones. The integrated concentrations in the mobile porosity are about equal for

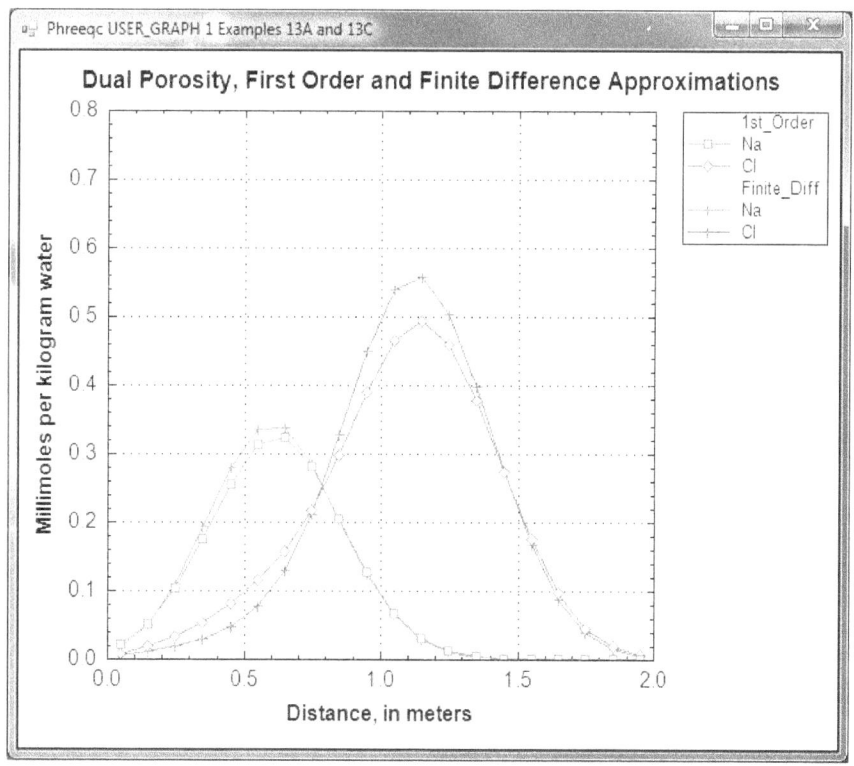

Figure 14. Results of simulations of transport with diffusion into spherical stagnant zones modeled by using finite difference and first-order exchange approximations.

the first-order exchange and the finite difference simulations. The exchange factor $f_{s \to 1} = 0.21$ for the first-order exchange approximation appears to provide adequate accuracy for this simulation. However, the first-order exchange approximation produces lower peaks and more tailing than the more exact solution obtained with finite differences. Discrepancies can also appear as deviations in the breakthrough curve (Van Genuchten, 1985). The first-order exchange model is probably least accurate when applied to simulating the transport behavior of spheres; other shapes of the stagnant area can give a better correspondence. It is clear that the linear exchange model is much easier to apply because any explicit model requires the preparation of extended lists of mixing factors (notice that a separate simulation with USER_PUNCH can serve that purpose, as illustrated in examples 8 and 21), which change when the discretization is adjusted. The calculation time for a finite difference model with multiple immobile-zone layers also may be considerably longer than for the single immobile-zone layer of the first-order exchange approximation.

Example 14—Advective Transport, Cation Exchange, Surface Complexation, and Mineral Equilibria

This example uses the phase-equilibrium, cation-exchange, and surface-complexation reaction capabilities of PHREEQC in combination with advective-transport capabilities to model the evolution of water in the Central Oklahoma aquifer. The geochemistry of the aquifer has been described by Parkhurst and others (1996). Two predominant water types occur in the aquifer: a calcium magnesium bicarbonate water with pH in the range of 7.0 to 7.5 in the unconfined part of the aquifer and a sodium bicarbonate water with pH in the range of 8.5 to 9.2 in the confined part of the aquifer. In addition, marine-derived sodium chloride brines exist below the aquifer and presumably in fluid inclusions and dead-end pore spaces within the aquifer. Large concentrations of arsenic, selenium, chromium, and uranium occur naturally within the aquifer. Arsenic is associated almost exclusively with the high-pH, sodium bicarbonate water type.

The conceptual model for the calculation of this example assumes that brines initially filled the aquifer. The aquifer contains calcite, dolomite, clays with cation-exchange capacity, and hydrous-ferric-oxide surfaces; initially, the cation exchanger and surfaces are in equilibrium with the brine. The aquifer is assumed to be recharged with rainwater that is concentrated by evaporation and equilibrates with calcite and dolomite in the vadose zone. This water then enters the saturated zone and reacts with calcite and dolomite in the presence of the cation exchanger and hydrous-ferric-oxide surfaces.

The calculations use the advective-transport capabilities of PHREEQC with just a single cell representing the saturated zone. A total of 200 pore volumes of recharge water are advected into the cell and, with each pore volume, the water is equilibrated with the minerals, cation exchanger, and the surfaces in the cell. The evolution of water chemistry in the cell represents the evolution of the water chemistry at a point near the top of the saturated zone of the aquifer.

Thermodynamic Data

The database *wateq4f.dat* contains thermodynamic data for the aqueous species of arsenic according to the compilation of Nordstrom and Archer (2003) and the surface complexation constants from Dzombak and Morel (1990). Unfortunately, these two sets of thermodynamic data are not internally consistent because the surface complexation constants were fit by using a different set of thermodynamic constants for arsenic aqueous species. To be consistent in the calculations, the thermodynamic data for both arsenic aqueous speciation and arsenic surface complexation are defined according to Dzombak and Morel (1990). The database *phreeqc.dat* was used for the calculations with the arsenic thermodynamic data defined with the SOLUTION_MASTER_SPECIES, SOLUTION_SPECIES, SURFACE_MASTER_SPECIES, and SURFACE_SPECIES data blocks within the input file.

Initial Conditions

Parkhurst and others (1996) provide data from which it is possible to estimate the moles of calcite, dolomite, and cation-exchange sites in the aquifer per liter of water. The weight percent ranges from 0 to 2 percent for calcite and 0 to 7 percent for dolomite, with dolomite much more abundant. Porosity is stated to be 0.22. Cation-exchange capacity for the clay ranges from 20 to 50 meq/100 g (milliequivalent per 100 grams), with an average clay content of 30 percent. For these example calculations, calcite was assumed to be present at 0.1 weight percent and dolomite, at 3 weight percent; by assuming a rock density of 2.7 kg/L, these percentages correspond to 0.1 mol/L for calcite and 1.6 mol/L for dolomite. The number of cation-exchange sites was estimated to be 1.0 eq/L.

The amount of arsenic on the surface was estimated from sequential extraction data on core samples (Mosier and others, 1991). Arsenic concentrations in the solid phases generally ranged from 10 to 20 ppm, which corresponds to 1.3 to 2.6 mmol/L arsenic. The number of surface sites were estimated from the amount of extractable iron in sediments, which ranged from 1.6 to 4.4 percent (Mosier and others, 1991). A content of 2 percent iron for the sediments corresponds to 3.4 mol/L of iron. However, most of the iron is in goethite and hematite, which have far fewer surface sites than hydrous ferric oxide. The fraction of iron in hydrous ferric oxide was arbitrarily assumed to be 0.1. Thus, a total of 0.34 mol of iron was assumed to be present as hydrous ferric oxide, and using a value of 0.2 for the number of sites per mole of iron, a total of 0.07 mol of sites per liter was used in the calculations. A gram formula weight of 89 g/mol was used to estimate that the mass of hydrous ferric oxide was 30 g/L (gram per liter). The specific surface area was assumed to be 600 m^2/g.

The brine that initially fills the aquifer was taken from Parkhurst and others (1996) and is given as solution 1 in the input file for this example (table 37). The pure-phase assemblage containing calcite and dolomite is defined with the **EQUILIBRIUM_PHASES** 1 data block. The brine is first equilibrated with calcite and dolomite and saved again as solution 1 (**SAVE** data block). The number of cation exchange sites is defined with **EXCHANGE** 1 and the number of surface sites is defined with **SURFACE** 1. The initial exchange and the initial surface composition are determined by equilibrium with the brine, after equilibration with calcite and dolomite (note that equilibration of exchangers and surfaces before mineral equilibration will yield different results because of buffering by the sorbed elements). The concentration of arsenic in the brine (0.025 µmol/kgw) was determined by trial and error to give a total of approximately 1.8 mmol arsenic on the surface, which is consistent with the sequential extraction data.

Table 37. Input file for example 14.

```
TITLE Example 14.--Transport with equilibrium_phases, exchange, and surface reactions
#
# Use phreeqc.dat
# Dzombak and Morel (1990) aqueous and surface complexation models for arsenic
# are defined here
#
SURFACE_MASTER_SPECIES
        Surf    SurfOH
SURFACE_SPECIES
        SurfOH = SurfOH
                log_k   0.0
        SurfOH  + H+ = SurfOH2+
                log_k   7.29
        SurfOH = SurfO- + H+
                log_k   -8.93
        SurfOH + AsO4-3 + 3H+ = SurfH2AsO4 + H2O
                log_k   29.31
        SurfOH + AsO4-3 + 2H+ = SurfHAsO4- + H2O
                log_k   23.51
        SurfOH + AsO4-3 = SurfOHAsO4-3
                log_k   10.58
SOLUTION_MASTER_SPECIES
        As      H3AsO4          -1.0    74.9216     74.9216
SOLUTION_SPECIES
        H3AsO4 = H3AsO4
                log_k           0.0
        H3AsO4 = AsO4-3 + 3H+
```

Table 37. Input file for example 14.—Continued

```
               log_k    -20.7
         H+ + AsO4-3 = HAsO4-2
               log_k    11.50
         2H+ + AsO4-3 = H2AsO4-
               log_k              18.46
SOLUTION 1 Brine
        pH      5.713
        pe      4.0     O2(g)     -0.7
        temp    25.
        units   mol/kgw
        Ca      .4655
        Mg      .1609
        Na      5.402
        Cl      6.642             charge
        C       .00396
        S       .004725
        As      .025 umol/kgw
END
USE solution 1
EQUILIBRIUM_PHASES 1
        Dolomite        0.0     1.6
        Calcite         0.0     0.1
SAVE solution 1
# prints initial condition to the selected-output file
SELECTED_OUTPUT
        -file ex14.sel
        -reset false
        -step
USER_PUNCH
        -head  m_Ca m_Mg m_Na umol_As pH mmol_sorbedAs
  10 PUNCH TOT("Ca"), TOT("Mg"), TOT("Na"), TOT("As")*1e6, -LA("H+"), SURF("As",
"Surf")*1000
END
PRINT
# skips print of initial exchange and initial surface to the selected-output file
        -selected_out false
EXCHANGE 1
        -equil with solution 1
        X       1.0
SURFACE 1
        -equil solution 1
# assumes 1/10 of iron is HFO
        SurfOH          0.07    600.    30.
END
SOLUTION 0 20 x precipitation
        pH      4.6
        pe      4.0     O2(g)     -0.7
        temp    25.
        units   mmol/kgw
        Ca      .191625
        Mg      .035797
        Na      .122668
        Cl      .133704
        C       .01096
```

Table 37. Input file for example 14.—Continued

```
        S        .235153         charge
EQUILIBRIUM_PHASES 0
        Dolomite         0.0     1.6
        Calcite          0.0     0.1
        CO2(g)          -1.5     10.
SAVE solution 0
END
PRINT
        -selected_out true
                         -status false
ADVECTION
        -cells 1
        -shifts 200
        -print_frequency 200
USER_GRAPH 1 Example 14
        -headings PV As(ppb) Ca(M) Mg(M) Na(M) pH
        -chart_title "Chemical Evolution of the Central Oklahoma Aquifer"
        -axis_titles "Pore volumes or shift number" "Log(Concentration, in ppb or
molal)" "pH"
        -axis_scale x_axis 0 200
        -axis_scale y_axis 1e-6 100 auto auto Log
  10 GRAPH_X STEP_NO
  20 GRAPH_Y TOT("As") * 74.92e6, TOT("Ca"), TOT("Mg"), TOT("Na")
  30 GRAPH_SY -LA("H+")
END
```

Recharge Water

The water entering the saturated zone of the aquifer was assumed to be in equilibrium with calcite and dolomite at a vadose-zone P_{CO_2} of $10^{-1.5}$ atm. The fourth simulation in the input set (the simulation following the third **END** statement) generates this water composition and stores it as solution 0 by using **SAVE** (table 37).

Advective-Transport Calculations

The **ADVECTION** data block (table 37) provides the necessary information to advect the recharge water into the cell representing the saturated zone. A total of 200 shifts is specified, which is equivalent to 200 pore volumes because only a single cell is used in this calculation.

The results of the calculations are plotted in figure 15. During the initial five pore volumes, the high concentrations of sodium, calcium, and magnesium decrease such that sodium is the dominant cation and calcium and magnesium concentrations are small. The pH increases to more than 9.0 and arsenic concentrations increase to approximately 100 ppb. Over the next 45 pore volumes, the pH gradually decreases, and the arsenic concentrations decrease to negligible concentrations. At about 100 pore volumes,

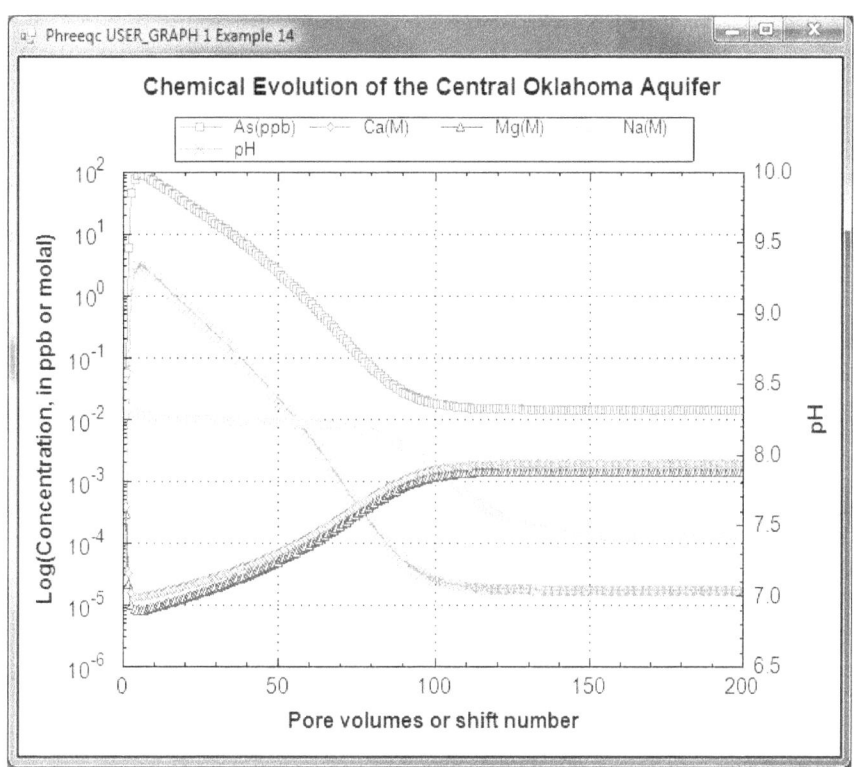

Figure 15. Results of transport simulation of the chemical evolution of groundwater due to calcium magnesium bicarbonate water inflow to an aquifer initially containing a brine, calcite and dolomite, a cation exchanger, and a surface that complexes arsenic.

the calcium and magnesium become the dominant cations, and the pH stabilizes at the pH of the infilling recharge water.

The advective-transport calculations produce three types of water that are similar to water types observed in the aquifer: the initial brine, a sodium bicarbonate water, and a calcium and magnesium bicarbonate water. The calculated pH values are consistent with observations of aquifer water. In the sodium dominated waters, the calculated pH is generally greater than 8.0 and sometimes greater than 9.0; in the calcium magnesium bicarbonate waters, the pH is slightly greater than 7.0. Sensitivity calculations indicate that the maximum pH depends on the amount of exchanger present. Decreasing the number of cation exchange sites decreases the maximum pH. Simulated arsenic concentrations are similar to values observed in the aquifer, where the maximum concentrations are from 75 to 150 ppb. Lower maximum pH values produce lower maximum arsenic concentrations. The stability constant for the surface complexation reactions have been taken directly from the literature; a decrease in the log K for the predominant arsenic complexation reaction tends to decrease the maximum arsenic concentration as well. In conclusion, the model results, which were based largely on measured values and literature thermodynamic data, can explain the trends in major ion chemistry, pH, and arsenic concentrations within the aquifer.

Example 15—1D Transport: Kinetic Biodegradation, Cell Growth, and Sorption

A test problem for advective-dispersive-reactive transport was developed by Tebes-Stevens and Valocchi (1997) and Tebes-Stevens and others (1998). Although based on relatively simple speciation chemistry, the solution to the problem demonstrates several interacting chemical processes that are common to many environmental problems: bacterially mediated degradation of an organic substrate; bacterial cell growth and decay; metal sorption; and aqueous speciation, including metal-ligand complexation. In this example, the test problem is solved with PHREEQC, which produces results almost identical to those of Tebes-Stevens and Valocchi (1997) and Tebes-Stevens and others (1998).

The test problem models the transport processes when a pulse of water containing NTA (nitrylotriacetate) and cobalt is injected into a column. The problem includes advection and dispersion in the column, aqueous equilibrium reactions, and kinetic reactions for NTA degradation, growth of biomass, and cobalt sorption.

Transport Parameters

The dimensions and hydraulic properties of the column are given in table 38.

Table 38. Hydraulic and physical properties of the column in example 15.
[m, meter; g/m^3, gram per cubic meter; g/L, gram per liter; m/h, meter per hour]

Property	Value
Length of column	10.0 m
Porosity, unitless	0.4
Bulk density	1.5e6 g/m^3
Grams of sediment per liter (from porosity and bulk density)	3.75e3 g/L
Pore-water velocity	1.0 m/h
Longitudinal dispersivity	0.05 m

Aqueous Model

Tebes-Stevens and Valocchi (1997) defined an aqueous model to be used for this test problem that includes the identity of the aqueous species and log *K*s of the species; activity coefficients were assumed to be 1.0. The database file in table 39 was constructed on the basis of their aqueous model. For the

PHREEQC simulation, NTA was defined as a new "element" in the **SOLUTION_MASTER_SPECIES** data block named "Nta". From this point on "NTA" will be referred to as "Nta" for consistency with the PHREEQC notation. The gram formula weight of Nta in **SOLUTION_MASTER_SPECIES** is immaterial if input units are moles in the **SOLUTION** data block, and is simply set to 1. The aqueous complexes of Nta are defined in the **SOLUTION_SPECIES** data block. Note that the activity coefficients of all aqueous species are defined with a large value for the a parameter (1×10^7) in the **-gamma** identifier, which forces the activity coefficients to be very nearly 1.0.

Table 39. Database for example 15.

```
SOLUTION_MASTER_SPECIES
C         CO2           2.0       61.0173        12.0111
Cl        Cl-           0.0       Cl             35.453
Co        Co+2          0.0       58.93          58.93
E         e-            0.0       0.0            0.0
H         H+            -1.       1.008          1.008
H(0)      H2            0.0       1.008
H(1)      H+            -1.       1.008
N         NH4+          0.0       14.0067        14.0067
Na        Na+           0.0       Na             22.9898
Nta       Nta-3         3.0       1.             1.
O         H2O           0.0       16.00          16.00
O(-2)     H2O           0.0       18.016
O(0)      O2            0.0       16.00
SOLUTION_SPECIES
2H2O = O2 + 4H+ + 4e-
        log_k   -86.08; -gamma  1e7    0.0
2 H+ + 2 e- = H2
        log_k   -3.15;  -gamma  1e7    0.0
H+ = H+
        log_k   0.0;    -gamma  1e7    0.0
e- = e-
        log_k   0.0;    -gamma  1e7    0.0
H2O = H2O
        log_k   0.0;    -gamma  1e7    0.0
CO2 = CO2
        log_k   0.0;    -gamma  1e7    0.0
Na+ = Na+
        log_k   0.0;    -gamma  1e7    0.0
Cl- = Cl-
        log_k   0.0;    -gamma  1e7    0.0
Co+2 = Co+2
        log_k   0.0;    -gamma  1e7    0.0
NH4+ = NH4+
        log_k   0.0;    -gamma  1e7    0.0
Nta-3 = Nta-3
        log_k   0.0;    -gamma  1e7    0.0
Nta-3 + 3H+ = H3Nta
        log_k   14.9;   -gamma  1e7    0.0
```

Table 39. Database for example 15.—Continued

```
Nta-3 + 2H+ = H2Nta-
        log_k    13.3;    -gamma  1e7   0.0
Nta-3 + H+ = HNta-2
        log_k    10.3;    -gamma  1e7   0.0
Nta-3 + Co+2 = CoNta-
        log_k    11.7;    -gamma  1e7   0.0
2 Nta-3 + Co+2 = CoNta2-4
        log_k    14.5;    -gamma  1e7   0.0
Nta-3 + Co+2 + H2O = CoOHNta-2 + H+
        log_k    0.5;     -gamma  1e7   0.0
Co+2 + H2O = CoOH+ + H+
        log_k    -9.7;    -gamma  1e7   0.0
Co+2 + 2H2O = Co(OH)2 + 2H+
        log_k    -22.9;   -gamma  1e7   0.0
Co+2 + 3H2O = Co(OH)3- + 3H+
        log_k    -31.5;   -gamma  1e7   0.0
CO2 + H2O = HCO3- + H+
        log_k    -6.35;   -gamma  1e7   0.0
CO2 + H2O = CO3-2 + 2H+
        log_k    -16.68;  -gamma  1e7   0.0
NH4+ = NH3 + H+
        log_k    -9.3;    -gamma  1e7   0.0
H2O = OH- + H+
        log_k    -14.0;   -gamma  1e7   0.0
END
```

Initial and Boundary Conditions

The background concentrations in the column are listed in table 40. The column contains no Nta or cobalt initially, but has a biomass of 1.36×10^{-4} g/L. A flux boundary condition is applied at the inlet of the column, and for the first 20 h (hours), a solution with Nta and cobalt enters the column; the concentrations in the pulse also are given in table 40. After 20 h, the background solution is introduced at the inlet until the experiment ends after 75 h. Na and Cl were not in the original problem definition but were added for charge balancing sorption reactions for PHREEQC (see Sorption Reactions).

Kinetic Degradation of Nta and Cell Growth

Nta is assumed to degrade in the presence of biomass and oxygen by the reaction:

$$HNta^{2-} + 1.62O_2 + 1.272H_2O + 2.424H^+ = 0.576C_5H_7O_2N + 3.12H_2CO_3 + 0.424NH_4^+.$$

PHREEQC requires kinetic reactants to be defined solely by the moles of each element that enter or leave the solution because of the reaction. Furthermore, the reactants should be charge balanced (no net charge should enter or leave the solution). The Nta reaction converts 1 mol $HNta^{2-}$ ($C_6H_7O_6N$) to 0.576 mol

Table 40. Concentration data for example 15.

[g/L, gram per liter; mol/L, mole per liter; Nta, nitrylotriacetate; ---, absent in pulse]

Constituent	Type	Pulse concentration	Background concentration
H^+	Aqueous	10.0e-6 mol/L	10.0e-6 mol/L
Total C	Aqueous	4.9e-7 mol/L	4.9e-7 mol/L
NH_4^+	Aqueous	0.0	0.0
O_2	Aqueous	3.125e-5 mol/L	3.125e-5 mol/L
Nta_3^-	Aqueous	5.23e-6 mol/L	0.0
Co^{2+}	Aqueous	5.23e-6 mol/L	0.0
Na	Aqueous	1.0e-3 mol/L	1.0e-3 mol/L
Cl	Aqueous	1.0e-3 mol/L	1.0e-3 mol/L
Biomass	Immobile	---	1.36e-4 g/L
$CoNta_{(ads)}$	Immobile	---	0.0
$Co_{(ads)}$	Immobile	---	0.0

$C_5H_7O_2N$, where the latter is chemically inert, and its concentration can be ignored. The difference in elemental mass contained in these two reactants provides the stoichiometry of the elements C, H, O, and N in the reaction. This stoichiometry is equal to the sum of the elements on the right-hand side of the equation, excluding $C_5H_7O_2N$, minus the sum of the elements on the left-hand side of the equation. The corresponding change in aqueous element concentrations per mole of $HNta^{2-}$ reaction is given in table 41 (positive coefficients indicate an increase in aqueous concentration, and negative coefficients indicate a decrease in aqueous concentration).

Table 41. Reaction stoichiometry for oxidation of Nta (nitrylotriacetate).

Component	Coefficient
Nta	-1.0
C	3.12
H	1.968
O	4.848
N	0.424

The following multiplicative Monod rate expression is used to describe the rate of Nta degradation:

$$R_{HNTA^{2-}} = -q_m X_m \left(\frac{c_{HNTA^{2-}}}{K_s + c_{HNTA^{2-}}} \right) \left(\frac{c_{O_2}}{K_a + c_{O_2}} \right), \tag{30}$$

where $R_{HNTA^{2-}}$ is the rate of HNta^{2-} degradation (mol L^{-1}h^{-1}, mole per liter per hour), q_m is the maximum specific rate of substrate utilization (mol/g cells/h), X_m is the biomass (g L^{-1}h^{-1}, gram per liter per hour), K_s is the half-saturation constant for the substrate Nta (mol/L), K_a is the half-saturation constant for the electron acceptor O$_2$ (mol/L), and c indicates concentration (mol/L). The rate of biomass production is dependent on the rate of substrate utilization and a first-order decay rate for the biomass:

$$R_{cells} = -YR_{HNTA^{2-}} - bX_m, \tag{31}$$

where R_{cells} is the rate of cell growth (g L^{-1}h^{-1}), Y is the microbial yield coefficient (g cells/mol Nta), and b is the first-order biomass decay coefficient (h^{-1}). The parameter values for these equations are listed in table 42.

Table 42. Kinetic rate parameters used in example 15.
[mol, mole; L, liter; g, gram; h, hour]

Parameter	Description	Parameter value
K_s	Half-saturation constant for donor	7.64e-7 mol/L
K_a	Half-saturation constant for acceptor	6.25e-6 mol/L
q_m	Maximum specific rate of substrate utilization	1.418e-3 mol Nta/g cells/h
Y	Microbial yield coefficient	65.14 g cells/mol Nta
b	First-order microbial decay coefficient	0.00208 h^{-1}

Sorption Reactions

Tebes-Stevens and Valocchi (1997) defined kinetic sorption reactions for Co^{2+} and CoNta$^-$ by the rate equation:

$$R_i = -k_m \left(c_i - \frac{s_i}{K_d} \right), \tag{32}$$

where i is either Co^{2+} or CoNta$^-$ (mol/L), s_i is the sorbed concentration (mol/g sediment), k_m is the mass transfer coefficient (h^{-1}), and K_d is the distribution coefficient (L/g, liter per gram). The values of the coefficients are given in table 43. The values of K_d were defined to give retardation coefficients of 20 and 3 for Co^{2+} and $CoNta^-$, respectively. Because the sorption reactions are defined to be kinetic, the initial moles of these reactants and the rates of reaction are defined with **KINETICS** and **RATES** data blocks; no surface definitions (**SURFACE**, **SURFACE_MASTER_SPECIES**, or **SURFACE_SPECIES**) are needed. Furthermore, all kinetic reactants are immobile, so that the sorbed species are not transported.

Table 43. Sorption coefficients for Co^{2+} and $CoNta^-$.
[h, hour; L/g, liter per gram]

Species	k_m	K_d
Co^{2+}	$1\ h^{-1}$	5.07e-3 L/g
$CoNta^-$	$1\ h^{-1}$	5.33e-4 L/g

When modeling with PHREEQC, kinetic reactants must be charge balanced. For sorption of Co^{2+} and $CoNta^-$, 1 mmol of NaCl was added to the solution definitions to have counter ions for the sorption process. The kinetic sorption reactions were then defined to remove or introduce (depending on the sign of the mole transfer) $CoCl_2$ and NaCoNta, which are charge balanced. To convert from moles sorbed per gram of sediment (s_i) to moles sorbed per liter of water, it is necessary to multiply by the grams of sediment per liter of water, 3.75×10^3 g/L.

Input File

Table 44 shows the input file derived from the preceding problem definition. Although rates have been given in units of mol $L^{-1}h^{-1}$, rates in PHREEQC are always mol/s (mole per second), and all rates have been adjusted to seconds in the definition of rate expressions in the input file. The density of water is assumed to be 1 kg/L.

Table 44. Input file for example 15.

```
DATABASE ex15.dat
TITLE Example 15.--1D Transport: Kinetic Biodegradation, Cell Growth, and Sorption
**********
PLEASE NOTE: This problem requires database file ex15.dat!!
**********
PRINT
        -reset false
        -echo_input true
                      -status false
SOLUTION 0 Pulse solution with NTA and cobalt
        units umol/L
```

Table 44. Input file for example 15.—Continued

```
          pH        6
          C         .49
          O(0)      62.5
          Nta       5.23
          Co        5.23
          Na        1000
          Cl        1000
SOLUTION 1-10 Background solution initially filling column
          units umol/L
          pH        6
          C         .49
          O(0)      62.5
          Na        1000
          Cl        1000
COPY solution 0 100 # for use later on, and in
COPY solution 1 101 # 20 cells model
END
RATES Rate expressions for the four kinetic reactions
#
          HNTA-2
          -start
10 Ks = 7.64e-7
20 Ka = 6.25e-6
30 qm = 1.407e-3/3600
40 f1 = MOL("HNta-2")/(Ks + MOL("HNta-2"))
50 f2 = MOL("O2")/(Ka + MOL("O2"))
60 rate = -qm * KIN("Biomass") * f1 * f2
70 moles = rate * TIME
80 PUT(rate, 1)    # save the rate for use in Biomass rate calculation
90 SAVE moles
          -end
#
          Biomass
          -start
10 Y = 65.14
20 b = 0.00208/3600
30 rate = GET(1)  # uses rate calculated in HTNA-2 rate calculation
40 rate = -Y*rate -b*M
50 moles = -rate * TIME
60 if (M + moles) < 0 then moles = -M
70 SAVE moles
          -end
#
          Co_sorption
          -start
10 km = 1/3600
20 kd = 5.07e-3
30 solids = 3.75e3
40 rate = -km*(MOL("Co+2") - (M/solids)/kd)
50 moles = rate * TIME
60 if (M - moles) < 0 then moles = M
70 SAVE moles
```

Table 44. Input file for example 15.—Continued

```
        -end
#
        CoNta_sorption
        -start
10 km = 1/3600
20 kd = 5.33e-4
30 solids = 3.75e3
40 rate = -km*(MOL("CoNta-") - (M/solids)/kd)
50 moles = rate * TIME
60 if (M - moles) < 0 then moles = M
70 SAVE moles
        -end
KINETICS 1-10 Four kinetic reactions for all cells
        HNTA-2
                -formula C -3.12 H -1.968 O -4.848 N -0.424 Nta 1.
        Biomass
                -formula        H 0.0
                -m              1.36e-4
        Co_sorption
                -formula CoCl2
                -m          0.0
                -tol 1e-11
        CoNta_sorption
                -formula NaCoNta
                -m          0.0
                -tol 1e-11
COPY kinetics 1 101 # to use with 20 cells
END
SELECTED_OUTPUT
        -file    ex15.sel
        -mol     Nta-3 CoNta- HNta-2 Co+2
USER_PUNCH
        -headings          hours  Co_sorb CoNta_sorb      Biomass
        -start
  10 punch TOTAL_TIME/3600 + 3600/2/3600
  20 punch KIN("Co_sorption")/3.75e3
  30 punch KIN("CoNta_sorption")/3.75e3
  40 punch KIN("Biomass")
USER_GRAPH 1 Example 15
        -headings 10_cells: Co+2 CoNTA- HNTA-2 pH
        -chart_title "Kinetic Biodegradation, Cell Growth, and Sorption: Dissolved Spe-
cies"
        -axis_titles "Time, in hours" "Micromoles per kilogram water" "pH"
        -axis_scale x_axis 0 75
        -axis_scale y_axis 0 4
        -axis_scale secondary_y_axis 5.799 6.8 0.2 0.1
        -plot_concentration_vs t
        -start
  10 x = TOTAL_TIME/3600 + 3600/2/3600
  20 PLOT_XY -1, -1, line_width = 0, symbol_size = 0
  30 PLOT_XY x, MOL("Co+2") * 1e6, color = Red, line_width = 0, symbol_size = 4
  40 PLOT_XY x, MOL("CoNta-") * 1e6, color = Green, line_width = 0, symbol_size = 4
```

Table 44. Input file for example 15.—Continued

```
  50 PLOT_XY x, MOL("HNta-2") * 1e6, color = Blue, line_width = 0, symbol_size = 4
  60 PLOT_XY x, -LA("H+"), y-axis = 2, color = Magenta, line_width = 0, symbol_size = 4
        -end
USER_GRAPH 2 Example 15
        -headings 10_cells: Co+2 CoNTA- Biomass
        -chart_title "Kinetic Biodegradation, Cell Growth, and Sorption: Sorbed Spe-
cies"
        -axis_titles "Time, in hours"  "Nanomoles per kilogram water" \
            "Biomass, in milligrams per liter"
        -axis_scale x_axis 0 75
        -axis_scale y_axis 0 2
        -axis_scale secondary_y_axis 0 0.4
        -plot_concentration_vs t
        -start
  10 x = TOTAL_TIME/3600 + 3600/2/3600
  20 PLOT_XY -1, -1, line_width = 0, symbol_size = 0
  30 PLOT_XY x, KIN("Co_sorption") / 3.75e3 * 1e9, color = Red, line_width = 0,
symbol_size = 4
  40 PLOT_XY x, KIN("CoNta_sorption") / 3.75e3 * 1e9, color = Green, line_width = 0, \
        symbol_size = 4
  50 PLOT_XY x, KIN("Biomass") * 1e3, y-axis = 2, color = Magenta, line_width = 0, \
        symbol_size = 4
        -end           -end
TRANSPORT First 20 hours have NTA and cobalt in infilling solution
        -cells                10
        -lengths              1
        -shifts               20
        -time_step            3600
        -flow_direction       forward
        -boundary_conditions  flux flux
        -dispersivities       .05
        -correct_disp         true
        -diffusion_coefficient 0.0
        -punch_cells          10
        -punch_frequency      1
        -print_cells          10
        -print_frequency      5

COPY solution 101 0 # initial column solution becomes influent
END
TRANSPORT Last 55 hours with background infilling solution
        -shifts                55
COPY cell 100 0 # for the 20 cell model...
COPY cell 101 1-20
END
USER_PUNCH
        -start
  10 punch TOTAL_TIME/3600 + 3600/4/3600
  20 punch KIN("Co_sorption")/3.75e3
  30 punch KIN("CoNta_sorption")/3.75e3
  40 punch KIN("Biomass")
        -end
```

Table 44. Input file for example 15.—Continued

```
USER_GRAPH 1
        -headings 20_cells: Co+2 CoNTA- HNTA-2 pH
        -start
  10 x = TOTAL_TIME/3600 + 3600/4/3600
  20 PLOT_XY -1, -1, line_width = 0, symbol_size = 0
  30 PLOT_XY x, MOL("Co+2") * 1e6, color = Red, symbol_size = 0
  40 PLOT_XY x, MOL("CoNta-") * 1e6, color = Green, symbol_size = 0
  50 PLOT_XY x, MOL("HNta-2") * 1e6, color = Blue, symbol_size = 0
  60 PLOT_XY x, -LA("H+"), y-axis = 2, color = Magenta, symbol_size = 0
        -end
USER_GRAPH 2
        -headings 20_cells: Co+2 CoNTA- Biomass
        -start
  10 x = TOTAL_TIME/3600 + 3600/4/3600
  20 PLOT_XY -1, -1, line_width = 0, symbol_size = 0
  30 PLOT_XY x, KIN("Co_sorption") / 3.75e3 * 1e9, color = Red, symbol_size = 0
  40 PLOT_XY x, KIN("CoNta_sorption") / 3.75e3 * 1e9, color = Green, symbol_size = 0
  60 PLOT_XY x, KIN("Biomass") * 1e3, y-axis = 2, color = Magenta, symbol_size = 0
        -end
TRANSPORT First 20 hours have NTA and cobalt in infilling solution
        -cells                  20
        -lengths                0.5
        -shifts                 40
        -initial_time           0
        -time_step              1800
        -flow_direction         forward
        -boundary_conditions    flux  flux
        -dispersivities         .05
        -correct_disp           true
        -diffusion_coefficient  0.0
        -punch_cells            20
        -punch_frequency        2
        -print_cells            20
        -print_frequency        10
COPY cell 101 0
END
TRANSPORT Last 55 hours with background infilling solution
        -shifts                 110
END
```

The 10-meter column was discretized with 10 cells of 1 meter each. The first two **SOLUTION** data blocks (table 44) define the infilling solution and the initial solution in cells 1 through 10. The solutions are copied to solution 100 and 101, to be used later in the 20-cell model.

The **RATES** data block defines the rate expressions for four kinetic reactions: HNta-2, Biomass, Co_sorption, and CoNta_sorption. The rate expressions are initiated with **-start**, defined with numbered Basic-language statements, and terminated with **-end**. The last statement of each expression is SAVE followed by a variable name. This variable is the number of moles of reaction over the time subinterval and is calculated from an instantaneous rate (mol/s) times the length of the time subinterval (s), which is given by the variable "TIME". Lines 30 and 20 in the first and second rate expressions and line 10 in the third and fourth rate expressions adjust parameters to units of seconds from units of hours. The function "MOL" returns the concentration of a species (mol/kgw), the function "M" returns the moles of the reactant for which the rate expression is being calculated, and "KIN" returns the moles of the specified kinetic reactant. The functions "PUT" and "GET" are used to save and retrieve a term that is common to both the HNta-2 and Biomass rate expressions (see also example 6).

The **KINETICS** data block defines the names of the rate expressions that apply to each cell; cells 1 through 10 are defined simultaneously in this example. For each rate expression that applies to a cell, the formula of the reactant (**-formula**) and the moles of the reactant initially present (**-m**, if needed to be different from the default of 1 mol) are defined. It is also possible to define a tolerance (**-tol**), in moles, for the accuracy of the numerical integration for a rate expression. Note that the HNta-2 rate expression generates a negative rate, so that elements with positive coefficients in the formula are removed from solution and negative coefficients add elements to solution. The biomass reaction adds "H 0.0", or zero moles of hydrogen; in other words, it does not add or remove anything from solution. The assimilation of carbon and nutrients that is associated with biomass growth is ignored in this simulation. Also, the kinetics block is copied to **KINETICS** 101, for use in the 20-cell model.

The **SELECTED_OUTPUT** data block punches the molalities of the aqueous species Nta-3, CoNta-, HNta-2 and Co+2 to the file *ex15.sel*. To each line in the file, the **USER_PUNCH** data block appends the time (in hours), the sorbed concentrations converted to mol/g sediment, and the biomass. **USER_GRAPH** plots the concentrations as symbols without lines to facilitate the comparison with the 20-cell model.

The first **TRANSPORT** data block defines the first 20 h of the experiment, during which Nta and cobalt are added at the column inlet. The column is defined to have 10 cells (**-cells**) of length 1 m (**-lengths**). The duration of the advective-dispersive transport simulation is 20 time steps (**-shifts**) of 3,600 seconds (**-time_step**). The direction of flow is forward (**-flow_direction**). Each end of the column is defined to have a flux boundary condition (**-boundary_conditions**). The dispersivity is 0.05 m (**-dispersivities**) and the diffusion coefficient is set to zero (**-diffusion_coefficient**). Data are written to the selected-output file only for cell 10 (**-punch_cells**) after each shift (**-punch_frequency**), and data are written to the output file only for cell 10 (**-print_cells**) after each fifth shift (**-print_frequency**).

At the end of the first advective-dispersive transport simulation, the initial column solution, which was stored as **SOLUTION** 101, is copied to **SOLUTION** 0, to become the influent for the next transport simulation. The second TRANSPORT data block defines the final 55 h of the experiment, during which Nta and cobalt are not present in the infilling solution. All parameters are the same as in the previous TRANSPORT data block; only the number of transport steps (**-shifts**) is increased to 55.

Grid Convergence

With advective-dispersive-reactive transport simulations, it is always necessary to check the numerical accuracy of the results. In general, analytical solutions will not be available for these complex simulations, so the only test of numerical accuracy is to refine the grid and time step, rerun the simulation, and compare the results. If simulations on two different grids give similar results, there is some assurance that the numerical errors are relatively small. If simulations on two different grids give significantly different results, the grid must be refined again and the process repeated. Unfortunately, doubling the grid size at least quadruples the number of solution calculations that must be made because the number of cells doubles and the time step is halved. If the cell size approaches the size of the dispersivity, it may require even more solution calculations because the number of mix steps in the dispersion calculation will increase as well.

To test grid convergence in this example, the number of cells in the column were doubled. All keyword data blocks that defined compositions for the range 1 through 10 were changed to 1 through 20. In addition, the parameters for advective-dispersive transport were adjusted to be consistent with the new number of cells. The final TRANSPORT data block in table 44 defines the 20-cell model. The number of cells and number of shifts are doubled; the cell length and time step are halved. To print information for the same location as the 10-cell model (the end of the column), the **-punch_cells** and **-print_cells** are set to cell 20. To print information at the same time in the simulation as the 10-cell model, **-punch_frequency** is set to every 2 shifts, **-print_frequency** is set to every 10 shifts, and the time step for going from the cell-midpoint to the column-end is halved on line 10 in USER_PUNCH.

Results

The distributions of aqueous and immobile constituents in the column at the end of 75 h are shown in figures 16 and 17 for the 10- and 20-cell models. In the experiment, two pore volumes of water with Nta and cobalt were introduced to the column over the first 20 h and then followed by 5.5 pore volumes of background water over the next 55 h. At 10 h, $HNta^{2-}$ begins to appear at the column outlet along with a rise in the pH (fig. 16). If Nta and cobalt were conservative and dispersion were negligible, the graph would

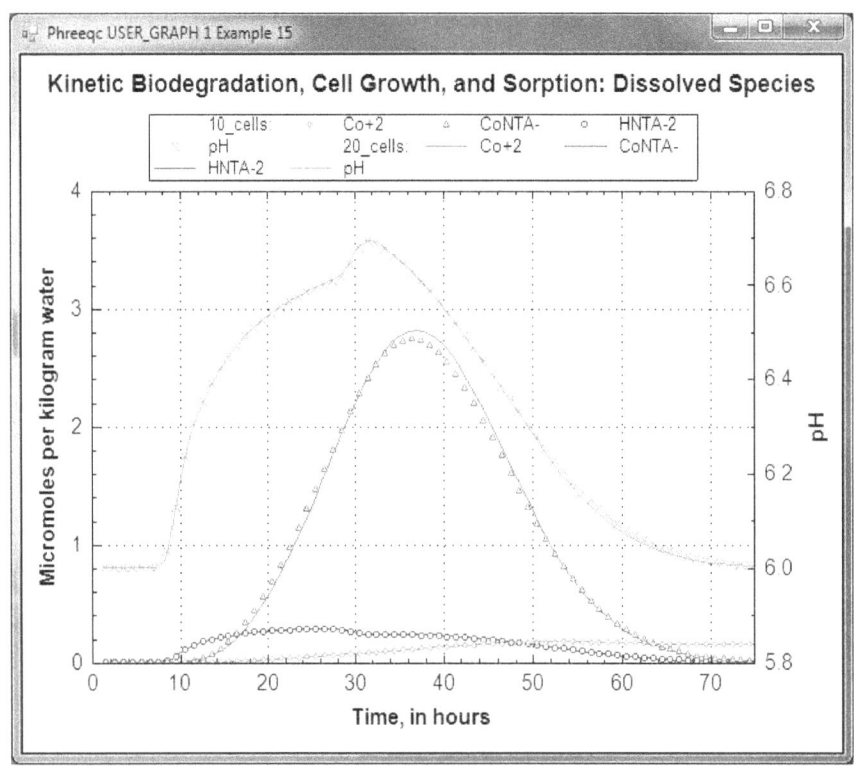

Figure 16. Dissolved concentrations and pH values at the outlet of the column for Nta and cobalt transport simulations with 10 (symbols) and 20 cells (lines).

show square pulses that increase at 10 h and decrease at 30 h. However, the movement of the Nta and cobalt is retarded relative to conservative movement by the sorption reactions, and small concentrations arrive early because of dispersion. The peak in Nta and cobalt concentrations occurs in the CoNta⁻ complex between 30 and 40 h. The peak in Co^{2+} concentration is even more retarded by its sorption reaction and does not show up until near the end of the experiment.

In figure 17, solid-phase concentrations are plotted against time for concentrations in the last cell of the column. The sorbed CoNta⁻ concentration peaks between 30 and 40 h and lags slightly behind the peak in the dissolved concentration of the CoNta⁻ complex. Initially, no Nta is present in the column and the biomass decreases slightly over the first 10 h because of the first-order decay rate for the biomass. When the Nta moves through the cells, the biomass increases because the Nta becomes available as substrate for microbes. After the peak in Nta has moved through the column, biomass concentrations level off and then begin to decrease because of decay. The K_d for cobalt sorption gives a greater retardation coefficient than the K_d for CoNta⁻ sorption, and the sorbed concentration of Co^{2+} appears to be still increasing at the end of the experiment.

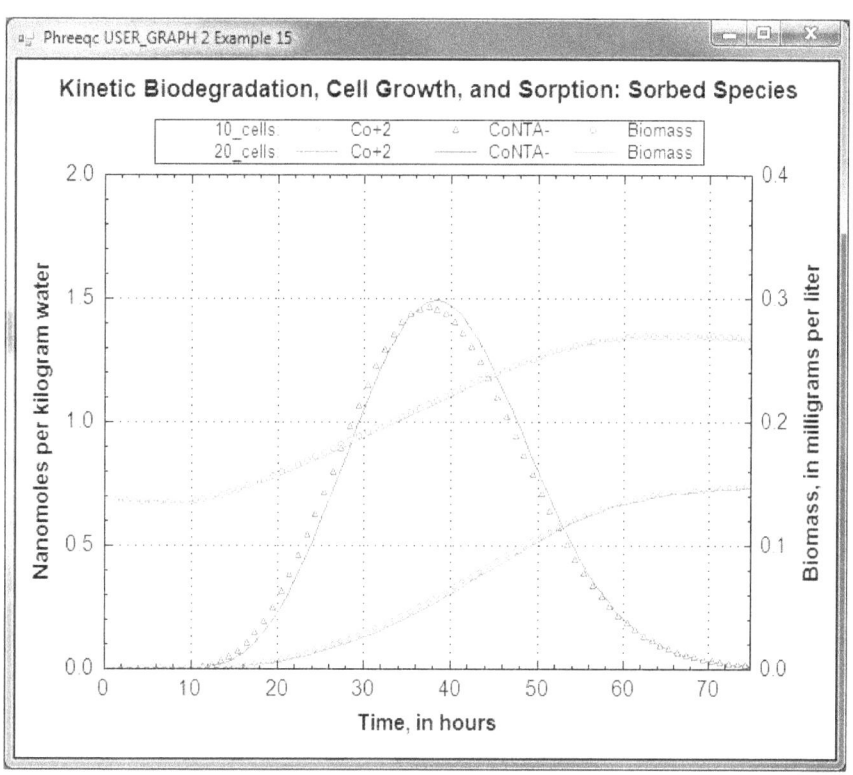

Figure 17. Concentrations of sorbed species and biomass at the outlet of the column for Nta and cobalt transport simulations with 10 (symbols) and 20 cells (lines).

Both the 10-cell and the 20-cell models give similar results, which indicates that the numerical errors in the advective-dispersive transport simulation are relatively small; furthermore, the results are very similar to results given by Tebes-Stevens and Valocchi (1997) and Tebes-Stevens and others (1998). However, Tebes-Stevens and Valocchi (1997) included another part to their test problem that increased the rate constants for the sorption reactions from 1 to 1,000 h^{-1}. The increased rate constants generate a stiff set of partial differential equations, which are equations that describe processes that occur on very different time scales. The stiff problem, with very fast sorption reactions, proved intractable for the explicit Runge-Kutta algorithm, but can be solved with the **-cvode** algorithm. With **-cvode**, the calculation of the slow sorption column takes about four times longer than with Runge-Kutta and calculation of the fast sorption column is about six times longer than for slow sorption. Another way to solve the stiff problem is to introduce the fast kinetic sorption reaction as an equilibrium process, which constitutes instantaneous rates. However, even with equilibrium sorption, grid convergence was computationally much more intensive; it was necessary to use 100 cells or more to arrive at a satisfactory solution. As an estimate of relative CPU times, the 20-cell model took 2.7 times the CPU time of the 10-cell model. A 200-cell model took approximately 600 times the CPU time of the 10-cell model.

Example 16—Inverse Modeling of Sierra Spring Waters

This example repeats the inverse modeling calculations of the chemical evolution of spring-water compositions in the Sierra Nevada that are described in a classic paper by Garrels and Mackenzie (1967). The same example is described in the manual for the inverse-modeling program NETPATH (Plummer and others, 1991 and 1994). The example uses two spring-water compositions, one from an ephemeral spring, which is less chemically evolved, and one from a perennial spring, which probably has had a longer residence time in the subsoil. The differences in composition between the ephemeral spring and the perennial spring are assumed to be caused by reactions between the water and the minerals and gases it contacts. The object of inverse modeling in this example is to find sets of minerals and gases that, when reacted in appropriate amounts, account for the differences in composition between the two solutions

NETPATH (Plummer and others, 1991 and 1994) and PHREEQC are both capable of performing inverse-modeling calculations. NETPATH has the advantage relative to PHREEQC that it provides isotope fractionation and carbon-14 dating, whereas PHREEQC, in inverse models, can only calculate isotope mole-balances. PHREEQC can calculate isotope fractionation in forward models (see example 20), but this requires a separate simulation and a more complicated setup than NETPATH. The major advantage of inverse modeling with PHREEQC relative to NETPATH is the capability to include uncertainties in the analytical data that are used in the calculation of inverse models. This capability produces inverse models that are more robust, that is, small changes in input data do not produce large differences in model-calculated mole transfers. Another advantage of PHREEQC is that any set of elements may be included in the inverse-modeling calculations, whereas NETPATH is limited to a selected, though relatively comprehensive, set of elements.

The analytical data for the two springs are given in table 45. The chemical compositions of minerals and gases postulated to react by Garrels and Mackenzie (1967) and their mole transfers are given in table 46. The selection of the identity and composition of the reactive phases, and the variation in space and time, is the most difficult part of inverse modeling. In general, the selection is made by knowledge of the

Table 45. Analytical data for spring waters in example 16.
[Analyses in millimoles per liter from Garrels and Mackenzie (1967)]

Spring	pH	SiO_2	Ca^{2+}	Mg^{2+}	Na^+	K^+	HCO_3^-	SO_4^{2-}	Cl^-
Ephemeral spring	6.2	0.273	0.078	0.029	0.134	0.028	0.328	0.010	0.014
Perennial spring	6.8	0.410	0.260	0.071	0.259	0.040	0.895	0.025	0.030

Table 46. Reactant compositions and mole transfers given by Garrels and Mackenzie (1967).
[Mole transfer in millimoles per kilogram water, positive numbers indicate dissolution and negative numbers indicate precipitation]

Reactant	Composition	Mole transfer
Halite	NaCl	0.016
Gypsum	$CaSO_4 \, 2H_2O$	0.015
Kaolinite	$Al_2Si_2O_5(OH)_4$	-0.033
Ca-Montmorillonite	$Ca_{0\,17}Al_{2\,33}Si_{3\,67}O_{10}(OH)_2$	-0.081
CO2gas	CO_2	0.427
Calcite	$CaCO_3$	0.115
Silica	SiO_2	0.0
Biotite	$KMg_3AlSi_3O_{10}(OH)_2$	0.014
Plagioclase	$Na_{0\,62}Ca_{0\,38}Al_{1\,38}Si_{2\,62}O_8$	0.175

flow system and the mineralogy along the flow path, although microscopic and chemical analysis of the aquifer material and isotopic composition of the water and minerals may provide additional insight for the selection of reactants. It is not necessary to know exactly which minerals are reacting, but it is necessary to have a comprehensive list of potential reactants.

The input file for this example is given in table 47. The SOLUTION_SPREAD data block is used to define the two spring waters. The INVERSE_MODELING data block defines the inverse-modeling calculations, including the solutions and phases to be used, the mole-balance equations, the uncertainty limits, whether all or only "minimal" models will be printed, and whether ranges of mole transfer that are consistent with the uncertainty limits will be calculated. A series of identifiers (sub-keywords preceded by a hyphen) specify the characteristics of the inverse model.

Table 47. Input file for example 16.

```
TITLE Example 16.--Inverse modeling of Sierra springs
SOLUTION_SPREAD
        -units mmol/L
# "\t" indicates tab
Number\t pH\t     Si\t     Ca\t     Mg\t     Na\t      K\t Alkalinity\t  S(6)\t     Cl
1\t      6.2\t  0.273\t  0.078\t  0.029\t  0.134\t  0.028\t      0.328\t  0.01\t   0.014
2\t      6.8\t  0.41 \t  0.26 \t  0.071\t  0.259\t  0.04 \t      0.895\t  0.025\t   0.03
INVERSE_MODELING 1
        -solutions 1 2
        -uncertainty 0.025
        -balances
                Ca       0.05      0.025
```

Table 47. Input file for example 16.—Continued

```
       -phases
              Halite
              Gypsum
              Kaolinite                   precip
              Ca-montmorillonite          precip
              CO2(g)
              Calcite
              Chalcedony                  precip
              Biotite                     dissolve
              Plagioclase                 dissolve
       -range
PHASES
Biotite
    KMg3AlSi3O10(OH)2 + 6H+ + 4H2O = K+ + 3Mg+2 + Al(OH)4- + 3H4SiO4
       log_k  0.0      # No log_k, Inverse modeling only
Plagioclase
    Na0.62Ca0.38Al1.38Si2.62O8 + 5.52 H+ + 2.48H2O =\
                0.62Na+ + 0.38Ca+2 + 1.38Al+3 + 2.62H4SiO4
       log_k  0.0      # No log_k, Inverse modeling only
END
```

The identifier **-solutions** selects the solutions to be used by solution number. Two or more solution numbers must be listed after the identifier. If only two solution numbers are given, the second solution is assumed to evolve from the first solution. If more than two solution numbers are given, the last solution listed is assumed to evolve from a mixture of the preceding solutions. The solutions to be used in inverse modeling are defined in the same way as any solutions used in PHREEQC models. Usually the analytical data are entered in the SOLUTION or SOLUTION_SPREAD data block, but solutions defined by batch-reaction calculations in the current or previous simulations may be used if they are saved with the SAVE keyword.

The **-uncertainty** identifier sets the default uncertainty limit for each analytical datum. In this example a fractional uncertainty limit of 0.025 (2.5 percent, default is 5 percent) is assumed for all of the analytical data except pH. By default, the uncertainty limit for pH is 0.05 unit. The uncertainty limit for pH can be set to an absolute value (standard units) with the **-balances** identifier. The uncertainty limit for other items in any of the solutions can be set explicitly to a fractional value, or to an absolute number (enter a negative number of moles or equivalents for alkalinity), also by using the **-balances** identifier.

By default, every inverse model includes mole-balance equations for every element in any of the phases included in **-phases** (except hydrogen and oxygen). If mole-balance equations are needed for elements not in the phases, that is, for elements with no source or sink (conservative mixing), the **-balances** identifier must be used to include those elements in the inverse-modeling equations (see example 17). In addition, the **-balances** identifier can be used to specify uncertainty limits for an element in each solution, as noted before. For demonstration, the uncertainty limit for calcium is set to 0.05 (5 percent) in solution 1 and 0.025 (2.5 percent) in solution 2 in example 16.

The phases to be used in the inverse-modeling calculations are defined with the **-phases** identifier, and may be constrained to dissolve only or precipitate only. In this example, kaolinite, Ca-montmorillonite, and chalcedony (SiO_2) are forced to precipitate only. Biotite and plagioclase are forced to dissolve (positive mole transfer) if they are present in an inverse model.

All of the phases used in inverse modeling must have been defined in the PHASES data block or as species in the EXCHANGE_SPECIES data block, either in the database file or in the input file. Thus, all phases defined in the database files are available for use in inverse modeling. Because biotite and plagioclase are not in the default database file *phreeqc.dat*, they are defined explicitly in the PHASES data block in the input file. For simplicity, the log Ks are set to zero for these phases, which does not affect inverse modeling because only the mineral stoichiometry is used. However, the saturation indices calculated for these phases will be spurious. The phases used in inverse modeling must have a charge-balanced reaction because of the charge-balance constraint for each solution. Each solution is adjusted to charge balance by adjusting the concentrations of the elements within their uncertainty limits while minimizing the objective function of the optimization method (see Parkhurst and Appelo, 1999, "Equations and Numerical Method for Inverse Modeling"). If a solution cannot be charge balanced, the solution will be noted in the output and the program will indicate that models cannot be found. Note that the reaction for plagioclase (table 47) is written on two lines, but interpreted to be on a single line because of the backslash "\" at the end of the first line. The **-range** identifier indicates that for each model the range of mole transfers will be calculated as constrained by the uncertainty limits.

In each inverse model, linear equations are solved for the mole balances of each element or valence state, the alkalinity balance, the electron balance, the water balance in the system, and the charge balance for each solution. The unknowns in the set of linear equations are the mole transfers for phases and redox reactions, and the concentration adjustments for each element in each solution, including alkalinity and pH (but excluding hydrogen and oxygen).

Results for the two inverse models found in this example are shown in table 48. The results begin with a listing of three columns for each solution showing concentrations in mol/kgw or pH of the original solution (column headed by Input), the added uncertainty (column Delta), and the sum of the two (column Input+Delta).

Table 48. Selected output for example 16.

Solution 1:

```
                    Input        Delta      Input+Delta
        pH      6.200e+00   +   1.246e-02   =   6.212e+00
        Al      0.000e+00   +   0.000e+00   =   0.000e+00
 Alkalinity     3.280e-04   +   5.500e-06   =   3.335e-04
      C(-4)     0.000e+00   +   0.000e+00   =   0.000e+00
       C(4)     7.825e-04   +   0.000e+00   =   7.825e-04
         Ca     7.800e-05   +  -3.900e-06   =   7.410e-05
         Cl     1.400e-05   +   0.000e+00   =   1.400e-05
       H(0)     0.000e+00   +   0.000e+00   =   0.000e+00
          K     2.800e-05   +  -7.000e-07   =   2.730e-05
         Mg     2.900e-05   +   0.000e+00   =   2.900e-05
         Na     1.340e-04   +   0.000e+00   =   1.340e-04
       O(0)     0.000e+00   +   0.000e+00   =   0.000e+00
      S(-2)     0.000e+00   +   0.000e+00   =   0.000e+00
       S(6)     1.000e-05   +   0.000e+00   =   1.000e-05
         Si     2.730e-04   +   0.000e+00   =   2.730e-04
```

Solution 2:

```
                    Input        Delta      Input+Delta
        pH      6.800e+00   +  -3.407e-03   =   6.797e+00
        Al      0.000e+00   +   0.000e+00   =   0.000e+00
 Alkalinity     8.951e-04   +  -1.796e-06   =   8.933e-04
      C(-4)     0.000e+00   +   0.000e+00   =   0.000e+00
       C(4)     1.199e-03   +   0.000e+00   =   1.199e-03
         Ca     2.600e-04   +   6.501e-06   =   2.665e-04
         Cl     3.000e-05   +   0.000e+00   =   3.000e-05
       H(0)     0.000e+00   +   0.000e+00   =   0.000e+00
          K     4.000e-05   +   1.000e-06   =   4.100e-05
         Mg     7.101e-05   +  -8.979e-07   =   7.011e-05
         Na     2.590e-04   +   0.000e+00   =   2.590e-04
       O(0)     0.000e+00   +   0.000e+00   =   0.000e+00
      S(-2)     0.000e+00   +   0.000e+00   =   0.000e+00
       S(6)     2.500e-05   +   0.000e+00   =   2.500e-05
         Si     4.100e-04   +   0.000e+00   =   4.100e-04
```

```
Solution fractions:                 Minimum        Maximum
  Solution   1    1.000e+00        1.000e+00      1.000e+00
  Solution   2    1.000e+00        1.000e+00      1.000e+00
```

```
Phase mole transfers:               Minimum        Maximum
        Halite    1.600e-05        1.490e-05      1.710e-05   NaCl
        Gypsum    1.500e-05        1.413e-05      1.588e-05   CaSO4:2H2O
     Kaolinite   -3.392e-05       -5.587e-05     -1.224e-05   Al2Si2O5(OH)4
Ca-Montmorillon  -8.090e-05       -1.100e-04     -5.154e-05   Ca0.165Al2.33Si3.67O10(OH)2
       CO2(g)     2.928e-04        2.363e-04      3.563e-04   CO2
       Calcite    1.240e-04        1.007e-04      1.309e-04   CaCO3
       Biotite    1.370e-05        1.317e-05      1.370e-05   KMg3AlSi3O10(OH)2
   Plagioclase    1.758e-04        1.582e-04      1.935e-04   Na0.62Ca0.38Al1.38Si2.62O8
```

Table 48. Selected output for example 16.—Continued

```
Redox mole transfers:

Sum of residuals (epsilons in documentation):         5.574e+00
Sum of delta/uncertainty limit:                       5.574e+00
Maximum fractional error in element concentration:    5.000e-02

Model contains minimum number of phases.
================================================================================


Solution 1:

                        Input          Delta      Input+Delta
            pH       6.200e+00  +   1.246e-02  =   6.212e+00
            Al       0.000e+00  +   0.000e+00  =   0.000e+00
    Alkalinity       3.280e-04  +   5.500e-06  =   3.335e-04
         C(-4)       0.000e+00  +   0.000e+00  =   0.000e+00
          C(4)       7.825e-04  +   0.000e+00  =   7.825e-04
            Ca       7.800e-05  +  -3.900e-06  =   7.410e-05
            Cl       1.400e-05  +   0.000e+00  =   1.400e-05
          H(0)       0.000e+00  +   0.000e+00  =   0.000e+00
             K       2.800e-05  +  -7.000e-07  =   2.730e-05
            Mg       2.900e-05  +   0.000e+00  =   2.900e-05
            Na       1.340e-04  +   0.000e+00  =   1.340e-04
          O(0)       0.000e+00  +   0.000e+00  =   0.000e+00
         S(-2)       0.000e+00  +   0.000e+00  =   0.000e+00
          S(6)       1.000e-05  +   0.000e+00  =   1.000e-05
            Si       2.730e-04  +   0.000e+00  =   2.730e-04

Solution 2:

                        Input          Delta      Input+Delta
            pH       6.800e+00  +  -3.407e-03  =   6.797e+00
            Al       0.000e+00  +   0.000e+00  =   0.000e+00
    Alkalinity       8.951e-04  +  -1.796e-06  =   8.933e-04
         C(-4)       0.000e+00  +   0.000e+00  =   0.000e+00
          C(4)       1.199e-03  +   0.000e+00  =   1.199e-03
            Ca       2.600e-04  +   6.501e-06  =   2.665e-04
            Cl       3.000e-05  +   0.000e+00  =   3.000e-05
          H(0)       0.000e+00  +   0.000e+00  =   0.000e+00
             K       4.000e-05  +   1.000e-06  =   4.100e-05
            Mg       7.101e-05  +  -8.980e-07  =   7.011e-05
            Na       2.590e-04  +   0.000e+00  =   2.590e-04
          O(0)       0.000e+00  +   0.000e+00  =   0.000e+00
         S(-2)       0.000e+00  +   0.000e+00  =   0.000e+00
          S(6)       2.500e-05  +   0.000e+00  =   2.500e-05
            Si       4.100e-04  +   0.000e+00  =   4.100e-04

Solution fractions:                    Minimum         Maximum
   Solution   1     1.000e+00        1.000e+00        1.000e+00
   Solution   2     1.000e+00        1.000e+00        1.000e+00

Phase mole transfers:                  Minimum         Maximum
        Halite       1.600e-05        1.490e-05        1.710e-05   NaCl
        Gypsum       1.500e-05        1.413e-05        1.588e-05   CaSO4:2H2O
     Kaolinite      -1.282e-04       -1.403e-04       -1.159e-04   Al2Si2O5(OH)4
        CO2(g)       3.061e-04        2.490e-04        3.703e-04   CO2
```

Table 48. Selected output for example 16.—Continued

```
        Calcite      1.106e-04      8.680e-05      1.182e-04    CaCO3
      Chalcedony     -1.084e-04     -1.473e-04     -6.906e-05   SiO2
         Biotite     1.370e-05      1.317e-05      1.370e-05    KMg3AlSi3O10(OH)2
      Plagioclase    1.758e-04      1.582e-04      1.935e-04    Na0.62Ca0.38Al1.38Si2.62O8

Redox mole transfers:

Sum of residuals (epsilons in documentation):          5.574e+00
Sum of delta/uncertainty limit:                        5.574e+00
Maximum fractional error in element concentration:     5.000e-02

Model contains minimum number of phases.
=============================================================================


Summary of inverse modeling:

        Number of models found: 2
        Number of minimal models found: 2
        Number of infeasible sets of phases saved: 20
        Number of calls to cl1: 62
```

Next, the relative fractions of each solution in the inverse model are printed. With only two solutions in the model, the fraction for each solution will be 1.0, usually. The fractions are derived from the mole balance on water, so if hydrated minerals consume or produce significant amounts of water or if evaporation is modeled (see example 17), the numbers may not sum to 1.0. In this example, all fractions are 1.0 because the amount of water from gypsum dissolution is too small to affect the four digits used for printing of the mixing fractions. The second and third columns of the block give the minimum and maximum fractional values that can be attained within the uncertainty limits. These two columns are printed if the **-range** identifier is used.

The next block of data lists the phase mole transfers. The first column contains the optimized transfers, which, when added to the adjusted concentrations in solution 1, exactly reproduce the adjusted concentrations in solution 2. Positive mole transfers indicate dissolution; negative mole transfers indicate precipitation. The second and third columns of the mole transfers are minimum and maximum values that can be attained within the uncertainty limits. These two columns are printed if the **-range** identifier is used. In general, these minima and maxima are highly correlated; that is, obtaining a maximum mole transfer of one phase may require a minimum mole transfer of another phase as well as maximum adjustments of element concentrations.

No redox mole transfers were calculated in this inverse model. Consequently, the block headed by "Redox mole transfers" is empty.

The next block of data prints the extent to which the analytical data were adjusted. If no adjustments were made, the three printed numbers would be zero. In table 48, the third number gives the "Maximum fractional error in an element concentration" as 0.05, which, in both models, is the maximum permitted error of 5 percent and applies to Ca in solution 1. The second number, the "Sum of delta/uncertainty limit", is the sum of all the numbers listed in the column labeled with "Delta", divided by the corresponding uncertainty limit. The first number, the "Sum of residuals", sums up the same values after multiplication with the fraction of the solution (in this example, because the fractions of the solutions are 1, the first and second numbers are the same). This "Sum of residuals" is minimized by the Simplex algorithm in PHREEQC. If no inverse model can be found with a proper subset of the solutions and phases in this model, the model is a "minimal model", and the statement "Model contains minimum number of phases" is printed.

Finally, a short summary of the calculations is printed. The summary includes the number of models, the number of minimal models (models with a minimum number of phases), the number of infeasible models that were tested, and the number of calls to the inequality equations solver, cl1 (calculation time is generally proportional to the number of calls to cl1).

The results of the example show that two inverse models exist using the phases suggested by Garrels and Mackenzie (1967). The main reactions are dissolution of calcite, plagioclase and carbon dioxide; kaolinite and Ca-montmorillonite precipitate in the first model, and kaolinite and chalcedony precipitate in the second model. Small amounts of halite, gypsum, and biotite dissolution are required in the models. The results of Garrels and Mackenzie (1967) fall within the range of mole transfers calculated in the first model of PHREEQC for all phases except carbon dioxide. The carbon dioxide mole transfer differs from Garrels and Mackenzie (1967) because they did not account for the dissolved carbon dioxide in the spring waters. Garrels and Mackenzie (1967) also ignored a small discrepancy in the mole balance for potassium. PHREEQC avoids this imbalance by adjusting concentrations in the two solutions. The PHREEQC calculations show that two inverse models can be found by adjusting concentrations by no more than the specified uncertainty limits (5 percent for Ca in solution 1). The PHREEQC calculations show the discrepancy in potassium can be accounted for by minor adjustments of concentrations, which indicates the discrepancy is not significant. This assessment of significance would be difficult without the uncertainty capabilities of inverse modeling with PHREEQC. The results of PHREEQC are concordant with the results of NETPATH.

Example 17—Inverse Modeling With Evaporation

Evaporation is handled in the same manner as other heterogeneous reactions for inverse modeling. To model evaporation (or dilution), it is necessary to include a phase with the composition "H2O". The important concept in modeling evaporation is the water mole-balance equation (see Parkhurst and Appelo, 1999, "Equations and Numerical Method for Inverse Modeling"). The moles of water in the initial solutions times their mixing fractions, plus water gained or lost by dissolution or precipitation of phases, plus water gained or lost through redox reactions, must equal the moles of water in the final solution. The equation is still approximate because it does not include the moles of water gained or lost in hydrolysis and complexation reactions in the solutions. The results of inverse modeling are compared with a forward model using Pitzer equations to calculate the sequence of salts that precipitate during evaporation.

This example uses data for the evaporation of Black Sea water that is presented in Carpenter (1978). Two analyses are selected, the initial Black Sea water and an evaporated water from which halite has precipitated. The hypothesis is that evaporation, precipitation of calcite, gypsum, and halite, and loss of carbon dioxide are sufficient to account for the changes in water composition of all of the major ions and bromide. The input file (table 49) contains the solution compositions in the **SOLUTION** data blocks. The total carbon in the solutions is unknown but is estimated by assuming that both solutions are in equilibrium with atmospheric carbon dioxide.

Table 49. Input file for example 17.

```
DATABASE ../database/pitzer.dat
TITLE Example 17.--Inverse modeling of Black Sea water evaporation
SOLUTION 1  Black Sea water
        units   mg/L
        density 1.014
        pH      8.0     # estimated
        Ca      233
        Mg      679
        Na      5820
        K       193
        S(6)    1460
        Cl      10340
        Br      35
        C       1       CO2(g) -3.5
SOLUTION 2  Composition during halite precipitation
        units   mg/L
        density 1.271
        pH      5.0     # estimated
        Ca      0.0
        Mg      50500
        Na      55200
```

Table 49. Input file for example 17.—Continued

```
        K          15800
        S(6)       76200
        Cl         187900
        Br         2670
        C          1           CO2(g)  -3.5
INVERSE_MODELING
        -solution 1 2
        -uncertainties .025
        -range
        -balances
                Br
                K
                Mg
        -phases
                H2O(g)   pre
                Calcite pre
                CO2(g)   pre
                Gypsum   pre
                Halite   pre
                Glauberite pre
                Polyhalite pre
END
```

The **INVERSE_MODELING** keyword defines the inverse model for this example. Solution 2, the solution during halite precipitation, evolves from solution 1, Black Sea water. Uncertainty limits of 2.5 percent are applied to all data. Water, calcite, carbon dioxide, gypsum, and halite are specified to be the potential reactants (**-phases**) that must precipitate, that is, must be removed from the aqueous phase.

Mole-balance equations for water, alkalinity, and electrons are always included in the inverse formulation. In addition, mole-balance equations are included for all the elements in the specified phases, in this case, for calcium, carbon, sulfur, sodium, and chloride. The **-balances** identifier is used to specify additional mole-balance equations for bromide, magnesium, and potassium. In the absence of alkalinity data, the calculated alkalinity of the solutions is controlled by the pH and the assumption that the solutions are in equilibrium with atmospheric carbon dioxide. Here, alkalinity is a minor contributor to charge balance.

Only one model is found in the inverse calculation. This model indicates that Black Sea water (solution 1) must be concentrated 88-fold to produce solution 2, as shown by the fractions of the two solutions in the inverse-model output (table 50). Thus, approximately 88 kg of water in Black Sea water is reduced to 1 kg of water in solution 2. Halite precipitates (19.75 mol) and gypsum precipitates (0.48 mol) during the evaporation process. Note that these mole transfers are relative to 88 kg of water. To find the loss per kilogram water in Black Sea water, it is necessary to divide by the mixing fraction of solution 1. The result is that 54.9 mol of water, 0.0004 mol of calcite, 0.0004 mol carbon dioxide, 0.0054 mol of gypsum, and 0.22 mol of halite have been removed per kilogram of Black Sea water. (This calculation could be accomplished by making solution 1 from solution 2, taking care to reverse the constraints on minerals from precipitation to dissolution.) All the other ions—magnesium, potassium, and bromide—are conservative within the 2.5-percent uncertainty limit that was specified. The inverse modeling shows that, with the given uncertainty limits, evaporation (loss of water), carbon dioxide outgassing, and calcite, halite, and gypsum precipitation can explain all of the changes in major ion composition.

Table 50. Selected output for example 17.

```
Solution 1: Black Sea water

                        Input            Delta        Input+Delta
            pH      8.000e+000   +    0.000e+000   =   8.000e+000
    Alkalinity      8.684e-004   +    0.000e+000   =   8.684e-004
            Br      4.401e-004   +    0.000e+000   =   4.401e-004
          C(4)      8.453e-004   +    0.000e+000   =   8.453e-004
            Ca      5.841e-003   +    0.000e+000   =   5.841e-003
            Cl      2.930e-001   +    8.006e-004   =   2.938e-001
             K      4.960e-003   +    1.034e-004   =   5.063e-003
            Mg      2.807e-002   +   -7.018e-004   =   2.737e-002
            Na      2.544e-001   +    0.000e+000   =   2.544e-001
          S(6)      1.527e-002   +    7.486e-005   =   1.535e-002

Solution 2: Composition during halite precipitation

                        Input            Delta        Input+Delta
            pH      5.000e+000   +    9.033e-013   =   5.000e+000
    Alkalinity      7.758e-006   +    0.000e+000   =   7.758e-006
            Br      3.785e-002   +    9.440e-004   =   3.880e-002
          C(4)      7.206e-006   +    0.000e+000   =   7.206e-006
            Ca      0.000e+000   +    0.000e+000   =   0.000e+000
            Cl      6.004e+000   +    1.501e-001   =   6.154e+000
             K      4.578e-001   +   -1.144e-002   =   4.464e-001
            Mg      2.354e+000   +    5.884e-002   =   2.413e+000
            Na      2.720e+000   +   -4.642e-002   =   2.674e+000
          S(6)      8.986e-001   +   -2.247e-002   =   8.761e-001

Solution fractions:                    Minimum           Maximum
    Solution   1    8.815e+001       8.780e+001        8.815e+001
    Solution   2    1.000e+000       1.000e+000        1.000e+000

Phase mole transfers:                  Minimum           Maximum
        H2O(g)      -4.837e+003      -4.817e+003      -4.817e+003   H2O
       Calcite      -3.827e-002      -3.923e-002      -3.716e-002   CaCO3
        CO2(g)      -3.624e-002      -3.737e-002      -3.497e-002   CO2
        Gypsum      -4.767e-001      -4.905e-001      -4.609e-001   CaSO4:2H2O
        Halite      -1.975e+001      -2.033e+001      -1.901e+001   NaCl
```

Table 50. Selected output for example 17.—Continued

```
Redox mole transfers:

Sum of residuals (epsilons in documentation):      1.943e+002
Sum of delta/uncertainty limit:                    7.820e+000
Maximum fractional error in element concentration: 2.500e-002


Model contains minimum number of phases.
================================================================================


Summary of inverse modeling:

             Number of models found: 1
             Number of minimal models found: 1
             Number of infeasible sets of phases saved: 11
             Number of calls to cl1: 29
```

The inverse model can be compared with a forward model that calculates the sequence of salts in evaporating seawater using Pitzer's equations for solute activities at high ionic strength (Hardie, 1991). The input file is in table 51. The input file allows precipitation (and possibly redissolution) of 12 phases, including carbonates, sulfates, and chlorides. At any point in the evaporation, if the number of moles of a precipitate is less than 1×10^{-5}, the number of moles is plotted as 1×10^{-5} by **USER_GRAPH**.

Table 51. Input file for example 17B.

```
DATABASE ../database/pitzer.dat
SOLUTION 1  Black Sea water
        units    mg/L
        density 1.014
        pH       8.0      # estimated
        Ca       233
        Mg       679
        Na       5820
        K        193
        S(6)     1460
        Cl       10340
        Br       35
        C        1       CO2(g) -3.5
EQUILIBRIUM_PHASES
 # carbonates...
 CO2(g) -3.5 10; Calcite 0 0
 # sulfates...
 Gypsum 0 0;     Anhydrite 0 0;  Glauberite 0 0;  Polyhalite 0 0
 Epsomite 0 0;   Kieserite 0 0;  Hexahydrite 0 0
 # chlorides...
 Halite 0 0;     Bischofite 0 0; Carnallite 0 0
USER_GRAPH Example 17B
```

Table 51. Input file for example 17B—Continued

```
-head H2O Na K Mg Ca Cl SO4 Calcite Gypsum Anhydrite Halite\
    Glauberite Polyhalite
-init false
-axis_scale x_axis 0 100
-axis_scale y_axis -5 1. 1
-axis_scale sy_axis -5 10 5 100
-axis_titles "Concentration factor" "Log(Molality)"  "Log(Moles of solid)"
-chart_title "Evaporating Black Sea water"
-start
10 graph_x 1 / tot("water")
20 graph_y log10(tot("Na")), log10(tot("K")), log10(tot("Mg")), log10(tot("Ca")),\
        log10(tot("Cl")), log10(tot("S"))
30 if equi("Calcite") > 1e-5 then graph_sy log10(equi("Calcite")) else graph_sy -5
35 if equi("Gypsum") > 1e-5 then graph_sy log10(equi("Gypsum")) else graph_sy -5
40 if equi("Anhydrite") > 1e-5 then graph_sy log10(equi("Anhydrite")) else graph_sy -5
50 if equi("Halite") > 1e-5 then graph_sy log10(equi("Halite")) else graph_sy -5
60 if equi("Glauberite") > 1e-5 then graph_sy log10(equi("Glauberite")) else graph_sy -5
70 if equi("Polyhalite") > 1e-5 then graph_sy log10(equi("Polyhalite")) else graph_sy -5
80 if STEP_NO > 20 THEN PRINT "x", "Na", "K", "Mg", "Ca", "Cl", "S"
90 if STEP_NO > 20 THEN PRINT 1 / tot("water"), (tot("Na")), (tot("K")), (tot("Mg")),\
                     (tot("Ca")), (tot("Cl")), (tot("S"))
-end
REACTION
 H2O -1; 0 36 3*4 6*1 2*0.25 0.176 4*0.05 5*0.03
INCREMENTAL_REACTIONS true
END
```

Results of the forward model at 90-fold concentration (fig. 18) show the same mineral precipitates that were included in the inverse model—calcite, gypsum, and halite. However, in the forward model, gypsum precipitates at lower concentration factors, but then transforms to anhydrite slightly before halite starts to precipitate. Similarly, glauberite [$Na_2Ca(SO_4)_2$] is calculated to precipitate when the water is 30 times concentrated, but redissolves at a concentration factor approaching 90. Polyhalite [$K_2MgCa_2(SO_4)_4$] is calculated to form in the forward model when the water is 80 times concentrated. If glauberite and polyhalite are included in an inverse-modeling calculation (in which solution 2 represents an 88 times concentrated water), neither mineral appears in a mole-balance model. This inverse modeling result for glauberite is consistent with the forward model, which indicates glauberite is redissolved before solution 2 is reached. Polyhalite is not found in the inverse model because precipitation would require the concentration of K^+ to decrease (the forward model, with polyhalite precipitation, predicts a K^+ concentration of 0.313 M at a concentration factor of 88), whereas the actual concentration in solution 2 (0.46 M) is slightly higher than expected from a concentration factor of 88 and no precipitation reactions (0.005 × 88 = 0.44 M). The inverse model result for polyhalite is consistent with most evaporation

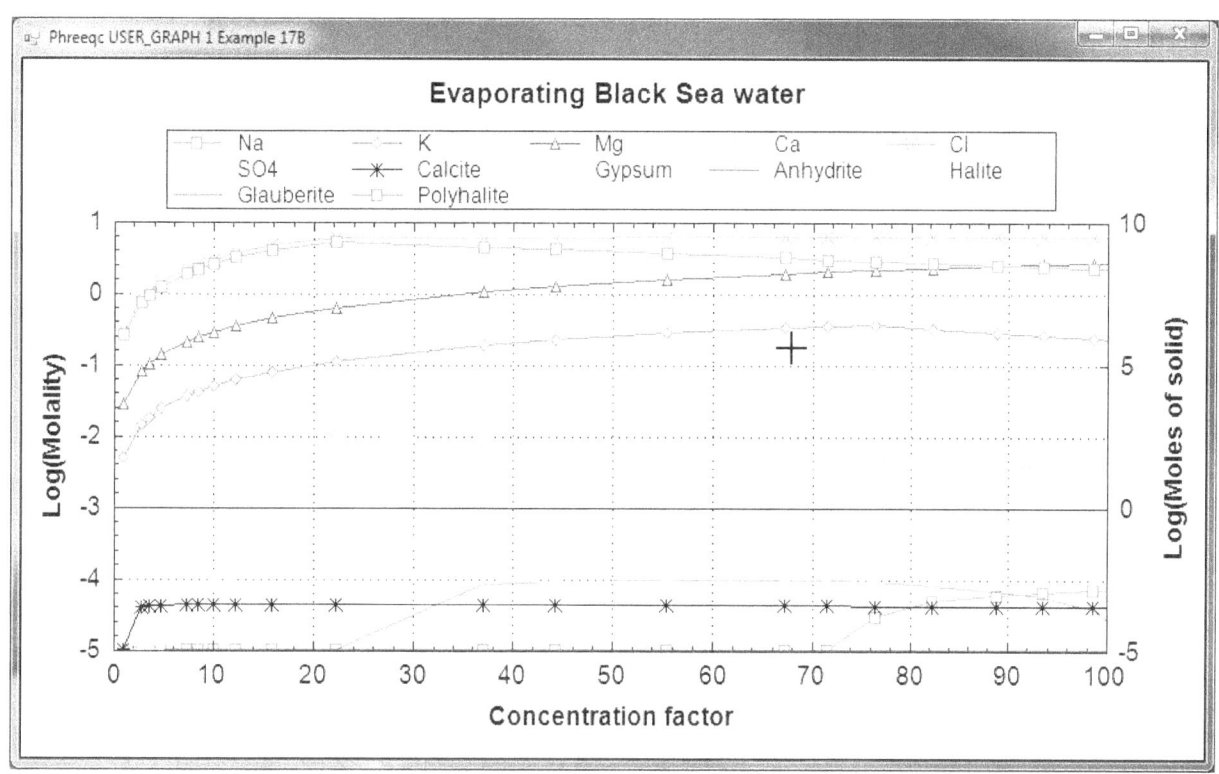

Figure 18. Concentrations in evaporating Black Sea water and precipitated salts, calculated with *pitzer.dat*.

experiments, which show that K^+ minerals first start to precipitate at somewhat higher concentration factors (McCaffrey and others, 1987; Zherebtsova and Volkova, 1966).

Example 18—Inverse Modeling of the Madison Aquifer

In this example, inverse modeling, including isotope mole-balance modeling, is applied to the evolution of water in the Madison aquifer in Montana. Plummer and others (1990) used mole-balance modeling to quantify the extent of dedolomitization at locations throughout the aquifer. In the dedolomitization process, anhydrite dissolution causes the precipitation of calcite and dissolution of dolomite. Additional reactions identified by mole-balance modeling include sulfate reduction, cation exchange, and halite and sylvite dissolution (Plummer and others, 1990). $\delta^{13}C$ and $\delta^{34}S$ data were used to corroborate the mole-balance models and carbon-14 was used to estimate groundwater ages (Plummer and others, 1990). Initial and final water samples were selected from a flow path that extends from north-central Wyoming northeast across Montana (Plummer and others, 1990, flow path 3). This pair of water samples was selected specifically because it was one of the few pairs that showed a relatively large discrepancy between previous mole-balance approaches and the mole-balance approach of PHREEQC, which includes uncertainties; results for most sample pairs were not significantly different between the two approaches. In addition, this pair of samples was selected because it was modeled in detail in Plummer and others (1990) to determine the sensitivity of mole-balance results to various model assumptions and was used as an example in the NETPATH manual (Plummer and others, 1994, example 6). Results of PHREEQC calculations are compared to NETPATH calculations. This example is also discussed in Parkhurst (1997).

Water Compositions and Reactants

The initial water for mole-balance modeling (solution 1, table 52) is the recharge water for flow path 3 (Plummer and others, 1990). This calcium-magnesium-bicarbonate water is typical of recharge water in a terrain containing calcite and dolomite. The final water (solution 2, table 52) is a sodium-calcium-sulfate water (with significant chloride concentration) (Plummer and others, 1990, "Mysse Flowing Well"). This water has a charge imbalance of +3.24 meq/kgw (milliequivalent per kilogram water) and contains measurable sulfide. An uncertainty limit of 5 percent was assigned to all chemical data, except iron, for the initial and final waters. The 5-percent uncertainty limit was chosen for the initial water because of spatial uncertainty in the location of a recharge water that is on the same flow path as the final water, and for the final water because it was near the smallest uncertainty limit necessary to obtain charge balance. Iron was assigned an uncertainty limit of 100 percent because of the small concentrations. An uncertainty limit of 0.1 unit was assigned to pH, which is a conservative estimate because of possible CO_2 degassing at this sampling site (L.N. Plummer, U.S. Geological Survey, written commun., 1996). $\delta^{13}C$ values increase from the initial water to the final water (-7.0 permil to -2.3 permil), as do $\delta^{34}S$ values (9.7 permil to 16.3 permil). Uncertainty limits for isotopic values of the initial solution were set to one-half the range in isotopic composition in four recharge waters from flow paths 3 and 4 (Plummer and others, 1990) (table 52). Similarly, uncertainty limits for isotopic values of the final water were set to one-half the range in isotopic composition in the samples from the distal end of flow path 3 (Plummer and others, 1990) (table 52).

Reactants considered by Plummer and others (1990) were dolomite, calcite, anhydrite, organic matter (CH_2O), goethite, pyrite, Ca/Na_2 cation exchange, halite, sylvite, and CO_2 gas. In their sensitivity calculations, Mg/Na_2 cation exchange and methane also were considered as potential reactants. The aquifer was considered to be a closed system with respect to CO_2, that is, no CO_2 is expected to be gained from or lost to a gas phase, and methane gain or loss was considered to be unlikely (Plummer and others, 1990). Therefore, CO_2 gas and methane were not included as reactants in the PHREEQC mole-balance modeling.

Table 52. Analytical data for solutions used in example 18.

[Charge balance is milliequivalents per kilogram water. All other data are in millimoles per kilogram water, except pH, $\delta^{13}C$, $\delta^{34}S$, and ^{14}C; Fe(2), ferrous iron; TDIC, total dissolved inorganic carbon; $\delta^{13}C$, carbon-13 of TDIC in permil relative to PDB (Pee Dee Belemnite standard); $\delta^{34}S(6)$, sulfur-34 of sulfate in permil relative to CDT (Canyon Diablo Troilite standard); $\delta^{34}S(-2)$, sulfur-34 of total sulfide in permil relative to CDT; ^{14}C, carbon-14 in percent modern carbon; $\pm$, indicates the uncertainty limit assigned in inverse modeling; --, not measured]

Analyte	Solution 1	Solution 2
Temperature, °C	9.9	63.0
pH	7.55	6.61
Ca	1.20	11.28
Mg	1.01	4.54
Na	0.02	31.89
K	0.02	2.54
Fe(2)	0.001	0.0004
TDIC	4.30	6.87
SO_4	0.16	19.86
H_2S	0	0.26
Cl	0.02	17.85
$\delta^{13}C$	-7.0±1.4 .	-2.3±0.2
$\delta^{34}S(6)$	9.7±0.9	16.3±1.5
$\delta^{34}S(-2)$	--	-22.1±7.0
^{14}C	52.3	0.8
Charge balance	+0.11	+3.24

The uncertainty limits for the isotopic compositions of dissolving phases were taken from data presented in Plummer and others (1990) with slight modifications as follows: $\delta^{13}C$ of dolomite, 1 to 5 permil; $\delta^{13}C$ of organic carbon, -30 to -20 permil; $\delta^{34}S$ of anhydrite, 11.5 to 15.5 permil. The $\delta^{13}C$ of precipitating calcite depends on the isotopic evolution of the solution and is affected by isotopic fractionation. The fractionation equations are not included in PHREEQC, so it is necessary to assume a compositional range of calcite that represents the average isotopic composition of the precipitating calcite. The average isotopic composition of precipitating calcite from NETPATH calculations was about -1.5 permil (Plummer and others, 1994) and an uncertainty limit of 1.0 permil was selected to account for uncertainties in fractionation factors. All carbon-14 modeling was done with NETPATH by using mole transfers from PHREEQC models. The $\delta^{34}S$ of precipitating pyrite was estimated to be -22 permil (Plummer and others, 1990) with an uncertainty limit of 2 permil; sensitivity analysis indicated that the isotopic value for the precipitating pyrite had little effect on mole transfers. The input file for PHREEQC is shown in table 53. Note that the log K values for sylvite, CH_2O, and the $Ca_{0.75}Mg_{0.25}/Na_2$ exchange reaction are set to zero in the **PHASES** and

EXCHANGE_SPECIES data blocks. The stoichiometry of each of these reactants is correct, which is all that is needed for mole-balance modeling; however, any saturation indices or forward modeling using these reactions would be incorrect because the log K values have not been properly defined.

Table 53. Input file for example 18.

```
TITLE Example 18.--Inverse modeling of Madison aquifer
SOLUTION 1 Recharge number 3
        units    mmol/kgw
        temp     9.9
        pe       0.
        pH       7.55
        Ca       1.2
        Mg       1.01
        Na       0.02
        K        0.02
        Fe(2)    0.001
        Cl       0.02
        S(6)     0.16
        S(-2)    0
        C(4)     4.30
        -i       13C     -7.0    1.4
        -i       34S      9.7    0.9
SOLUTION 2 Mysse
        units    mmol/kgw
        temp     63.
        pH       6.61
        pe       0.
        redox    S(6)/S(-2)
        Ca       11.28
        Mg       4.54
        Na       31.89
        K        2.54
        Fe(2)    0.0004
        Cl       17.85
        S(6)     19.86
        S(-2)    0.26
        C(4)     6.87
        -i       13C      -2.3    0.2
        -i       34S(6)   16.3    1.5
        -i       34S(-2) -22.1    7
INVERSE_MODELING 1
        -solutions 1 2
        -uncertainty 0.05
        -range
        -isotopes
                13C
                34S
```

Table 53. Input file for example 18.—Continued

```
        -balances
                Fe(2)      1.0
                pH         0.1
        -phases
                Dolomite            dis      13C      3.0      2
                Calcite             pre      13C      -1.5     1
                Anhydrite           dis      34S      13.5     2
                CH2O                dis      13C      -25.0    5
                Goethite
                Pyrite              pre      34S      -22.     2
                CaX2                pre
                Ca.75Mg.25X2        pre
                MgX2                pre
                NaX
                Halite
                Sylvite
PHASES
   Sylvite
        KCl = K+ + Cl-
        -log_k  0.0
   CH2O
        CH2O + H2O = CO2 + 4H+ + 4e-
        -log_k  0.0
EXCHANGE_SPECIES
        0.75Ca+2 + 0.25Mg+2 + 2X- = Ca.75Mg.25X2
        log_k    0.0
END
```

Mole-balance calculations included equations for all elements in the reactive phases (listed under the identifier **-phases**) and for $\delta^{34}S$ and $\delta^{13}C$. NETPATH calculations included isotopic fractionation equations to calculate the $\delta^{13}C$ of the final water, thus accounting for fractionation during precipitation, whereas PHREEQC calculated only a mole-balance equation on $\delta^{13}C$. The adjusted concentrations (original data plus calculated δs) from the PHREEQC results were rerun with NETPATH to obtain carbon-14 ages and to consider the fractionation effects of calcite precipitation. One NETPATH calculation used the charge-balancing option to identify the effects of charge-balance errors. The charge-balance option of NETPATH adjusts the concentrations of all cationic elements by a fraction, f, and of all anionic elements by a fraction $\frac{1}{f}$ to achieve charge balance for the solution. (The charge-balance option of NETPATH was improved in version 2.13 to produce exact charge balance; previous versions produced only approximate charge balance.)

For all NETPATH calculations (including the ones that used PHREEQC-adjusted concentrations), carbon dioxide was included as a potentially reactive phase, but the $\delta^{34}S$ of anhydrite was adjusted to produce zero mole transfer of carbon dioxide. The $\delta^{13}C$ of dolomite and organic matter were adjusted within their uncertainty limits to reproduce the $\delta^{13}C$ of the final solution as nearly as possible.

Madison Aquifer Results and Discussion

The predominant reactions determined by mole-balance modeling are dedolomitization, ion exchange, halite dissolution, and sulfate reduction, as listed in table 54 for the various modeling options discussed next. The driving force for dedolomitization is dissolution of anhydrite (about 20 mmol/kgw, table 54), which causes calcite precipitation and dolomite dissolution. Some of the calcium from anhydrite dissolution and (or) magnesium from dolomite dissolution is taken up by ion-exchange sites, which release sodium to solution. About 15 mmol/kgw of halite dissolves. Sulfate and iron oxyhydroxide reduction by organic matter leads to precipitation of pyrite.

Plummer and others (1990) realized that the stoichiometry of the exchange reaction was not well defined and considered two variations on these reactions in the sensitivity analysis of the mole-balance model. Pure Ca/Na_2 exchange and pure Mg/Na_2 exchange were considered as potential reactants (NETPATH *A* and *B*, table 54). When PHREEQC was run with these two reactants, a model was found with Mg/Na_2 (PHREEQC *B*), but no model was found with pure Ca/Na_2 exchange. This difference between NETPATH and PHREEQC results is attributed to the charge imbalance of the solutions. Solution 2 (table 52) has a charge imbalance of 3.24 meq/kgw, which is more than 3 percent relative to the sum of cation and anion equivalents. This is not a large percentage error, but the absolute magnitude in milliequivalents is large relative to some of the mole transfers of the mole-balance models. When using the revised mole-balance equations in PHREEQC with Ca/Na_2 exchange as the only exchange reaction, it is not possible simultaneously to attain mole balance on elements and isotopes, produce charge balance for each solution, and keep uncertainty terms within the specified uncertainty limits. The exchange reaction with the largest calcium component for which a model could be found was about $Ca_{0.75}Mg_{0.25}/Na_2$ (PHREEQC *C*). This exchange reaction was then used in NETPATH to find NETPATH *C*. NETPATH *C'* was calculated by using the charge-balance option of NETPATH with all phases and constraints the same as in NETPATH *C*.

One consistent difference between the NETPATH models without the charge-balance option (NETPATH *A*, *B*, and *C*) and the PHREEQC models is that the amount of organic-matter oxidation and the mole transfers of goethite and pyrite are larger in the PHREEQC models. These differences are attributed to the effects of charge balance on the mole transfers. It has been noted that charge-balance errors frequently manifest themselves as erroneous mole transfers of single component reactants, such as carbon dioxide or

Table 54. Mole-balance results for example 18.

[Results are in millimoles per kilogram water, unless otherwise noted. [14]C, carbon-14 in percent modern carbon (pmc); $\delta^{13}C$, carbon-13 in permil rPDB (Pee Dee Belemnite); $\delta^{34}S$, sulfur-34 in permil CDT (Canyon Diablo Troilite); CH_2O represents organic matter; Positive numbers for mineral mass transfer indicate dissolution; negative numbers indicate precipitation; For exchange reactions, positive numbers indicate a decrease in calcium and (or) magnesium and an increase in sodium in solution. --, reactant not included in model]

Result	Ca/Na$_2$	Mg/Na$_2$		Ca$_{0.75}$Mg$_{0.25}$/Na		
	NETPATH A	NETPATH B	PHREEQC B	NETPATH C	NETPATH C' Charge balanced	PHREEQC C
Ca/Na$_2$ exchange	8.3	--	--	--	--	--
Ca$_{0.75}$Mg$_{0.25}$/Na$_2$ exchange	--	--	--	8.3	7.6	7.7
Mg/Na$_2$ exchange	--	8.3	7.7	--	--	--
Dolomite [CaMg(CO$_3$)$_2$]	3.5	11.8	11.2	5.6	5.3	5.4
Calcite (CaCO$_3$)	-5.3	-21.8	-23.9	-9.4	-12.3	-12.1
Anhydrite (CaSO$_4$)	20.1	20.1	22.9	20.1	22.5	22.5
CH$_2$O	0.8	0.8	4.1	0.8	4.3	3.5
Goethite (FeOOH)	0.1	0.1	1.0	0.1	1.0	0.8
Pyrite (FeS$_2$)	-0.1	-0.1	-1.0	-0.1	-1.0	-0.8
Halite (NaCl)	15.3	15.3	15.3	15.3	15.8	15.3
Sylvite (KCl)	2.5	2.5	2.5	2.5	2.5	2.5
Carbon dioxide (CO$_2$)	0.0	0.0	--	0.0	0.0	--
[14]C, reaction adjusted	12.5	0.6	0.4	5.9	3.8	3.8
Apparent age (year)	22,700	-2,200	-5,400	16,500	13,000	12,900
$\delta^{34}S$, Anhydrite	15.6	15.6	12.8	15.6	12.5	13.4
$\delta^{13}C$, Dolomite	3.6	1.0	3.0	1.9	5.0	5.0
$\delta^{13}C$, CH$_2$O	-25.0	-30.0	-21.4	-25.0	-20.0	-20.0
Calculated $\delta^{13}C$, final water	-2.3	-2.2	-3.0	-2.3	-4.3	-3.3
Calculated $\delta^{34}S$, final water	15.8	15.8	16.1	15.8	15.9	16.0

organic matter (Plummer and others, 1994). Except for differences in mole transfers in organic matter, goethite, and pyrite, the Mg/Na$_2$ models are similar (NETPATH B and PHREEQC B). However, both models imply a negative carbon-14 age which is impossible, as noted by Plummer and others (1990).

The PHREEQC model most similar to the pure Ca/Na$_2$ exchange model (NETPATH A) is the Ca$_{0.75}$Mg$_{0.25}$/Na$_2$ model (PHREEQC C). This model has larger mole transfers of carbonate minerals and organic matter than the Ca/Na$_2$ model, which decreases the reaction-adjusted carbon-14 activity and produces a younger groundwater age, 12,900 (PHREEQC C) compared with 22,700 (NETPATH A). This large change in the calculated age can be attributed to differences in the reactions involving carbon. Two effects can be noted: the change in the exchange reaction and the adjustments for charge-balance errors. The effect of the change in exchange reaction can be estimated from the differences between NETPATH A, which contains pure Ca/Na$_2$ exchange, and NETPATH C, which contains Ca$_{0.75}$Mg$_{0.25}$/Na$_2$ exchange, but neither model includes corrections for charge imbalances in the solution compositions. The increase in magnesium in the exchange reaction causes larger mole transfers of calcite and dolomite and decreases the calculated age from 22,700 to 16,500 yr. The effects of charge-balance errors are estimated by the differences between NETPATH C and C', which differ only in that the NETPATH charge-balance option was used in NETPATH C'. Charge balancing the solutions produces larger mole transfers of organic matter and calcite and decreases the calculated age from 16,500 to 13,000 yr. The mole transfers and calculated age for NETPATH C' are similar to PHREEQC C but differ slightly because the uncertainty terms in the PHREEQC model have been calculated to achieve not only charge balance but also to reproduce as closely as possible the observed $\delta^{13}C$ of the final solution.

One advantage of the revised mole-balance formulation in PHREEQC is that much of the sensitivity analysis that was formerly accomplished by setting up and running multiple models can now be done by including uncertainty limits for all chemical and isotopic data simultaneously. For example, one run of the revised mole-balance formulation determines that no pure Ca/Na$_2$ model can be found even if any or all of the chemical data were adjusted by as much as plus or minus 10 percent. This kind of information would be difficult and time consuming to establish with previous mole-balance formulations. Another improvement is the explicit inclusion of charge-balance constraints. In this example, including the charge-balance constraint requires a change in the exchange reaction and adjustments to solution composition, which have the combined effect of lowering the estimated maximum age of the groundwater by about 10,000 yr. If Mg/Na$_2$ exchange is the sole exchange reaction, the age would be modern. Thus, the estimated range in age is large, 0 to 13,000 yr. However, because the calcium to magnesium ratio in solution is approximately 2.5:1 and the cation-exchange constants for calcium and magnesium are approximately equal (Appelo and Postma, 2005), the combined exchange reaction with a dominance for calcium is more plausible, which gives more credence to the older age. Furthermore, comparisons with other carbon-14 ages in the aquifer and with groundwater flow-model ages also indicate that the older end of the age range is more reasonable.

Sorption of heavy metals and organic pollutants on natural materials can be described by linear, Freundlich, or Langmuir isotherms. All three isotherms can be calculated by PHREEQC, as shown in this example for Cd^{+2} sorbing on a loamy soil (Christensen, 1984; Appelo and Postma, 2005). A more mechanistic approach, also illustrated here, is to model the distribution of Cd^{+2} over the sorbing components in the soil, in this case, in and on organic matter, clay minerals, and iron oxyhydroxides.

Figure 19 shows the measured sorbed concentration (μg/g soil, microgram per gram) as a function of solute concentration (μg/L water, microgram per liter), and three isotherms defined by:

$$\frac{(\text{LinearCd}^{+2})}{(\text{Cd}^{+2})} = K_{linear}, \tag{33}$$

$$\frac{(\text{FreundlichCd}^{+2})}{(\text{Cd}^{+2})^n} = K_{Freundlich}, \text{ and} \tag{34}$$

$$\frac{(\text{LangmuirCd}^{+2})}{(\text{Cd}^{+2})} = \frac{(\text{Langmuir\_total\_sites})}{K_{Langmuir} + (\text{Cd}^{+2})}, \tag{35}$$

where the brackets indicate concentrations in μg/g soil and μg/L water, and Linear, Freundlich, and Langmuir refer to the sorption sites that will be used in definitions for the **SURFACE** keyword.

The three isotherms correspond to reactions and mass-action equations that can be entered in PHREEQC as follows:

$$\text{Linear} + \text{Cd}^{+2} = \text{LinearCd}^{+2}; \qquad \frac{[\text{LinearCd}^{+2}]}{[\text{Cd}^{+2}]} = K_{linear, \text{PHREEQC}}[\text{Linear}], \tag{36}$$

$$\text{Freundlich} + n\text{Cd}^{+2} = \text{FreundlichCd}^{+2}; \quad \frac{[\text{FreundlichCd}^{+2}]}{[\text{Cd}^{+2}]^n} = K_{Freundlich, \text{PHREEQC}}[\text{Freundlich}], \tag{37}$$

and

$$\text{Langmuir} + \text{Cd}^{+2} = \text{LangmuirCd}^{+2}; \qquad \frac{[\text{LangmuirCd}^{+2}]}{[\text{Cd}^{+2}]} = K_{Langmuir, \text{PHREEQC}}[\text{Langmuir}], \tag{38}$$

where the square brackets indicate activities (unitless).

The Langmuir isotherm (equation 35) and its PHREEQC formulation (equation 38) are equivalent for sorption of Cd on 1 g soil when (Langmuir_total_sites) / 112.4×10^6 = moles$_{Langmuir\ sites,\ \text{PHREEQC}}$, and

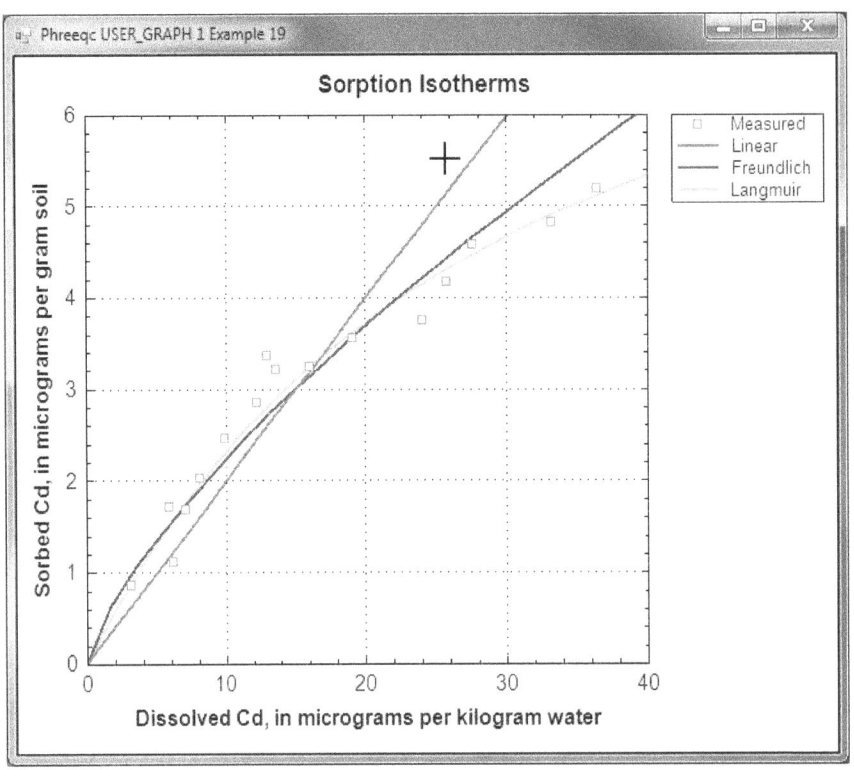

Figure 19. Measured sorption of Cd on loamy soil and three isotherms calculated by PHREEQC.

$(K_{Langmuir} / 112.4{\times}10^6)^{-1} = K_{Langmuir,\ \mathrm{PHREEQC}}$, where $112.4{\times}10^6$ µg/mol (microgram per mole) is the atomic weight of Cd. Thus, the Langmuir isotherm represents a valid chemical reaction that can be defined in PHREEQC. That is not the case for the linear and the Freundlich isotherms because the equations do not account for the decrease of the "free" sites when Cd^{+2} sorbs. However, in PHREEQC the number of sorption sites can be made large; for example $moles_{linear,\ \mathrm{PHREEQC}} = 10^{100}$ mol, so that a decrease of a few moles of free sites by surface complexation will be negligible relative to the total number of free sites. In activity terms, the effect of the large number of sites is that [Linear] = 1 at all times in equation 36. By reducing the constant in the mass-action equation by the same large number, the required distribution is satisfied. Thus, $K_{linear,\ \mathrm{PHREEQC}} = K_{linear} / moles_{linear,\ \mathrm{PHREEQC}}$. Similarly for the Freundlich equation, $K_{Freundlich,\ \mathrm{PHREEQC}} = K_{Freundlich} / moles_{Freundlich,\ \mathrm{PHREEQC}}$. The remaining problem with the Freundlich equation (equation 37) is that the stoichiometric coefficient for Cd^{+2} is n on the left-hand side of the equation, but is 1 on the right-hand side of the equation. This inconsistency can be ignored in PHREEQC by adding **-no_check** to the reaction equation.

The coefficients in the linear, Freundlich, and Langmuir isotherm equations that fit the measurements are taken from Appelo and Postma (2005). For the linear isotherm K_{linear} = 0.2 (L water / g soil), and for the Freundlich and Langmuir isotherms:

$$\frac{\text{Freundlich}Cd^{+2}}{(Cd^{+2})^{0.722}} = 0.421 \frac{\text{L water}}{\text{g soil}} \text{, and} \tag{39}$$

$$\frac{\text{Langmuir}Cd^{+2}}{Cd^{+2}} = \frac{9.50 \frac{\mu g}{\text{g soil}}}{30.9 \frac{\mu g}{\text{L water}}{Cd^{+2}} + 1} . \tag{40}$$

The PHREEQC constants then become as follows:

$$\log(K_{linear,\ \text{PHREEQC}}) = \log(0.2) - \log(10^{100}) = -100.7, \tag{41}$$

$$\log(K_{Freundlich,\ \text{PHREEQC}}) = \log(0.421) + (0.722 - 1) \times \log(112.4\times10^6) - \log(10^{100}) = -102.61, \text{ and} \tag{42}$$

$$\log(K_{Langmuir,\ \text{PHREEQC}}) = \log(112.4\times10^6 / 30.9) = 6.56. \tag{43}$$

The input file (table 55) defines the sorption reactions with keywords **SURFACE_MASTER_SPECIES** and **SURFACE_SPECIES**. To avoid accumulation of charge in the solution, the sorption formula is specified to be *surface*$CdCl_2$ with **-mole_balance** for each of the reactions, where *surface* is Linear, Freundlich, or Langmuir. The keyword **SURFACE** defines the moles of sites, and the identifier **-no_edl** specifies that the calculations will be done without the charge/potential term in the mass-action equations for the sorbed species. Next a 1 mM (millimolar) $CaCl_2$ is entered with keyword **SOLUTION**, followed by a reaction of 0.7 mmol $CdCl_2$ in 20 steps (**REACTION** data block). **USER_GRAPH** plots the isotherms, and the experimental data are added to the plot with **-plot_tsv_file** *ex19_meas.tsv*.

Table 55. Input file for example 19.

```
TITLE Example 19.--Linear, Freundlich and Langmuir isotherms for
      Cd sorption on loamy sand. Calculates Example 7.1
      from Appelo and Postma, 2005. Data from Christensen, 1984.
SURFACE_MASTER_SPECIES
      Linear Linear
      Freundlich Freundlich
      Langmuir Langmuir
SURFACE_SPECIES
  Linear = Linear
  Linear + Cd+2 = LinearCd+2
      -log_k -100.7          # log10(0.2) - 100
      -mole_balance LinearCdCl2
  Freundlich = Freundlich
```

Table 55. Input file for example 19.—Continued

```
   Freundlich + 0.722 Cd+2 = FreundlichCd+2
       -log_k -102.61          # log10(0.421) + (0.722 - 1) * log10(112.4e6) - 100
       -no_check
       -mole_balance FreundlichCdCl2
   Langmuir = Langmuir
   Langmuir + Cd+2 = LangmuirCd+2
       -log_k 6.56             # log10(112.4 / 30.9e-6)
       -mole_balance LangmuirCdCl2
SURFACE 1
       Linear 1e100 1 1
       Freundlich 1e100 1 1
       Langmuir 8.45e-8 1 1  # 9.5 / 112.4e6
       -no_edl
SOLUTION 1
       pH   6
       Ca   1
       Cl   2
REACTION 1
       CdCl2 1
       0.7e-6 in 20
USER_GRAPH Example 19
       -headings Linear Freundlich Langmuir
       -chart_title "Sorption Isotherms"
       -axis_titles "Dissolved Cd, in micrograms per kilogram water" \
                    "Sorbed Cd, in micrograms per gram soil"
       -plot_tsv_file ex19_meas.tsv
       -axis_scale x_axis 0 40
       -axis_scale y_axis 0 6
       -initial_solutions true
   -start
10 x = act("Cd+2") * 112.4e6
20 PLOT_XY x, mol("LinearCd+2")*112e6, color = Green, symbol = None, line_width = 2
30 PLOT_XY x, mol("FreundlichCd+2")*112e6, color = Blue, symbol = None, line_width = 2
40 PLOT_XY x, mol("LangmuirCd+2")*112e6, color = Orange, symbol = None, line_width = 2
   -end
PRINT
     -reset false
END
```

Christensen (1984) measured the exchange capacity of the soil ($CEC = 56$ meq/kg, milliequivalent per kilogram), the iron-oxyhydroxide content (2,790 ppm Fe), and organic matter content ($OM = 0.7$ weight percent), which can be inserted in a deterministic PHREEQC model. The database *phreeqc.dat* contains the exchange reaction of Cd^{+2} and Ca^{+2} (the predominant cation in the experiment) with the exchange site X^-, and the surface complexation reactions of Cd^{+2} on hydrous ferric oxide (Hfo). Christensen measured the iron content by extracting the soil with dithionite, which reduces and dissolves all amorphous and crystalline iron oxyhydroxides. It is assumed that only 10 percent of analyzed iron has the reactivity represented by Hfo in the *phreeqc.dat* database. The measured exchange capacity is entered as moles X^-, although part of the exchange capacity is located in organic matter, which is modeled separately. Organic matter binds Cd^{+2} more strongly than clay minerals, so the overestimation of the clay exchange capacity is assumed to be negligible. However, the assumption is checked (and corrected) in the model results, from which the contribution of organic matter to the *CEC* is estimated by summing the Ca-complexes on humic acid and the calcium excess in the double layer.

Complexation on organic matter can be modeled with WHAM, the Windermere Humic Acid Model of Tipping and Hurley (Tipping and Hurley, 1992; Tipping, 1998; Lofts and Tipping, 2000). The WHAM model was designed for humic and fulvic acids in surface water, but it has been applied to Cd sorption on organic matter in soils (Shi and others, 2007). Complexation constants are defined for protons and cations on a number of monodentate and bidentate binding sites, which can be entered simply as SURFACE_SPECIES in PHREEQC. Furthermore, an empirical parameter for the charge-potential relation of the (probably spherical) humic molecules is incorporated into the Gouy Chapman relation that is used in PHREEQC by adjusting the specific surface area with a function that depends on the ionic strength of the solution (Appelo and Postma, 2005).

The input file (table 56) starts with the SOLUTION_MASTER_SPECIES for the complexing sites on humic acid and the SURFACE_SPECIES for protons, Cd^{+2} and Ca^{+2}. Next the SURFACE is defined, with number of sites for 3.5 mg humic acid (1 g soil with 0.7 percent OM, is about 3.5 mg organic carbon, taken all as humic acid initially, but corrected to an active fraction by comparing model and measured results). In the WHAM model, the humic acid has a total charge of -7.1 meq/g (milliequivalent per gram), distributed over four monoprotic carboxylic sites (*nHA* carrying a charge of -2.84 meq/g humic acid), four monoprotic phenolic sites (*nHB*, charge = -1.42 meq/g humic acid) and twelve diprotic sites that combine carboxylic and phenolic charges (charge = -2.84 meq/g humic acid). Hfo, assumed to represent 10 percent of analyzed iron, is part of the same SURFACE. Keyword EXCHANGE defines the moles of X⁻ in 1 g soil, the SOLUTION is 1 mM $CaCl_2$ with a very small Cd^{+2} concentration to avoid a zero division in USER_GRAPH. USER_GRAPH prints the contribution of organic matter to the *CEC* and the distribution coefficient for Cd^{+2} (L/kg, liter per kilogram), and plots the sorption of Cd^{+2} by organic matter, clay exchange, and iron oxyhydroxides (fig. 20).

Table 56. Input file for example 19B.

```
TITLE Example 19B.--Cd sorption on X, Hfo and OC in loamy soil
PRINT
        -reset false
        -user_print true
SURFACE_MASTER_SPECIES
# Monodentate 60%
  H_a  H_aH;  H_b  H_bH;  H_c  H_cH;  H_d  H_dH
  H_e  H_eH;  H_f  H_fH;  H_g  H_gH;  H_h  H_hH
# Bidentate 40%
  H_ab H_abH2;  H_ad H_adH2;  H_af H_afH2;   H_ah H_ahH2
  H_bc H_bcH2;  H_be H_beH2;  H_bg H_bgH2;   H_cd H_cdH2
  H_cf H_cfH2;  H_ch H_chH2;  H_de H_deH2;   H_dg H_dgH2
SURFACE_SPECIES
  H_aH = H_aH; log_k 0;  H_bH = H_bH; log_k 0;   H_cH = H_cH; log_k 0;  \
      H_dH = H_dH; log_k 0;
  H_eH = H_eH; log_k 0;  H_fH = H_fH; log_k 0;   H_gH = H_gH; log_k 0;  \
      H_hH = H_hH; log_k 0;

  H_abH2 = H_abH2; log_k 0;  H_adH2 = H_adH2; log_k 0;  H_afH2 = H_afH2; log_k 0;
  H_ahH2 = H_ahH2; log_k 0;  H_bcH2 = H_bcH2; log_k 0;  H_beH2 = H_beH2; log_k 0;
  H_bgH2 = H_bgH2; log_k 0;  H_cdH2 = H_cdH2; log_k 0;  H_cfH2 = H_cfH2; log_k 0;
  H_chH2 = H_chH2; log_k 0;  H_deH2 = H_deH2; log_k 0;  H_dgH2 = H_dgH2; log_k 0;
# Protons
  H_aH = H_a- + H+; log_k  -1.59
  H_bH = H_b- + H+; log_k  -2.70
  H_cH = H_c- + H+; log_k  -3.82
  H_dH = H_d- + H+; log_k  -4.93

  H_eH = H_e- + H+; log_k  -6.88
  H_fH = H_f- + H+; log_k  -8.72
  H_gH = H_g- + H+; log_k  -10.56
  H_hH = H_h- + H+; log_k  -12.40


  H_abH2 = H_abH- + H+; log_k -1.59;  H_abH- = H_ab-2 + H+; log_k -2.70
  H_adH2 = H_adH- + H+; log_k -1.59;  H_adH- = H_ad-2 + H+; log_k -4.93
  H_afH2 = H_afH- + H+; log_k -1.59;  H_afH- = H_af-2 + H+; log_k -8.72
  H_ahH2 = H_ahH- + H+; log_k -1.59;  H_ahH- = H_ah-2 + H+; log_k -12.40
  H_bcH2 = H_bcH- + H+; log_k -2.70;  H_bcH- = H_bc-2 + H+; log_k -3.82
  H_beH2 = H_beH- + H+; log_k -2.70;  H_beH- = H_be-2 + H+; log_k -6.88
  H_bgH2 = H_bgH- + H+; log_k -2.70;  H_bgH- = H_bg-2 + H+; log_k -10.56
  H_cdH2 = H_cdH- + H+; log_k -3.82;  H_cdH- = H_cd-2 + H+; log_k -4.93
  H_cfH2 = H_cfH- + H+; log_k -3.82;  H_cfH- = H_cf-2 + H+; log_k -8.72
  H_chH2 = H_chH- + H+; log_k -3.82;  H_chH- = H_ch-2 + H+; log_k -12.40
  H_deH2 = H_deH- + H+; log_k -4.93;  H_deH- = H_de-2 + H+; log_k -6.88
  H_dgH2 = H_dgH- + H+; log_k -4.93;  H_dgH- = H_dg-2 + H+; log_k -10.56
# Calcium
  H_aH + Ca+2 = H_aCa+ + H+; log_k  -3.20
  H_bH + Ca+2 = H_bCa+ + H+; log_k  -3.20
  H_cH + Ca+2 = H_cCa+ + H+; log_k  -3.20
  H_dH + Ca+2 = H_dCa+ + H+; log_k  -3.20

  H_eH + Ca+2 = H_eCa+ + H+; log_k  -6.99
  H_fH + Ca+2 = H_fCa+ + H+; log_k  -6.99
```

Table 56. Input file for example 19B.—Continued

```
  H_gH + Ca+2 = H_gCa+ + H+; log_k  -6.99
  H_hH + Ca+2 = H_hCa+ + H+; log_k  -6.99

  H_abH2 + Ca+2 = H_abCa + 2H+; log_k -6.40
  H_adH2 + Ca+2 = H_adCa + 2H+; log_k -6.40
  H_afH2 + Ca+2 = H_afCa + 2H+; log_k -7.45
  H_ahH2 + Ca+2 = H_ahCa + 2H+; log_k -10.2
  H_bcH2 + Ca+2 = H_bcCa + 2H+; log_k -6.40
  H_beH2 + Ca+2 = H_beCa + 2H+; log_k -10.2
  H_bgH2 + Ca+2 = H_bgCa + 2H+; log_k -10.2
  H_cdH2 + Ca+2 = H_cdCa + 2H+; log_k -6.40
  H_cfH2 + Ca+2 = H_cfCa + 2H+; log_k -10.2
  H_chH2 + Ca+2 = H_chCa + 2H+; log_k -10.2
  H_deH2 + Ca+2 = H_deCa + 2H+; log_k -10.2
  H_dgH2 + Ca+2 = H_dgCa + 2H+; log_k -10.2
# Cadmium
  H_aH + Cd+2 = H_aCd+ + H+; log_k  -1.52
  H_bH + Cd+2 = H_bCd+ + H+; log_k  -1.52
  H_cH + Cd+2 = H_cCd+ + H+; log_k  -1.52
  H_dH + Cd+2 = H_dCd+ + H+; log_k  -1.52

  H_eH + Cd+2 = H_eCd+ + H+; log_k  -5.57
  H_fH + Cd+2 = H_fCd+ + H+; log_k  -5.57
  H_gH + Cd+2 = H_gCd+ + H+; log_k  -5.57
  H_hH + Cd+2 = H_hCd+ + H+; log_k  -5.57

  H_abH2 + Cd+2 = H_abCd + 2H+; log_k -3.04
  H_adH2 + Cd+2 = H_adCd + 2H+; log_k -3.04
  H_afH2 + Cd+2 = H_afCd + 2H+; log_k -7.09
  H_ahH2 + Cd+2 = H_ahCd + 2H+; log_k -7.09
  H_bcH2 + Cd+2 = H_bcCd + 2H+; log_k -3.04
  H_beH2 + Cd+2 = H_beCd + 2H+; log_k -7.09
  H_bgH2 + Cd+2 = H_bgCd + 2H+; log_k -7.09
  H_cdH2 + Cd+2 = H_cdCd + 2H+; log_k -3.04
  H_cfH2 + Cd+2 = H_cfCd + 2H+; log_k -7.09
  H_chH2 + Cd+2 = H_chCd + 2H+; log_k -7.09
  H_deH2 + Cd+2 = H_deCd + 2H+; log_k -7.09
  H_dgH2 + Cd+2 = H_dgCd + 2H+; log_k -7.09

END
SURFACE 1
# 1 g soil = 0.7% Organic Matter ~ 3.5 mg Organic Carbon.
# 7.1 meq charge per g OC
# For Psi vs I (= ionic strength) dependence, adapt specific surface area in PHRC:
# SS = 159300 - 220800/(I)^0.09 + 91260/(I)^0.18
# Example: SS = 46514 m2/g for I = 0.003 mol/l
#
# 3.5 mg OC, 0.025 meq total charge, distributed over the sites:
# charge on 4 nHA sites: -2.84 / 4 * 3.5e-3 / 1e3 (eq)
  H_a  2.48e-06 46.5e3 3.50e-03
  H_b  2.48e-06; H_c  2.48e-06; H_d  2.48e-06
# charge on 4 nHB sites: 0.5 * charge on nHA sites
```

Table 56. Input file for example 19B.—Continued

```
  H_e  1.24e-06; H_f  1.24e-06; H_g  1.24e-06; H_h  1.24e-06
# charge on 12 diprotic sites: -2.84 / 12 * 3.5e-3 / 1e3
  H_ab 8.28e-07; H_ad 8.28e-07; H_af 8.28e-07; H_ah 8.28e-07
  H_bc 8.28e-07; H_be 8.28e-07; H_bg 8.28e-07; H_cd 8.28e-07
  H_cf 8.28e-07; H_ch 8.28e-07; H_de 8.28e-07; H_dg 8.28e-07
        -Donnan
# 1 g soil = 2.79 mg Fe = 0.05 mmol Fe = 4.45 mg FeOOH
# 10% has ferrihydrite reactivity
  Hfo_w 1e-6 600 4.45e-4
  Hfo_s 0.025e-6
        -equilibrate 1
EXCHANGE 1
  X 55.7e-6
        -equilibrate 1
SOLUTION 1
  pH   6.0
  Ca   1
  Cl   2
  Cd   1e-6
REACTION 1
  CdCl2 1
  2e-6 in 20
USER_GRAPH Example 19
        -headings Cd_HumicAcids CdX2 Cd_Hfo TOTAL
        -chart_title "Deterministic Sorption Model"
        -axis_titles "Dissolved Cd, in micrograms per kilogram water" \
                     "Sorbed Cd, in micrograms per gram soil"
        -plot_tsv_file ex19_meas.tsv
        -axis_scale x_axis 0 40
        -axis_scale y_axis 0 6
        -initial_solutions true
  -start
10 H_Cd = SURF("Cd", "H") + EDL("Cd", "H")
20 print CHR$(10) + " ug Cd/L =", tot("Cd") * 112.4e6, " ug Cd/g = ", H_Cd * 112.4e6 \
       ," Kd (L/kg) = ", H_Cd*1e3/tot("Cd"), " ug Cd/g in DL =", \
       EDL("Cd", "H") * 112.4e6
30 print "Excess meq Ca in DL =", EDL("Ca", "H")*2 - EDL("water", "H") * tot("Ca")*2
40 print "Excess meq Cl in DL =", EDL("Cl", "H")  - EDL("water", "H") * tot("Cl")
50 print "Surface charge      =", EDL("Charge", "H")
55 af_OM = 1 / 9
60 H_Ca = (SURF("Ca", "H") + EDL("Ca", "H")) * af_OM
70 print 'Total Ca in/on organic matter =', H_Ca, ' CEC on OM =' H_Ca*200/TOT("X"),\
       '%.'
80 x = TOT("Cd") * 112.4e6
90 H_Cd = H_Cd * 112.4e6 * af_OM
100 CdX2 = mol("CdX2") * 112.4e6 * 0.96
110 Hfo_Cd = (mol("Hfo_wOCd+") + mol("Hfo_sOCd+")) * 112.4e6
120 PLOT_XY x, H_Cd, color = Green, line_width = 2, symbol = None
130 PLOT_XY x, CdX2, color = Brown, line_width = 2, symbol = None
140 PLOT_XY x, Hfo_Cd, color = Black, line_width = 2, symbol = None
150 PLOT_XY x, H_Cd + CdX2 + Hfo_Cd, color = Red, line_width = 2, symbol = None
  -end
END
```

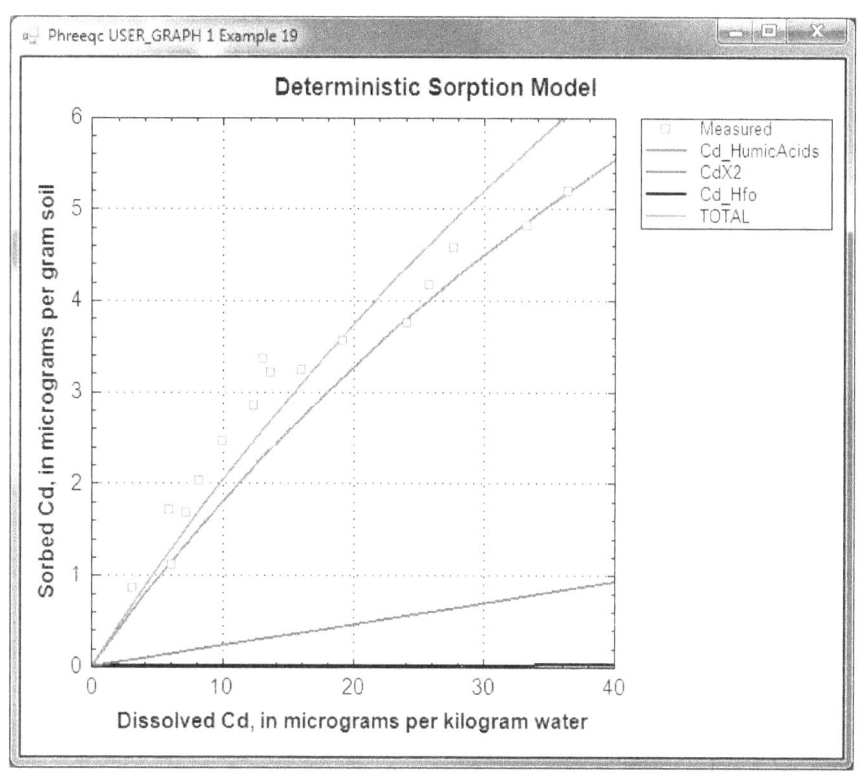

Figure 20. Measured sorption of Cd on loamy soil and calculated sorption on iron oxyhydroxides ("Hfo"), clay minerals ("X"), and organic matter ("HumicAcids"). TOTAL represents the sum of all sorbed Cd.

In line with many experimental results, figure 20 illustrates that Cd in soils is mainly bound to organic matter. The contribution of clay exchange is about 20 percent, and the contribution of iron-oxyhydroxide surface complexation is negligible. Actually, the sorbed concentration of Cd^{+2} on 3.5 mg humic acids would be nine times higher, but the concentration was reduced by multiplying with an active fraction of organic matter to match the measured data. The smaller active fraction is understandable because a soil contains lignin and a variety of other, less reactive compounds than the pure humic acid that formed the basis for the Tipping and Hurley database. Shi and others (2007) found an average active fraction of 0.6 for organic carbon concentrations that ranged from 0.18 to 7.15 weight percent in New Jersey soils, whereas Weng and others (2002) and Bonten and others (2008) used an active fraction of 0.3 for soil organic matter in the NICA-Donnan model.

Earlier it was assumed that the organic-matter contribution to the CEC was negligible. The assumption is checked by calculating the Ca exchanged on organic matter relative to the amount exchanged on clays. The amount of calcium exchanged on organic matter is printed in the output file for the largest cadmium concentration as:

```
Total Ca in/on organic matter =  1.1810e-006   CEC on OM =  4.2407e+000 %.
```

Although the correction is so small as to be unnecessary, the CEC attributed to clays should be reduced by a factor of 0.96 to account for the 4 percent of the exchange capacity that is located in or on organic matter.

Example 20—Distribution of Isotopes Between Water and Calcite

The database *iso.dat* implements the approach to isotope reactions described by Thorstenson and Parkhurst (2002, 2004), in which minor isotopes are treated as individual thermodynamic components. The aqueous and solid species of minor isotopes have slightly different equilibrium constants than those of the major isotopes, which account for fractionation processes. The treatment of isotopes in gases requires a separate species for each isotopic variant of a gas; for example, the isotopic variants of carbon dioxide are CO_2, $C^{18}OO$, $C^{18}O_2$, $^{13}CO_2$, $^{13}C^{18}OO$, and $^{13}C^{18}O_2$. Similarly, every isotopic variant of a mineral must be included as a component of a solid solution to represent completely the isotopic composition of the solid. The equilibrium constants in *iso.dat* are derived from empirical fractionation factors, from statistical mechanical theory, or, where no data are available (the most common case), by assuming no fractionation. However, the database is a framework that can be expanded as additional isotopic thermodynamic data become available.

Example 20 considers the fractionation of oxygen-18, carbon-13, and carbon-14 between solution and calcite under two conditions: (1) in a system where a solution and isotopic solid solution equilibrate with sequential additions of calcite of a specified isotopic composition, hereafter referred to as marine calcite. This system is termed "closed" in the sense that all the masses of elements and isotopes that enter the system remain in the system; and (2) in a system where a solution and isotopic solid solution equilibrate with sequential additions of marine calcite, but the isotopic solid solution is removed from the system following equilibration. This system is termed "open" in the sense that masses of elements and isotopes are removed from the reaction system at each step. Both systems are isolated from the unsaturated zone, which precludes any further addition of carbon dioxide. The open system is equivalent to calculations in NETPATH that implement isotopic exchange, and the results of the PHREEQC open-system calculation are compared to NETPATH calculations for carbon-13 and carbon-14.

Example 20 is divided into two parts: first, the isotopic composition of a marine calcite solid solution in equilibrium with a solution of 0.0 permil oxygen-18 and 0.0 permil carbon-13 is estimated; second, the isotopic evolution of the solution and solid solution are calculated for the closed- and open-reaction systems. In addition to carbon-13 and oxygen-18, carbon-14 is included in the isotopic-evolution calculations, which allows carbon-14 to partition into the solid; however, the calculations do not consider the radioactive decay of carbon-14.

Calculation of the Composition of a Marine Calcite Isotopic Solid Solution

The input file that estimates the isotopic composition of calcite in equilibrium with a specified solution isotopic composition is given in table 57. The first line contains the DATABASE keyword and indicates that the *iso.dat* database file must be used for this calculation. The *iso.dat* database has thousands of aqueous species, many of which are negligibly small for most calculations. The PRINT data block is used to exclude species from the printed distribution of species for an element or isotope if the species concentration is less than a fraction of 10^{-6} relative to the total concentration of the element or isotope.

Table 57. Input file for example 20A.

```
DATABASE ../database/iso.dat
TITLE Example 20A.--Calculate carbonate solid solution
PRINT
            -censor_species          1e-006
SOLUTION 1 # water to find composition of marine carbonate
            pH        8.2
            Na        1         charge
            Ca        10        Calcite         0
            C         2
            [13C]     0         # permil
            [14C]     0         # pmc
            D         0         # permil
            [18O]     0         # permil
END
SOLID_SOLUTION 1 No [14C]
Calcite
            -comp     Calcite                         0
            -comp     CaCO2[18O](s)                   0
            -comp     CaCO[18O]2(s)                   0
            -comp     CaC[18O]3(s)                    0
            -comp     Ca[13C]O3(s)                    0
            -comp     Ca[13C]O2[18O](s)               0
            -comp     Ca[13C]O[18O]2(s)               0
            -comp     Ca[13C][18O]3(s)                0
END
RUN_CELLS
            -cells 1
USER_PRINT
-start
            10 PRINT pad("Component", 20), "Mole fraction"
```

Table 57. Input file for example 20A.—Continued

```
                   20 t = LIST_S_S("Calcite", count, name$, moles)
                   30 for i = 1 to count
                   40   PRINT pad(name$(i),20), moles(i)/t
                   50 next i
   -end
   END
```

The **SOLUTION** data block defines an aqueous solution with isotopic composition of 0.0 permil carbon-13 and 0.0 permil oxygen-18, which are approximately the values in seawater. The solution contains dissolved inorganic carbon at a concentration of 2.0 mmol/kgw; the calcium concentration is determined by equilibrium (saturation index 0.0) with pure carbon-12, oxygen-16 calcite; and the sodium concentration is determined by charge balance. In the next simulation, a solid solution is defined that has eight components, but the number of moles of each component is zero. Thus, when reacting, components may precipitate, but no components initially are available to dissolve.

In the next simulation **RUN_CELLS** causes the solution and all of the reactants numbered 1 to react, namely **SOLUTION** 1 reacts with **SOLID_SOLUTIONS** 1. The solution is initially in equilibrium with a pure carbon-12, oxygen-16 calcite; however, a solid solution of all of the isotopic variants of calcite will be slightly less soluble than this pure solid. Thus, a solid solution should form, but the mole transfers to the solid should be small enough not to affect the isotopic composition of the solution substantially.

The composition of the isotopic solid solution is printed to the output file, but the printout only contains three significant digits. **USER_PRINT** is used to print the mole fractions with greater precision. The function LIST_S_S is used to retrieve the total number of moles of components (*t*, line 20) in the "Calcite" solid solution (as defined in **SOLID_SOLUTIONS**) and the name and number of moles of each component (*name$* and *moles*, line 20). From these data, the mole fraction of each component is calculated and printed to the **USER_PRINT** section of the output file.

The mole fractions are interpreted as the number of moles of each component in a mole of the solid solution. The mole fractions calculated by the simulation are shown in table 58. The "Calcite" component represents the pure carbon-12, oxygen-16 component, which is approximately 98 percent of the solid; all other components are approximately 1 percent mole fraction or less.

"Calcite" is the name of the entire solid solution and also is the name of one of the components in the solid solution. Using "Calcite" for the solid solution name allows functions defined in *iso.dat* to be used to list isotope ratios and isotope alphas in the output file. The "Isotope Ratios" part of the printout (not shown) gives the isotopic compositions of the solution and the solid in terms of permil. The isotopic composition of oxygen-18 [R(18)] and carbon-13 [R(13)] in the solution are virtually unchanged from 0 permil. The oxygen-18 composition of the solid [R(18O) Calcite] is approximately 29 permil and the carbon-13 composition of the solid [R(13C) Calcite] is approximately 2 permil.

Table 58. Mole fractions of isotopic components in a marine carbonate solid solution.

```
Component               Mole fraction
Calcite                  9.8283e-001
Ca[13C]O3(s)             1.1011e-002
CaCO2[18O](s)            6.0825e-003
Ca[13C]O2[18O](s)        6.8147e-005
CaCO[18O]2(s)            1.2548e-005
Ca[13C]O[18O]2(s)        1.4058e-007
CaC[18O]3(s)             8.6284e-009
Ca[13C][18O]3(s)         9.6671e-011
```

Calculation of Open- and Closed-System Isotopic Evolution of Calcite and Solution

The concept for the isotope-evolution calculations is that water that has reacted with carbon dioxide in the unsaturated zone moves into the saturated zone and comes into contact with marine calcite with the composition defined by the previous calculation without further contact with carbon dioxide gas. The marine calcite reacts incrementally with the solution (0.0005 mol per increment), and after a few increments, a solid solution forms. In the open-system calculation, the solution moves on to react with pristine marine calcite, leaving the previously formed solid solution behind. In the closed system, the entire system—previously formed solid solution, increment of marine calcite, and solution—reacts to a new equilibrium. Although the isotopic compositions of the unsaturated zone solution are substantially different from the marine calcite, with sufficient reaction of the marine calcite, the solution will ultimately approach a composition that is in equilibrium with the pure marine calcite for both the open- and closed-system paths.

The input file for the open- and closed-system calculations is given in table 59. As in the previous calculations, *iso.dat* is used as the database and the species concentrations printed in the distribution of species are censored at 10^{-6}. Carbon-14 is included in the calculations, which results in very small concentrations of carbon-14 atoms in solution. These small concentrations can lead to instabilities in the numerical method used to solve the constitutive equations. The **KNOBS** data block sets a smaller step size and an alternative equation scaling than the default, which helps in the solution of the equations. Alternatively, results are linear for such small concentrations of carbon-14, and a larger concentration of carbon-14 (say 10 or 100 times larger) could be used to increase numerical stability, provided the results are reduced in the same proportion.

Table 59. Input file for example 20B.

```
DATABASE ../database/iso.dat
TITLE Example 20B.--Isotope evolution.
PRINT
                -censor_species        1e-006
KNOBS
                -diagonal_scale
                -step 10
                -pe    5
#
# Open system calculation
#
SOLID_SOLUTION 1 With [14C]
Calcite
                -comp    Calcite              0
                -comp    CaCO2[18O](s)        0
                -comp    CaCO[18O]2(s)        0
                -comp    CaC[18O]3(s)         0
                -comp    Ca[13C]O3(s)         0
                -comp    Ca[13C]O2[18O](s)    0
                -comp    Ca[13C]O[18O]2(s)    0
                -comp    Ca[13C][18O]3(s)     0
                -comp    Ca[14C]O3(s)         0
                -comp    Ca[14C]O2[18O](s)    0
                -comp    Ca[14C]O[18O]2(s)    0
                -comp    Ca[14C][18O]3(s)     0
END
REACTION 1
                Calcite             9.8283e-001
                Ca[13C]O3(s)        1.1011e-002
                CaCO2[18O](s)       6.0825e-003
                Ca[13C]O2[18O](s)   6.8147e-005
                CaCO[18O]2(s)       1.2548e-005
                Ca[13C]O[18O]2(s)   1.4058e-007
                CaC[18O]3(s)        8.6284e-009
                Ca[13C][18O]3(s)    9.6671e-011
                0.0005 mole
END
USER_PRINT
10 PRINT "Calcite added: ", GET(0) * RXN
USER_GRAPH 1 Example 20
        -headings Open--Dissolved Open--Calcite
        -chart_title "Oxygen-18"
        -axis_titles "Marine calcite reacted, in moles" "Permil"
        -axis_scale x_axis 0 0.05 a a
        -axis_scale y_axis -10 30 a a
    -start
10 PUT(GET(0) + 1, 0)
20 PLOT_XY RXN*GET(0),ISO("R(18O)"), color=Red, line_w=2, symbol=None
30 PLOT_XY RXN*GET(0),ISO("R(18O)_Calcite"), color=Green, line_w=2, symbol=None
    -end
END
USER_GRAPH 2 Example 20
```

Table 59. Input file for example 20B.—Continued

```
            -headings Open--Dissolved Open-Calcite
            -chart_title "Carbon-13"
            -axis_titles "Marine calcite reacted, in moles" "Permil"
            -axis_scale x_axis 0 0.05 a a
            -axis_scale y_axis -25 5.0 a a
            -plot_tsv  ex20-c13.tsv
    -start
10 PLOT_XY RXN*GET(0),ISO("R(13C)"), color=Red, line_w=2, symbol=None
20 PLOT_XY RXN*GET(0),ISO("R(13C)_Calcite"), color=Green, line_w=2, symbol=None
    -end
END
USER_GRAPH 3 Example 20
        -headings Open--Dissolved Open--Calcite
        -chart_title "Carbon-14"
        -axis_titles "Marine calcite reacted, in moles" "Percent modern carbon"
        -axis_scale x_axis 0 0.05 a a
        -axis_scale y_axis 0 100 a a
        -plot_tsv  ex20-c14.tsv
    -start
10 PLOT_XY RXN*GET(0),ISO("R(14C)"), color=Red, line_w=2, symbol=None
20 PLOT_XY RXN*GET(0),ISO("R(14C)_Calcite"), color=Green, line_w=2, symbol=None
    -end
END
SOLUTION 1
                pH      5       charge
                pe      10
                C       2       CO2(g)  -1.0
                [13C]   -25     # permil
                [14C]   100     # pmc
                [18O]   -5      # permil
SELECTED_OUTPUT
                -reset false
                -file ex20_open
USER_PUNCH
-start
10 FOR i = 1 to 100
20 PUNCH EOL$ + "USE solution 1"
30 PUNCH EOL$ + "USE solid_solution 1"
40 PUNCH EOL$ + "USE reaction 1"
50 PUNCH EOL$ + "SAVE solution 1"
60 PUNCH EOL$ + "END"
70 NEXT i
-end
END
PRINT
                -selected_output false
END
INCLUDE$ ex20_open
END
#
# Closed system calculation
#
```

Table 59. Input file for example 20B.—Continued

```
USER_GRAPH 1 Oxygen-18
        -headings Closed--Dissolved Closed--Calcite
    -start
10 PUT(GET(1) + 1, 1)
20 PLOT_XY RXN*GET(1),ISO("R(18O)"), color=Blue, line_w=0, symbol=Circle
30 PLOT_XY RXN*GET(1),ISO("R(18O)_Calcite"), color=Black, line_w=0, symbol=Circle
    -end
END
USER_GRAPH 2 Carbon-13
        -headings Closed--Dissolved Closed--Calcite
    -start
10 PLOT_XY RXN*GET(1),ISO("R(13C)"), color=Blue, line_w=2, symbol=None
20 PLOT_XY RXN*GET(1),ISO("R(13C)_Calcite"), color=Black, line_w=2, symbol=None
    -end
END
USER_GRAPH 3 Carbon-14
        -headings Closed--Dissolved Closed--Calcite
    -start
10 PLOT_XY RXN*GET(1),ISO("R(14C)"), color=Blue, line_w=2, symbol=None
20 PLOT_XY RXN*GET(1),ISO("R(14C)_Calcite"), color=Black, line_w=2, symbol=None
    -end
END
USER_PRINT
10 PRINT "Calcite added: ", GET(1), GET(1)*0.0005, RXN
SOLUTION 1
                pH        5        charge
                pe        10
                C         2        CO2(g)  -1.0
                [13C]     -25      # permil
                [14C]     100      # pmc
                [18O]     -5       # permil
END
INCREMENTAL_REACTIONS true
# Alternative to redefinition of REACTION 1
#REACTION_MODIFY 1
#               -steps
#                       0.05
#               -equal_increments1
#               -count_steps      100
REACTION 1
                Calcite                 9.8283e-001
                Ca[13C]O3(s)            1.1011e-002
                CaCO2[18O](s)           6.0825e-003
                Ca[13C]O2[18O](s)       6.8147e-005
                CaCO[18O]2(s)           1.2548e-005
                Ca[13C]O[18O]2(s)       1.4058e-007
                CaC[18O]3(s)            8.6284e-009
                Ca[13C][18O]3(s)        9.6671e-011
                0.05 mole in 100 steps
RUN_CELLS
                -cells 1
END
```

As in the previous calculation, a solid solution with zero moles of components is defined in the SOLID_SOLUTIONS data block; however, this time, four carbon-14 solid components also are defined. A reaction is defined by using the composition of the calcite isotopic solid solution of the previous calculation. The components are present in the proportions given by the mole fractions, but the reaction adds only 0.0005 mol of marine calcite per increment. Next, three USER_GRAPH data blocks are defined to display the isotopic composition of oxygen-18, carbon-13, and carbon-14 in the solid and in solution as the reactions proceed.

The SOLUTION definition represents an unsaturated zone water. The carbon concentration is determined by equilibrium with a soil-zone partial pressure of 0.1 atm for carbon dioxide gas. Carbon-13 is set to -25 permil to represent an organic matter source of the carbon dioxide. Carbon-14 of the soil gas is assumed to be 100 pmc. Oxygen-18 is arbitrarily set to -5 permil, which is typical of a temperate climate.

The open-system simulation requires a series of calculations where an increment of marine calcite is added to the solution and equilibrated with a solid solution, followed by saving the solution composition. This sequence is repeated 100 times. It is possible to use an editor to create the input file by cutting and pasting 100 copies of the sequence of statements. For compactness, USER_PUNCH is used to write 100 copies of the sequence into the file *ex20_open*, which is then included in the input file with an INCLUDES statement. The PRINT data block (or a new definition of USER_PUNCH) is necessary to avoid repeatedly writing USER_PUNCH output to file for each subsequent calculation.

The closed-system calculation is similar to the open-system calculation. USER_GRAPH data blocks are used to set the parameters for the closed-system plots, which will be added to the open-system plots. The initial unsaturated-zone water is redefined with SOLUTION. However, in the closed system, the compositions of both the solution and the solid solution need to be saved after each increment. By saving the compositions at each reaction step, the closed-system definition is actually much simpler than the open system definition. The reaction can be defined as 0.05 mol in 100 steps, and RUN_CELLS can be used to bring together all the reactants numbered 1 (REACTION 1, SOLUTION 1, and SOLID_SOLUTIONS 1) for the series of 100 calculations. RUN_CELLS automatically saves the compositions of all reactants after each reaction step.

The isotopic compositions of the open and closed systems as a function of marine calcite reacted are shown in figure 21. The isotopic compositions begin at the composition of the unsaturated-zone water.

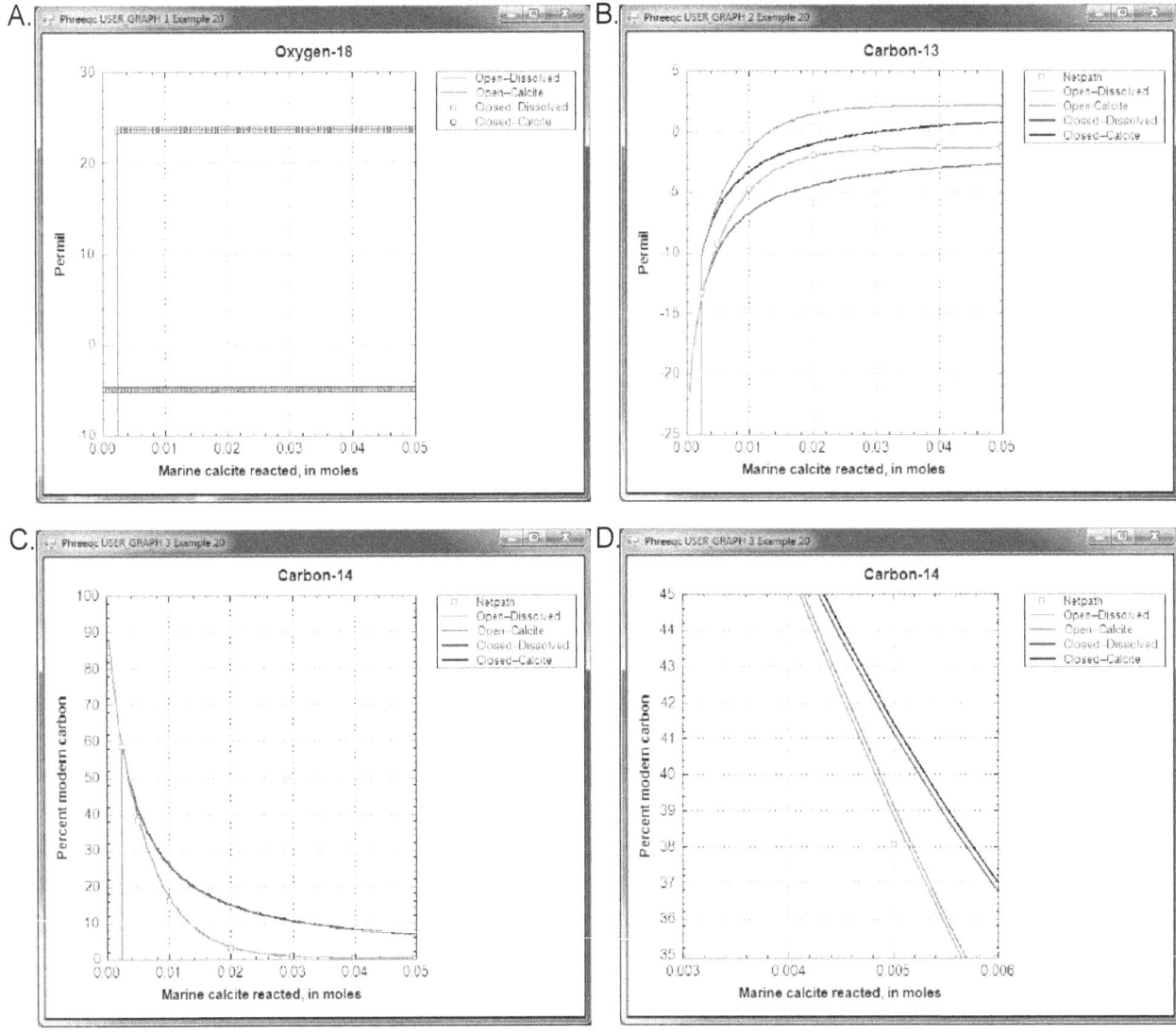

Figure 21. USER_GRAPH plots of example 20 open- and closed-system isotopic evolution for calcite and water: (A) oxygen-18, (B) carbon-13, (C) carbon-14, and (D) detail of carbon-14.

Initially, the water is undersaturated with the isotopic solid solution and no solid is present. The vertical lines on panels A, B, and C indicate the point at which a solid solution begins to form.

Oxygen-18 is essentially constant in the water and solid through the course of the reactions because the water phase dominates. There are 55.5 mol of oxygen in the water, whereas only 0.05 mol of marine calcite react. Thus, oxygen-18 in the water remains essentially equal to the initial composition of -5.0 permil, although, in detail, there is a slight trend to heavier compositions. The isotopic solid solution in equilibrium with the water is approximately 24 permil. Open- and closed-system calculations give identical results for oxygen-18.

As more and more marine calcite reacts, the carbon-13 isotopic composition of the solid approaches 2 permil, which is the composition of the marine calcite (fig. 21, panel B). The open system approaches the value more quickly than the closed system because only the isotopes in the water are carried forward to the next step and the solid solution (containing some of the lighter isotopes) is left behind. In the closed system, more reaction is needed to approach the marine calcite composition because all the light isotopes from the initial solution remain in the system and must be diluted by more heavy isotopes from the marine calcite. The carbon in the water is about 3.5 permil lighter than the carbon in the solid solution for both the open- and closed-system calculations.

Initially, all carbon-14 is in the unsaturated-zone water. As marine calcite reacts, the carbon-14 partitions into the solid and asymptotically approaches the point when all carbon-14 is in the solid solution. As with carbon-13, the open-system calculation approaches this point more quickly than the closed system calculation. Fractionation does occur for carbon-14 as it does for carbon-13; the detailed plot in figure 21, panel D, shows that the solid solution is slightly heavier than the solution, but the effect is insignificantly small. Radioactive decay of carbon-14 has been ignored in these calculations; all of the effects of decreasing carbon-14 concentrations in these calculations are the result of partitioning carbon-14 into the solid. If the rate of partitioning carbon-14 into solids is fast relative to the half-life of carbon-14 (approximately 6,000 yr), then it would be an important consideration in carbon-14 groundwater dating. If substantial loss of carbon-14 to the solid were erroneously attributed to radioactive decay, the resulting groundwater ages would be too old.

NETPATH was used to model the effects of isotopic exchange with the marine calcite on the evolution of carbon-13 and carbon-14 from the point that the solid solution forms. The fractionation factors of Deines and others (1974) are used in the *iso.dat* database and, consequently, the Deines option in NETPATH was used to maintain consistency in the calculations. NETPATH uses an analytical solution to the fractionation equations, whereas PHREEQC calculations used numerical integration with 0.0005 mole reaction steps. NETPATH results and PHREEQC results are very similar (fig. 21, panels B and C), which indicates both that the isotopic approach in PHREEQC is workable and that the reactions steps were sufficiently small to obtain accurate results.

This example illustrates how PHREEQC version 3 can simulate a diffusion experiment, as is now often performed for assessing the properties of a repository for nuclear waste in a clay formation. A sample is cut from a core of clay, enveloped in filters, and placed in a diffusion cell (see Van Loon and others, 2004, for details). Solutions with tracers are circulated at the surfaces of the filters, the tracers diffuse into and out of the clay, and the solutions are sampled and analyzed regularly in time. The concentration changes are interpreted with Fick's diffusion equations to obtain transport parameters for modeling the rates of migration of elements away from a waste repository. Transport in clays is mainly diffusive because of the low hydraulic conductivity, and solutes are further retarded by sorption (cations) and by exclusion from part of the pore space (anions).

For calculating diffusion, one needs to account for the different diffusion coefficients of the tracers, the effect of the filters, and the properties of the clay. Figure 22 presents a schematic diagram of the processes in the clay (Appelo and others, 2010). The pores in the clay are lined with clay minerals that have a negative surface charge. The charge is partly neutralized by cations that are bound to the surface and partly by the electrostatic double layer that extends some distance into the pore. The double layer contains an excess of cations (counter-ions, in general) and a deficit of anions (co-ions, in general). In swelling clay

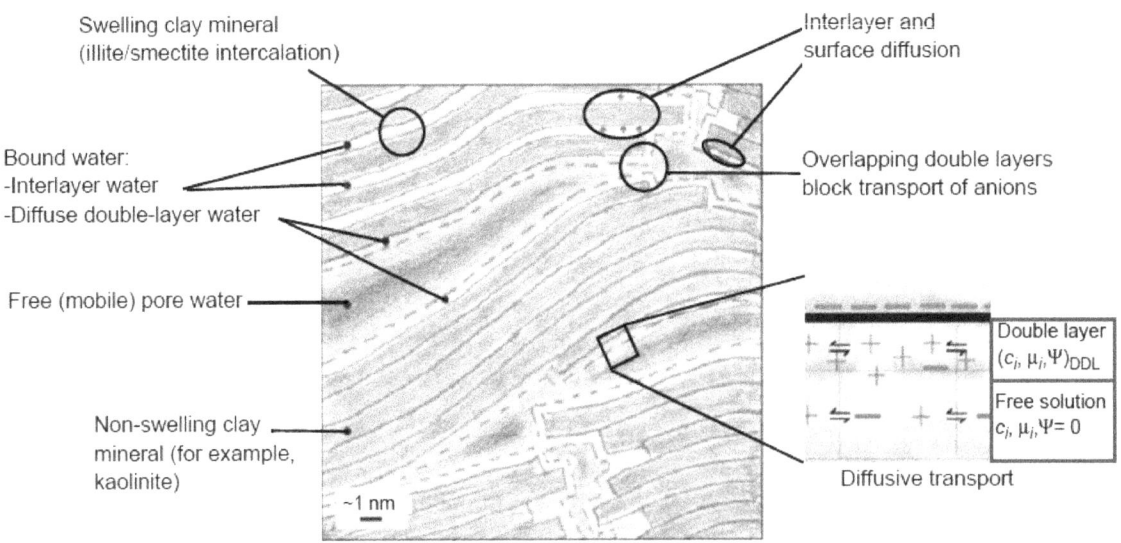

Figure 22. A diagram of the pore space in Opalinus Clay, showing three water types with associated diffusion domains. The right-hand side presents a pore as simplified in PHREEQC (Modified from Nagra, 2002; Appelo and others, 2010).

minerals, such as montmorillonite, cations in the interlayer space also neutralize part of the negative charge. In comparison with concentration gradients that drive diffusion in free (uncharged) pore water, the gradient for a cation is magnified in the double layer, whereas the gradient for an anion is diminished, and the gradient for a neutral species remains the same. The charge in the double layer lowers the dielectric permittivity of the water, which enhances the ion-association of cations and anions into neutral species. Also, the viscosity of water may be higher than in free pore water. Where double layers overlap in pore constrictions, anions are impeded and forced to take longer paths, whereas cations and neutral species pass through unhindered.

PHREEQC has capabilities to model many of these diffusion processes. An averaged double layer composition can be calculated by using the identifier **-Donnan** in keyword SURFACE, which, in essence, neutralizes the surface charge. Solute species can be assigned an enrichment factor in the Donnan pore space to emulate the additional concentration change as a result of different ion association. For diffusion, the viscosity can be set differently with respect to free pore water. In this example, all these properties are adjusted for the tracers tritium (HTO), Na^+, Cs^+ and Cl^- to simulate experimental results.

The experiments to be modeled were done in a radial diffusion cell, shown in figure 23. The radial cell enables measurement of diffusion parallel to the bedding plane of the clay (Van Loon and others, 2004). A solution with tracers is circulated at the surface of the inner filter, and another solution with the same major ions, but without tracers, contacts the surface of the outer filter. The latter solution is removed regularly and analyzed for the tracer that has diffused through the filters and the clay. Meanwhile, the outer solution is replaced by fresh solution, thus, maintaining tracer concentrations near zero in the outer solution. In the simulations, low solubility phases were defined to maintain small concentrations of tracers in the outer solution. The moles of precipitates of these phases give the accumulated mass that has diffused through the column.

For a linear column, PHREEQC calculates diffusion automatically by using the parameters entered with identifier **-multi_D** in keyword TRANSPORT. However, diffusion in the experiment is radial, and the filters have different properties than the clay. Also, the boundary solutions for the default (linear) column have a constant composition, whereas we want to know the concentration changes in these solutions during the diffusion experiment. All these experimental details can be simulated for a stagnant column by defining mixing factors among the individual cells with keyword MIX.

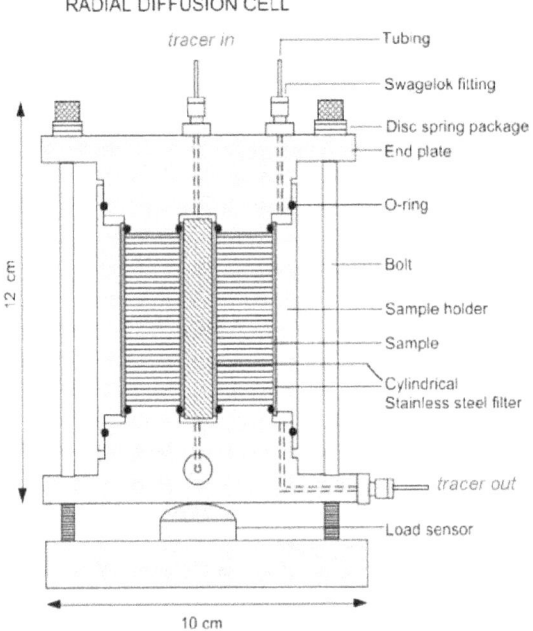

Figure 23. Radial diffusion cell used for analyzing diffusion parallel to the bedding plane of clay (Van Loon and others, 2004).

The mixing factors can be derived from Fick's diffusion equations, $F = -D_e \nabla C$ and $\frac{\partial C}{\partial t} = -\nabla \cdot F$, which transform to finite differences for an arbitrarily shaped cell j:

$$c_j^{t2} = c_j^{t1} + D_w \Delta t \sum_{i \neq j}^{n} \frac{\varepsilon_{ij}}{G_{ij}h_{ij}} \frac{A_{ij}}{V_j} (c_i^{t1} - c_j^{t1})f_{bc}, \tag{44}$$

where c_j^{t1} is the concentration in cell j at the current time (mol/m^3), c_j^{t2} is the concentration in cell j after the time step, D_w is the tracer diffusion coefficient (m^2/s), Δt is the time step (s), i is an adjacent cell, ε_{ij} is the porosity over the interface of cells i and j (unitless), G_{ij} is the geometrical factor that corrects for tortuosity of the porous medium (unitless), h_{ij} is the distance between midpoints of the cells (m), A_{ij} is the shared surface area of cells i and j (m^2), V_j is the volume of cell j (m^3), and f_{bc} is a factor for boundary cells (unitless). The summation is for all cells (up to n) adjacent to j. When h_{ij} is equal for all cells, a central difference algorithm is obtained that has second-order accuracy $[O(h)^2]$ (Appelo and others, 2008). It is therefore advantageous to make the grid regular. However, the same accuracy is achievable for a heterogeneous domain, even with widely variable gridsize, if the harmonic mean of the parameters in $\dfrac{\varepsilon_{ij}}{G_{ij}h_{ij}}$

is used. These parameters together, translate the tracer diffusion coefficient D_w into the effective diffusion coefficient D_e.

The harmonic mean can be derived in general (omitting the activity coefficient to simplify the formulas) as follows. The fluxes inside cells i and j and over the interface of the two cells must be the same and are given by:

$$J_{ij} = -\frac{\varepsilon_{ij}}{G_{ij}}D_w\frac{c_j - c_i}{h_{ij}},$$ (45)

$$J_i = -\frac{\varepsilon_i}{G_i}D_w\frac{c_{ij} - c_i}{h_i/2}, \text{ and}$$ (46)

$$J_j = -\frac{\varepsilon_j}{G_j}D_w\frac{c_j - c_{ij}}{h_j/2},$$ (47)

where h is the cell length (m), and c_{ij} is the concentration at the interface. Substituting $\varepsilon_i/G_i = g_i$, in equation 46 and similarly in equation 47, the two equations can be combined while eliminating the concentration c_{ij} to give:

$$J_{ij} = J_i = J_j = -\frac{2}{h_j/g_j + h_i/g_i}D_w(c_j - c_i).$$ (48)

Multiplying the flux by the surface area, the time step, and the boundary factor, and dividing by the volume of the cell for which the concentration change is calculated (here, cell j), the mixing factor is obtained:

$$mixf_{ij} = \frac{2}{h_j/g_j + h_i/g_i}D_w\frac{A_{ij}\Delta t}{V_j}f_{bc},$$ (49)

where f_{bc} is 2 for cells in contact with a constant concentration cell, and 1 otherwise. When calculating $mixf$, V_j is set to 10^{-3} m^3, and for D_w the default diffusion coefficient is used, entered with identifier **-multi_D**. In multicomponent-diffusion mode, PHREEQC adapts the volume V_j (entered as 10^{-3} m^3 in equation 49), to the actual volume of water in cell j, and multiplies the mixing factor for each solute species a by the ratio of the tracer diffusion coefficient for a and the default diffusion coefficient ($D_{w,\,a}/D_w$).

To avoid numerical oscillations, it is necessary that $mixf < 0.5$, which can be realized by limiting the time step. This maximum permissible time step is usually determined by the cell with the smallest volume (in the model) and the fluxes of the proton, which normally has the highest tracer diffusion coefficient. However, if the proton concentration is sufficiently buffered by alkalinity or other species, and the solutions are uniform except for the tracers, the tracer with the highest D_w may be selected for calculating the maximum permissible time step. If, nevertheless, time steps are too large, PHREEQC warns that negative concentrations are calculated (and the program stops because the system reached an infeasible state), the time step can be subdivided. For example, **-time_step** 5e2 3 will subdivide the time step of 500 s in three equal ones of 166.7 s.

The input file in table 60 defines the physical and chemical properties of the clay pore space and writes the mixing factors for diffusional transport in the filters and the clay. The filters used by Van Loon have a geometrical factor of 4 for all the tracers, whether charged or not (Glaus and others, 2008). In the clay, the geometrical factor for HTO is 6.2. For cations, the geometrical factor appears to be 2 to 4 times smaller than for tritium. However, in this example (example 21), the same geometrical factor is used for both HTO and the cations; the smaller apparent geometrical factors are modeled by subdividing the pore space into free pore water and double-layer water. The concentrations of cations are higher in double-layer water than in free pore water, and hence, the concentration gradients of cations are also higher, which, as noted previously enhances their diffusion. In models that do not account for this physical aspect of the clay pore space, the faster diffusion is treated by diminishing the geometrical factor.

For anions, the geometrical factor is about 1.5 times larger than for tritium. This larger factor is related to narrowing of the pores, where overlapping double layers decrease the accessible porosity for anions and obstruct their passage (as noted above). The model accounts for the observed, smaller accessible porosity for anions relative to tritium and cations by the calculation of anion exclusion in the double layer.

The file starts with the tracer species, where, for the monovalent tracers $^{22}Na^+$ (="Na_tr$^+$") and Cs^+, an enrichment factor for the double layer is entered with **-erm_ddl** in the SOLUTION_SPECIES data block. The enrichment is related to increased complexation of the polyvalent cations in the low dielectric permittivity of the double layer, and thus, more charge must be counterbalanced by monovalent cations. Sorption of Cs^+ is much stronger than that of Na^+, which is modeled by two surface complexes and one exchange reaction with large constants. The constants are based on the measured adsorption isotherm for Opalinus Clay, but may be generally applicable because they are associated primarily with strong sorption sites on illite. Next, the file writes SOLUTION 0-2 for a regular column, followed by SOLUTION 3,

Table 60. Input file for example 21.

```
TITLE Diffusion through Opalinus Clay in a radial diffusion cell,
      Appelo and others, 2010, GCA, v. 74, p. 1201-1219.
SOLUTION_MASTER_SPECIES
#   element   species   alk gfw_formula element_gfw
    Hto       Hto       0.0  20               20
    Na_tr     Na_tr+    0.0  22               22
    Cl_tr     Cl_tr-    0.0  36               36
    Cs        Cs+       0.0  132.905          132.905
SOLUTION_SPECIES
    Hto = Hto;                log_k 0;  -gamma 1e6 0;                 -dw 2.236e-9
    Na_tr+ = Na_tr+;          log_k 0;  -gamma 4.0 0.075;             -dw 1.33e-9;  -erm_ddl 1.23
    Cl_tr- = Cl_tr-;          log_k 0;  -gamma 3.5 0.015;             -dw 1.31e-9 # dw = dw(water) / 1.55 = 2.03e-9 / 1.55
    Cs+ = Cs+;                log_k 0;  -gamma 3.5 0.015;             -dw 2.07e-9;  -erm_ddl 1.23
SURFACE_MASTER_SPECIES
    Su_fes Su_fes-    # Frayed Edge Sites
    Su_ii Su_ii-      # Type II sites of intermediate strength
    Su_ Su_-          # Double layer, planar sites are modeled with EXCHANGE
SURFACE_SPECIES
    Su_fes- = Su_fes-; log_k 0
    Na+ + Su_fes- = NaSu_fes; log_k 10
    Na_tr+ + Su_fes- = Na_trSu_fes; log_k 10
    K+ + Su_fes- = KSu_fes; log_k 12.4
    Cs+ + Su_fes- = CsSu_fes; log_k 17.14

    Su_ii- = Su_ii-; log_k 0
    Na+ + Su_ii- = NaSu_ii; log_k 10
    Na_tr+ + Su_ii- = Na_trSu_ii; log_k 10
    K+ + Su_ii- = KSu_ii; log_k 12.1
    Cs+ + Su_ii- = CsSu_ii; log_k 14.6

    Su_- = Su_-; log_k 0

EXCHANGE_SPECIES
    Na_tr+ + X- = Na_trX; log_k 0.0;  -gamma 4.0 0.075
    Cs+ + X- = CsX;       log_k 2.04; -gamma 3.5 0.015

SOLUTION 0-2 column with only cell 1, two boundary solutions 0 and 2.
    Na 1; Cl 1
END
```

Table 60. Input file for example 21.—Continued

```
KNOBS; -diagonal_scale true # -tolerance 1e-20 # because of low concentrations

SOLUTION 3 tracer solution
    pH 7.6; pe 14 O2(g) -1.0; temp 23
    Na 240; K 1.61; Mg 16.9; Ca 25.8; Sr 0.505
    Cl 300; S(6) 14.1; Fe(2) 0.0; Alkalinity 0.476
# uncomment tracer concentrations and kg water 1 by 1...
    Hto 1.14e-6;        -water 0.2
#   Cl_tr 2.505e-2; -water 0.502
#   Cs 1; Na_tr 1.87e-7; -water 1.02
SELECTED_OUTPUT
    -file radial; -reset false
USER_PUNCH
    # Define symbols and pi...
1   nl$ = EOL$                      # newline
2   x$  = CHR$(35)                  # cross '#'
3   sc$ = CHR$(59)                  # semicolon ';'
4   pi  = 2 * ARCTAN(1e10)          # 3.14159...

    # Define experimental parameters...
10  height = 0.052                  # length of the clay cylinder / m
20  r_int = 6.58e-3                 # inner radius of clay cylinder / m
30  r_ext = 25.4e-3                 # outer radius
40  thickn_filter1 = 1.8e-3         # tracer-in filter thickness / m
50  thickn_filter2 = 1.6e-3         # tracer-out filter thickness / m
60  por_filter1 = 0.418             # porosity
70  por_filter2 = 0.367
80  G_filter1 = 4.18                # geometrical factor. (for filters, G = por / 10)
90  G_filter2 = 3.67
100 V_end = 0.2                     # volume of the tracer-out solution / L
110 thickn_clay = r_ext - r_int     # clay thickness / m
120 por_clay = 0.159
130 rho_b_eps = 2.7 * (1 - por_clay) / por_clay  # clay bulk density / porosity / (kg/L)
140 CEC = 0.12 * rho_b_eps          # CEC / (eq/L porewater)
150 A_por = 37e3 * rho_b_eps        # pore surface area / (m2/L porewater)

160 DIM tracer$(4), exp_time(4), scale_y1$(4), scale_y2$(4), profile_y1$(4), profile_y2$(4)
170 DATA 'Hto', 'Cl_tr', 'Na_tr', 'Cs'
180 READ tracer$(1), tracer$(2), tracer$(3), tracer$(4)
    # experimental times (seconds) for HTO, 36Cl, 22Na and Cs, respectively,
```

Table 60. Input file for example 21.—Continued

```
       # in order of increasing times...
200    DATA 86400 * 20, 86400 * 40, 86400 * 45, 86400 * 1000
210    READ exp_time(1), exp_time(2), exp_time(3), exp_time(4)
       # scale y1-axis (flux) (not used)...
230    DATA '1', '1', '1', '1'
240    READ scale_y1$(1), scale_y1$(2), scale_y1$(3), scale_y1$(4)
       # scale y2-axis (mass) (not used)...
260    DATA '1', '1', '1', '1'
270    READ scale_y2$(1), scale_y2$(2), scale_y2$(3), scale_y2$(4)
       # scale max of the profile y axes...
280    DATA '0 1.2e-9', '0 2.5e-5', '0 2e-10', '0 auto'
290    READ profile_y1$(1), profile_y1$(2), profile_y1$(3), profile_y1$(4)
300    DATA '0 1.2e-9', '0 2.5e-5', '0 6e-10', '0 auto'
310    READ profile_y2$(1), profile_y2$(2), profile_y2$(3), profile_y2$(4)

       # Define model parameters...
350    Dw = 2.5e-9                               # default tracer diffusion coefficient /  (m2/s)
360    nfilt1 = 1                                # number of cells in filter 1
370    nfilt2 = 1                                # number of cells in filter 2
380    nclay = 11                                # number of clay cells
390    f_free = 0.117                            # fraction of free pore water (0.01 - 1)
400    f_DL_charge = 0.45                        # fraction of CEC charge in electrical double layer
410    tort_n = -0.99                            # exponent in Archie's law, -1.045 without filters
420    G_clay = por_clay^tort_n                  # geometrical factor
430    interlayer_D$ = 'false'                   # 'true' or 'false' for interlayer diffusion
440    G_IL = 700                                # geometrical factor for clay interlayers
450    punch_time = 60 * 60 * 6                  # punch time / seconds
460    profile$ = 'true'                         # 'true' or 'false' for c/x profile visualization
470    IF nfilt1 = 0 THEN thickn_filter1 = 0
480    IF nfilt2 = 0 THEN thickn_filter2 = 0

       # See which tracer is present...
490    IF tot("Hto") > 1e-10 THEN tracer = 1 ELSE \
       IF tot("Cl_tr") > 1e-10 THEN tracer = 2 ELSE tracer = 3

       # Define clay pore water composition...
520    sol$ = nl$ +  ' pH 7.6' + sc$ +' pe 14 O2(g) -1.0' + sc$ +' temp 23'
530    sol$ = sol$ + nl$ + ' Na 240' + sc$ +' K 1.61' + sc$ +' Mg 16.9' + sc$ +' Ca 25.8' + sc$ +' Sr 0.505'
540    sol$ = sol$ + nl$ + ' Cl 300' + sc$ +' S(6) 14.1' + sc$ +' Fe(2) 0.0' + sc$ +' Alkalinity 0.476'
```

Table 60. Input file for example 21.—Continued

```
    # Define phases in which the tracers precipitate...
550 tracer_phases$ = nl$ + 'PHASES '
560 tracer_phases$ = tracer_phases$ + nl$ + ' Hto = Hto' + sc$ +' log_k -15'
570 tracer_phases$ = tracer_phases$ + nl$ + ' Na_tr = Na_tr' + nl$ + ' A_Hto' + nl$ + ' log_k -14'
580 tracer_phases$ = tracer_phases$ + nl$ + ' Na_trCl = Na_tr+ + Cl-' + sc$ +' log_k -14'
590 tracer_phases$ = tracer_phases$ + nl$ + ' NaCl_tr = Na+ + Cl_tr-' + sc$ +' log_k -13'
600 DIM tracer_equi$(4)
610 FOR i = 1 TO 4
620     tracer_equi$(i) = nl$ + 'A_' + tracer$(i) + ' 0 0'
630 NEXT i

    # Write solutions for the cells...
650 punch nl$ + 'PRINT ' + sc$ + ' -reset false' + sc$ + ' -echo_input true' + sc$ + ' -user_print true'
660 IF nfilt1 = 0 THEN GOTO 800
670 punch nl$ + x$ + ' filter cells at tracer-in side...'
680 r1 = r_int - thickn_filter1
690 xf1 = thickn_filter1 / nfilt1
700 FOR i = 1 TO nfilt1
710     num$ = TRIM(STR$(i + 3)) + sc$
720     V_water = 1e3 * height * por_filter1 * pi * (SQR(r1 + xf1) - SQR(r1))
730     punch nl$ + 'SOLUTION ' + num$ + ' -water ' + STR$(V_water)
740     punch sol$ + nl$
750     r1 = r1 + xf1
760 NEXT i

800 punch nl$ + nl$ + x$ + ' cells in Opalinus Clay...'
810 r1 = r_int
820 x = thickn_clay / nclay
830 FOR i = 1 TO nclay
840     num$ = TRIM(STR$(i + 3 + nfilt1)) + sc$
850     V_water = 1e3 * height * por_clay * pi * (SQR(r1 + x) - SQR(r1))
860     punch nl$ + 'SOLUTION ' + num$ + ' -water ' + STR$(V_water * f_free)
870     punch sol$
880     IF f_free = 1 and tracer = 1 THEN GOTO 960
890     punch nl$ + 'SURFACE ' + num$ + ' -equil ' + num$
900     punch nl$ + ' Su_ ' + TRIM(STR$(f_DL_charge * CEC * V_water)) + STR$(A_por) + ' ' + STR$(V_water)
910     punch nl$ + ' Su_ii ' + TRIM(STR$(7.88e-4 * rho_b_eps * V_water))
920     punch nl$ + ' Su_fes ' + TRIM(STR$(7.4e-5 * rho_b_eps * V_water))
930     IF f_free < 1 THEN punch nl$ + ' -Donnan ' + TRIM(STR$((1 - f_free) * 1e-3 / A_por))
940     punch nl$ + 'EXCHANGE ' + num$ + ' -equil ' + num$
```

Table 60. Input file for example 21.—Continued

```
 950   punch nl$ + ' X ' + TRIM(STR$((1 - f_DL_charge) * CEC * V_water)) + nl$
 960   r1 = r1 + x
 970   NEXT i

1000  IF nfilt2 = 0 THEN GOTO 1200
1010  punch nl$ + nl$ + x$ + ' tracer-out filter cells...'
1020  r1 = r_ext
1030  xf2 = thickn_filter2 / nfilt2
1040  FOR i = 1 TO nfilt2
1050    num$ = TRIM(STR$(i + 3 + nfilt1 + nclay)) + sc$
1060    V_water = 1e3 * height * por_filter2 * pi * (SQR(r1 + xf2) - SQR(r1))
1070    punch nl$ + 'SOLUTION ' + num$ + ' -water ' + STR$(V_water)
1080    punch sol$ + nl$
1090    r1 = r1 + xf2
1100  NEXT i

1200  punch nl$ + x$ + ' outside solution...'
1210  num$ = TRIM(STR$(4 + nfilt1 + nclay + nfilt2)) + sc$
1220  punch nl$ + 'SOLUTION ' + num$ + ' -water ' + STR$(V_end)
1230  punch sol$
1240  punch nl$ + 'END'

#     Write phases in which the tracers precipitate...
1300  punch nl$ + tracer_phases$
1310  punch nl$ + 'EQUILIBRIUM_PHASES ' + num$ + tracer_equi$(tracer)
1312  If tracer = 3 THEN punch nl$ + tracer_equi$(tracer + 1)
1320  punch nl$ + 'END'

#     Define mixing factors for the diffusive flux between cells 1 and 2:
#       J_12 = -2 * Dw / (x_1 / g_1 + x_2 / g_2) * (c_2 - c_1)
#     Multiply with dt * A / (V = 1e-3 m3).    (Actual volumes are given with SOLUTION; -water)
#     Use harmonic mean: g_1 = por_1 / G_1, g_2 = por_2 / G_2, x_1 = Delta(x_1), etc.
1400  IF nfilt1 > 0 THEN gf1 = por_filter1 / G_filter1
1410  IF nfilt2 > 0 THEN gf2 = por_filter2 / G_filter2
1420  g = por_clay / G_clay
#     Find max time step = 0.5 * V_water * dx * G_factor / (Dw * por * A * fbc)
#       V_water = por * pi * height * ((r + dr)^2 - r^2)
#         A = por * pi * height * r * 2
#     At the inlet of the tracers, fbc = 2...
1500  IF nfilt1 = 0 THEN GOTO 1530
```

Table 60. Input file for example 21.—Continued

```
1510    r1 = r_int - thickn_filter1
1520    ff = (SQR(r1 + xf1) - SQR(r1)) * xf1 * G_filter1 / (r1 * 2) / 2
1530    ff1 = (SQR(r_int + x) - SQR(r_int)) * x * G_clay / (r_int * 2) / 2
        # Perhaps the clay has very small cells...
1540    IF nfilt1 = 0 THEN ff = ff1 ELSE IF ff1 * 2 < ff THEN ff = ff1 * 2
        # Or at the filter1-clay transition, fbc = 1...
1550    IF nfilt1 > 0 THEN ff1 = (SQR(r_int + x) - SQR(r_int)) * (xf1 / gf1 + x / g) / (2 * r_int * 2)
1560    IF nfilt1 > 0 AND ff1 < ff THEN ff = ff1
        # Perhaps filter2 has very small cells...
1570    IF nfilt2 > 0 THEN ff1 = (SQR(r_ext + xf2) - SQR(r_ext)) * xf2 * G_filter2 / (r_ext * 2)
1580    IF nfilt2 > 0 AND ff1 < ff THEN ff = ff1
1590    dt_max = 0.5 * ff / Dw
        # Check with punch times, set shifts...
1610    IF punch_time < dt_max THEN dt = punch_time ELSE dt = dt_max
1620    punch_fr = 1
1630    IF dt < punch_time THEN punch_fr = ceil(punch_time / dt)
1640    dt = punch_time / punch_fr
1650    shifts = ceil(exp_time(tracer) / dt)
        # Write mixing factors...
1700    punch nl$ + nl$ + x$ + ' mixing factors...'
1710    r1 = r_int
1720    IF nfilt1 > 0 THEN r1 = r_int - thickn_filter1
1730    A = height * 2 * pi
1740    FOR i = 0 TO nfilt1 + nclay + nfilt2
1750    IF i = 0 OR i = nfilt1 + nclay + nfilt2 THEN fbc = 2 ELSE fbc = 1
1760    IF i > nfilt1 OR nfilt1 = 0 THEN GOTO 1810
1770    IF i < nfilt1 THEN mixf = Dw * fbc / (xf1 / gf1) * dt * A * r1 / 1e-3
1780    IF i = nfilt1 THEN mixf = 2 * Dw / (xf1 / gf1 + x / g) * dt * A * r1 / 1e-3
1790    IF i < nfilt1 THEN r1 = r1 + xf1 ELSE r1 = r1 + x
1800    GOTO 1880
1810    IF i > nfilt1 + nclay THEN GOTO 1860
1820    mixf = Dw * fbc / (x / g) * dt * A * r1 / 1e-3
1830    IF i = nfilt1 + nclay AND nfilt2 > 0 THEN mixf = 2 * Dw / (xf2 / gf2 + x / g) * dt * A * r1 / 1e-3
1840    IF i < nfilt1 + nclay THEN r1 = r1 + x ELSE r1 = r1 + xf2
1850    GOTO 1880
1860    mixf = Dw * fbc / (xf2 / gf2) * dt * A * r1 / 1e-3
1870    r1 = r1 + xf2
1880    punch nl$ + 'MIX ' + TRIM(STR$(i + 3)) + sc$ + STR$(i + 4) + STR$(mixf)
1890    NEXT i
1900    punch nl$ + 'END'
```

Table 60. Input file for example 21.—Continued

```
      # Write TRANSPORT....
2000  punch nl$ + 'TRANSPORT'
2010  stag = 2 + nfilt1 + nclay + nfilt2
2020  punch nl$ + '    -warnings true'
2030  punch nl$ + '    -shifts ' + TRIM(STR$(shifts))
2040  punch nl$ + '    -flow diff' + sc$ + ' -cells 1' + sc$ + ' -bcon 1 2' + sc$ + ' -stag ' + TRIM(STR$(stag))
2050  punch nl$ + '    -time ' + STR$(dt)
2060  punch nl$ + '    -multi_D true ' + STR$(Dw) + STR$(por_clay) + ' 0.0 ' + TRIM(STR$(-tort_n))
2070  punch nl$ + '    -interlayer_D ' + interlayer_D$ + ' 0.001 0.0 ' + TRIM(STR$(G_IL))
2080  punch nl$ + '    -punch_fr ' + TRIM(STR$(punch_fr)) + sc$ + ' -punch_c ' + TRIM(STR$(2 + stag))

      # Write USER_GRAPH....
2180  FOR i = 0 to 1
2190    punch nl$ + 'USER_GRAPH ' + TRIM(STR$(tracer + i)) + ' Example 21' + nl$
2200    punch nl$ + '    -chart_title "' + tracer$(tracer + i) + ' Diffusion to Outer Cell"'
2210    punch nl$ + '    -plot_tsv_file ex21_' + tracer$(tracer + i) + '_rad.tsv'
2220    punch nl$ + '    -axis_scale x_axis 0 ' + TRIM(STR$(exp_time(tracer + i) / (3600 * 24)))
2230    punch nl$ + '    -axis_titles "Time, in days" "Flux, in moles per square meter per second" \
                        "Accumulated mass, in moles"'
2240    punch nl$ + '    -plot_concentration_vs time'
2250    punch nl$ + '    10 days = total_time / (3600 * 24)'
2260    punch nl$ + '    20 a = equi("A_' + tracer$(tracer + i) + '")'
2270    punch nl$ + '    30 IF get(1) = 0 AND total_time > 0 THEN put(total_time, 1)'
2280    punch nl$ + '    40 dt = get(1)'
2290    A = 2 * pi * r_ext * height
2300    i$ = TRIM(STR$(2 + i))
2310    punch nl$ + '    50 plot_xy days - dt / (2 * 3600 * 24), (a - get(' + i$ + ')) / dt /' + STR$(A) + \
                      ', color = Green, symbol = None'
2320    punch nl$ + '    60 put(a, ' + i$ + ')'
2330    punch nl$ + '    70 plot_xy days, equi("A_' + tracer$(tracer + i) + \
                      '"), y_axis = 2, color = Red, symbol = None'
2340    IF tracer < 3 THEN GOTO 2360
2350  NEXT i
2360  punch nl$ + 'END'

2400  IF profile$ = 'true' THEN GOSUB 3000
2410  IF tracer < 3 THEN END # finished for Hto and Cl

      # Continue with Cs....
```

Table 60. Input file for example 21.—Continued

```
2420 IF profile$ = 'false' THEN punch nl$ + 'USER_GRAPH ' + TRIM(STR$(tracer)) + sc$ + ' -detach' ELSE \
                               punch nl$ + 'USER_GRAPH ' + TRIM(STR$(tracer + 4)) + sc$ + ' -detach'
2440 tracer = tracer + 1
2450 punch nl$ + 'TRANSPORT'
2460 shifts = ceil((exp_time(tracer) - exp_time(tracer - 1))/ dt)
2480 punch nl$ + ' -shifts ' + TRIM(STR$(shifts))
2490 punch nl$ + ' -punch_fr ' + TRIM(STR$(punch_fr)) + sc$ + ' -punch_c ' + TRIM(STR$(2 + stag))
2500 punch nl$ + 'END'
2510 IF profile$ = 'true' THEN GOSUB 3000
2520 END # finished...

     # Write TRANSPORT and USER_GRAPH for concentration profile...
3000 punch nl$ + 'TRANSPORT'
3010 punch nl$ + ' -shifts 0'
3020 punch nl$ + ' -punch_fr 2' + sc$ + ' -punch_c 3-' + TRIM(STR$(2 + stag))
     # Write USER_GRAPH...
3030 punch nl$ + 'USER_GRAPH ' + TRIM(STR$(tracer)) + sc$ + ' -detach'
3040 punch nl$ + 'USER_GRAPH ' + TRIM(STR$(tracer + 4)) + ' Example 21' + nl$
3050 punch nl$ + ' -chart_title "' + tracer$(tracer) + ' Concentration Profile: Filter1 | Clay | Filter2"'
3060 REM punch nl$ + ' -plot_tsv_file + tracer$(tracer) + ' _prof.tsv'
3070 punch nl$ + ' -axis_scale x_axis 0 ' + TRIM(STR$((thickn_filter1 + thickn_clay + thickn_filter2) * 1e3))
3080 punch nl$ + ' -axis_scale y_axis ' + profile_y1$(tracer)
3090 punch nl$ + ' -axis_scale sy_axis ' + profile_y2$(tracer)
3100 punch nl$ + ' -axis_titles ' + '"Distance, in millimeters" "Free pore-water molality" "Total molality"'
3110 punch nl$ + ' -headings ' + tracer$(tracer) + '_free ' + tracer$(tracer) + '_tot'
3120 punch nl$ + ' -plot_concentration_vs x'
3130 punch nl$ + ' -initial_solutions true'
3140 punch nl$ + ' 10 IF cell_no = 3 THEN xval = 0 ELSE xval = get(14)'
3150 punch nl$ + ' 20 IF (' + TRIM(STR$(nfilt1)) + ' = 0 OR cell_no > ' + TRIM(STR$(nfilt1 + 3)) + ') THEN
GOTO 60'
3160 punch nl$ + ' 30 IF (cell_no = 4) THEN xval = xval + 0.5 * ' + TRIM(STR$(xf1))
3170 punch nl$ + ' 40 IF (cell_no > 4 AND cell_no < ' + TRIM(STR$(nfilt1 + 4)) + \
     ') THEN xval = xval + ' + TRIM(STR$(xf1))
3180 punch nl$ + ' 50 GOTO 200'
3190 punch nl$ + ' 60 IF (cell_no = ' + TRIM(STR$(4 + nfilt1)) + ') THEN xval = xval + 0.5 * ' + \
     TRIM(STR$(xf1)) + ' + 0.5 * ' + TRIM(STR$(x))
3200 punch nl$ + ' 70 IF (cell_no = ' + TRIM(STR$(4 + nfilt1)) + ' AND cell_no < ' + \
     TRIM(STR$(4 + nfilt1 + nclay)) + ') THEN xval = xval + ' + TRIM(STR$(x)) + ' ELSE GOTO 90'
3210 punch nl$ + ' 80 GOTO 200'
3220 punch nl$ + ' 90 IF (cell_no = ' + TRIM(STR$(4 + nfilt1 + nclay)) + ') THEN xval = xval + 0.5 * ' + \
```

Table 60. Input file for example 21.—Continued

```
        TRIM(STR$(x)) + ' ' + 0.5 * ' ' + TRIM(STR$(xf2))
3230 punch nl$ + ' 100 IF (cell_no > ' + TRIM(STR$(4 + nfilt1 + nclay)) + ' AND cell_no <= ' + \
        TRIM(STR$(3 + nfilt1 + nclay + nfilt2)) + ') THEN xval = xval + ' + TRIM(STR$(xf2)) + \
3240 punch nl$ + ' 110 IF (cell_no = ' + TRIM(STR$(4 + nfilt1 + nclay + nfilt2)) + \
        ') THEN xval = xval + 0.5 * ' + TRIM(STR$(xf2))
3250 punch nl$ + ' 200 y1 = TOT("' + tracer$(tracer) + '")'
3260 punch nl$ + ' 210 plot_xy xval * 1e3, y1, color = Blue, symbol = Plus'
3270 punch nl$ + ' 220 IF cell_no = 3 THEN put(y1, 15)'
3280 punch nl$ + ' 230 IF (cell_no < ' + TRIM(STR$(4 + nfilt1)) + ' OR cell_no > ' + \
        TRIM(STR$(3 + nfilt1 + nclay)) + ') THEN GOTO 400'
3290 punch nl$ + ' 240 y2 = SYS("' + tracer$(tracer) + '") / (tot("water") + edl("water"))'
     # Remove REM if total conc's per kg solid must be plotted (and adapt axis_titles)...
3310 punch nl$ + ' 250 REM y2 = y2 / ' + TRIM(STR$(rho_b_eps)) + x$ + ' conc / kg solid'
3320 punch nl$ + ' 260 plot_xy xval * 1e3, y2, symbol = Circle, y_axis = 2'
3330 punch nl$ + ' 270 IF (cell_no > ' + TRIM(STR$(5 + nfilt1)) + ') THEN GOTO 400'
3340 punch nl$ + ' 280 IF ' + TRIM(STR$(nfilt1)) + ' THEN plot_xy ' + TRIM(STR$(thickn_filter1 * 1e3)) + \
        ', get(15), color = Black, symbol = None'
3350 punch nl$ + ' 290 IF ' + TRIM(STR$(nfilt2)) + ' THEN plot_xy ' + \
        TRIM(STR$((thickn_filter1 + thickn_clay) * 1e3)) + ', get(15), color = Black, symbol = None'
3360 punch nl$ + ' 300 put(0, 15)'
3370 punch nl$ + ' 400 put(xval, 14)'
3380 punch nl$ + 'END'
3390 RETURN
END
PRINT
  -selected_output false; -status false
INCLUDE$ radial
END
```

which forms the start of the stagnant column and circulates at the inner filter of the diffusion cell. This solution contains the tracers and the amounts of water used in the experiment. The lines can be uncommented one by one to run the file successively with the different tracers and amounts of water. When SOLUTION 3 is calculated, USER_PUNCH is processed to write a SELECTED_OUTPUT file named *radial*, which contains the SOLUTION definitions for the cells in the filters and the clay, the mixing factors, and the TRANSPORT and USER_GRAPH data blocks. The Basic lines in USER_PUNCH do the following tasks:

- Lines 1–4 define variables that facilitate printing of special symbols used in printing the output—end of line, comment sign, and semicolon—and define π, which is used to write definitions for the radial configuration of the experimental cell.

- Lines 10–150 define the dimensions of the experimental cell and properties of the filters and the clay that have been measured and, thus, should be considered as constant.

- Lines 160–310 fill arrays with the names of the tracers, the experimental times, and the scale factors for the graphs.

- Lines 350–480 give model parameters that may be varied to simulate the experiments, and can be changed to check the diffusion model. Typically, for checking the numerics, the number of cells for the filters (variables `nfilt1` and `nfilt2`) and for the clay (`nclay`), and the time step (`punch_time`) can be altered without affecting the calculated results. It also is interesting to set `nfilt1` and/or `nfilt2` to zero and inspect the effects that the filters have on the fluxes. (The program will fail if `nclay` is set to zero.) Values of the other model parameters were derived from the geometrical factors obtained by fitting the measured tracer diffusion curves (Appelo and others, 2010). By adjusting these parameters, the Basic program and the functioning of PHREEQC can be verified. For example, `f_free` sets the fraction of free pore water, partitioning the pore space in free and double layer water. Values may range from 0.01 to 1. The variable has little effect on tritium which, as an uncharged species, diffuses equally quickly in free and double-layer water (if the latter is given the viscosity of free pore water). However, the variable has a major effect on the diffusion of $^{36}Cl^-$, which diffuses more quickly in the free pore water. The parameter `f_DL_charge` partitions the Cation Exchange Capacity (CEC) over the double layer and exchange sites. Increasing its value does not affect the diffusion of tritium, but decreases the flux of Cl^- and increases the flux of Na^+.

- Line 490 checks which tracer is present to set the experimental times and scale factors for the graphs.

- Lines 520–540 define clay pore-water composition.

- Lines 550–630 define the solid phases in which the tracers precipitate in the outer solution. The moles of this phase will record the amounts that have diffused. The phases have such a low solubility that the tracer concentration is essentially zero.

- Lines 650–1100 write the solutions for the filter cells and the clay, with radially increasing amounts of water. For the clay, keyword **SURFACE** is used to define the moles of the surface sites of Su_, Su_ii and Su_fes, for the double layer, and the sites on illite that sorb the alkaline cations with intermediate and high strength, respectively. The fixed sites of the Cation Exchange Capacity (CEC) are defined with **EXCHANGE**.

- Lines 1200–1320 write the external **SOLUTION** and the **EQUILIBRIUM_PHASES** in which the tracers are captured.

- Lines 1400–1900 calculate and write the mixing factors as explained above in equations 44–49. First, the maximum time step is derived from one of the following: the innermost filter cell, at the transition of the inner filter to the clay, the innermost clay cell, or the outer filter cell. The time step is decreased when it is larger than desired by `punch_time`. With this time step, the mixing factors are calculated for each cell and written to the file, taking care of the heterogeneities at physical boundaries and the radial outline of the field.

- Lines 2000–2520 write data blocks for **TRANSPORT** and **USER_GRAPH**. The experimental data (L.R. Van Loon, Paul Scherrer Institute, written commun., 2011) are plotted ("**-plot_tsv_file** *file_name*") together with the calculated accumulated mass in the outer solution and the flux, obtained by dividing the mass that has accumulated by the time interval and the outer surface area of the clay.

- Lines 3000–3390 contain the subroutine to write **TRANSPORT** and **USER_GRAPH** data blocks for plotting concentration profiles in the filters and the clay.

The experiments with the tracers ^{22}Na and Cs can be calculated together, because the tracer solutions contained identical amounts of water. Thus, when the ^{22}Na profile has been plotted after 45 days, another **TRANSPORT** block is written to calculate diffusion of Cs during 1,000 days in total.

Following the **END** after **USER_PUNCH**, the file *radial* is loaded in the input file with **INCLUDE$** and then processed. Before the file is included, writing to file *radial* is suspended (**PRINT**; -selected_output false), thus avoiding rewriting the file over and over again, each time a solution is calculated.

Three transport calculations with different tracers can be run with the input file (1) HTO, (2) ^{36}Cl, and (3) ^{22}Na and Cs together. To perform one of these three calculations, uncomment the appropriate line defining the tracer(s) in **SOLUTION** 3. Note that the length of the Cs experiment is about 1,000 days compared to less than 50 days for the other experiments; consequently, the Cs calculation is 20 times longer than the others.

The results of the three calculations are shown in figures 24 and 25. As shown, the arrival times of the tracers that accumulate in the outer solution are delayed by the storage in pore water and the sorption on minerals in the clay. The delay increases in the order $^{36}Cl^- < HTO < {}^{22}Na^+ < Cs^+$ (fig. 24). The total storage can be obtained from the graphs by extrapolating the straight-line segment of the accumulated mass to time zero (fig. 24) and reading the value from the secondary Y-axis (a negative number, because mass is lost). The flux, the derivative of the mass with time, shows that the accumulation of HTO, and of Cs^+ in particular, decreases during the experiment (fig. 24) because the concentration is diminishing in the tracer solution. The volume of the solution with HTO, the first experiment performed, was relatively small (0.2 L) and insufficient to maintain a constant concentration for the inner solution. Sorption of Cs^+ is so strong (more than 99.5 percent of Cs^+ resides in the solid phase) that 1 L of inner solution simply contains insufficient mass to fill all the sorption sites on the clay. Figure 25 shows on the primary Y axis the concentration profile at the end of the diffusion periods, including concentrations in the inner filter, the free pore water of the clay, and the outer filter. The figure shows on the secondary Y axis the total moles, which includes moles sorbed and moles dissolved in free and double layer water, expressed per kilogram of total water. The two concentrations are the same for HTO (the concentration of HTO is the same in free and in double layer water), the free pore-water concentration of Cl^- is twice the total concentration (the concentration of Cl^- in the double layer is smaller than in free pore water), and the free pore-water concentrations of Na^+, and particularly of Cs^+, are smaller than the total concentrations because the cations have a higher concentration in the double layer than in free pore water, and because cations are sorbed on the surface and exchange sites.

The model simulates the experimental results very well (fig. 24), except for Cs^+. The calculated arrival time of Cs^+ is almost 100 days later than observed and then the mass accumulates too slowly. This behavior of Cs^+ has been found in many similar experiments. It has been modeled by increasing the diffusion coefficient and decreasing the sorption capacity for Cs^+ relative to batch experiments; these adjustments can be easily incorporated in this example.

The diffusion of Cs^+ can be increased by setting interlayer diffusion to true in Line 430:

```
430  interlayer_D$ = 'true'
```

(It is of interest to see the different effects of interlayer diffusion on $^{22}Na^+$ and on Cs^+.)

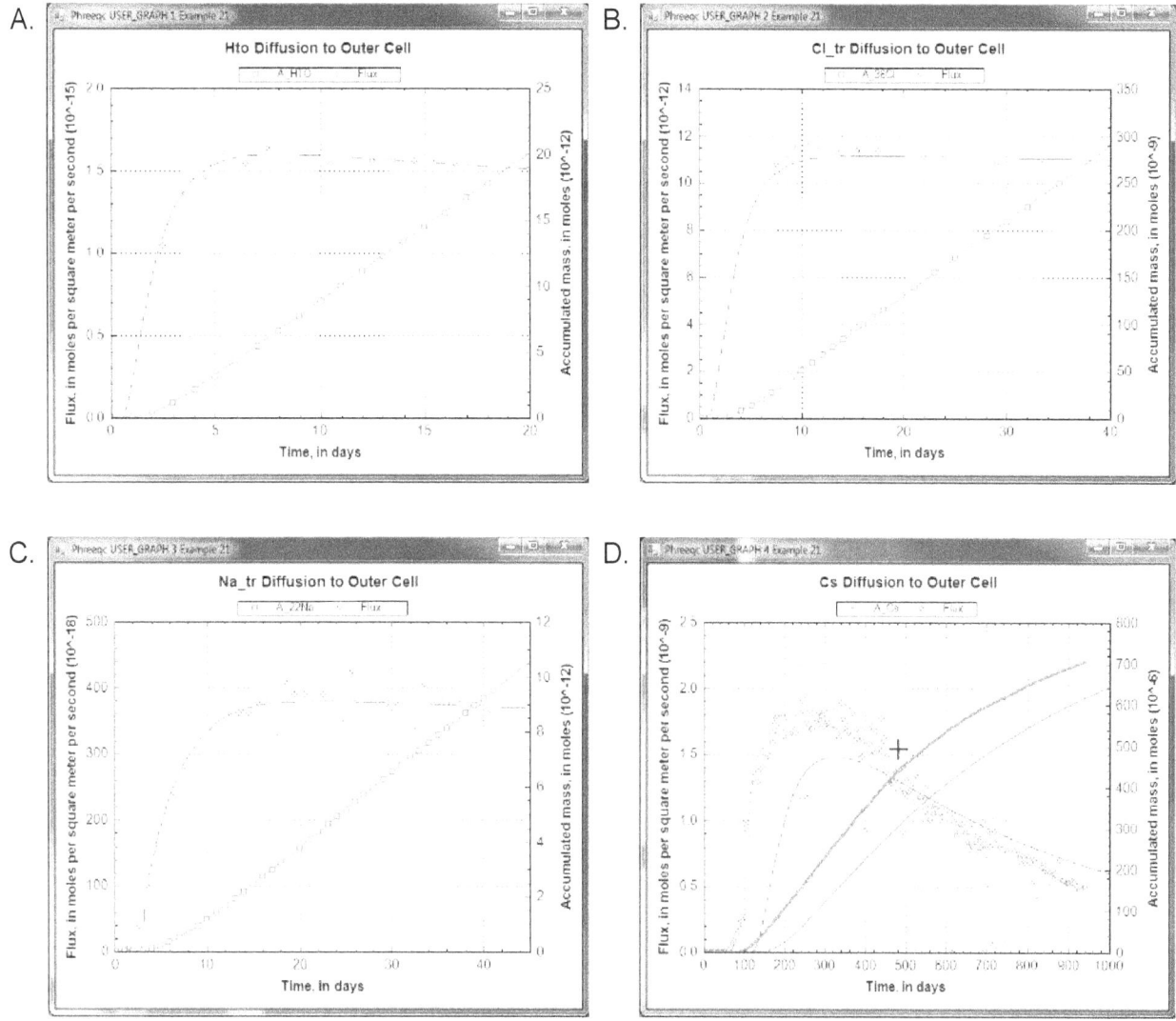

Figure 24. Mass outflow (red) and corresponding flux (green) by diffusion through the radial cell for (A) HTO, (B) ^{36}Cl$^-$, (C) ^{22}Na$^+$, and (D) Cs$^+$. Lines are modeled; symbols indicate measured data.

The results, shown in figure 26, illustrate that the arrival time of Cs$^+$ can be matched by increasing diffusion, but that the mass accumulates too quickly. Another option for reducing the delay of the tracer arrival is to decrease the sorption of Cs$^+$, either by decreasing the moles of surface and exchange sites or by decreasing the complexation constants, or both. By adjusting both the sorption and the diffusion coefficient, it may be possible to simulate the experimental data for Cs$^+$. However, it remains difficult to explain why the sorption capacity is different between batch and diffusion experiments for Cs$^+$, but not for the other cations.

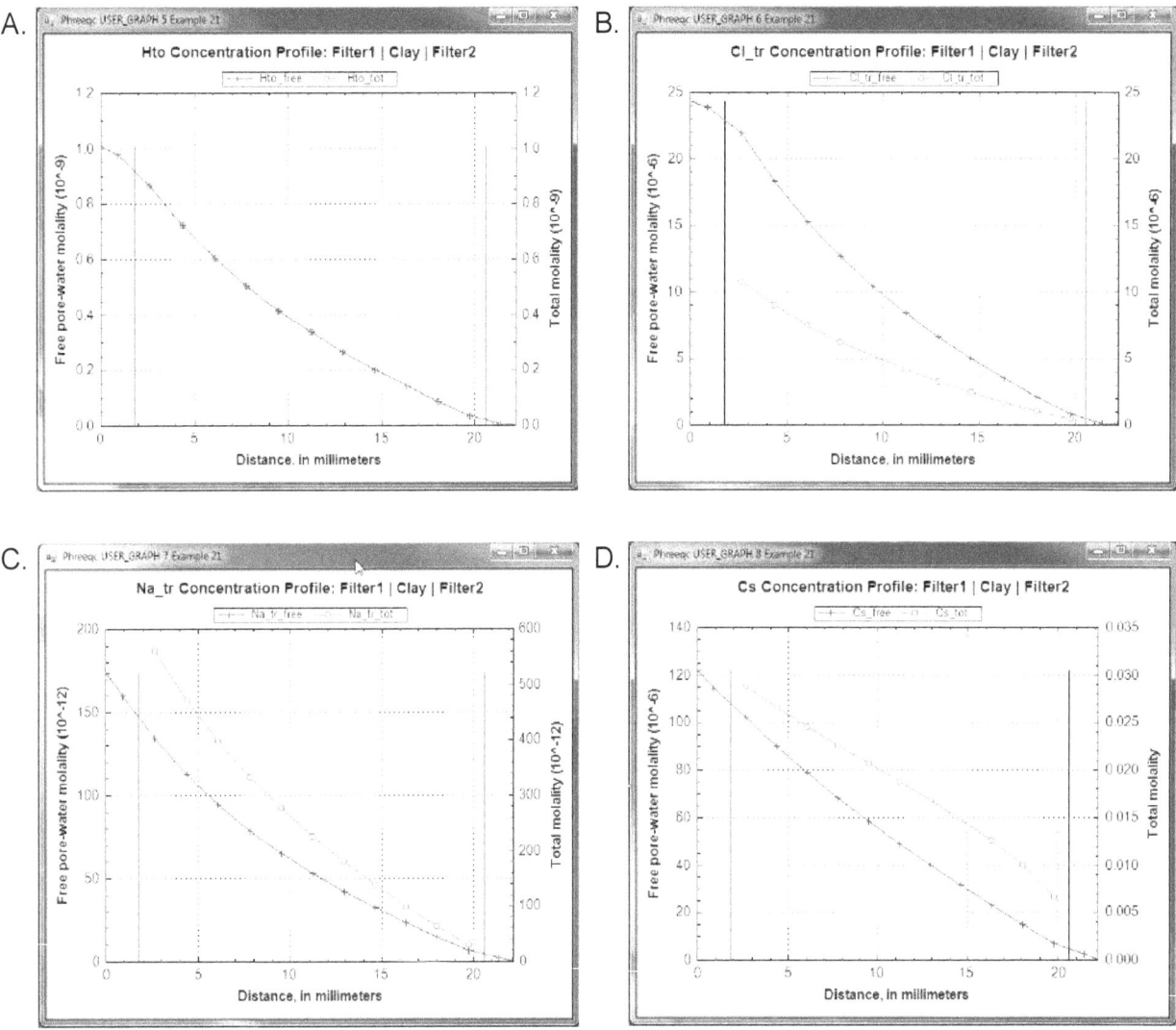

Figure 25. Concentration profiles in the filters and free pore water in the clay (blue), and total moles per kilogram water in the double-layer water in the clay (red) at the end of the calculations in the radial cell for (A) HTO, (B) ^{36}Cl$^-$, (C) ^{22}Na$^+$, and (D) Cs$^+$. Vertical lines indicate the positions of the filters.

Alternatively, and in line with the heterogeneous distribution of Cs$^+$ in the clay after the experiment, the relatively fast arrival time can be modeled with a dual-porosity structure in which the pore space is subdivided into continuous and stagnant pores that can exchange by diffusion. The continuous pores allow Cs$^+$ to move more rapidly through the clay than is calculated for a homogeneous medium, depending on the proportion, the flow velocity, and the diffusional exchange with the stagnant pores. With equation 49, and some effort, the dual-porosity structure can be introduced in the Basic program. Otherwise, the C program that is given as supplementary information in Appelo and others (2010) can be used. Similarly to the Basic program, the C program writes a complete PHREEQC input file for diffusion of Cs$^+$, but in a dual porosity

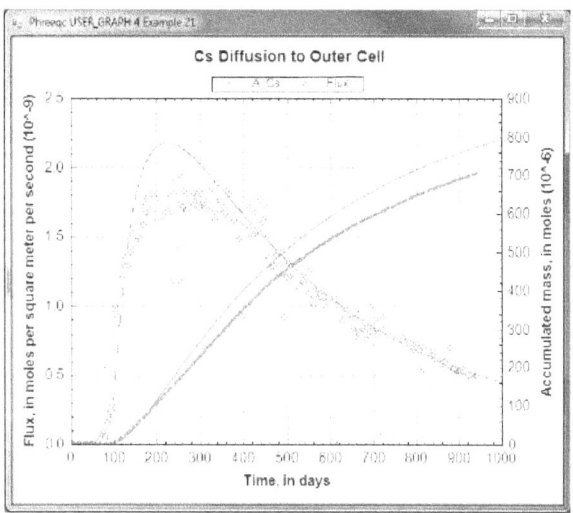

Figure 26. Mass outflow (red) and corresponding flux (green) of Cs⁺ in the diffusion cell when interlayer diffusion is included.

clay; in addition, the program permits distributing the surface and exchange sites differently between the stagnant and continuous pores.

Example 22—Modeling Gas Solubilities: CO$_2$ at High Pressures

The solubility of gas i is given by $m_i = K_H \dfrac{\varphi_i P_i}{\gamma_i}$, where m is the molality, γ is the activity coefficient in water, K_H is the equilibrium constant, P is the partial pressure, and φ is the fugacity coefficient (the activity coefficient in the gas phase). PHREEQC calculates the fugacity coefficient with the Peng-Robinson equation of state (Peng and Robinson, 1976) from the critical pressure and temperature, and the acentric factor of the gas in a gas mixture to obtain the limiting volume and the attraction factor in the Van der Waals equation. The fugacity coefficient is close to 1 when the total pressure of the gas phase is less than about 10 atm, and it can be neglected in the solubility calculation. At higher pressures, the effect can be substantial, as shown for CO_2 in figure 27. At low pressures, the concentration of CO_2 increases near-linearly with pressure. At 25 °C and pressures higher than 62 atm, the concentration increases more gradually because the fugacity coefficient drops rapidly.

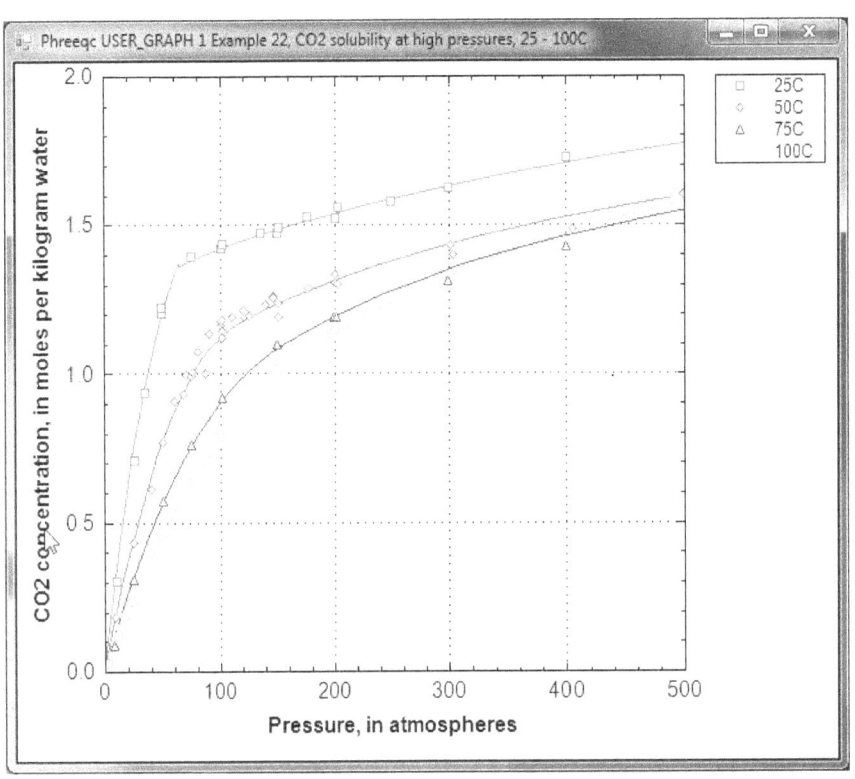

Figure 27. Solubility of CO_2 gas as a function of gas pressure at 25, 50, 75 and 100 °C. Data points are from compilations by Duan and others (2006) and Spycher and others (2003); lines are calculated with PHREEQC.

The aqueous concentrations for a point in figure 27 can be calculated by initial equilibrium in the **SOLUTION**:

```
SOLUTION
  pH 3 charge
  -pressure 1000
  C(4) 100 CO2(g) 3 # 1000 atm
END
```

or with **EQUILIBRIUM_PHASES**:

```
SOLUTION
  -pressure 1000
EQUILIBRIUM_PHASES
  CO2(g) 3 # 1000 atm
END
```

For these two options the solutions or the equilibrium phases need to be redefined for each pressure, either by explicitly repeating the keywords in the file, or by writing an input file in Basic and reading it in with **INCLUDES**.

Another option is to use a fixed-volume GAS_PHASE and let the pressure increase by adding CO_2 with keyword REACTION (table 61). This option permits other gases to be included; for example, H_2O, for which the pressure and fugacity coefficient also will be calculated. Furthermore, the pressure of the solution does not need to be defined as in the options above because it will change with the pressure of the gas phase.

Table 61. Input file for example 22.

```
TITLE Example 22.--Compare experimental CO2 solubilities at high CO2 pressure
    with Peng-Robinson calc'ns with fixed-volume gas_phase, 25, 50, 75, 100 oC.
SOLUTION 1
GAS_PHASE 1
    -fixed_volume
    CO2(g)  0
    H2O(g)  0
REACTION
 CO2 1; 0 27*1
INCREMENTAL_REACTIONS true

USER_GRAPH 1 Example 22, CO2 solubility at high pressures, 25 - 100C
 -plot_tsv_file co2.tsv
 -axis_titles "Pressure, in atmospheres" \
            "CO2 concentration, in moles per kilogram water"
 -axis_scale x_axis 0 500
 -axis_scale y_axis 0 2
 -connect_simulations false
 10 graph_x PR_P("CO2(g)")
 20 graph_y TOT("C(4)")
 -end
USER_GRAPH 2 Example 22, P-Vm of CO2 gas, 25 - 100C
 -headings 25C
 -axis_titles "Molar volume of CO2 gas, in liters per mole" \
            "CO2 pressure, in atmospheres"
 -axis_scale x_axis 0 1
 -axis_scale y_axis 0 500
 -connect_simulations false
 10 plot_xy gas_vm, gas_p symbol = None
 -end
END

USE solution 1
USE gas_phase 1
USE reaction 1
REACTION_TEMPERATURE 2
 50
USER_GRAPH 2
 -headings 50C
END
USE solution 1
USE gas_phase 1
USE reaction 1
```

Table 61. Input file for example 22.—Continued

```
REACTION_TEMPERATURE 3
 75
USER_GRAPH 2
 -headings 75C
END

USE solution 1
USE gas_phase 1
USE reaction 1
REACTION_TEMPERATURE 4
 100
USER_GRAPH 2
 -headings 100C
END
```

Figure 28 shows a plot of the gas pressure as a function of the molar volume at four temperatures, 25, 50, 75, and 100 °C. At 50 °C and higher, the graph displays a smooth increase of the gas pressure with decreasing volume. However, at 25 °C the curve flattens out between 0.07 and 0.15 L/mol. The constant pressure for decreasing volume indicates that CO_2 gas becomes liquid when it is compressed to less than

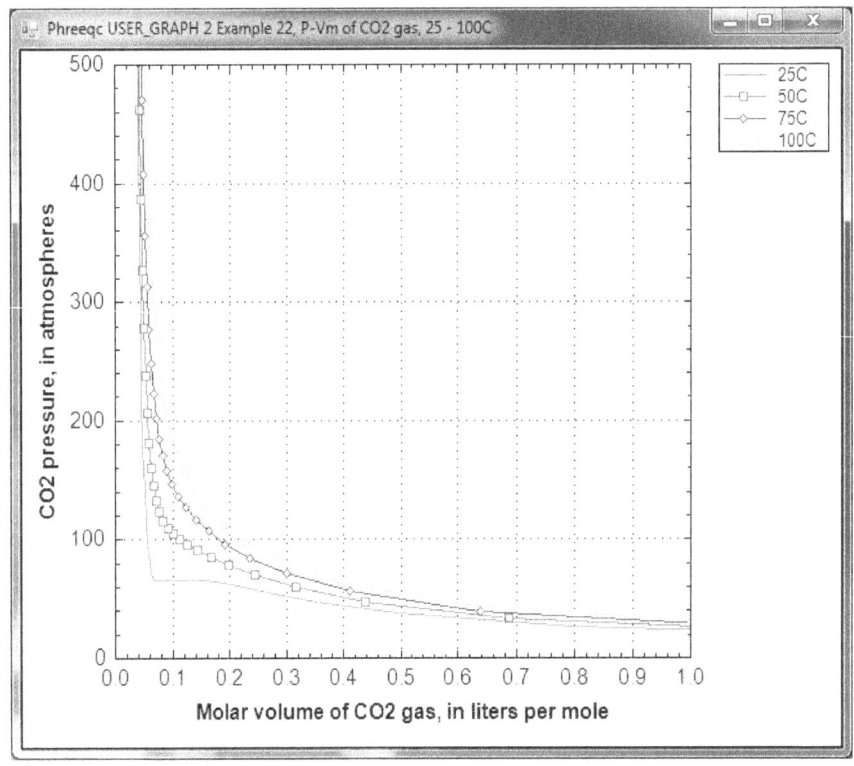

Figure 28. Pressure-volume curves of CO_2 gas at 25, 50, 75 and 100 °C. The calculated points are omitted at 25 °C to show more clearly how the curve flattens when the gas liquefies.

0.141 L/mol at 25 °C. At 0.068 L/mol all the gas has been liquefied, and further decrease of the molar volume then results again in pressure increase. The gas/liquid transition appears in the cubic Van der Waals equation (the basis of Peng-Robinson's equation), when it has three roots at temperatures below the critical temperature of the gas (31 °C for CO_2 gas).

When the pressure is calculated from the molar volume of the gas or the gas mixture, PHREEQC checks whether the equation has three roots. If so, it will test whether another root can be found at a larger molar volume; the root with the maximum pressure will be used as the solution of the equation. Thus, the gas pressure may remain constant (or, possibly even decrease slightly because the parameters in a gas-mixture may change) even when the molar volume decreases. The gas liquefaction may result in small kinks in the calculated solubility curve (calculated by equilibrating a solution with a GAS_PHASE), and one is visible in figure 27 at 65 atm and 25 °C.

Allison, J.D., Brown, D.S., and Novo-Gradac, K.J., 1991, MINTEQA2/PRODEFA2—A geochemical assessment model for environmental systems—Version 3.0 user's manual: Athens, Georgia, Environmental Research Laboratory, Office of Research and Development, U.S. Environmental Protection Agency, 106 p.

Appelo, C.A.J., 1994a, Cation and proton exchange, pH variations, and carbonate reactions in a freshening aquifer: Water Resources Research, v. 30, no. 10, p. 2793–2805.

Appelo, C.A.J., 1994b, Some calculations on multicomponent transport with cation exchange in aquifers: Ground Water, v. 32, no. 6, p. 968–975.

Appelo, C.A.J., and Postma, Dieke, 2005, Geochemistry, groundwater and pollution (2nd ed.): Leiden, The Netherlands, A.A. Balkema, 649 p.

Appelo, C.A.J., Van Loon, L.R., and Wersin, P., 2010, Multicomponent diffusion of a suite of tracers (HTO, Cl, Br, I, Na, Sr, Cs) in a single sample of Opalinus clay: Geochimica et Cosmochimica Acta, v. 74, no. 4, p. 1201–1219.

Appelo, C.A.J., and Wersin, P., 2007, Multicomponent diffusion modeling in clay systems with application to the diffusion of tritium, iodide, and sodium in Opalinus clay: Environmental Science and Technology, v. 41, no. 14, p. 5002–5007.

Appelo, C.A.J., Vinsot, A., Mettler, S., Wechner, S., 2008, Obtaining the porewater composition of a clay rock by modeling the in- and out-diffusion of anions and cations from an in-situ experiment: Journal of Contaminant Hydrology, v. 101, p. 67–76.

Atkins, P.W., and de Paula, Julio, 2002, Atkin's physical chemistry (7th ed.): Oxford, United Kingdom, Oxford University Press, 1,149 p.

Ball, J.W., and Nordstrom, D.K., 1991, WATEQ4F—User's manual with revised thermodynamic data base and test cases for calculating speciation of major, trace and redox elements in natural waters: U.S. Geological Survey Open-File Report 90–129, 185 p.

Blount, C.W., and Dickson, F.W., 1973, Gypsum-anhydrite equilibria in systems $CaSO_4$-H_2O and $CaSO_4$-$NaCl$-H_2O: American Mineralogist, v. 58, p. 323–331.

Bonten, L.T.C., Groenenberg, J.E., Weng, Liping, and van Riemsdijk, W.H., 2008, Use of speciation and complexation models to estimate heavy metal sorption in soils: Geoderma, v. 146, no. 1–2, p. 303–310.

Borkovec, Michal, and Westall, John, 1983, Solution of the Poisson-Boltzmann equation for surface excesses of ions in the diffuse layer at the oxide-electrolyte interface: Journal of Electroanalytical Chemistry, v. 150, p. 325–337.

Bradley, D.J., and Pitzer, K.S., 1979, Thermodynamics of electrolytes—12. Dielectric properties of water and Debye-Hueckel parameters to 350 °C and 1 kbar: Journal of Physical Chemistry, v. 83, p. 1599–1603.

Carpenter, A.B., 1978, Origin and chemical evolution of brines in sedimentary basins: Oklahoma Geological Survey Circular 79, p. 60–77.

Charlton, S.R., and Parkhurst, D.L., 2002, PHREEQCI—A graphical user interface to the geochemical model PHREEQC: U.S. Geological Survey Fact Sheet FS–031–02, April 2002, 2 p.

Charlton, S.R., and Parkhurst, D.L., 2011, Modules based on the geochemical model PHREEQC for use in scripting and programming languages: Computers and Geosciences, v. 37, no. 10, p. 1653–1663.

Christensen, T.H., 1984, Cadmium soil sorption at low concentration—I. Effect of time, cadmium load, pH, and calcium: Water, Air, and Soil Pollution: v. 21, no. 1–4, p. 105–114.

Clegg, S.L., and Whitfield, Michael, 1991, Activity coefficients in natural waters, *in* Pitzer, K.S., ed., Activity coefficients in electrolyte solutions (2nd ed.): Boca Raton, Fla., CRC Press, p. 279–434.

Clegg, S.L., and Whitfield, Michael, 1995, A chemical model of seawater including dissolved ammonia and the stoichiometric dissociation constant for ammonia in estuarine water and seawater from –2 to 40 °C: Geochimica et Cosmochimica Acta, v. 59, no. 12, p. 2403–2421.

Cohen, S.D., and Hindmarsh, A.C., 1996, CVODE, A stiff/nonstiff ODE solver in C: Computers in Physics, v. 10, no. 2, p. 138–143.

Crank, John, 1975, The mathematics of diffusion (2nd ed.): Clarendon Press, 414 p.

Daveler, S.A., and Wolery, T.J., 1992, EQPT, A data file preprocessor for the EQ3/6 software package—User's guide and related documentation (Version 7.0): Lawrence Livermore National Laboratory, UCRL-MA-110662 PT II, 89 p., *http://www.wipp.energy.gov/library/CRA/2009_CRA/ references/Others/Daveler_Wolery_1992_EQPT_A_Data_File_Processor.pdf*, accessed September 24, 2010.

Davis, J.A., and Kent, D.B, 1990, Surface complexation modeling in aqueous geochemistry: Reviews in Mineralogy and Geochemistry, v. 23, p. 177–260.

Deines, Peter, Langmuir, Donald, and Harmon, R.S., 1974, Stable carbon isotope ratios and the existence of a gas phase in the evolution of carbonate ground waters: Geochimica et Cosmochimica Acta, v. 38, no. 7, p. 1147–1164.

Drummond, S.E., 1981, Boiling and mixing of hydrothermal fluids—Chemical effects on mineral precipitation: University Park, Penn., Pennsylvania State University, Ph.D. thesis, 760 p.

Duan, Zhenhao, Sun, Rui, Zhu, Chen, and Chou, I.-M., 2006, An improved model for the calculation of CO_2 solubility in aqueous solutions containing Na^+, K^+, Ca^{2+}, Mg^{2+}, Cl^-, and SO_4^{2-}: Marine Chemistry, v. 98, no. 2-4, p. 131–139.

Dzombak, D.A., and Morel, F.M.M., 1990, Surface complexation modeling—Hydrous ferric oxide: New York, John Wiley, 393 p.

Garrels, R.M., and Mackenzie, F.T., 1967, Origin of the chemical composition of some springs and lakes, *in* Equilibrium concepts in natural water systems: American Chemical Society, Advances in Chemistry Series no. 67, p. 222–242.

Glaus, M.A., Rossé, Roger, Van Loon, L.R., and Yaroshchuk, A.E., 2008, Tracer diffusion in sintered stainless steel filters—Measurement of effective diffusion coefficients and implications for diffusion studies with compacted clays: Clays and Clay Minerals, v. 56, no. 6, p. 677–685.

Glynn, P.D., 1990, Modeling solid-solution reactions in low-temperature aqueous systems, chap. 6 *of* Melchior, D.C., and Bassett, R.L., eds., Chemical modeling of aqueous systems II: Washington, D.C., American Chemical Society Symposium Series 416, chap. 6, p. 74–86.

Glynn, P.D., 1991, MBSSAS—A code for the computation of Margules parameters and equilibrium relations in binary solid-solution aqueous-solution systems: Computers and Geosciences, v. 17, no. 7, p. 907–966.

Glynn, P.D., and Parkhurst, D.L., 1992, Modeling non-ideal solid-solution aqueous-solution reactions in mass-transfer computer codes, *in* Kharaka, Y.K., and Maest, A.S., eds., Proceedings of the 7th International Symposium on Water-Rock Interaction, July 13–18, 1992: Park City, Utah, Rotterdam, Balkema, p. 175–179.

Glynn, P.D., and Reardon, E.J., 1990, Solid-solution aqueous-solution equilibria—Thermodynamic theory and representation: American Journal of Science, v. 290, p. 164–201.

Grathwohl, Peter, 1998, Diffusion in natural porous media—Contaminant transport, sorption/desorption and dissolution kinetics: Norwell, Mass., Kluwer Academic Publishers, 207 p.

Grenthe, Ingmar, Plyasunov, A.V., and Spahiu, Kastriot, 1997, Estimations of medium effects on thermodynamic data, *in* Grenthe, Ingmar, and Puigdomenech, Ignasi, eds., Modelling in aquatic chemistry, chap. IX: Paris, OECD Nuclear Energy Agency, p. 325–426.

Hardie, L.A., 1991, On the significance of evaporites: Annual Review of Earth and Planetary Sciences. v. 19, p. 131–168.

Helgeson, H.C., Brown, T.H., Nigrini, Andrew, and Jones, T.A., 1970, Calculation of mass transfer in geochemical processes involving aqueous solutions: Geochimica et Cosmochimica Acta, v. 34, no. 5, p. 569–592.

Helgeson, H.C., Garrels, R.M., and Mackenzie, F.T., 1969, Evaluation of irreversible reactions in geochemical processes involving minerals and aqueous solutions—II. Applications: Geochimica et Cosmochimica Acta, v. 33, no. 4, p. 455–481.

Hiemstra, T., and Van Riemsdijk, W.H., 1996, A surface structural approach to ion adsorption—The charge distribution (CD) model: Journal of Colloid Interface Science, v. 179, no. 2, p. 488–508.

Johnson, J.W., Oelkers, E.H., and Helgeson, H.C., 1992, SUPCRT92—A software package for calculating the standard molal thermodynamic properties of minerals, gases, aqueous species, and reactions from 1 to 5000 bar and 0 to 1000 °C: Computers and Geosciences, v. 18, no. 7, p. 899–947.

Lofts, S., and Tipping, E., 2000, Solid-solution metal partitioning in the Humber rivers—Application of WHAM and SCAMP: Science of the Total Environment, v. 251–252, p. 381–399.

McCaffrey, M.A., Lazar, B., Holland, H.D., 1987, The evaporation path of seawater and the coprecipitation of Br^- and K^+ in halite: Journal of Sedimentary Research, v. 57, no. 5, p. 928–938.

Laliberté, Marc, 2009, A model for calculating the heat capacity of aqueous solutions, with updated density and viscosity data: Journal of Chemical and Engineering Data, v. 54, p. 1725–1760.

Millero, F.J., 2000, The equation of state of lakes: Aquatic Geochemistry, v. 6, no. 1, p. 1–17.

Mosier, E.L., Papp, C.S.E., Motooka, J.M., Kennedy, K.R., and Riddle, G.O., 1991, Sequential extraction analyses of drill core samples, Central Oklahoma aquifer: U.S. Geological Survey Open-File Report 91–347, 42 p.

Nagra, 2002, Projekt Opalinuston—Synthese der geowissenschaftlichen Untersuchungsergebnisse: Nagra, Wettingen, Switzerland, Nagra Technical Report NTB 02–03, 659 p.

Nordstrom, D.K., and Archer, D.G., 2003, Arsenic thermodynamic data and environmental geochemistry, *in* Welch, A.H., and Stollenwerk, K.G., eds., Arsenic in ground water—Geochemistry and occurrence: Boston, Kluwer Academic Publishers, p. 1–25.

Nordstrom, D.K., Plummer, L.N.,Wigley, T.M.L.,Wolery, T.J., Ball, J.W., Jenne, E.A., Bassett, R.L., Crerar, D.A., Florence, T.M., Fritz, B., Hoffman, M., Holdren, G.R., Jr., Lafon, G.M., Mattigod, S.V., McDuff, R.E., Morel, F., Reddy, M.M., Sposito, G., and Thrailkill, J., 1979, A comparison of computerized chemical models for equilibrium calculations in aqueous systems, *in* Jenne, E.A., ed., Chemical modeling in aqueous systems—Speciation, sorption, solubility, and kinetics: Series 93, American Chemical Society, p. 857–892.

Parkhurst, D.L., 1995, User's guide to PHREEQC—A computer program for speciation, reaction-path, advective-transport, and inverse geochemical calculations: U.S. Geological Survey Water-Resources Investigations Report 95–4227, 143 p.

Parkhurst, D.L., 1997, Geochemical mole-balance modeling with uncertain data: Water Resources Research, v. 33, no. 8, p. 1957–1970.

Parkhurst, D.L., and Appelo, C.A.J., 1999, User's guide to PHREEQC (Version 2)—A computer program for speciation, batch-reaction, one-dimensional transport, and inverse geochemical calculations: U.S. Geological Survey Water-Resources Investigations Report 99–4259, 312 p.

Parkhurst, D.L., and Charlton, S.R., 2008, NetpathXL—An Excel® interface to the program NETPATH: U.S. Geological Survey Techniques and Methods, book 6, chap. A26, 11 p.

Parkhurst, D.L., Christenson, Scott, and Breit, G.N., 1996, Ground-water-quality assessment of the central Oklahoma aquifer, Oklahoma—Geochemical and geohydrologic investigations: U.S. Geological Survey Water-Supply Paper 2357–C, 101 p.

Parkhurst, D.L., Kipp, K.L., and Charlton, S.R., 2010, PHAST Version 2—A program for simulating groundwater flow, solute transport, and multicomponent geochemical reactions: U.S. Geological Survey Techniques and Methods, book 6, chap. A35, 235 p.

Parkhurst, D.L., Thorstenson, D.C., and Plummer, L.N., 1980, PHREEQE—A computer program for geochemical calculations: U.S. Geological Survey Water-Resources Investigations Report 80–96, 195 p. (Revised and reprinted August, 1990.)

Peng, D.-Y., and Robinson, D.B., 1976, A new two-constant equation of state: Industrial and Engineering Chemistry Fundamentals, v. 15, p. 59–64.

Pitzer, K.S., 1973, Thermodynamics of electrolytes—1. Theoretical basis and general equations: Journal of Physical Chemistry, v. 77, no. 2, p. 268–277.

Plummer, L.N., Busby, J.F., Lee, R.W., and Hanshaw, B.B., 1990, Geochemical modeling of the Madison aquifer in parts of Montana, Wyoming, and South Dakota: Water Resources Research, v. 26, no. 9, p. 1981–2014.

Plummer, L.N., and Busenberg, Eurybiades, 1987, Thermodynamics of aragonite-strontianite solid solutions—Results from stoichiometric dissolution at 25 and 76 °C: Geochimica et Cosmochimica Acta, v. 51, p. 1393–1411.

Plummer, L.N., Parkhurst, D.L., Fleming, G.W., and Dunkle, S.A., 1988, A computer program incorporating Pitzer's equations for calculation of geochemical reactions in brines: U.S. Geological Survey Water-Resources Investigations Report 88–4153, 310 p.

Plummer, L.N., Prestemon, E.C., and Parkhurst, D.L., 1991, An interactive code (NETPATH) for modeling NET geochemical reactions along a flow PATH: U.S. Geological Survey Water-Resources Investigations Report 91–4078, 227 p.

Plummer, L.N., Prestemon, E.C., and Parkhurst, D.L., 1994, An interactive code (NETPATH) for modeling NET geochemical reactions along a flow PATH, version 2.0: U.S. Geological Survey Water-Resources Investigations Report 94–4169, 130 p.

Plummer, L.N., Wigley, T.M.L., and Parkhurst, D.L., 1978, The kinetics of calcite dissolution in CO_2-water systems at 5 to 60 °C and 0.0 to 1.0 atm CO_2: American Journal of Science, v. 278, p. 179–216.

Post, Vincent, 2012, Phreeqc for Windows, *http://pfw.antipodes.nl/*, accessed May 31, 2012.

Rahnemaie, Rasoul, Hiemstra, Tjisse, and Van Riemsdijk, W.H., 2006, Inner- and outer-sphere complexation of ions at the goethite-solution interface: Journal of Colloid and Interface Science, v. 297, no. 2, p. 379–388.

Redlich, Otto, and Meyer, D.M., 1964, The molal volumes of electrolytes: Chemical Reviews, v. 64, no. 3, p. 221–227.

Robie, R.A., Hemingway, B.S., and Fisher, J.R., 1978, Thermodynamic properties of minerals and related substances at 298.15 K and 1 bar (105 pascals) pressure and at higher temperatures: U.S. Geological Survey Bulletin 1452, 456 p.

Robinson, R.A., and Stokes, R.H., 2002, Electrolyte solutions: Dover Publications, 559 p.

Schofield, R.K., 1947, Calculation of surface areas from measurements of negative adsorption: Nature, v. 160, p. 408–410.

Shi, Zhenqing, Allen, H.E., Di Toro, D.M., Lee, S-Z., Flores Meza, D.M., and Lofts, Steve, 2007, Predicting cadmium adsorption on soils using WHAM VI: Chemosphere, v. 69, no. 4, p. 605–612.

Singer, P.C., and Stumm, W., 1970, Acid mine drainage—The rate determining step: Science, v. 167, p. 1121–1123.

Spycher, Nicholas, Pruess, Karsten, and Ennis-King, Jonathan, 2003, CO_2-H_2O mixtures in the geological sequestration of CO_2—I. Assessment and calculation of mutual solubilities from 12 to 100°C and up to 600 bar: Geochimica et Cosmochimica Acta, v. 67, p. 3015–3031.

Stachowicz, Monika, Hiemstra, Tjisse, and van Riemsdijk, W.H., 2006, Surface speciation of As(III) and As(V) in relation to charge distribution: Journal of Colloid and Interface Science, v. 302, no. 1, p. 62–75.

Thorstenson, D.C., and Parkhurst, D.L., 2002, Calculation of individual isotope equilibrium constants for implementation in geochemical models: U.S. Geological Survey Water-Resources Investigations Report 2002–4172, 129 p.

Thorstenson, D.C., and Parkhurst, D.L., 2004, Calculation of individual isotope equilibrium constants for geochemical reactions: Geochimica et Cosmochimica Acta, v. 68, no. 11, p. 2449–2465.

Tipping, E., 1998, Humic ion-binding model VI—An improved description of the interactions of protons and metal ions with humic substances: Aquatic Geochemistry, v. 4, no. 1, p. 3–47.

Tipping, E., and Hurley, M.A., 1992, A unifying model of cation binding by humic substances: Geochimica et Cosmochimica Acta, v. 56, no. 10, p. 3627–3641.

Tournassat, C., and Appelo, C.A.J., 2011, Modelling approaches for anion-exclusion in compacted Na-bentonite: Geochimica et Cosmochimica Acta, v. 75, no. 13, p. 3698–3710.

Truesdell, A.H., and Jones, B.F., 1974, WATEQ, A computer program for calculating chemical equilibria of natural waters: Journal of Research of the U.S. Geological Survey, v. 2, no. 2, p. 233–274.

Tebes-Stevens, Caroline, and Valocchi, A.J., 1997, Reactive transport simulation with equilibrium speciation and kinetic biodegradation and adsorption/desorption reactions: A Workshop on Subsurface Reactive Transport Modeling, Pacific Northwest National Laboratory, Richland, Washington, October 29–November 1, 1997.

Tebes-Stevens, C., Valocchi, A.J., VanBriesen, J.M., Rittmann, B.E., 1998, Multicomponent transport with coupled geochemical and microbiological reactions—Model description and example simulations: Journal of Hydrology, v. 209, p. 8–26.

U.S. Environmental Protection Agency, 1998, MINTEQA2/PRODEFA2, A geochemical assessment model for environmental systems—User manual supplement for version 4.0: Athens, Georgia, National Exposure Research Laboratory, Ecosystems Research Division, 76 p. Revised September 1999.

Van Genuchten, M.Th., 1985, A general approach for modeling solute transport in structured soils: International Association of Hydrogeologists Memoirs, v. 17, no. 2, p. 513–526.

Van Loon, L.R., Soler, J.M., Müller, W., and Bradbury, M.H., 2004, Anisotropic diffusion in layered argillaceous rocks—A case study with Opalinus clay: Environmental Science and Technology, v. 38, no. 21, p. 5721–5728.

Van Loon, L.R., Glaus, M.A., and Müller, W., 2007, Anion exclusion effects in compacted bentonites: Towards a better understanding of anion diffusion: Applied Geochemistry, v. 22, p. 2536–2552.

Vinograd, J.R., and McBain, J.W., 1941, Diffusion of electrolytes and of the ions in their mixtures: Journal of the American Chemical Society, v. 63, p. 2008–2015.

Wagner, W., and Pruss, A., 2002, The IAPWS formulation 1995 for the thermodynamic properties of ordinary water substance for general and scientific use: Journal of Physical and Chemical Reference Data, v. 31, p. 387–535.

Waite, T.D., Davis, J.A., Payne, T.E., Waychunas, G.A., and Xu, N., 1994, Uranium(VI) adsorption to ferrihydrite—Application of a surface complexation model: Geochimica et Cosmochimica Acta, v. 58, no. 24, p. 5465–5478.

Weng, L., Temminghoff, E.J.M., Lofts, S., Tipping, E., and Van Riemsdijk, W.H., 2002, Complexation with dissolved organic matter and solubility control of heavy metals in a sandy soil: Environmental Science and Technology, v. 36, p. 4804–4810.

Williamson, M.A., and Rimstidt, J.D., 1994, The kinetics and electrochemical rate-determining step of aqueous pyrite oxidation: Geochimica et Cosmochimica Acta, v. 58, no. 24, p. 5443–5454.

Wolery, T.J., 1979, Calculation of chemical equilibrium between aqueous solution and minerals—The EQ3/6 software package: Livermore, Calif., Lawrence Livermore National Laboratory Report UCRL-52658, 41 p.

Wolery, T.J., Jackson, K.J., Bourcier, W.L., Bruton, C.J., Viani, B.E., Knauss, K.G., and Delany, J.M., 1990, Current status of the EQ3/6 software package for geochemical modeling, chap. 8 *in* Melchior, D.C., and Bassett, R.L., eds., Chemical modeling of aqueous systems II: Washington, D.C., American Chemical Society Symposium Series 416, p. 104–116.

Zherebtsova, I.K., and Volkova, N.N., 1966, Experimental study of behavior of trace elements in the process of natural solar evaporation of Black Sea water and Sasyk-Sivash brine: Geochemistry International, v. 3, p. 656–670.

Appendix A. Keyword Data Blocks for Programmers

A number of keywords are intended to be used when scripting or programming with an IPhreeqc module (Charlton and Parkhurst, 2011). The IPhreeqc module provides a set of methods that expose the full capabilities of PHREEQC to other programs and allows retrieval of specified values (as defined in SELECTED_OUTPUT and USER_PUNCH).

One use of an IPhreeqc module is to incorporate geochemical reactions into a transport model. The strategy is to specify reactants for a set of cells in an IPhreeqc module and perform geochemical reactions for a time step with the RUN_CELLS data block. The solution concentrations are retrieved from the module, transported by the transport model, and returned to the IPhreeqc module for renewed geochemical calculations. A series of keyword data blocks have been written that facilitate transferring and modifying concentration data. These data blocks end with the suffixes **_MODIFY** and **_RAW**. In this reactive-transport strategy, SOLUTION_MODIFY could be used to update solution concentrations in the IPhreeqc module after a transport step. SOLUTION_MODIFY is related to SOLUTION_RAW, which is an input data block that is written by the DUMP operation and contains a complete description of a solution composition. SOLUTION_MODIFY has the same format as SOLUTION_RAW and allows one or more data items of the solution composition to be modified and also allows new data items (moles of another element in solution, for example) to be added.

Keyword equivalents to SOLUTION_MODIFY and SOLUTION_RAW data blocks exist for equilibrium-phase assemblages, exchange assemblages, gas phases, solid-solution assemblages, surface assemblages, kinetic-reaction assemblages, stoichiometric reactions, reaction pressures, and reaction temperatures. A complete list of **_MODIFY** and **_RAW** data blocks is given in table A1. There is no keyword **MIX_RAW** or **MIX_MODIFY** because the MIX data block can be used. Similarly, **REACTION_TEMPERATURE_MODIFY** and **REACTION_TEMPERATURE_MODIFY** data blocks are not needed; the REACTION_PRESSURE and REACTION_TEMPERATURE data blocks can be used to modify temperature definitions. The set of **_MODIFY** keywords provides capabilities to change any data item in any reactant. Although transport is expected to apply primarily to solutions, it would be possible to transport exchange assemblages analogously to solutions by transporting the elements that define the composition of exchangers and then updating the exchange composition of a cell by using EXCHANGE_MODIFY with the results of the transport calculations.

Table A1. List of keyword data blocks for scripting and programming.

Keyword data block	Function
EQUILIBRIUM_PHASES_MODIFY	Modify the definition of an equilibrium-phase assemblage
EQUILIBRIUM_PHASES_RAW	Complete description of an equilibrium-phase assemblage as written by DUMP
EXCHANGE_MODIFY	Modify the definition of an exchange assemblage
EXCHANGE_RAW	Complete description of an exchange assemblage as written by DUMP
GAS_PHASE_MODIFY	Modify the definition of a gas phase
GAS_PHASE_RAW	Complete description of a gas phase as written by DUMP
KINETICS_MODIFY	Modify the definition of a kinetic reactant assemblage
KINETICS_RAW	Complete description of a kinetic reactant assemblage as written by DUMP
REACTION_MODIFY	Modify the definition of an irreversible reaction
REACTION_RAW	Complete description of a REACTION definition as written by DUMP
REACTION_PRESSURE_RAW	Complete description of a REACTION_PRESSURE definition as written by DUMP
REACTION_TEMPERATURE_RAW	Complete description of a REACTION_TEMPERATURE definition as written by DUMP
SOLID_SOLUTIONS_MODIFY	Modify the definition of a solid-solution assemblage
SOLID_SOLUTIONS_RAW	Complete description of a solid-solution assemblage as written by DUMP
SOLUTION_MODIFY	Modify the definition of a solution composition
SOLUTION_RAW	Complete description of a solution as written by DUMP
SURFACE_MODIFY	Modify the definition of a surface-assemblage composition
SURFACE_RAW	Complete description of a surface-assemblage composition as written by DUMP

The _RAW data blocks (and possibly MIX) are written by the DUMP operation and are intended to be used without modification. The DUMP operation can be used to save the state of a calculation. The _RAW data blocks that are written by DUMP can be read by PHREEQC and the entire chemical state of the calculation will be restored to the point where DUMP was executed. DUMP also can be used to transfer data among IPhreeqc modules by dumping the data in one module and reading the data in another.

The use of the _MODIFY data blocks is complicated and subject to errors. Any item of data can be changed, but some are used for internal calculations, and changing the value externally has no effect. Some items cannot be reasonably changed, and others are interrelated so that a change to one may require a change to another as well. In particular, the number of moles of elements, including H and O, and the charge balance should not be changed independently because doing so may cause unforeseen pe and (or) pH changes. Varying just the number of moles of a single element (sodium for example) would be equivalent to adding or removing sodium metal from solution, which also would produce unexpected redox and pH reactions.

The _MODIFY data blocks are not intended for general use, but can be used by program developers. As such, developers are largely responsible for their use, and only limited support is provided. In the following description of input for _MODIFY keywords, a subset of the complete set of data in the _RAW data blocks is presented. Each subset is somewhat arbitrary, but has, in principle, the data items that could reasonably be changed with appropriate care.

EQUILIBRIUM_PHASES_MODIFY

This keyword data block is used to modify the definition of a previously defined equilibrium-phase assemblage. New phases may be added and the quantity of each phase may be changed. The format of the data block is the same as the EQUILIBRIUM_PHASES_RAW data block, except that the data block need not be complete. The **EQUILIBRIUM_PHASES_MODIFY** data block can be used selectively to change data items. The Example data block lists a subset of identifiers that can be used in the **EQUILIBRIUM_PHASES_MODIFY** data block.

Example data block

```
Line 0:      EQUILIBRIUM_PHASES_MODIFY 1 Added Barite
Line 1:      -component      Barite
Line 2:          -add_formula           BaCl2
Line 3:          -si                    0
Line 4:          -moles                 10
Line 5:          -force_equality        0
Line 6:          -dissolve_only         1
Line 7:          -precipitate_only      0
```

Explanation

Line 0: **EQUILIBRIUM_PHASES_MODIFY** *number* [*description*]

EQUILIBRIUM_PHASES_MODIFY is the keyword for the data block.

number—Positive integer to identify the equilibrium-phase assemblage to modify.

description—Optional comment that describes the equilibrium-phase assemblage.

Line 1: **-component** *name*

-component—Identifier that indicates information will be defined for a phase in the equilibrium-phase assemblage. The identifier **-component** is required to precede the other identifiers for a phase definition. Optionally, **component** or **-c[omponent]**.

name—Name of a phase that has been defined in a PHASES data block.

Line 2: **-add_formula** *alternative formula*

-add_formula—Defines a reactant other than the phase that reacts to produce equilibrium; see *alternative formula* in the description of the EQUILIBRIUM_PHASES data block. Optionally, **add_formula** or **-a[dd_formula]**.

alternative formula—Phase name or chemical formula.

Line 3: **-si** *saturation index*

> **-si**—Defines the target saturation index for the phase; see *saturation index* in the description of the EQUILIBRIUM_PHASES data block. Optionally, **si** or -s[i].
>
> *saturation index*—Target saturation index for the phase.

Line 4: **-moles** *moles*

> **-moles**—Defines the amount of the phase that can react; see *amount* in the description of the EQUILIBRIUM_PHASES data block. Optionally, **moles** or -m[oles].
>
> *moles*—Moles of phase or *alternative formula*.

Line 5: **-force_equality** (*1* or *0*)

> **-force_equality**—Defines whether the equation for phase equilibrium is an inequality or equality constraint in the set of equations to be solved; see **-force_equality** in the description of the EQUILIBRIUM_PHASES data block. Optionally, **force_equality** or -f[orce_equality].
>
> (*1* or *0*)—A value of 1 indicates true. A value of 0 indicates false.

Line 6: **-dissolve_only** (*1* or *0*)

> **-dissolve_only**—Defines whether the phase is required only to dissolve; see **-dissolve_only** in the description of the EQUILIBRIUM_PHASES data block. Optionally, **dissolve_only** or -di[ssolve_only].
>
> (*1* or *0*)—A value of 1 indicates true. A value of 0 indicates false.

Line 7: **-precipitate_only** (*1* or *0*)

> **-precipitate_only**—Defines whether the phase is required only to precipitate; see **-precipitate_only** in the description of the EQUILIBRIUM_PHASES data block. Optionally, **precipitate_only** or -p[recipitate_only].
>
> (*1* or *0*)—A value of 1 indicates true. A value of 0 indicates false.

Notes

The **EQUILIBRIUM_PHASES_MODIFY** data block allows modification of a preexisting equilibrium-phase assemblage. The most common uses are to add a new phase to the phase assemblage and to change the amount of a phase that is present in the assemblage. It also is possible to change the target saturation index for a phase, whether the phase can only dissolve or can only precipitate, and whether an equality or inequality constraint is included for the phase in the set of equations that are solved.

EQUILIBRIUM_PHASES_MODIFY modifies only the data items specifically defined in the data block. Any data items in the equilibrium-phases definition not modified by the data block remain unchanged.

Related keywords

EQUILIBRIUM_PHASES and EQUILIBRIUM_PHASES_RAW.

EQUILIBRIUM_PHASES_RAW

This keyword data block is written by the DUMP operation. It is intended to be used to reinitialize simulations at the point that the DUMP command was executed or to transfer equilibrium-phase compositions between IPhreeqc modules.

Notes

The **EQUILIBRIUM_PHASES_RAW** data block contains a complete listing of the internal data structures that define an equilibrium-phase assemblage. **EQUILIBRIUM_PHASES_RAW** data blocks are written by DUMP and read by PHREEQC without user modification. The formats should be considered dynamic because future modifications to PHREEQC could result in additional data that are included in the **EQUILIBRIUM_PHASES_RAW** data block.

Related keywords

EQUILIBRIUM_PHASES and EQUILIBRIUM_PHASES_MODIFY.

EXCHANGE_MODIFY

This keyword data block is used to modify the definition of a previously defined exchange assemblage. New exchangers may be added and the quantity and composition of each exchanger may be altered. The format of the data block is the same as the EXCHANGE_RAW data block except that the data block need not be complete. The **EXCHANGE_MODIFY** data block can be used selectively to change data items. The Example data blocks list a subset of identifiers that can be used in **EXCHANGE_MODIFY** data blocks.

Example data block 1

```
Line 0:        EXCHANGE_MODIFY 1 Added Y
Line 1:        -component      Y
Line 2:            -totals
Line 3:                Na      1.0
Line 3a:                Y       1.0
Line 4:        -exchange_gammas            true
```

Explanation 1

Line 0: **EXCHANGE_MODIFY** *number* [*description*]

EXCHANGE_MODIFY is the keyword for the data block.

number—Positive integer to identify the exchange assemblage to modify.

description—Optional comment that describes the exchange assemblage.

Line 1: **-component** *exchange site*

-component—Identifier that indicates information for an exchanger in the exchange assemblage will be added or modified. The identifier **-component** is required to precede the other identifiers that define the exchanger. Optionally, **component** or **-c**[**omponent**].

exchange site—Exchange site formula.

Line 2: **-totals**

-totals—Identifier begins a block of data containing the moles of elements in the exchanger. Optionally, **totals** or **-t**[**otals**].

Line 3: *element moles*

element—An element name or an exchange site name.

moles—Moles of element or exchange site present in the exchanger; it is a positive number.

Line 4: **-exchange_gammas** (*1* or *0*)

-**exchange_gammas**—This identifier selects whether exchange activity coefficients are assumed to be equal to aqueous activity coefficients when using the Pitzer or SIT aqueous model. Optionally, **exchange_gammas** or -**ex**[**change_gammas**]. The option has no effect when using ion-association aqueous models.

(*1* or *0*)—A value of 1 indicates true. A value of 0 indicates false.

Notes 1

The **EXCHANGE_MODIFY** data block allows modification of a preexisting exchange assemblage, but care is needed for any modifications because data items have interdependencies. The most common uses are to add a new exchanger to an exchange assemblage and to change the amount of an exchanger that is present in the assemblage. However, it is important that the -**totals** data block defines an exchange composition, such that the number of equivalents of exchange sites is balanced by an equal number of equivalents of ions. The totals are assumed to balance in charge, and any charge in the formulas of -**totals** will be ignored. If, for example, 1 mol of X, with master species X$^-$, and 2 moles of Na, with master species Na$^+$, are defined, it will be equivalent to adding 1 mol of NaX and 1 mol of Na metal, and unexpected pH and redox conditions will result. It also is possible to change the choice of activity coefficient option (-**exchange_gammas**) when using the Pitzer or SIT aqueous models.

EXCHANGE_MODIFY modifies only the data items specifically defined in the data block. Any data items in the exchanger definition not modified by the data block remain unchanged. Note specifically that the -**totals** data block modifies the elements as defined in the data block, but any element not listed in the data block will remain unchanged in the exchange composition.

Example data block 2

```
Line 0:       EXCHANGE_MODIFY 1 Added Y
Line 1:       -component       NaY
          Line 2:        -totals
Line 3:                   Na      0.002
Line 3a:                  Y       0.002
Line 4:            -phase_name    Na-Mont
Line 5:            -phase_proportion   0.1
Line 6:       -exchange_gammas            true
```

Explanation 2

Line 0: **EXCHANGE_MODIFY** *number* [*description*]

EXCHANGE_MODIFY is the keyword for the data block.

number—Positive integer to identify the exchange assemblage to modify.

description—Optional comment that describes the exchange assemblage.

Line 1: **-component** *exchange formula*

-component—Defines the formula for the exchange site; see *exchange formula* in the description of the EXCHANGE data block. Optionally, **formula** or **-f[ormula]**.

exchange formula—Exchange site formula.

Line 2: **-totals**

-totals—Same as Explanation 1.

Line 3: *element moles*.

element moles—Same as Explanation 1.

Line 4: (**-phase_name** or **-rate_name**) *name*

-phase_name—Defines the name of an equilibrium phase to which the number of exchange sites is related. See **equilibrium_phase** in the description of the EXCHANGE data block. Optionally, **phase_name** or **-p[hase_name]**.

-rate_name—Defines the name of a kinetic reactant to which the number of exchange sites is related. See **kinetic_reactant** in the description of the EXCHANGE data block. Optionally, **rate_name** or **-r[ate_name]**.

name—Name of an equilibrium phase or a kinetic reactant.

Line 5: **-phase_proportion** *exchange_per_mole*

-phase_proportion—Defines the number of moles of exchange formula per mole of phase or kinetic reactant; see *exchange_per_mole* in the description of the EXCHANGE_MODIFY data block. Optionally, **phase_proportion** or **-phase_p[roportion]**.

exchange_per_mole—Number of moles of the exchange formula per mole of phase or kinetic reactant, unitless.

Line 6: **-exchange_gammas** (*1* or *0*).

-exchange_gammas—Same as Explanation 1

Notes 2

This example shows how to add a new exchanger that is related to an equilibrium phase. As in Example data block 1, it is important that the **-totals** data block defines a charge-balanced exchange composition. If charge balance is not preserved or the number of moles of exchange sites is inconsistent with the number of moles of the phase, unexpected pH and redox conditions will result. Moreover, the number of moles of the exchange site (Y in this example) must be consistent with the number of moles of the equilibrium phase, as defined by the **-phase_proportion** and **-formula** identifiers. Use of **-phase_name** or **-rate_name** for an exchanger numbered *n* requires that an EQUILIBRIUM_PHASES *n* or KINETICS *n* has been defined.

Related keywords

EXCHANGE and EXCHANGE_RAW.

EXCHANGE_RAW

This keyword data block is written by the DUMP operation. It is intended to be used to reinitialize simulations at the point that the DUMP command was executed, or to transfer exchange-assemblage compositions between IPhreeqc modules.

Notes

The **EXCHANGE_RAW** data block contains a complete listing of the internal data structures that define an exchange assemblage. **EXCHANGE_RAW** data blocks are written by DUMP and read by PHREEQC without user modification. The formats should be considered dynamic because future modifications to PHREEQC could result in additional data that are included in the **EXCHANGE_RAW** data block.

Related keywords

EXCHANGE and EXCHANGE_MODIFY.

GAS_PHASE_MODIFY

This keyword data block is used to modify the definition of a previously defined gas phase. New gas components may be added, and the quantity of each gas component may be changed. In addition, it is possible to change the type of the gas phase between fixed pressure and fixed volume, and to change the total pressure of a fixed-pressure gas phase and the volume of a fixed-volume gas phase. The format of the data block is the same as the GAS_PHASE_RAW data block except that the data block need not be complete. The **GAS_PHASE_MODIFY** data block can be used selectively to change data items. This Example data block lists a subset of identifiers that can be used in **GAS_PHASE_MODIFY** data blocks.

Example data block

```
Line 0:    GAS_PHASE_MODIFY 1 Add CO2(g) component
Line 1:       -type              0
Line 2:       -total_p           1
Line 3:       -volume            1.5
Line 4:       -component     CO2(g)
Line 5:            -moles    1.4305508698401e-005
```

Explanation

Line 0: **GAS_PHASE_MODIFY** *number* [*description*]

GAS_PHASE_MODIFY is the keyword for the data block.

number—Positive integer to identify the gas phase to modify.

description—Optional comment that describes the gas phase.

Line 1: **-type** (*1* or *0*)

-type—Identifier that indicates the type of the gas phase, either fixed pressure or fixed volume. Optionally, **type** or **-t[ype]**.

(*1* or *0*)—1 indicates a fixed-volume gas phase, and 0 indicates a fixed-pressure gas phase.

Line 2: **-total_p** *pressure*

-total_p—Defines the pressure for a fixed-pressure gas phase. Optionally, **pressure**, **total_p**, **-p[ressure]**, or **-to[tal_p]**.

pressure—Total pressure of a fixed-pressure gas phase (atmospheres).

Line 3: **-volume** *volume*

-volume—Defines the volume for a fixed-volume gas phase. Optionally, **volume** or **-v[olume]**.

volume—Total volume of a fixed-volume gas phase (liter).

Line 4: **-component** *component_name*

> **-component**—Begins a data block that adds gas components or modifies the number of moles of gas components. Optionally, **component** or **-c[omponent]**.

> *component_name*—Name of a gas component as defined in a PHASES data block.

Line 5: **-moles** *moles*

> *moles*—Number of moles of the gas component in the gas phase; it is a positive number.

Notes

The **GAS_PHASE_MODIFY** data block allows modification of a preexisting gas phase. The most common uses are to change the pressure (**-total_p**) of a fixed-pressure gas phase, to change the volume (**-volume**) of a fixed-volume gas phase, to add a new gas component to the gas phase, and to change the number of moles of a gas component that is present in the gas phase. It also is possible to change the gas phase from a fixed-pressure gas phase to a fixed-volume gas phase, and the reverse.

The **GAS_PHASE_MODIFY** data block modifies only the data items specifically defined in the data block. Any data items in the gas-phase definition not modified by the data block remain unchanged. In particular, if a single new component is defined, any other gas components already defined in the gas phase are unchanged.

Related keywords

GAS_PHASE and GAS_PHASE_RAW.

GAS_PHASE_RAW

This keyword data block is written by the DUMP operation. It is intended to be used to reinitialize simulations at the point that the DUMP command was executed, or to transfer gas-phase compositions between IPhreeqc modules.

Notes

The **GAS_PHASE_RAW** data block contains a complete listing of the internal data structures that define a gas phase. **GAS_PHASE_RAW** data blocks are written by DUMP and read by PHREEQC without user modification. The formats should be considered dynamic because future modifications to PHREEQC could result in additional data that are included in the **GAS_PHASE_RAW** data block.

Related keywords

GAS_PHASE and GAS_PHASE_MODIFY.

KINETICS_MODIFY

This keyword data block is used to modify the definition of a previously defined assemblage of kinetic reactants. New reactants may be added, and the quantity of each reactant may be changed. Other parameters from the **KINETICS** data block also may be changed. The format of the data block is the same as for the **KINETICS_RAW** data block, except that the data block need not be complete. The **KINETICS_MODIFY** data block can be used selectively to change data items. The Example data block lists a subset of identifiers that can be used in **KINETICS_MODIFY** data blocks.

Example data block

```
Line 0:    KINETICS_MODIFY 2 Revising kinetics
Line 1:        -component   Calcite
Line 2:            -tol                    1e-08
Line 3:            -m                      1
Line 4:            -m0                     1
Line 5:            -namecoef
Line 6:                 CaCO3    1
Line 7:            -d_params
Line 8:                 1 1 1 1
Line 9:        -equal_increments    1
Line 10:       -count          2
Line 11:       -steps
Line 12:            1.0     1.3    2.0
Line 13:       -step_divide      1
Line 14:       -rk              3
Line 15:       -bad_step_max     500
Line 16:       -use_cvode        0
Line 17:       -cvode_steps      100
Line 18:       -cvode_order      5
```

Explanation

Line 0: **KINETICS_MODIFY** *number* [*description*]

KINETICS_MODIFY is the keyword for the data block.

number—Positive integer to identify the kinetic-reaction assemblage to modify.

description—Optional comment that describes the kinetic-reaction assemblage.

Line 1: **-component** *name*

-component—Identifier that indicates information for a kinetic reactant in the assemblage will be added or modified. The identifier **-component** is required to precede the other identifiers related to the reactant (as indicated by the indentation in the Example data block). Optionally, **component** or **-c[omponent]**.

name—Name of a kinetic reactant for which a rate expression has been defined in RATES.

Line 2: **-tol** *tolerance*

-tol—Defines the tolerance for errors in integrating the kinetic rate equation; see **-tol** in the description of the KINETICS data block. Optionally, **tol** or **-t[ol]**.

tolerance—Error tolerance for integrating the kinetic rate equation.

Line 3: **-m** *moles*

-m—Defines the current number of moles of the reactant; see **-m** in the description of the KINETICS data block. Optionally, **m** or **-m**.

moles—Moles of reactant.

Line 4: **-m0** *moles*

-m0—Defines an initial number of moles of the reactant; see **-m0** in the description of the KINETICS data block. Optionally, **m0** or **-m0**.

moles—Initial number of moles associated with the reactant.

Line 5: **-namecoef**

-namecoef—Marks the beginning of a data block that defines the stoichiometry of the kinetic reaction; see **-formula** in the description of the KINETICS data block. Optionally, **namecoef** or **-n[amecoef]**.

Line 6: *chemical_formula, stoichiometric_coefficient*

chemical_formula—A chemical formula that is added or removed by the kinetic reaction; see **-formula** in the description of the KINETICS data block. Multiple line 7s may be included to define the complete stoichiometry of the kinetic reaction.

stoichiometric_coefficient—Stoichiometric coefficient of the *chemical_formula*. The coefficient may be positive or negative.

Line 7: **-d_params**

-d_params—Begins a data block that defines parameters for the rate equation associated with the kinetic reactant; see **-parms** in the description of the KINETICS data block. Optionally, **d_params** or **-d[_params]**.

Line 8: *list_of_parameters*

list_of_parameters—List of parameters for the rate equation.

Line 9: **-equal_increments** (*1* or *0*)

> **-equal_increments**—Specifies whether the time step for the kinetic reaction will be split into a number of equal steps; see **-steps** for example 2 in the description of the **KINETICS** data block. Optionally, **equal_increments**, **-e[qual_increments]**, or **-e[qualincrements]**.

> (*1* or *0*)—A value of 1 indicates true. A value of 0 indicates false.

Line 10: **-count** *count*

> **-count**—Specifies the number of equal steps, if **-equal_increments** is 1. If **-equal_increments** is 0, *count* is ignored. Optionally, **count** or **-co[unt]**.

> *count*—Integer number of equal time increments for kinetic steps. If *count* is greater than 0, then the time step, given in the first value for **-steps**, will be split into *count* equal time steps.

Line 11: **-steps**

> **-steps**—Begins a data block that defines a series of times over which to integrate the rate equations; see **-steps** in the description of the **KINETICS** data block. If **-equal_increments** is greater than 1, the first time step will be divided into equal increments according to the value given for **-count**, and the other time steps will be ignored. Optionally, **steps** or **-steps**.

Line 12: *list_of_steps*

> *list_of_steps*—List of time steps for integrating the rate equation.

Line 13: **-step_divide** *step_divide*

> **-step_divide**—Specifies whether to begin integrating with a reduced time substep; see **-step_divide** in the description of the **KINETICS** data block. Optionally, **step_divide** or **-s[tep_divide]**.

> *step_divide*—Parameter greater than or equal to 1 that is used to reduce the initial time substep of integration.

Line 14: **-rk** (**1, 2, 3**, or **6**)

> **-rk**—Designates the preferred number of time subintervals to use when integrating rates with the Runge-Kutta method; see **-runge_kutta** in the description of the **KINETICS** data block. Optionally, **rk** or **-r[k]**.

> (**1, 2, 3**, or **6**)—Preferred number of time subintervals.

Line 15: **-bad_step_max** *tries*

> **-bad_step_max**—Defines the maximum number of attempts at integrating a set of kinetic reactions over a time step; see **-bad_step_max** in the description of the KINETICS data block. Optionally, **bad_step_max** or **-b[ad_step_max]**.

> *tries*—The maximum number of integration attempts.

Line 16: **-use_cvode** (*1* or *0*)

> **-use_cvode**—Specifies whether to use the explicit Runge-Kutta method or the implicit CVODE method (Cohen and Hindmarsh, 1996) to integrate the kinetic rate equations; see **-cvode** in the description of the KINETICS data block. Optionally, **use_cvode** or **-u[se_cvode]**.

> (*1* or *0*)—A value of 1 indicates true. A value of 0 indicates false.

Line 17: **-cvode_steps** *steps*

> **-cvode_steps**—Specifies the maximum number of steps that will be taken during one invocation of CVODE; see **-cvode_steps** in the description of the KINETICS data block. Optionally, **cvode_steps** or **-cvode_[steps]**.

> *steps*—Maximum number of steps.

Line 18: **-cvode_order** (*1, 2, 3, 4,* or *5*)

> **-cvode_order**—Specifies the number of terms to use in the extrapolation of rates when using the CVODE method; see **-cvode_order** in the description of the KINETICS data block. Optionally, **cvode_order** or **-cvode_o[rder]**.

> (*1, 2, 3, 4,* or *5*)—Number of terms used in the extrapolation of rates.

Notes

The **KINETICS_MODIFY** data block allows modification of a preexisting assemblage of kinetic reactions. The most common uses of the data block are to add a new kinetic reactant to the assemblage and to change the amount of a kinetic reactant that is present in the assemblage. It also is possible to change the other parameters of the assemblage of kinetic reactants, including the choice of integration methods, the parameters of the integration method, and the time steps associated with the assemblage.

KINETICS_MODIFY modifies only the data items specifically defined in the data block. Any data items in the assemblage of kinetic reactants not modified by the data block remain unchanged.

Related keywords

KINETICS and KINETICS_RAW.

KINETICS_RAW

This keyword data block is written by the **DUMP** operation. It is intended to be used to reinitialize simulations at the point that the **DUMP** command was executed or to transfer the definition of an assemblage of kinetic reactants between IPhreeqc modules.

Notes

The **KINETICS_RAW** data block contains a complete listing of the internal data structures that define an assemblage of kinetic reactants. **KINETICS_RAW** data blocks are written by **DUMP** and read by PHREEQC without user modification. The formats should be considered dynamic because future modifications to PHREEQC could result in additional data that are included in the **KINETICS_RAW** data block.

Related keywords

KINETICS and **KINETICS_MODIFY**.

REACTION_MODIFY

This keyword data block is used to modify the definition of a previously defined irreversible reaction. New reactants may be added, and the stoichiometry of reactants may be changed. The format of the data block is the same as REACTION_RAW data block, except that the data block need not be complete. The **REACTION_MODIFY** data block can be used selectively to change data items. The Example data block lists a subset of identifiers that can be used in **REACTION_MODIFY** data blocks.

Example data block

```
Line  0:  REACTION_MODIFY 1              Adding calcite and MgSO4
Line  1:         -units              uMol
Line  2:         -reactant_list
Line  3:             Calcite         2
Line  3a:            MgSO4           0.2
Line  4:         -steps
Line  5:             .1
Line  6:         -equal_increments   1
Line  7:         -count_steps        10
```

Explanation

Line 0: **REACTION_MODIFY** *number* [*description*]

REACTION_MODIFY is the keyword for the data block.

number—Positive integer to identify the stoichiometric reaction to modify.

description—Optional comment that describes the stoichiometric reaction.

Line 1: **-units** *units*

-units—Units are moles, millimoles, micromoles, or an abbreviation of these units. Optionally, **units** or **-u[nits]**.

Line 2: **-reactant_list**

-reactant_list—Begins a data block that defines reactants in the stoichiometric reaction. Optionally, **reactant_list** or **-r[eactant_list]**.

Line 3: (*phase name* or *formula*), *relative stoichiometry*

(*phase name* or *formula*)—The name of a phase defined in a PHASES data block or a chemical formula; see (*phase name* or *formula*) in the description of the REACTION data block.

relative stoichiometry—Amount of this reactant relative to other reactants; see *relative stoichiometry* in the description of the REACTION data block.

Line 4: **-steps**

-steps—Begins a data block that defines the steps for the reaction. Optionally, **steps** or **-s[teps]**.

Line 5: *list of reaction amounts*

list of reaction amounts—Defines the amounts of the stoichiometric reaction to add to a solution or mixture; see *reaction amounts* in the description of the REACTION data block.

Line 6: **-equal_increments** (*1* or *0*)

-equal_increments—Defines whether a single reaction amount is divided into one or more equal steps as specified by **-count_steps**; see "**in**" in the description of the REACTION data block. Optionally, **equal_increments** or **-eq[ual_increments]**.

(*1* or *0*)—A value of 1 indicates true, and the first *reaction amount* is split into equal increments. A value of 0 indicates false, and the *list of reaction amounts* determines the amounts of the stoichiometric reaction added to a solution or mixture.

Line 7: **-count_steps** *n*

-count_steps—If **-equal_increments** is equal to 1, the value of *n* determines the number of equal increments for the reaction; see "**in**" in the description of the REACTION data block. Optionally, **count_steps** or **-c[ount_steps]**.

n—Number of equal steps for the stoichiometric reaction.

Notes

The **REACTION_MODIFY** data block allows modification of a preexisting stoichiometric reaction definition. The most common uses of the data block are to add a new reactant to the list of reactants, to change the relative coefficients of reactants, and to change the reactant amounts. It also is possible to change the units for the reaction and whether the reaction is added in equal increments or as defined by a list of reaction increments.

REACTION_MODIFY modifies only the data items specifically defined in the data block. Any data items in the stoichiometric reaction definition not modified by the data block remain unchanged. If **-steps** is defined in the data block, the previous list of reaction amounts is removed and is replaced by the new definitions.

Related keywords

REACTION and REACTION_RAW.

REACTION_RAW

This keyword data block is written by the DUMP operation. It is intended to be used to reinitialize simulations at the point that the DUMP command was executed or to transfer reactions defined by REACTION data blocks between IPhreeqc modules.

Notes

The **REACTION_RAW** data block contains a complete listing of the internal data structures that define a stoichiometric reaction (REACTION data block definition). **REACTION_RAW** data blocks are written by DUMP and read by PHREEQC without user modification. The formats should be considered dynamic because future modifications to PHREEQC could result in additional data that are included in the **REACTION_RAW** data block.

Related keywords

REACTION and REACTION_MODIFY.

REACTION_PRESSURE_RAW

This keyword data block is written by the DUMP operation. It is intended to be used to reinitialize simulations at the point that the DUMP command was executed, or to transfer reaction-pressure specifications defined by REACTION_PRESSURE data blocks between IPhreeqc modules.

Notes

The **REACTION_PRESSURE_RAW** data block contains a complete listing of the internal data structure that defines reaction temperatures (as defined in REACTION_PRESSURE data blocks). **REACTION_PRESSURE_RAW** data blocks are written by DUMP and can be read directly by PHREEQC. The format of the REACTION_PRESSURE data block is sufficiently simple, so that a **REACTION_PRESSURE_RAW** data block is not needed; simply use REACTION_PRESSURE to define new reaction pressures.

Related keywords

REACTION_PRESSURE.

REACTION_TEMPERATURE_RAW

This keyword data block is written by the DUMP operation. It is intended to be used to reinitialize simulations at the point that the DUMP command was executed or to transfer reaction-temperature specifications defined by REACTION_TEMPERATURE data blocks between IPhreeqc modules.

Notes

The **REACTION_TEMPERATURE_RAW** data block contains a complete listing of the internal data structure that defines reaction temperatures (as defined in REACTION_TEMPERATURE data blocks). **REACTION_TEMPERATURE_RAW** data blocks are written by DUMP and can be read directly by PHREEQC. The format of the REACTION_TEMPERATURE data block is sufficiently simple, so that a **REACTION_TEMPERATURE_MODIFY** data block is not needed; simply use REACTION_TEMPERATURE to define new reaction temperatures.

Related keywords

REACTION_TEMPERATURE.

SOLID_SOLUTIONS_MODIFY

This keyword data block is used to modify the definition of a previously defined solid-solution assemblage. New solid solutions and new components of solid solutions may be added. The number of moles of each solid-solution component may be changed. The format of the data block is the same as the SOLID_SOLUTIONS_RAW data block except that the data block need not be complete. The **SOLID_SOLUTIONS_MODIFY** data block can be used selectively to change data items. The Example data block lists a subset of identifiers that can be used in **SOLID_SOLUTIONS_MODIFY** data blocks.

Example data block

```
Line 0:      SOLID_SOLUTION_MODIFY 1
Line 1:      -solid_solution      Calcite_SS
Line 2:          -component       calcite
Line 3:              -moles          0.2
Line 2a:         -component       rhodochrosite
Line 3a:             -moles          0.2
Line 2b:         -component       siderite
Line 3b:             -moles          0.001
```

Explanation

Line 0: **SOLID_SOLUTIONS_MODIFY** *number* [*description*]

SOLID_SOLUTIONS_MODIFY is the keyword for the data block.

number—Positive integer to identify the solid-solution assemblage to modify.

description—Optional comment that describes the solid-solution assemblage.

Line 1: **-solid_solution** *name*

-solid_solution—Identifier that indicates a solid solution will be defined. The identifier **-solid_solution** is required to precede the other identifiers that define the solid-solution composition. Optionally, **solid_solution** or **-s[olid_solution]**.

name—Name of the solid solution.

Line 2: **-component** *component*

-component—Identifier begins a data block that modifies a solid-solution component. The identifier **-component** is required to precede the modifications to the component. Optionally, **component** or **-c[omponent]**.

component—The solid solution component to be added or modified. If *component* is already in the solid solution, then the number of moles is changed; if *component* is not in the solid solution, then it is added.

Line 3: **-moles** *moles*

-moles—Identifier for the number of moles of a solid-solution component. Optionally, **moles** or **-m**[**oles**].

moles—Number of moles of *component* in solid solution *name*.

Notes

The **SOLID_SOLUTIONS_MODIFY** data block allows modification of a preexisting solid-solution assemblage. The most common uses are to change the amount of a component in a solid solution, to add a new component to an ideal solid solution, and to add a new solid solution to a solid-solution assemblage.

SOLID_SOLUTIONS_MODIFY modifies only the data items specifically defined in the data block. Any data items in the solid-solution assemblage definition not modified by the data block remain unchanged.

Related keywords

SOLID_SOLUTIONS and SOLID_SOLUTIONS_RAW.

SOLID_SOLUTIONS_RAW

This keyword data block is written by the DUMP operation. It is intended to be used to reinitialize simulations at the point that the DUMP command was executed or to transfer solid-solution compositions between IPhreeqc modules.

Notes

The **SOLID_SOLUTIONS_RAW** data block contains a complete listing of the internal data structures that define a solid-solution assemblage. **SOLID_SOLUTIONS_RAW** data blocks are written by DUMP and read by PHREEQC without user modification. The formats should be considered dynamic because future modifications to PHREEQC could result in additional data that are included in the **SOLID_SOLUTIONS_RAW** data block.

Related keywords

SOLID_SOLUTIONS and SOLID_SOLUTIONS_MODIFY.

SOLUTION_MODIFY

This keyword data block is used to modify the definition of a previously defined solution composition. New elements may be added to the solution, and the number of moles of elements may be changed. However, care is needed to preserve the correct relations between the number of moles of elements, equivalents of charge imbalance, and moles of hydrogen and oxygen. Initial estimates for log activity of master species are updated automatically if the number of moles of an element is adjusted; alternatively, new initial estimates for log activity of master species may be specified explicitly with **-activities**. The format of the data block is the same as the SOLUTION_RAW data block except that the data block need not be complete. The **SOLUTION_MODIFY** data block can be used selectively to change data items. The Example data block lists a subset of identifiers that can be used in **SOLUTION_MODIFY** data blocks.

Example data block

```
Line 0:    SOLUTION_MODIFY 1 Modified solution description
Line 1:        -temp              25
Line 2:        -total_h           111.01243359981
Line 3:        -total_o           55.506216800086
Line 4:        -cb                -3.657928589893e-10
Line 5:        -totals
Line 6:                Cl         0.0020000000000003
Line 6a:               Fe(2)      3.9999998897204e-007
Line 6b:               Fe(3)      1.1027960586156e-014
Line 6d:               Na         0.0020000000000003
Line 6c:               S          0.02012
```

The following lines optionally set initial estimates of master variables.

```
Line 7:        -pH      6.61
Line 8:        -pe      0
Line 9:        -mu      0.07
Line 10:       -ah2o    0.9985
Line 11:       -activities
Line 12a:              Cl         -3.0
Line 12b:              Fe(2)      -7.6
Line 12c:              Fe(3)      -23.4
Line 12d:              Na         -3.0
Line 12e:              S          -3.9
```

Explanation

Line 0: **SOLUTION_MODIFY** *number* [*description*]

SOLUTION_MODIFY is the keyword for the data block.

number—Positive integer to identify the solution to modify.

description—Optional comment that describes the solution.

Line 1: **-temp** *tc*

-temp *tc*—Sets the temperature for the solution; *tc* is in degrees Celsius.

Line 2: **-total_h** *moles*

-total_h *moles*—Total moles of hydrogen in the solution. A solution with one kilogram of water, contains approximately 111.0 moles of hydrogen; however, the value must be known precisely to set the correct redox conditions.

Line 3: **-total_o** *moles*

-total_o *moles*—Total moles of oxygen in the solution. A solution with one kilogram of water, contains approximately 55.5 moles of oxygen; however, the value must be known precisely to set the correct redox conditions.

Line 4: **-cb** *equivalents*

-cb *equivalents*—Equivalents of charge imbalance in the solution. Although physical solutions are nearly perfectly balanced in charge, analytical errors can result in apparent charge imbalances for solutions. Surface reactions without explicit definition of the double layer composition also can impart charge imbalances to a solution. The value of the charge imbalance must be known precisely to set the correct redox conditions.

Line 5: **-totals**

-totals—Identifier begins a block of data that modifies the number of moles of elements or element valence states in the solution.

Line 6: (*Element* or *element valence state*) *moles*

Element or *element valence state*—An element or element valence state that has been defined in a SOLUTION_MASTER_SPECIES data block. If a redox element is specified (that is, a redox element name without following parentheses), all valence states of the element are removed from the solution and *moles* is the total number of moles of the redox element in solution, inclusive of all valence states. If an element valence state is defined (that is, a redox element name with following parentheses), then, if present, the element (no parentheses) is removed from solution, and the element valence state (with parentheses) is added to the solution.

moles—Number of moles of the element or element valence state in the solution.

Line 7: **-pH** *pH*

> **-pH** *pH*—Sets the initial estimate of the pH, which is used in solving the system of nonlinear equations that define solution equilibrium.

Line 8: **-pe** *pe*

> **-pe** *pe*—Sets the initial estimate of the pe, which is used in solving the system of nonlinear equations that define solution equilibrium.

Line 9: **-mu** *mu*

> **-mu** *mu*—Sets the initial estimate of the ionic strength, which is used in solving the system of nonlinear equations that define solution equilibrium.

Line 10: **-ah2o** *ah2o*

> **-ah2o** *ah2o*—Sets the initial estimate of the activity of water, which is used in solving the system of nonlinear equations that define solution equilibrium.

Line 11: **-activities**

> **-activities**—Identifier begins a data block that sets initial estimates of the activities of the master species, which are used in solving the system of nonlinear equations that define solution equilibrium. For redox elements, the activity of only one redox state is used in solving the equilibrium equations. The activity used is usually that of the primary master species; however, it is possible that the equations have been rewritten so that the primary master species has been replaced by one of the secondary master species, in which case, the initial estimate of the secondary master species is the one that is used to begin solving the nonlinear equations that define solution equilibrium. When the number of moles of an element are changed, log activities for that element [and (or) element valence states] automatically will be adjusted by log of the new number of moles divided by the old number of moles. Use of **-activities** overrides this automatic adjustment for each element in the **-activities** list.

Line 12: *master species*, *log activity*

> *master species*—An element or element valence state master species that has been defined in a **SOLUTION_MASTER_SPECIES** data block.

> *log activity*—Initial estimate of the log activity of the master species, which is used in solving the system of nonlinear equations that define solution equilibrium.

Notes

The **SOLUTION_MODIFY** data block allows modification of a preexisting solution definition. Lines 1 through 6 in the Example data block set the temperature and the number of moles of elements and element redox states for the solution. Care is needed to maintain concordance between the number of moles of elements (including hydrogen and oxygen) and the charge imbalance. Inconsistent alteration of the moles of elements and equivalents of charge imbalance will result in unexpected pH and redox reactions.

Lines 7 through 12 set initial estimates of unknowns for the equations that define solution equilibrium but do not affect the actual solution composition or the ultimate values of the numerical solution to the equations. The definitions on lines 7 through 12 could affect whether a numerical solution is found or the number of iterations needed to find a numerical solution. The initial estimates of unknowns can make a substantial difference in the number of iterations to solve a chemical system and, consequently, in run times, especially for transport calculations. If the number of moles of an element is adjusted (**-totals**), by default, the log activity associated with that element is automatically adjusted by the log of the ratio of the new moles to the old moles. Logic is included to adjust valence states as well as total moles of a redox element. Use of **-activities** for an element supersedes the automatic adjustment for log activities for that element.

The most common use of a **SOLUTION_MODIFY** data block is expected to be in transport models that use IPhreeqc modules. The transport code is used to transport elemental concentrations conservatively, and then the moles of elements in the solutions in the IPhreeqc module are updated with the transport results by using a series of SOLUTION_MODIFY data blocks. Geochemical reactions are calculated in the IPhreeqc module, and the new compositions of solutions are retrieved by extracting data defined by SELECTED_OUTPUT or by the output from a DUMP data block. The reacted solution concentrations then are used to begin a new conservative transport step in the transport model. Care is needed to convert the moles of an element in an IPhreeqc solution (as written with DUMP) to the concentration of the element used in the transport code, and, conversely, to convert the concentrations from transport to the number of moles in an IPhreeqc solution.

SOLUTION_MODIFY modifies only the data items specifically defined in the data block. Any data items in the solution definition not modified by the data block remain unchanged. Modification of the number of moles of redox elements requires special care. A redox element is defined simply by the element name and represents the total moles of that element, including all valence states. A redox element valence state is defined by the element name followed by a valence state in parentheses. When a redox element is defined in **-totals**, all element valence states are removed from the solution, and only the total concentration remains. When a redox element valence state is defined in **-totals**, then the redox element (total moles), if it is present, is removed, but other valence-state concentrations are retained.

Related keywords

SOLUTION and SOLUTION_RAW.

SOLUTION_RAW

This keyword data block is written by the DUMP operation. It is intended to be used to reinitialize simulations at the point that the DUMP command was executed, or to transfer solution compositions between IPhreeqc modules.

Notes

The **SOLUTION_RAW** data block contains a complete listing of the internal data structures that define a solution. **SOLUTION_RAW** data blocks are written by DUMP and read by PHREEQC without user modification. The formats should be considered dynamic because future modifications to PHREEQC could result in additional data that are included in the **SOLUTION_RAW** data block.

Related keywords

SOLUTION and SOLUTION_MODIFY.

SURFACE_MODIFY

This keyword data block is used to modify the definition of a previously defined surface-assemblage composition. New surfaces and surface-site types may be added, and the quantity of each surface-site type may be changed. In addition, it is possible to change the elemental composition of the surface and the elemental composition in the diffuse layer of the surface. The format of the data block is the same as the SURFACE_RAW data block except that the data block need not be complete. The **SURFACE_MODIFY** data block can be used selectively to change data items. This Example data block lists a subset of identifiers that can be used in **SURFACE_MODIFY** data blocks.

Example data block

```
Line 0:    SURFACE_MODIFY      1 Surface assemblage after simulation 2.
Line 1:        -only_counter_ions      0
Line 2:        -thickness              1e-008
Line 3:        -debye_lengths          0
Line 4:        -DDL_viscosity          1
Line 5:        -DDL_limit              0.8
Line 6:        -component                      Hfo_sOH

Line 7:            -charge_balance              0.0010927661324556
Line 8:            -Dw                          1e-009
Line 9:            -totals
Line 10:               Ca                       0.0008229045119665
Line 10a:              H                        0.0094469571085231
Line 10b:              Hfo_s                    0.01
Line 10c:              O                        0.01
Line 6a:        -component                      Hfo_wOH

Line 7a:           -charge_balance             -0.0010124274171143
Line 9a:           -totals
Line 10d:              Ca                       1.1196909316487e-007
Line 10e:              H                        0.00094211950362998
Line 10f:              Hfo_w                    0.0010000000015016
Line 10g:              O                        0.0029934399595643
Line 10h:              S                        0.00050535150978271
Line 11:        -charge_component               Hfo

Line 12:           -specific_area               600
Line 13:           -grams                       1
Line 14:           -charge_balance             -1.5437177225538e-012
Line 15:           -capacitance0                1
```

```
Line 16:            -capacitance1          5
Line 17:            -diffuse_layer_totals
Line 18:                   Ca              6.7655957651525e-007
Line 18a:                  H               0.66607460244918
Line 18b:                  Na              2.2662527289495e-006
Line 18c:                  O               0.33320521533783
Line 18d:                  S               4.1978356284299e-005
```

Explanation

Line 0: **SURFACE_MODIFY** *number* [*description*]

> **SURFACE_MODIFY** is the keyword for the data block.

> *number*—Positive integer to identify the surface assemblage to modify.

> *description*—Optional comment that describes the surface assemblage.

Line 1: **-only_counter_ions** (*1* or *0*)

> **-only_counter_ions**—Excludes co-ions in diffuse layer. See Line 7 in Explanation 1 of the SURFACE data block.

> (*1* or *0*)—A value of 1 indicates true. A value of 0 indicates false.

Line 2: **-thickness** *thickness*

> **-thickness**—Thickness of diffuse layer. See **-Donnan** and **-diffuse_layer** in the SURFACE data block. Optionally, **thickness** or **-t**[**hickness**].

> *thickness*—Thickness of diffuse layer, m.

Line 3: **-debye_lengths** *lengths*

> **-debye_lengths**—A factor for the Debye length is defined to determine the thickness of the diffuse layer. See **-Donnan** in the SURFACE data block. Optionally, **debye_lengths** or **-de**[**bye_lengths**].

> *lengths*—Factor used to calculate the thickness of the diffuse layer, unitless.

Line 4: **-DDL_viscosity** *fraction*

> **-DDL_viscosity**—When considering multicomponent diffusion in the diffuse layer of surfaces, *fraction* is the viscosity in the diffuse layer relative to the viscosity in the free pore space. See **-Donnan** in the SURFACE data block. Optionally, **DDL_viscosity** or **-DD**[**L_viscosity**].

> *fraction*—Viscosity in the diffuse layer divided by the viscosity in the free pore space, unitless.

Line 5: **-DDL_limit** *limit*

-DDL_limit—If **debye_lengths** are used to define the thickness of the diffuse layer, *limit* is the maximum fraction of the total water that can be in the diffuse layer. See **-Donnan** in the SURFACE data block. Optionally, **DDL_limit** or **-DDL_l[imit]**.

limit—Maximum fraction of water in the diffuse layer, unitless.

Line 6: **-component** *surface site formula*

-component—Identifier that indicates that information for a surface site will be defined. The identifier **-component** is required to precede the other identifiers that define information for the surface site. Optionally, **component** or **-c[omponent]**.

surface site formula—Surface site formula. Surface site names have an underscore that is preceded by the surface name. The formula is a charge-balanced formula for the surface site.

Line 7: **-charge_balance** *equiv*

-charge_balance—Sum of the charge for all the species of the surface site. Optionally, **charge_balance** or **-charge_b[alance]**.

equiv—Surface charge for the surface site defined by **-component**, eq.

Line 8: **-Dw** *diffusion coefficient*

-Dw—Diffusion coefficient for the surface site, which is used only if **-multi_D** is **true** in a TRANSPORT simulation. Optionally, **Dw** or **-D[w]**.

diffusion coefficient—Diffusion coefficient for the surface site, m^2/s.

Line 9: **-totals**

-totals—Identifier begins a block of data containing the moles of elements related to the surface site. Optionally, **totals** or **-t[otals]**.

Line 10: *element moles*

element—An element name or a surface site name.

moles—Moles of element or surface site, mol.

Line 11: **-charge_component**

-charge_component—Identifier that indicates that information on surface charge for a surface will be defined. The identifier **-charge_component** is required to precede the other identifiers that define information for the surface charge. Optionally, **charge_component** or **-ch[arge_component]**.

surface name—Name of the surface. Surface site names have an underscore; the surface name precedes the underscore.

Line 12: **-specific_area** *area*

> **-specific_area**—Specific area for the surface. See the SURFACE data block. Optionally, **specific_area** or **-s[specific_area]**.
>
> *area*—Specific area for the surface, m^2/g, or, if the surface is related to an equilibrium phase or kinetic reactant, m^2/mol.

Line 13: **-grams** *grams*

> **-grams**—Mass of the material containing the surface. See the SURFACE data block. Optionally, **grams** or **-g[rams]**.
>
> *grams*—Mass of material containing the surface, g.

Line 14: **-charge_balance** *equiv*

> **-charge_balance**—Charge of the surface, including the charge in the diffuse layer. If **-Donnan** or **-diffuse_layer** are defined for the SURFACE, then the charge balance for the surface will be near zero, such that the charge in the diffuse layer will be equal to the sum of the charge for all species for all surface sites for that surface. Optionally, **charge_balance** or **-c[harge_balance]**.
>
> *equiv*—Charge for the surface, eq.

Line 15: **-capacitance0** *c*

> **-capacitance0**—Capacitance for the 0-1 planes in the CD-MUSIC formulation. See **-capacitances** in the SURFACE data block. Optionally, **capacitance0** or **-ca[pacitance0]**.
>
> *c*—Capacitance for the 0-1 planes, F/m^2.

Line 16: **-capacitance1** *c*

> **-capacitance1**—Capacitance for the 1-2 planes in the CD-MUSIC formulation. See **-capacitances** in the SURFACE data block. Optionally, **capacitance1** or **-capacitance1**.
>
> *c*—Capacitance for the 1-2 planes, F/m^2.

Line 17: **-diffuse_layer_totals**

> **-diffuse_layer_totals**—Identifier begins a block of data containing the moles of elements in the diffuse layer for a surface. Values are present only if **-Donnan** or **-diffuse_layer** are defined for the surface. Optionally, **diffuse_layer_totals** or **-d[iffuse_layer_totals]**.

Line 18: *element, moles*

    element—An element name.

    moles—Moles of element in the diffuse layer for the surface, mol.

Notes

The **SURFACE_MODIFY** data block allows modification of a preexisting surface assemblage, but care is needed for any modifications because data items have interdependencies. Although it is possible to add a new surface or surface site to a surface assemblage, defining consistent data for the new item is difficult. It is important that the **-totals** data block is consistent with the **-charge_balance** for surface components and that the **-diffuse_layer_totals** data block is consistent with the **-charge_balance** for surface charge components. If charge balance is not preserved, unexpected pH and redox conditions will result.

One use of the **SURFACE_MODIFY** data block might be in transporting surfaces, for example in conjunction with a sediment transport model. Another use might be for diffusive transport in the diffuse layer of surfaces. These transport calculations need to transport the charge [component **-charge_balance** and (or) charge-component **-charge_balance**], elements [component **-totals** and (or) charge-component **-diffuse_layer_totals**], and possibly other quantities, such as **-grams** and **-specific area**. SURFACE_MODIFY data blocks could be used to update surface-assemblage definitions after the transport calculations.

A **SURFACE_MODIFY** data block modifies only the data items specifically defined in the data block. Any data items in the surface assemblage definition not modified by the data block remain unchanged. Note specifically, that the **-totals** data block modifies all elements included in the data block, but the number of moles of any element related to the surface site that is not listed in the data block will remain unchanged.

Related keywords

SURFACE and SURFACE_RAW.

SURFACE_RAW

This keyword data block is written by the DUMP operation. It is intended to be used to reinitialize simulations at the point that the DUMP command was executed or to transfer surface-assemblage compositions between IPhreeqc modules.

Notes

The **SURFACE_RAW** data block contains a complete listing of the internal data structures that define a surface assemblage. **SURFACE_RAW** data blocks are written by DUMP and read by PHREEQC without user modification. The formats should be considered dynamic because future modifications to PHREEQC could result in additional data that are included in the **SOLUTION_RAW** data block.

Related keywords

SURFACE and SURFACE_MODIFY.

www.ingramcontent.com/pod-product-compliance
Lightning Source LLC
Chambersburg PA
CBHW081427170526
45166CB00008B/2121